CLASSICAL MECHANICS

Point Particles and Relativity

Springer
New York
Berlin
Heidelberg
Hong Kong
London
Milan
Paris
Tokyo

THEORETICAL PHYSICS

Greiner
Quantum Mechanics
An Introduction 3rd Edition

Greiner
Quantum Mechanics
Special Chapters

Greiner · Müller
Quantum Mechanics
Symmetries 2nd Edition

Greiner
Relativistic Quantum Mechanics
Wave Equations 2nd Edition

Greiner · Reinhardt
Field Quantization

Greiner · Reinhardt
Quantum Electrodynamics
2nd Edition

Greiner · Schramm · Stein
Quantum Chromodynamics

Greiner · Maruhn
Nuclear Models

Greiner · Müller
Gauge Theory of Weak Interactions
2nd Edition

Greiner
Classical Mechanics
Point Particles and Relativity

Greiner
Classical Mechanics
Systems of Particles and Hamiltonian Dynamics

Greiner
Classical Electrodynamics

Greiner · Neise · Stöcker
Thermodynamics and Statistical Mechanics

Walter Greiner

CLASSICAL MECHANICS

Point Particles and Relativity

Foreword by D. Allan Bromley

With 317 Figures

Walter Greiner
Institut für Theoretische Physik
Johann Wolfgang Goethe-Universität
Robert Mayer Strasse 10
Postfach 11 19 32
D-60054 Frankfurt am Main
Germany
greiner@th.physik.uni-frankfurt.de

Library of Congress Cataloging-in-Publication Data
Greiner, Walter, 1935-
Classical mechanics : point particles and relativity / Walter Greiner.
p. cm.-- (Classical theoretical physics)
Includes bibliographical references and index.
ISBN 0-387-95586-0 (softcover : alk. paper)
1. Mechanics--Problems, exercises, etc. 2. Relativity (Physics)--Problems, exercises, etc. I. Title II. Series.

QC125.2 .G74 2003
531--dc21 2002030570

ISBN 0-387-95586-0 Printed on acid-free paper.

Translated from the German *Mechanik: Teil 2*, by Walter Greiner, published by Verlag Harri Deutsch, Thun, Frankfurt am Main, Germany, © 1989.

Printed in the United States of America.

9 8 7 6 5 4 3 2 1 SPIN 10892857

www.springer-ny.com

Springer-Verlag New York Berlin Heidelberg
A member of BertelsmannSpringer Science+Business Media GmbH

Foreword

More than a generation of German-speaking students around the world have worked their way to an understanding and appreciation of the power and beauty of modern theoretical physics—with mathematics, the most fundamental of sciences—using Walter Greiner's textbooks as their guide.

The idea of developing a coherent, complete presentation of an entire field of science in a series of closely related textbooks is not a new one. Many older physicians remember with real pleasure their sense of adventure and discovery as they worked their ways through the classic series by Sommerfeld, by Planck, and by Landau and Lifshitz. From the students' viewpoint, there are a great many obvious advantages to be gained through the use of consistent notation, logical ordering of topics, and coherence of presentation; beyond this, the complete coverage of the science provides a unique opportunity for the author to convey his personal enthusiasm and love for his subject.

These volumes on classical physics, finally available in English, complement Greiner's texts on quantum physics, most of which have been available to English-speaking audiences for some time. The complete set of books will thus provide a coherent view of physics that includes, in classical physics, thermodynamics and statistical mechanics, classical dynamics, electromagnetism, and general relativity; and in quantum physics, quantum mechanics, symmetries, relativistic quantum mechanics, quantum electro- and chromodynamics, and the gauge theory of weak interactions.

What makes Greiner's volumes of particular value to the student and professor alike is their completeness. Greiner avoids the all too common "it follows that...," which conceals several pages of mathematical manipulation and confounds the student. He does not hesitate to include experimental data to illuminate or illustrate a theoretical point, and these data, like the theoretical content, have been kept up to date and topical through frequent revision and expansion of the lecture notes upon which these volumes are based.

Moreover, Greiner greatly increases the value of his presentation by including something like one hundred completely worked examples in each volume. Nothing is of greater importance to the student than seeing, in detail, how the theoretical concepts and tools under study are applied to actual problems of interest to working physicists. And, finally, Greiner adds brief biographical sketches to each chapter covering the people responsible for the development of the theoretical ideas and/or the experimental data presented. It was Auguste Comte (1789–1857) in his *Positive Philosophy* who noted, "To understand a science it is necessary to know its history." This is all too often forgotten in modern physics teaching, and the bridges that Greiner builds to the pioneering figures of our science upon whose work we build are welcome ones.

Greiner's lectures, which underlie these volumes, are internationally noted for their clarity, for their completeness, and for the effort that he has devoted to making physics an integral whole. His enthusiasm for his sciences is contagious and shines through almost every page.

These volumes represent only a part of a unique and Herculean effort to make all of theoretical physics accessible to the interested student. Beyond that, they are of enormous value to the professional physicist and to all others working with quantum phenomena. Again and again, the reader will find that, after dipping into a particular volume to review a specific topic, he or she will end up browsing, caught up by often fascinating new insights and developments with which he or she had not previously been familiar.

Having used a number of Greiner's volumes in their original German in my teaching and research at Yale, I welcome these new and revised English translations and would recommend them enthusiastically to anyone searching for a coherent overview of physics.

D. Allan Bromley
Henry Ford II Professor of Physics
Yale University
New Haven, Connecticut, USA

Preface

Theoretical physics has become a many faceted science. For the young student, it is difficult enough to cope with the overwhelming amount of new material that has to be learned, let alone obtain an overview of the entire field, which ranges from mechanics through electrodynamics, quantum mechanics, field theory, nuclear and heavy-ion science, statistical mechanics, thermodynamics, and solid-state theory to elementary-particle physics; and this knowledge should be acquired in just eight to ten semesters, during which, in addition, a diploma or master's thesis has to be worked on or examinations prepared for. All this can be achieved only if the university teachers help to introduce the student to the new disciplines as early as possible, in order to create interest and excitement that in turn set free essential new energy.

At the Johann Wolfgang Goethe University in Frankfurt am Main, we therefore confront the student with theoretical physics immediately, in the first semester. Theoretical Mechanics I and II, Electrodynamics, and Quantum Mechanics I—An Introduction are the courses during the first two years. These lectures are supplemented with many mathematical explanations and much support material. After the fourth semester of studies, graduate work begins, and Quantum Mechanics II—Symmetries, Statistical Mechanics and Thermodynamics, Relativistic Quantum Mechanics, Quantum Electrodynamics, Gauge Theory of Weak Interactions, and Quantum Chromodynamics are obligatory. Apart from these, a number of supplementary courses on special topics are offered, such as Hydrodynamics, Classical Field Theory, Special and General Relativity, Many-Body Theories, Nuclear Models, Models of Elementary Particles, and Solid-State Theory.

This volume of lectures, *Classical Mechanics: Point Particles and Relativity*, deals with the first and more elementary part of the important field of classical mechanics. We have tried to present the subject in a manner that is both interesting to the student and easily accessible. The main text is therefore accompanied by many exercises and examples that have been worked out in great detail. This should make the book useful also for students wishing to study the subject on their own.

Beginning the education in theoretical physics at the first university semester, and not as dictated by tradition after the first one and a half years in the third or fourth semester, has brought along quite a few changes as compared to the traditional courses in that discipline.

Especially necessary is a greater amalgamation between the actual physical problems and the necessary mathematics. Therefore, we treat in the first semester vector algebra and analysis, the solution of ordinary, linear differential equations, Newton's mechanics of a mass point culminating in the discussion of Kepler's laws (planetary motion), elements of astronomy, addressing modern research issues like the dark matter problem, and the mathematically simple mechanics of special relativity.

Many explicitly worked-out examples and exercises illustrate the new concepts and methods and deepen the interrelationship between physics and mathematics. As a matter of fact, this first-semester course in theoretical mechanics is a precursor to theoretical physics. This changes significantly the content of the lectures of the second semester addressed in the volume *Classical Mechanics: System of Particles and Hamiltonian Dynamics*.

The new mathematical tools are explained and exercised in many physical examples. In the lecturing praxis, the deepening of the exhibited material is carried out in a three-hour-per-week *theoretica*, that is, group exercises where eight or ten students solve the given exercises under the guidance of a tutor.

Biographical and historical footnotes anchor the scientific development within the general context of scientific progress and evolution. In this context, I thank the publishers Harri Deutsch and F. A. Brockhaus (*Brockhaus Enzyklopädie*, F.A. Brockhaus, Wiesbaden—marked by [BR]) for giving permission to extract the biographical data of physicists and mathematicians from their publications.

We should also mention that in preparing some early sections and exercises of our lectures we relied on the book *Theory and Problems of Theoretical Mechanics*, by Murray R. Spiegel, McGraw-Hill, New York, 1967.

Over the years, we enjoyed the help of several students and collaborators, in particular, H. Angermüller, P. Bergmann, H. Betz, W. Betz, G. Binnig (Nobel prize 1986), J. Briechle, M. Bundschuh, W. Caspar, C. v. Charewski, J. v. Czarnecki, R. Fickler, R. Fiedler, B. Fricke (now professor at Kassel University), C. Greiner (now professor at JWG-University, Frankfurt am Main), M. Greiner, W. Grosch, R. Heuer, E. Hoffmann, L. Kohaupt, N. Krug, P. Kurowski, H. Leber, H. J. Lustig, A. Mahn, B. Moreth, R. Mörschel, B. Müller (now professor at Duke University, Durham, N.C.), H. Müller, H. Peitz, J. Rafelski (now professor at University of Arizona, Tuscon), G. Plunien, J. Reinhardt, M. Rufa, H. Schaller, D. Schebesta, H. J. Scheefer, H. Schwerin, M. Seiwert, G. Soff (now professor at Technical University Dresden), M. Soffel (now professor at Technical University Dresden), E. Stein (now professor at Maharishi University, Vlodrop, Netherlands), K. E. Stiebig, E. Stämmler, H. Stock, H. Störmer (Nobel prize 1998), J. Wagner, and R. Zimmermann. They all made their way in science and society, and meanwhile work as professors at universities, as leaders in industry, and in other places. We particularly acknowledge the recent help of Dr. Sven Soff and Dr. Stefan Scherer during the preparation of the English manuscript. The figures were drawn by Mrs. A. Steidl.

The English manuscript was copy-edited by Kristen Cassereau and the production of the book was supervised by Timothy Taylor of Springer-Verlag New York, Inc.

Walter Greiner
Johann Wolfgang Goethe-Universität
Frankfurt am Main, Germany

Contents

Examples

PART I

VECTOR CALCULUS

1 Introduction and Basic Definitions

Physical quantities that are completely determined by the specification of one numerical value and a unit are called

scalars (e.g., mass, temperature, energy, wavelength).

Quantities that for a complete description besides the numerical value and the physical unit still need the specification of their direction are called

vectors (e.g., force, velocity, acceleration, torque).

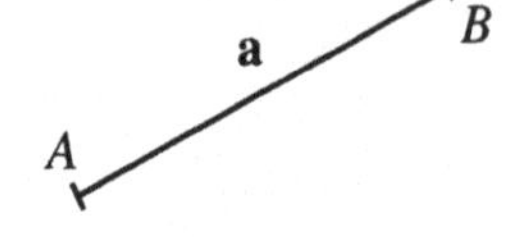

Vector **a** pointing from A to B.

A vector may be represented geometrically by an oriented distance, i.e., by a distance associated with a direction, such that holds; for example: Let A be the initial point and B the endpoint of the vector **a** (compare figure).

The *magnitude* of the vector is then represented by the length of the distance AB. A vector is frequently described symbolically by a Latin letter with a small arrow attached to elucidate the vector character. Other possible representations make use of German letters or emphasize the quantity by bold printing.

The magnitude of a vector **a** is written as: $|\mathbf{a}| = a$.

Definition: Two vectors **a** and **b** are called *equal* if

1. $|\mathbf{a}| = |\mathbf{b}|$,

2. $\mathbf{a} \uparrow\uparrow \mathbf{b}$ (aligned; parallel).

We then write $\mathbf{a} = \mathbf{b}$.

The vectors **a** and **b** are equal.

That means: All distances of equal length and equal orientation are representations of the same vector on equal footing. Hence, the specific location of the vector in space is being disregarded.

A vector with *opposite direction but equal magnitude* of **a** is denoted as $-\mathbf{a}$. Oppositely equal vectors have the same length ($|\mathbf{a}| = |-\mathbf{a}|$) and are located on parallel straight lines but have opposite orientations; that is, they are antiparallel ($\mathbf{a} \uparrow\downarrow -\mathbf{a}$). If, for instance, $\mathbf{a} = \overrightarrow{AB}$, then $-\mathbf{a} = \overrightarrow{BA}$.

Addition: If two vectors **a** and **b** are added, the initial point of the one vector is brought by a parallel shift to coincide with the endpoint of the other one. The sum $\mathbf{a}+\mathbf{b}$, also called the *resultant*, then corresponds to the distance from the initial point of the first vector to the endpoint of the second one. This sum may also be found as the diagonal of the parallelogram formed by **a** and **b** (compare the figure).

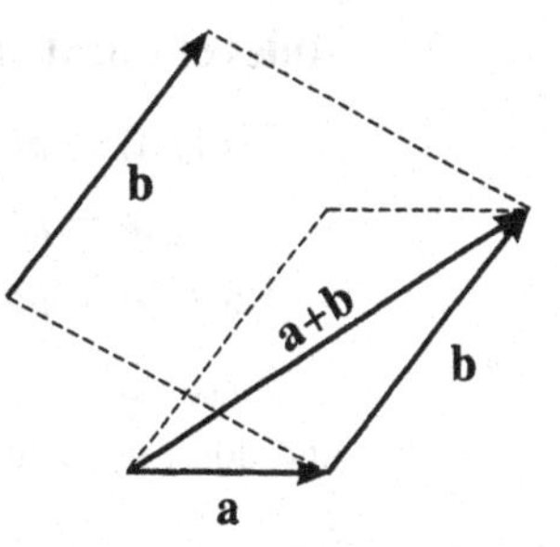

Addition of the vectors **a** and **b**.

Rules of calculation: There hold

$$\mathbf{a}+\mathbf{b}=\mathbf{b}+\mathbf{a} \qquad \text{(commutation law)}$$

and

$$(\mathbf{a}+\mathbf{b})+\mathbf{c}=\mathbf{a}+(\mathbf{b}+\mathbf{c}) \qquad \text{(association law)},$$

as is seen immediately (compare the figures).

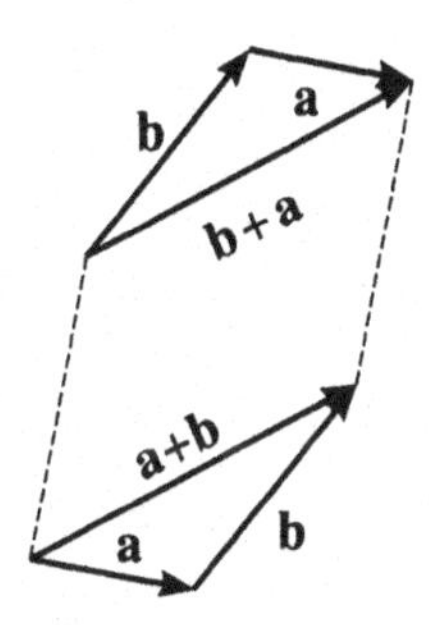

Illustration of the commutativity of the addition of vectors.

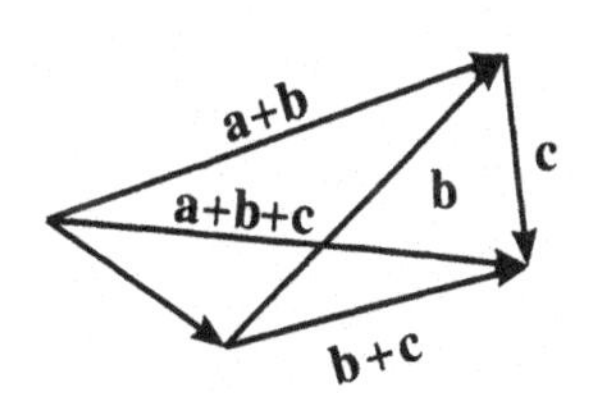

Illustration of the associativity of the addition of vectors.

Subtraction: The difference of two vectors **a** and **b** is defined as

$$\mathbf{a}-\mathbf{b}=\mathbf{a}+(-\mathbf{b}).$$

Zero (Null) vector: The vector difference $\mathbf{a}-\mathbf{a}$ is denoted as zero vector (or null vector):

$$\mathbf{a}-\mathbf{a}=\mathbf{0} \quad \text{or} \quad \mathbf{a}-\mathbf{a}=0.$$

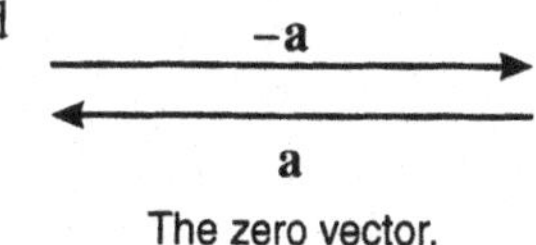

The zero vector.

The zero vector has magnitude 0; it is orientationless.

Multiplication of a vector by a scalar: The product $p\mathbf{a}$ of a vector **a** by a scalar p, where p is a real number, is understood as the vector having the same orientation as **a** and the magnitude $|p\mathbf{a}|=|p|\cdot|\mathbf{a}|$.

Rules of calculation:

$$q(p\mathbf{a}) = p(q\mathbf{a}) = qp\mathbf{a} \quad \text{(where } p \text{ and } q \text{ are real)},$$
$$(p+q)\mathbf{a} = p\mathbf{a} + q\mathbf{a},$$
$$p(\mathbf{a}+\mathbf{b}) = p\mathbf{a} + p\mathbf{b}.$$

These rules are immediately intelligible and don't need any further explanation.

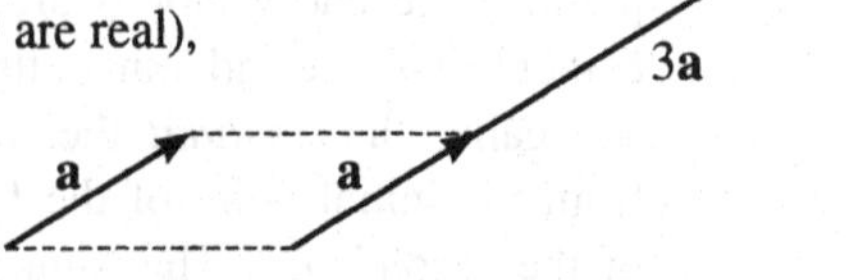

The multiplication of a vector **a** by a scalar p (in this case, $p = 3$).

2 The Scalar Product

The physical quantities force and path are oriented quantities and are represented by the vectors **F** and **s**. The *mechanical work* W performed by a force **F** along a straight path **s** is

$$W = Fs\cos\varphi = |\mathbf{F}|\,|\mathbf{s}|\cos\varphi,$$

where φ is the angle enclosed by **F** and **s**. W by itself, although originating from two vectors, is a *scalar* quantity. With a view on physical applications of this kind, we therefore define:

The *scalar product* $\mathbf{a}\cdot\mathbf{b}$ of two vectors is understood as

$$\mathbf{a}\cdot\mathbf{b} = |\mathbf{a}|\cdot|\mathbf{b}|\cdot\cos\varphi,$$

where φ is the angle enclosed by **a** and **b**. $\mathbf{a}\cdot\mathbf{b}$ is a *real number*. Expressed by words, the scalar product is defined as follows: $\mathbf{a}\cdot\mathbf{b} = |\mathbf{a}|$ multiplied by the projection of **b** onto **a**, or vice versa.

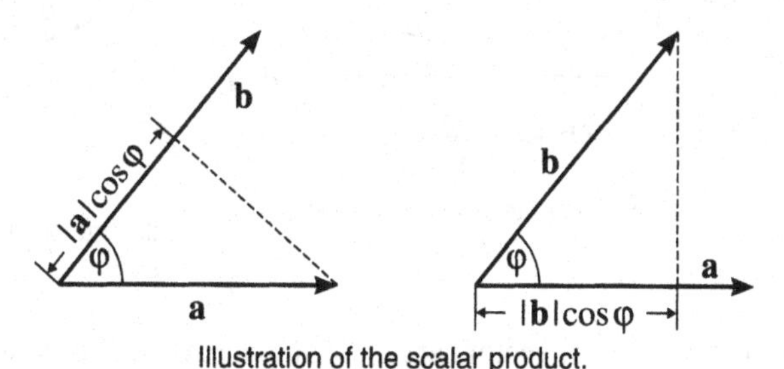

Illustration of the scalar product.

The visual meaning of the scalar product:

magnitude of the projection of **b** onto **a** multiplied by $|\mathbf{a}|$, or

magnitude of the projection of **a** onto **b** multiplied by $|\mathbf{b}|$.

Properties of the scalar product: $\mathbf{a}\cdot\mathbf{b}$ takes its maximum value for φ equal to zero ($\cos 0 = 1$, $\mathbf{a}$ parallel to $\mathbf{b}$)

$$\mathbf{a}\cdot\mathbf{b} = |\mathbf{a}|\cdot|\mathbf{b}|.$$

For $\varphi = \pi$ the scalar product takes its minimum value ($\cos\pi = -1$, $\mathbf{a}$ antiparallel to $\mathbf{b}$), namely

$$\mathbf{a}\cdot\mathbf{b} = -|\mathbf{a}|\cdot|\mathbf{b}|.$$

For $\varphi = \pi/2$, $\mathbf{a}\cdot\mathbf{b} = 0$ holds, even if $\mathbf{a}$ and $\mathbf{b}$ are nonzero ($\cos\pi/2 = 0$, $\mathbf{a}$ perpendicular to $\mathbf{b}$); thus

$$\mathbf{a}\cdot\mathbf{b} = 0 \qquad \text{if} \qquad \mathbf{a}\perp\mathbf{b}.$$

Rules of calculation: The following are true:

$$\mathbf{a}\cdot\mathbf{b} = \mathbf{b}\cdot\mathbf{a} \qquad \text{(commutativity);}$$

$$\mathbf{a}\cdot(\mathbf{b}+\mathbf{c}) = \mathbf{a}\cdot\mathbf{b}+\mathbf{a}\cdot\mathbf{c} \qquad \text{(distributivity);}$$

$$p(\mathbf{b}\cdot\mathbf{c}) = (p\,\mathbf{b})\cdot\mathbf{c} \qquad \text{(associativity).}$$

The first and last rules are immediately intelligible; the second rule is illustrated in the figure below.

If $\mathbf{b}$, $\mathbf{c}$, $\mathbf{a}$ are not coplanar, the rule of distributivity may easily be visualized by a triangle located in space. The vector $\mathbf{a}$ may easily be visualized by a pencil or a pointing rod (compare the figures!).

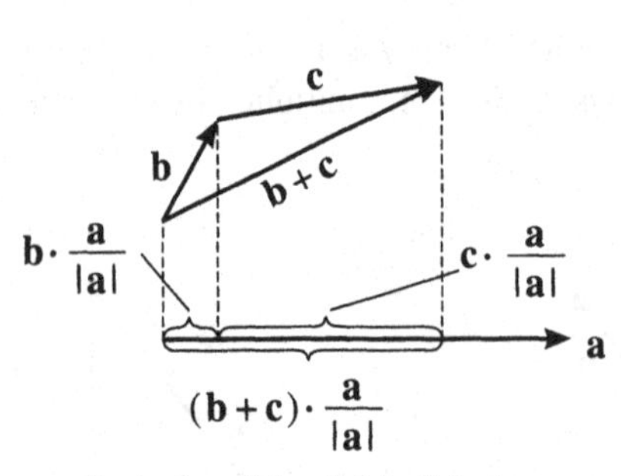

Illustration of the distributivity law.

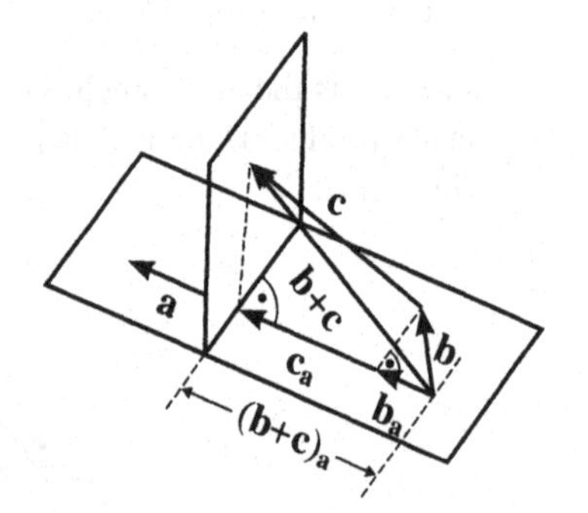

Illustration of the distributivity law in space.

Unit vectors: Unit vectors are understood as vectors of magnitude 1. If $\mathbf{a} \neq \mathbf{0}$, then

$$\mathbf{e} = \frac{\mathbf{a}}{|\mathbf{a}|}$$

is a unit vector pointing along $\mathbf{a}$. Actually, the magnitude of $\mathbf{e}$ equals 1 since $|\mathbf{e}| = \big|\mathbf{a}/|\mathbf{a}|\big| = |\mathbf{a}|/|\mathbf{a}| = 1$. A possibility frequently used in physics is to assign a direction to a scalarly formulated equation by the unit vector. For example, the gravitational force has the magnitude

$$F = \gamma \frac{mM}{r^2}.$$

It is acting along the connecting line between the two masses M and m, hence

$$\mathbf{F} = -\gamma \frac{mM}{r^2} \frac{\mathbf{r}}{|\mathbf{r}|}.$$

$\mathbf{F}$ is the force applied by the mass M to the mass m. Its direction is given by $-\mathbf{e}_r = -\mathbf{r}/|\mathbf{r}|$. Hence it is acting toward the mass M.

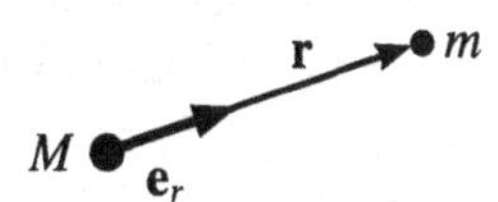

The unit vector pointing from the big mass to the small mass is $\mathbf{e}_r = \mathbf{r}/|\mathbf{r}|$.

Cartesian unit vectors: The unit vectors pointing along the positive x-, y-, and z-axes of a Cartesian coordinate frame are defined as follows:

$\mathbf{e}_1$ (in x-direction) or also $\mathbf{i}$;

$\mathbf{e}_2$ (in y-direction) or also $\mathbf{j}$;

$\mathbf{e}_3$ (in z-direction) or also $\mathbf{k}$.

There exist two kinds of Cartesian coordinate frames, namely right-handed frames and left-handed frames (compare the figures below).

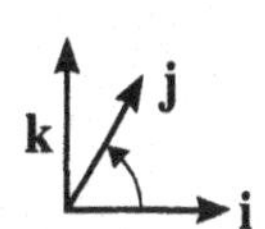

right-handed system: **k** points into the direction of a right-handed screw when $\mathbf{i} \mapsto \mathbf{j}$ is rotated along the shortest possible way.

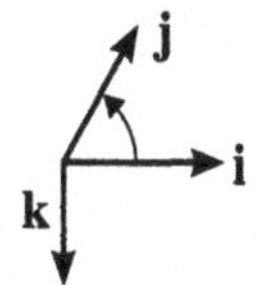

left-handed system: **k** points into the direction of a left-handed screw when $\mathbf{i} \mapsto \mathbf{j}$ is rotated along the shortest possible way.

We shall always use only right-handed frames in these lectures!

Orthonormality relations: $\mathbf{i}, \mathbf{j}, \mathbf{k}$ or $\mathbf{e}_1, \mathbf{e}_2, \mathbf{e}_3$ will be used in the following always concurrently, depending on convenience.

We now consider the properties of the Cartesian unit vectors with respect to formation of scalar products: Since the enclosed angle is each a right one, the following relations hold:

$$\begin{aligned} &\mathbf{i}\cdot\mathbf{i} = \mathbf{j}\cdot\mathbf{j} = \mathbf{k}\cdot\mathbf{k} = 1 \quad \text{(because of } \varphi = 0\text{, hence } \cos 0 = 1\text{);} \\ &\mathbf{i}\cdot\mathbf{j} = \mathbf{i}\cdot\mathbf{k} = \mathbf{j}\cdot\mathbf{k} = 0 \quad \text{(because of } \varphi = \pi/2\text{, hence } \cos \pi/2 = 0\text{).} \end{aligned} \tag{2.1}$$

These relations are combined by defining

$$\mathbf{e}_\mu \cdot \mathbf{e}_\nu = \delta_{\mu\nu}, \qquad \text{where} \quad \delta_{\mu\nu} = \begin{cases} 0 & \text{for } \nu \neq \mu, \\ 1 & \text{for } \nu = \mu, \end{cases}$$

and is called the *Kronecker symbol.*[1] For the three-dimensional space, μ and ν are running from 1 to 3, $\mathbf{e}_1 = \mathbf{i}$, $\mathbf{e}_2 = \mathbf{j}$, $\mathbf{e}_3 = \mathbf{k}$.

[1] *Leopold Kronecker*, b. Dec. 7, 1823, Liegnitz (Legnica)—d. Dec. 29, 1891, Berlin. Kronecker was a rich private person who moved to Berlin in 1855. He taught for many years at the university there, without having a chair. Only in 1883, after retirement of his teacher and friend *Kummer*, he took a professorship. His most important publications concern arithmetics, theory of ideals, number theory, and elliptic functions. Kronecker was the leading representative of the *Berlin School*, which claimed the necessity of arithmetization of the entire mathematics.

3 Component Representation of a Vector

The vector **a**, which is uniquely represented by the sum of vectors—in our example by the sum of the vectors **b**, **c**, **d**, **f**—is called the *linear combination* of the vectors (e.g., **b**, **c**, **d**, and **f**). The term "vectors" and their "linear combination" thus graphically form a closed polygon, the vector polygon. One may, of course, conclude from given vectors **b**, **c**, **d** on the linear combination that yields the arbitrary (but fixed) vector **a**.

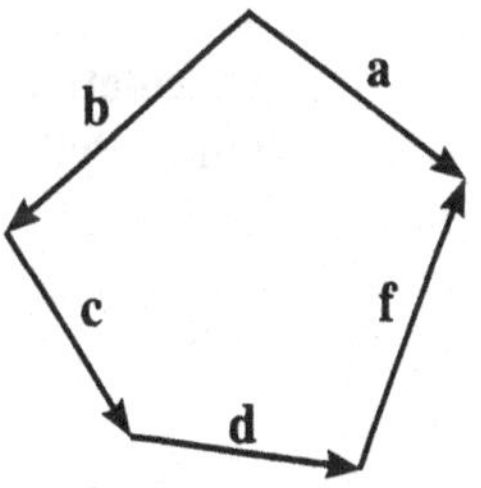

The vector polygon.

According to the definition introduced above, the vector **a** then must be a linear combination of the vectors **b**, **c**, **d**; thus

$$\mathbf{a} = q_1\mathbf{b} + q_2\mathbf{c} + q_3\mathbf{d}.$$

q_1, q_2, and q_3 are denoted as components of the vector **a** with respect to **b**, **c**, **d**. The vectors **b**, **c**, **d** must be linearly independent, that is, none of the three vectors may be represented by the other two vectors. Otherwise not every arbitrary vector **a** could be combined out of the three basic vectors **b**, **c**, **d**. If, for example, **d** could be expressed by **b** and **c**, hence $\mathbf{d} = \alpha\mathbf{b} + \beta\mathbf{c}$, then $\mathbf{a} = (q_1 + q_3\alpha)\mathbf{b} + (q_2 + q_3\beta)\mathbf{c}$ would always be confined to lie in the plane spanned by **b** and **c**. But an arbitrary vector **a** in general does not lie in this plane (e.g., points out of this plane). One says: *The base* **b**, **c** *is incomplete for arbitrary vectors* **a**. In the three-dimensional space one therefore always needs three basic vectors that are linearly independent (i.e., cannot be expressed by each other).

Component representation of a vector in Cartesian coordinates: Any vector of the three-dimensional space may be represented as a linear combination of the Cartesian unit vectors **i**, **j**, **k**. This representation leads to simple and transparent calculations, due to the orthogonality relations. One then has

$$\mathbf{a} = a_x\mathbf{i} + a_y\mathbf{j} + a_z\mathbf{k},$$

where $a_x = \mathbf{a} \cdot \mathbf{i}$, $a_y = \mathbf{a} \cdot \mathbf{j}$, and $a_z = \mathbf{a} \cdot \mathbf{k}$ are the projections of **a** onto the axes of the frame. The unit vectors **i**, **j**, **k** (or $\mathbf{e}_1, \mathbf{e}_2, \mathbf{e}_3$) are also called *base vectors.*

Besides the representation as a sum of vectors along the unit vectors, the vector **a** still may be represented as

$$\mathbf{a} = (a_x, a_y, a_z) \qquad \text{(row notation)},$$

$$\mathbf{a} = \begin{pmatrix} a_x \\ a_y \\ a_z \end{pmatrix} \qquad \text{(column notation)}.$$

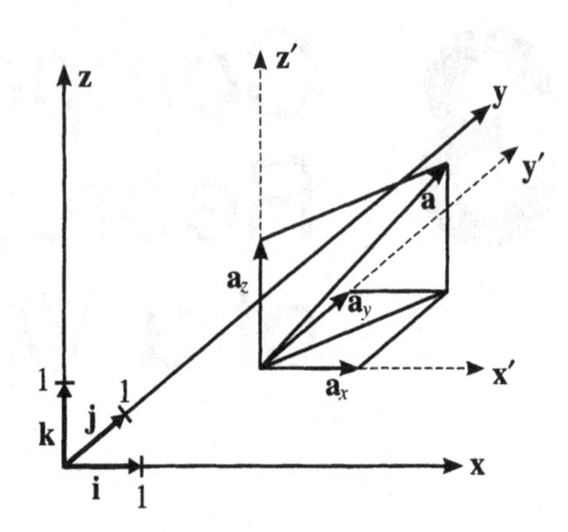

The components of a vector are obtained by parallel projection.

If the base vectors are known, it is sufficient to know the three components.

Calculation of the magnitude of a vector from the components: According to the theorem of Pythagoras, the magnitude of a vector **a** is calculated from its Cartesian components as follows:

$$|\mathbf{a}| = \sqrt{a_x^2 + a_y^2 + a_z^2}.$$

Addition of vectors expressed by components: One has

$$\begin{aligned}\mathbf{a} + \mathbf{b} &= \sum_{i=1}^{3} a_i\mathbf{e}_i + \sum_{i=1}^{3} b_i\mathbf{e}_i = \sum_{i=1}^{3}(a_i + b_i)\mathbf{e}_i \\ &= (a_1 + b_1)\mathbf{e}_1 + (a_2 + b_2)\mathbf{e}_2 + (a_3 + b_3)\mathbf{e}_3 \\ &= (a_1 + b_1, a_2 + b_2, a_3 + b_3)\,.\end{aligned}$$

Here both commutativity as well as associativity of vector addition have been used repeatedly. Thus, the components of the sum vector are the sums of the corresponding components of the individual vectors.

The scalar product in component representation: One has

$$\begin{aligned}\mathbf{a} \cdot \mathbf{b} &= (a_x\mathbf{i} + a_y\mathbf{j} + a_z\mathbf{k}) \cdot (b_x\mathbf{i} + b_y\mathbf{j} + b_z\mathbf{k}), \\ &= a_xb_x\mathbf{i} \cdot \mathbf{i} + a_xb_y\mathbf{i} \cdot \mathbf{j} + a_xb_z\mathbf{i} \cdot \mathbf{k} + a_yb_x\mathbf{j} \cdot \mathbf{i} + a_yb_y\mathbf{j} \cdot \mathbf{j} + a_yb_z\mathbf{j} \cdot \mathbf{k} \\ &\quad + a_zb_x\mathbf{k} \cdot \mathbf{i} + a_zb_y\mathbf{k} \cdot \mathbf{j} + a_zb_z\mathbf{k} \cdot \mathbf{k}\,.\end{aligned}$$

Taking into account the orthonormality relations (2.1), we then get

$$\mathbf{a} \cdot \mathbf{b} = a_x b_x + a_y b_y + a_z b_z \,. \tag{3.1}$$

Finally, setting for the indices $x \mathrel{\hat{=}} 1$, for $y \mathrel{\hat{=}} 2$, and for $z \mathrel{\hat{=}} 3$, then one can write

$$\mathbf{a} \cdot \mathbf{b} = \sum_{i=1}^{3} a_i b_i \,. \tag{3.2}$$

Hence, the scalar product of two vectors may be evaluated simply by multiplying the corresponding components of the vectors by each other and summing over the three products.

Problem 3.1: Addition and subtraction of vectors

A DC-10 "flies" north-west at 930 km/h relative to ground. A strong breeze blows from the west with 120 km/h relative to ground.

What are the velocity and direction of flight of the plane, assuming that there is no wind deflection?

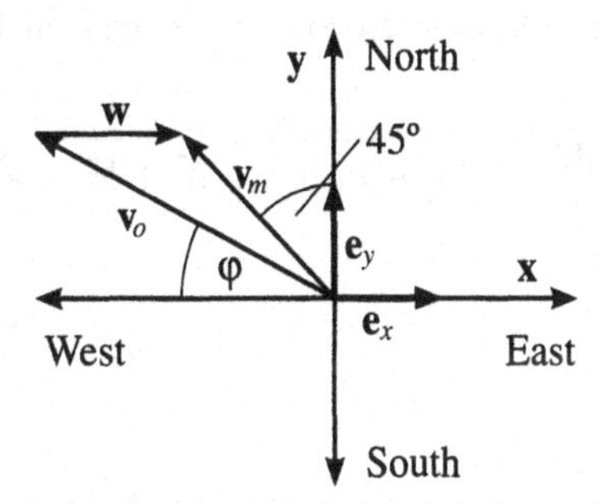

The relative directions of wind and airplane velocity.

Solution Let

$|\mathbf{v}_m| = 930$ km/h, the velocity of the plane in the wind,

$|\mathbf{v}_0| =$ the velocity of the plane without wind,

$|\mathbf{w}| =$ the wind velocity.

Now we can write

$$\begin{aligned}
\mathbf{w} &= 120\,\mathbf{e}_x \,, \\
\mathbf{v}_m &= -930\cos(45^\circ)\mathbf{e}_x + 930\sin(45^\circ)\mathbf{e}_y \\
&= -657.61\mathbf{e}_x + 657.61\mathbf{e}_y \,, \\
\mathbf{v}_0 &= \mathbf{v}_m - \mathbf{w} = -777.61\mathbf{e}_x + 657.61\mathbf{e}_y \\
&\Rightarrow\; v_0 = |\mathbf{v}_0| = 1018.39\,\text{km/h}, \\
\tan\varphi &= \frac{|v_{0y}|}{|v_{0x}|} = 0.846 \\
&\Rightarrow\; \varphi = 40.2^\circ .
\end{aligned}$$

The position vector: A point P in space may be uniquely fixed by specifying the vector beginning at the origin of the coordinate frame and pointing to the point P as endpoint.

The components of this vector, the position vector, then correspond to the coordinates (x, y, z) of the point P. Thus, for the position vector, which is mostly abbreviated by $\mathbf{r}$, there holds

$$\mathbf{r} = x\mathbf{i} + y\mathbf{j} + z\mathbf{k}, \quad \text{or: } \mathbf{r} = (x, y, z);$$
$$|\mathbf{r}| = \sqrt{x^2 + y^2 + z^2}.$$

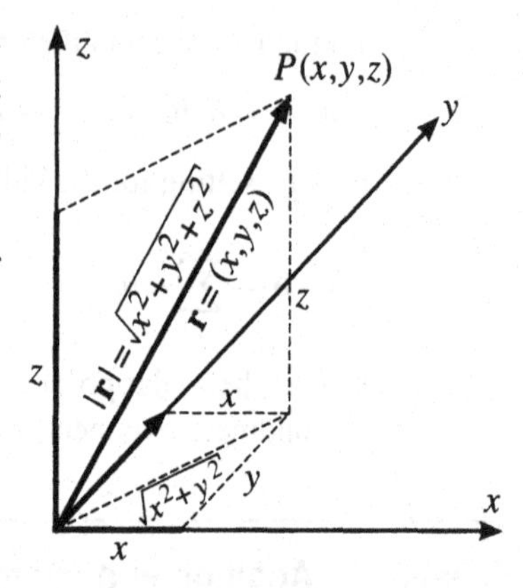

The position vector and its coordinates.

The angle between two vectors: From the knowledge of the two possibilities for representing the scalar product

$$\mathbf{a} \cdot \mathbf{b} = |\mathbf{a}|\,|\mathbf{b}| \cos\varphi = a_x b_x + a_y b_y + a_z b_z,$$

one obtains the following relation for the angle enclosed by $\mathbf{a}$ and $\mathbf{b}$:

$$\cos\varphi = \frac{\mathbf{a} \cdot \mathbf{b}}{|\mathbf{a}|\,|\mathbf{b}|} = \frac{a_x b_x + a_y b_y + a_z b_z}{\sqrt{a_x^2 + a_y^2 + a_z^2}\sqrt{b_x^2 + b_y^2 + b_z^2}}.$$

4 The Vector Product (Axial Vector)

One may define a further product between vectors. Here a new vector arises that is defined as follows.

Definition: The vector product of two vectors **a** and **b** is the vector

$$\mathbf{a} \times \mathbf{b} = (|\mathbf{a}| \cdot |\mathbf{b}| \sin\varphi)\mathbf{n}, \tag{4.1}$$

where **n** is the unit vector being perpendicular to the plane fixed by **a** and **b**, and pointing out of the plane as a right-handed helix when rotating the first vector of the product into the second vector. Note that the rotation has to be performed along the shortest path.

The magnitude of the vector product is equal to the area of the parallelogram spanned by **a** and **b**, as is seen from the figure.

Geometrical interpretation of the absolute value of the vector product as area.

$$F = |\mathbf{a} \times \mathbf{b}| = |\mathbf{a}| \cdot |\mathbf{b}| \sin\varphi = ab \sin\varphi,$$

Properties of the vector product: $\mathbf{a} \times \mathbf{b}$ takes its maximum magnitude for $\varphi = \pi/2$, $\sin(\pi/2) = 1$, **a** perpendicular to **b**, $|\mathbf{a} \times \mathbf{b}| = |\mathbf{a}|\,|\mathbf{b}|$.

$\mathbf{a} \times \mathbf{b}$ vanishes for $\varphi = 0$ ($\sin 0 = 0$, **a** parallel to **b**).

$$\mathbf{a} \times \mathbf{b} = 0 \begin{cases} \text{if } \mathbf{a} \uparrow\downarrow \mathbf{b} \text{ or} & (\uparrow\downarrow \text{ means antiparallel}) \\ \text{if } \mathbf{a} \uparrow\uparrow \mathbf{b}. & (\uparrow\uparrow \text{ means parallel}) \end{cases}$$

The formula also includes the special case $\mathbf{a} = \mathbf{b}$, thus

$$\mathbf{a} \times \mathbf{a} = 0.$$

Notations:

$\odot$ represents a vector perpendicular to the drawing plane and pointing out of the plane (arrowhead).

$\otimes$ represents a vector perpendicular to the drawing plane and pointing into the plane (arrowbase).

Rules of calculation: The vector product has the following properties:

$$\begin{array}{lll} \text{I.} & \mathbf{a}\times\mathbf{b} = -\mathbf{b}\times\mathbf{a} & \text{(no commutativity);} \\ \text{II.} & \mathbf{a}\times(\mathbf{b}+\mathbf{c}) = \mathbf{a}\times\mathbf{b}+\mathbf{a}\times\mathbf{c} & \text{(distributivity);} \\ \text{III.} & \mathbf{a}\times(\mathbf{b}\times\mathbf{c}) \neq (\mathbf{a}\times\mathbf{b})\times\mathbf{c} & \text{(no associativity);} \\ \text{IV.} & p(\mathbf{a}\times\mathbf{b}) = (p\mathbf{a})\times\mathbf{b} = \mathbf{a}\times(p\mathbf{b}). & \end{array} \tag{4.2}$$

Rule I follows immediately from the definition of the vector product (compare with the figure).

Rule III: The vector on the left side lies in the plane spanned by the vectors **b** and **c**; the vector on the right side is in the plane spanned by **a** and **b**. The subsequent example also shows that associativity does not hold. One has $\mathbf{e}_1\times(\mathbf{e}_2\times\mathbf{e}_2) = 0$, but $(\mathbf{e}_1\times\mathbf{e}_2)\times\mathbf{e}_2 = -\mathbf{e}_1$.

Rule II: The proof is given in two steps:

1. Let **a** be perpendicular ($\perp$) on **b** and **c**, that is, $\mathbf{a}\cdot\mathbf{b} = \mathbf{a}\cdot\mathbf{c} = 0$. Then $\mathbf{a}\times(\mathbf{b}+\mathbf{c}) = \mathbf{a}\times\mathbf{b}+\mathbf{a}\times\mathbf{c}$. The proof for that may be read off immediately from the two figures. $\mathbf{a}\times\mathbf{b}$ stands $\perp$ on **b** and **a**, is rotated against **b** by 90°, and is longer than **b** by the factor $|\mathbf{a}|$. The situation is similar for $\mathbf{a}\times\mathbf{c}$ and $\mathbf{a}\times(\mathbf{b}+\mathbf{c})$. The parallelogram of the vectors $\mathbf{a}\times\mathbf{b}$, $\mathbf{a}\times\mathbf{c}$, $\mathbf{a}\times(\mathbf{b}+\mathbf{c})$ emerges from that of the vectors **b**, **c**, $(\mathbf{b}+\mathbf{c})$ by a rotation about **a** by 90° and subsequent stretching by $|\mathbf{a}|$.

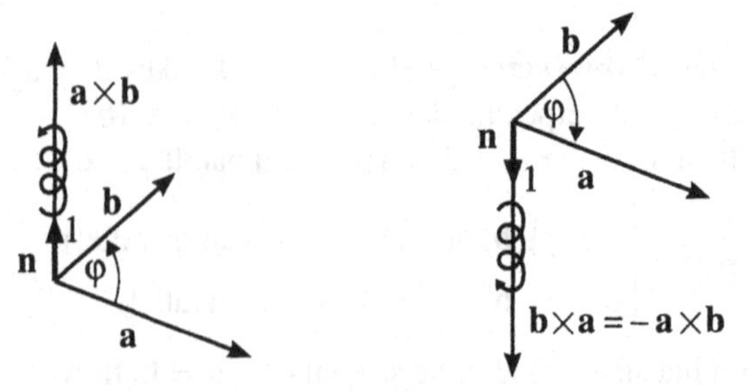

Illustration of the calculational rule I. The direction of rotation is shown.

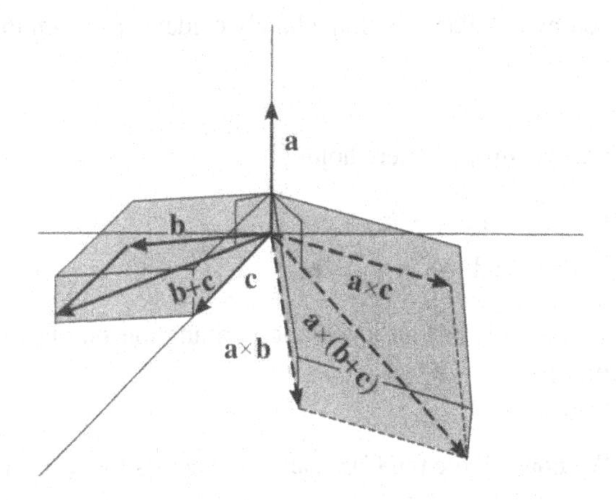

Perspective view of the special case: $\mathbf{a} \perp \mathbf{b}$ and $\mathbf{a} \perp \mathbf{c}$.

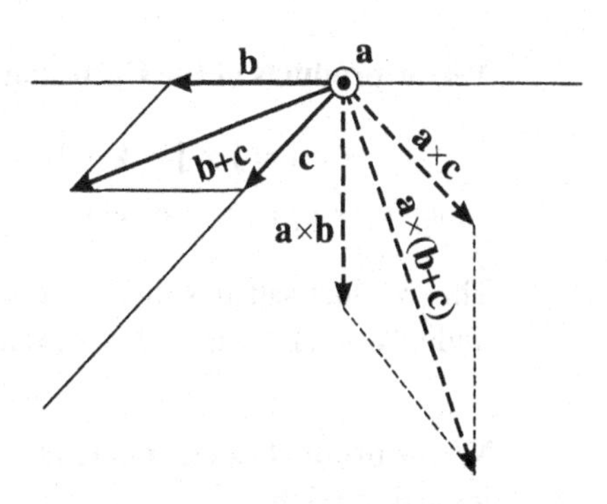

Plan view from top of the special case: $\mathbf{a} \perp \mathbf{b}$ and $\mathbf{a} \perp \mathbf{c}$.

2. We now decompose in the general case

$$\mathbf{b} = \mathbf{b}_\perp + \mathbf{b}_{||},$$
$$\mathbf{c} = \mathbf{c}_\perp + \mathbf{c}_{||},$$

that is, $\mathbf{b}$ and $\mathbf{c}$ into components $\perp$ and $||$ to $\mathbf{a}$ (compare with the figure).

Then, on the one hand, the following holds:

$$\mathbf{a} \times \mathbf{b} = (|\mathbf{a}| \cdot |\mathbf{b}| \cdot \sin\varphi)\,\frac{\mathbf{a} \times \mathbf{b}}{|\mathbf{a} \times \mathbf{b}|};$$

and, on the other hand,

$$\mathbf{a} \times \mathbf{b}_\perp = (|\mathbf{a}| \cdot |\mathbf{b}_\perp|)\frac{\mathbf{a} \times \mathbf{b}}{|\mathbf{a} \times \mathbf{b}|}$$
$$= (|\mathbf{a}| \cdot |\mathbf{b}| \cdot \sin\varphi)\,\frac{\mathbf{a} \times \mathbf{b}}{|\mathbf{a} \times \mathbf{b}|},$$

The general case: the vectors $\mathbf{b}$ and $\mathbf{c}$ are both decomposed into components parallel ($||$) and perpendicular ($\perp$) to $\mathbf{a}$.

and therefore

$$\mathbf{a} \times \mathbf{b} = \mathbf{a} \times \mathbf{b}_\perp. \tag{4.3}$$

This holds for any arbitrary vector $\mathbf{b}$. Therefore, one immediately concludes $\mathbf{a} \times \mathbf{c} = \mathbf{a} \times \mathbf{c}_\perp$. We may then conclude

$$\begin{aligned} \mathbf{a} \times (\mathbf{b} + \mathbf{c}) &= \mathbf{a} \times (\mathbf{b} + \mathbf{c})_\perp = \mathbf{a} \times (\mathbf{b}_\perp + \mathbf{c}_\perp) \\ &= \mathbf{a} \times \mathbf{b}_\perp + \mathbf{a} \times \mathbf{c}_\perp && \text{(because of the special case in 1)} \\ &= \mathbf{a} \times \mathbf{b} + \mathbf{a} \times \mathbf{c} && \text{(because of (4.3))} \end{aligned}$$

q.e.d.

Rule IV: The rule for multiplication by a scalar p is immediately evident if we remind ourselves of the meaning of $p\mathbf{a}$.

Vector products of the Cartesian unit vectors: There holds

$$\mathbf{i}\times\mathbf{i}=\mathbf{j}\times\mathbf{j}=\mathbf{k}\times\mathbf{k}=0, \quad \text{and} \quad \mathbf{i}\times\mathbf{j}=\mathbf{k},$$

$$\text{or} \quad \mathbf{e}_1\times\mathbf{e}_1=\mathbf{e}_2\times\mathbf{e}_2=\mathbf{e}_3\times\mathbf{e}_3=0, \quad \text{and} \quad \mathbf{e}_1\times\mathbf{e}_2=\mathbf{e}_3. \tag{4.4}$$

This product satisfies the cyclic permutability. For an anticyclic permutation one has to multiply by the factor -1, for example, $\mathbf{j}\times\mathbf{i}=-\mathbf{k}$.

Vector product in components: We now denote the Cartesian unit vectors by $\mathbf{e}_1$, $\mathbf{e}_2$, $\mathbf{e}_3$ instead of $\mathbf{i}$, $\mathbf{j}$, $\mathbf{k}$.

Let

$$\mathbf{a}=a_1\mathbf{e}_1+a_2\mathbf{e}_2+a_3\mathbf{e}_3=\sum_{i=1}^{3}a_i\mathbf{e}_i \quad \text{and} \quad \mathbf{b}=\sum_{i=1}^{3}b_i\mathbf{e}_i$$

be two arbitrary vectors. When forming the vector product of the two vectors $\mathbf{a}=\sum_{i=1}^{3}a_i\mathbf{e}_i$ and $\mathbf{b}=\sum_{i=1}^{3}b_i\mathbf{e}_i$, one obtains

$$\begin{aligned}\mathbf{a}\times\mathbf{b}&=(a_1\mathbf{e}_1+a_2\mathbf{e}_2+a_3\mathbf{e}_3)\times(b_1\mathbf{e}_1+b_2\mathbf{e}_2+b_3\mathbf{e}_3)\\&=a_1b_2\mathbf{e}_3-a_2b_1\mathbf{e}_3+a_2b_3\mathbf{e}_1-a_3b_2\mathbf{e}_1+a_3b_1\mathbf{e}_2-a_1b_3\mathbf{e}_2\\&=(a_2b_3-a_3b_2)\mathbf{e}_1+(a_3b_1-a_1b_3)\mathbf{e}_2+(a_1b_2-a_2b_1)\mathbf{e}_3.\end{aligned} \tag{4.5}$$

It is now practical to introduce the determinant notation.

Determinants: A rectangular array of numbers is called a *matrix* (see the figure).

$$\begin{array}{c}\text{column}\\ \downarrow\end{array}$$

$$\begin{pmatrix}a_{11} & a_{12} & \dots & a_{1q}\\ a_{21} & a_{22} & \dots & a_{2q}\\ \dots & \dots & \dots & \dots\\ a_{p1} & a_{p2} & \dots & a_{pq}\end{pmatrix} \quad \leftarrow \text{row}$$

For the case $q=p$, the matrix is called *quadratic*. One then can assign a numerical value D to it, called a *determinant*. It is defined as follows:

I. $$\det(a_{11})\equiv|a_{11}|=a_{11};$$

II. $$\det\begin{pmatrix}a_{11} & a_{12}\\ a_{21} & a_{22}\end{pmatrix}\equiv\begin{vmatrix}a_{11} & a_{12}\\ a_{21} & a_{22}\end{vmatrix}=a_{11}a_{22}-a_{12}a_{21}; \tag{4.6}$$

$$\textbf{III.}\ \det\begin{pmatrix} a_{11} & a_{12} & a_{13} \\ a_{21} & a_{22} & a_{23} \\ a_{31} & a_{32} & a_{33} \end{pmatrix} \equiv \begin{vmatrix} a_{11} & a_{12} & a_{13} \\ a_{21} & a_{22} & a_{23} \\ a_{31} & a_{32} & a_{33} \end{vmatrix}$$

$$= a_{11}\begin{vmatrix} a_{22} & a_{23} \\ a_{32} & a_{33} \end{vmatrix} - a_{12}\begin{vmatrix} a_{21} & a_{23} \\ a_{31} & a_{33} \end{vmatrix} + a_{13}\begin{vmatrix} a_{21} & a_{22} \\ a_{31} & a_{32} \end{vmatrix}.$$

The evaluation of the 3×3 determinants may be simplified by using the so-called Sarrus rule.[1] The procedure is: Establish an additional auxiliary matrix by writing the first two columns of the original matrix once again to its right side, and form the product terms, involving signs, according to the following scheme.

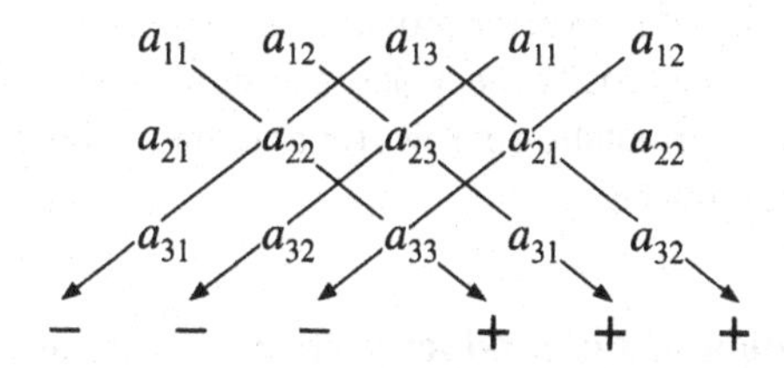

Multiple-row determinants may be reduced to determinants of lower order by expansion with respect to a row or column (formation of subdeterminants), analogous to (eq. (4.6), III). We will see this method at work in Example 4.4 on the Laplace expansion theorem.

Rules of calculation: The most important rules for calculations involving determinants are

1. If two rows or columns of the quadratic matrix are identical or proportional to each other, then the determinant of this matrix = 0.
2. When permuting any two neighboring rows or columns, the sign of the determinant changes.
3. The determinant of the matrix reflected at the main diagonal (also called the transposed matrix) is equal to the original determinant.

[1] *Pierre Frédéric Sarrus*, b. 1798—d. 1861, Saint Affriques. Sarrus was professor of mathematics in Strasbourg from 1826 untill 1856. He dealt mainly with the numerical solution of equations with several unknowns (1832), with multiple integrals (1842), and with the determination of comet orbits (1843). The rule for evaluating three-row determinants is named after him.

4. The expansion theorem that has been used in equation (4.6), III, with respect to the first row, holds for the first column in the same way.

These rules may easily be checked explicitly in the cases quoted above (eq. (4.6), I, II, III). The cases I–III are the most important ones in the present context. The properties of the 3×3 determinants will be outlined and discussed in more detail in the context of Problem 4.3. The rules hold, however, in general for arbitrary determinants.

The vector product (4.5) may now be written as a three-row determinant:

$$\begin{aligned} \mathbf{a} \times \mathbf{b} &= \begin{vmatrix} \mathbf{e}_1 & \mathbf{e}_2 & \mathbf{e}_3 \\ a_1 & a_2 & a_3 \\ b_1 & b_2 & b_3 \end{vmatrix} \\ &= \mathbf{e}_1(a_2b_3 - a_3b_2) + \mathbf{e}_2(a_3b_1 - a_1b_3) + \mathbf{e}_3(a_1b_2 - a_2b_1) \,. \end{aligned} \tag{4.7}$$

If the two vectors of the cross product are equal, then the two lower rows of the determinant are also equal, and the vector product vanishes.

Further, one may easily check based on equation (4.6), III, that the sign of the determinant changes under permutation of rows (or columns). This corresponds to the anticommutativity of the vector product.

Representation of the product vector: As we already stated in the definition of the vector product, the magnitude of the product vector may be visualized by a distance but better by the area of the parallelogram formed by the vectors. This vector is not determined by its length and orientation only (such vectors are called *polar vectors*) but is called an *axial vector*. To understand this difference, we consider space reflections: We thereby change from the components a_1, a_2, a_3 to the new base vectors $a_1' = -a_1$, $a_2' = -a_2$, $a_3' = -a_3$. The vector $\mathbf{a}$ is thus reflected at the origin. Under a space reflection, which is also called a *parity transformation*, a polar vector changes its sign: $\mathbf{a} \to -\mathbf{a}$. An axial vector, on the contrary, remains unchanged: $\mathbf{a} \times \mathbf{b} = (-\mathbf{a}) \times (-\mathbf{b})$.

The invention of these new vectors is necessary since certain physical quantities need a handedness for a complete description. The handedness is taken into account by an axial vector. Such kinds of quantities are, for instance, the angular velocity and the angular momentum. One should get straight in one's mind that a handedness remains unchanged under a space reflection!

An axial vector may, however, also be represented by an oriented distance.

The double-vector product: The vector product $\mathbf{a} \times (\mathbf{b} \times \mathbf{c})$ is called the double-vector product. To evaluate it, we denote the components of $\mathbf{b} \times \mathbf{c}$ as follows: Let

$(\mathbf{b} \times \mathbf{c})_x$ be the x-component,

$(\mathbf{b} \times \mathbf{c})_y$ be the y-component, and

$(\mathbf{b} \times \mathbf{c})_z$ be the z-component.

For the x-component of the double-vector product, it then follows that

$$\begin{aligned}(\mathbf{a}\times(\mathbf{b}\times\mathbf{c}))_x &= a_y(\mathbf{b}\times\mathbf{c})_z - a_z(\mathbf{b}\times\mathbf{c})_y\\ &= a_y(b_x c_y - b_y c_x) - a_z(b_z c_x - b_x c_z).\end{aligned}$$

We add $a_x b_x c_x - a_x b_x c_x = 0$ and obtain

$$\begin{aligned}(\mathbf{a}\times(\mathbf{b}\times\mathbf{c}))_x &= b_x(a_x c_x + a_y c_y + a_z c_z) - c_x(a_x b_x + a_y b_y + a_z b_z)\\ &= b_x(\mathbf{a}\cdot\mathbf{c}) - c_x(\mathbf{a}\cdot\mathbf{b}).\end{aligned}$$

Analogous considerations for the y- and z-components of $\mathbf{a}\times(\mathbf{b}\times\mathbf{c})$ yield the

Graßmann expansion theorem: One has

$$\mathbf{a}\times(\mathbf{b}\times\mathbf{c}) = (\mathbf{a}\cdot\mathbf{c})\mathbf{b} - (\mathbf{a}\cdot\mathbf{b})\mathbf{c},$$

while

$$(\mathbf{a}\times\mathbf{b})\times\mathbf{c} = (\mathbf{a}\cdot\mathbf{c})\mathbf{b} - (\mathbf{b}\cdot\mathbf{c})\mathbf{a}. \tag{4.8}$$

This is another proof of the fact that the vector product is not associative (see (4.2), III).

Problem 4.1: Vector product

(a) The vector $(1, a, b)$ is perpendicular to the two vectors $(4, 3, 0)$ and $(5, 1, 7)$. Find a and b.

(b) Evaluate in Cartesian coordinates the vector product $\mathbf{a}\times\mathbf{b}$ for $\mathbf{a} = (1, 7, 0)$ and $\mathbf{b} = (1, 1, 1)$.

(c) Show that

$$(\mathbf{a}\times\mathbf{b})^2 = a^2b^2 - (\mathbf{a}\cdot\mathbf{b})^2 .$$

Solution (a) It must hold that $(1, a, b)\cdot(4, 3, 0) = 0$ and $(1, a, b)\cdot(5, 1, 7) = 0$. This yields the two equations

$$4 + 3a = 0 \quad \text{and} \quad 5 + a + 7b = 0 \quad \Rightarrow \quad a = -\frac{4}{3}, b = -\frac{11}{21}.$$

(b)

$$\begin{aligned}(\mathbf{a}\times\mathbf{b})_x &= (a_y b_z - a_z b_y) = 7;\\ (\mathbf{a}\times\mathbf{b})_y &= (a_z b_x - a_x b_z) = -1;\\ (\mathbf{a}\times\mathbf{b})_z &= (a_x b_y - a_y b_x) = -6.\end{aligned}$$

(c)

$$\begin{aligned}(\mathbf{a}\times\mathbf{b})^2 &= (|\mathbf{a}|\cdot|\mathbf{b}|\cdot\sin\varphi\cdot\mathbf{e}_n)^2 = |\mathbf{a}|^2|\mathbf{b}|^2\sin^2\varphi(\mathbf{e}_n)^2\\ &= |\mathbf{a}|^2|\mathbf{b}|^2(1-\cos^2\varphi)(\mathbf{e}_n)^2\\ &= a^2b^2 - (\mathbf{a}\cdot\mathbf{b})^2\end{aligned}$$

Here $\varphi : \sphericalangle(\mathbf{a}, \mathbf{b})$ and $\mathbf{e}_n$ is the unit vector along $\mathbf{a}\times\mathbf{b}$.

Problem 4.2: Proof of theorems on determinants

The most important theorems on determinants are as follows:

(a) Under permutation of rows and columns (reflection at the main diagonal), the value of a determinant remains unchanged.

(b) Under permutation of two arbitrary neighboring rows, the sign of the determinant changes.

(c) If all elements of a row contain a common factor c, then it may be pulled out of the determinant.

(d) If two rows of a determinant are proportional to each other, then the determinant = 0.

(e) The value of a determinant remains unchanged when adding a multiple of any row to another row.

Check these rules for a general 3×3 determinant.

Solution From the lecture we know the definition of the 3-determinant:

$$D = \begin{vmatrix} a_{11} & a_{12} & a_{13} \\ a_{21} & a_{22} & a_{23} \\ a_{31} & a_{32} & a_{33} \end{vmatrix}$$

$$= a_{11}(a_{22}a_{33} - a_{23}a_{32}) - a_{12}(a_{21}a_{33} - a_{23}a_{31}) + a_{13}(a_{21}a_{32} - a_{22}a_{31})$$

$$= a_{11}a_{22}a_{33} - a_{11}a_{23}a_{32} - a_{12}a_{21}a_{33} + a_{12}a_{23}a_{31} + a_{13}a_{21}a_{32} - a_{13}a_{22}a_{31}. \quad \textbf{(4.9)}$$

(a) Permutation of rows and columns of D (reflection at the main diagonal) leads to

$$\widetilde{D} = \begin{vmatrix} a_{11} & a_{21} & a_{31} \\ a_{12} & a_{22} & a_{32} \\ a_{13} & a_{23} & a_{33} \end{vmatrix}$$

$$= a_{11}(a_{22}a_{33} - a_{32}a_{23}) - a_{21}(a_{12}a_{33} - a_{32}a_{13}) + a_{31}(a_{12}a_{23} - a_{22}a_{13})$$

$$= a_{11}a_{22}a_{33} - a_{11}a_{32}a_{23} - a_{21}a_{12}a_{33} + a_{21}a_{32}a_{13} + a_{31}a_{12}a_{23} - a_{31}a_{22}a_{13}$$

$$= a_{11}(a_{22}a_{33} - a_{23}a_{32}) - a_{12}(a_{21}a_{33} - a_{23}a_{31}) + a_{13}(a_{21}a_{32} - a_{22}a_{31}).$$

A comparison with D (see above) yields

$$\widetilde{D} = D. \quad \textbf{(4.10)}$$

(b) Permutation of, for example, the second and third rows of D yields

$$D' = \begin{vmatrix} a_{11} & a_{12} & a_{13} \\ a_{31} & a_{32} & a_{33} \\ a_{21} & a_{22} & a_{23} \end{vmatrix}$$

$$= a_{11}(a_{32}a_{23} - a_{33}a_{22}) - a_{12}(a_{31}a_{23} - a_{33}a_{21}) + a_{13}(a_{31}a_{22} - a_{32}a_{21})$$

$$= a_{11}a_{32}a_{23} - a_{11}a_{33}a_{22} - a_{12}a_{31}a_{23} + a_{12}a_{33}a_{21} + a_{13}a_{31}a_{22} - a_{13}a_{32}a_{21}.$$

By means of 4.9, one immediately concludes that

$$D' = -D. \quad \textbf{(4.11)}$$

This means: When we permute the second and third rows, the determinant changes its sign. Similarly, one may check that for permuting other rows. From (a) the same result follows for the columns: When we permute neighboring columns, the determinant also changes its sign.

(c) We investigate

$$D'' = \begin{vmatrix} a_{11} & a_{12} & a_{13} \\ ca_{21} & ca_{22} & ca_{23} \\ a_{31} & a_{32} & a_{33} \end{vmatrix}$$

$$= a_{11}(ca_{22}a_{33} - ca_{23}a_{32}) - a_{12}(ca_{21}a_{33} - ca_{23}a_{31}) + a_{13}(ca_{21}a_{32} - ca_{22}a_{31})$$

$$= c\,[a_{11}(a_{22}a_{33} - a_{23}a_{32}) - a_{12}(a_{21}a_{33} - a_{23}a_{31}) + a_{13}(a_{21}a_{32} - a_{22}a_{31})]$$

and compare with 4.9. Obviously,

$$D'' = cD. \tag{4.12}$$

(d) For example, let the third row be proportional to the second row; thus

$$\widetilde{D}' = \begin{vmatrix} a_{11} & a_{12} & a_{13} \\ a_{21} & a_{22} & a_{23} \\ \lambda a_{21} & \lambda a_{22} & \lambda a_{23} \end{vmatrix}$$

$$= a_{11}(\lambda a_{22}a_{23} - \lambda a_{23}a_{22}) - a_{12}(\lambda a_{21}a_{23} - \lambda a_{23}a_{21}) + a_{13}(\lambda a_{21}a_{22} - \lambda a_{22}a_{21})$$

$$= 0. \tag{4.13}$$

Similarly, one may check the assertion for the proportionality of other rows. From (a), it follows immediately that the determinant also vanishes if two columns are proportional to each other.

(e) We add, for example, a multiple of the first row to the second row. Then

$$\widetilde{D}'' = \begin{vmatrix} a_{11} & a_{12} & a_{13} \\ a_{21} + \lambda a_{11} & a_{22} + \lambda a_{12} & a_{23} + \lambda a_{13} \\ a_{31} & a_{32} & a_{33} \end{vmatrix}$$

$$\begin{aligned}
&= a_{11}\Big[(a_{22} + \lambda a_{12})a_{33} - (a_{23} + \lambda a_{13})a_{32}\Big] \\
&\quad - a_{12}\Big[(a_{21} + \lambda a_{11})a_{33} - (a_{23} + \lambda a_{13})a_{31}\Big] \\
&\quad + a_{13}\Big[(a_{21} + \lambda a_{11})a_{32} - (a_{22} + \lambda a_{12})a_{31}\Big] \\
&= a_{11}a_{22}a_{33} + \lambda a_{11}a_{12}a_{33} - a_{11}a_{23}a_{32} - \lambda a_{11}a_{13}a_{32} \\
&\quad - a_{12}a_{21}a_{33} - \lambda a_{12}a_{11}a_{33} + a_{12}a_{23}a_{31} + \lambda a_{12}a_{13}a_{31} \\
&\quad + a_{13}a_{21}a_{32} + \lambda a_{13}a_{11}a_{32} - a_{13}a_{22}a_{31} - \lambda a_{13}a_{12}a_{31} \\
&= a_{11}(a_{22}a_{33} - a_{23}a_{32}) - a_{12}(a_{21}a_{33} - a_{23}a_{31}) + a_{13}(a_{21}a_{32} - a_{22}a_{31}).
\end{aligned}$$

A comparison with 4.9 yields the assertion

$$\widetilde{D}'' = D. \tag{4.14}$$

Problem 4.3: Determinants

Calculate using the theorems on determinants:

$$\text{(a)}\quad \begin{vmatrix} x & x+1 & x+2 \\ 0 & 1 & 2 \\ 3 & 3 & 3 \end{vmatrix} \qquad \text{(b)}\quad \begin{vmatrix} a & d & xa+yd \\ b & e & xb+ye \\ c & f & xc+yf \end{vmatrix} \qquad \text{(c)}\quad \begin{vmatrix} 4 & 5 & 22 \\ 8 & 11 & 44 \\ 3 & 7 & 1 \end{vmatrix}$$

Solution (a) We form the linear combination

$$\alpha \cdot (2.\ \text{row}) + \beta \cdot (3.\ \text{row}) \quad \text{with} \quad \alpha = 1, \beta = \frac{x}{3}$$

and obtain $(x, x+1, x+2)$, thus just the first row. From (i) and (ii) of (4.6) it follows that the determinant is always equal to zero.

(b) The third column is a linear combination of the first and second columns with the factors x and y:

$$x\begin{pmatrix} a \\ b \\ c \end{pmatrix} + y\begin{pmatrix} d \\ e \\ f \end{pmatrix} = \begin{pmatrix} xa+yd \\ xb+yc \\ xc+yf \end{pmatrix}.$$

From this it follows that the determinant becomes zero.

(c) We expand with respect to the first row:

$$4\begin{vmatrix} 11 & 44 \\ 7 & 1 \end{vmatrix} - 5\begin{vmatrix} 8 & 44 \\ 3 & 1 \end{vmatrix} + 22\begin{vmatrix} 8 & 11 \\ 3 & 7 \end{vmatrix} = 4(-297) - 5(-124) + 22(23) = -62.$$

Example 4.4: Laplace expansion theorem

Let $A = (a_{ik})$ be a $n \times n$ matrix, and S_{ik} be the submatrices of A obtained by erasing the ith row and the kth column of the matrix A. The matrices S_{ik} thus are $(n-1) \times (n-1)$ matrices. For each i with $1 \le i \le n$, it holds that

$$\det A = \sum_{k=1}^{n} (-1)^{i+k} a_{ik} \det S_{ik} \qquad \text{(expansion with respect to } i\text{th row)}$$

and also

$$\det A = \sum_{k=1}^{n} (-1)^{i+k} a_{ki} \det S_{ki} \qquad \text{(expansion with respect to } i\text{th column).}$$

We check the theorem explicitly for 3-determinants and expand at first the general 3×3 determinant:

Expansion of $\det A = \begin{vmatrix} a_{11} & a_{12} & a_{13} \\ a_{21} & a_{22} & a_{23} \\ a_{31} & a_{32} & a_{33} \end{vmatrix}$ with respect to the first row yields

$$\det A = (-1)^{1+1} a_{11} S_{11} + (-1)^{1+2} a_{12} S_{12} + (-1)^{1+3} a_{13} S_{13} \tag{4.15}$$

$$= a_{11} \begin{vmatrix} a_{22} & a_{23} \\ a_{32} & a_{33} \end{vmatrix} - a_{12} \begin{vmatrix} a_{21} & a_{23} \\ a_{31} & a_{33} \end{vmatrix} + a_{13} \begin{vmatrix} a_{21} & a_{22} \\ a_{31} & a_{32} \end{vmatrix}. \tag{4.16}$$

Expansion of the 3-determinant with respect to the second column yields

$$\det A = (-1)^{1+2} a_{12} S_{12} + (-1)^{2+2} a_{22} S_{22} + (-1)^{3+2} a_{32} S_{32}. \tag{4.17}$$

The first term on the right side is identical with the second term of 4.15. The last two terms of 4.17 read explicitly

$$a_{22} \begin{vmatrix} a_{11} & a_{13} \\ a_{31} & a_{33} \end{vmatrix} - a_{32} \begin{vmatrix} a_{11} & a_{13} \\ a_{21} & a_{23} \end{vmatrix} = a_{22}(a_{11}a_{33} - a_{13}a_{31}) - a_{32}(a_{11}a_{23} - a_{13}a_{21}). \tag{4.18}$$

The sum of the first and third terms of 4.15 or 4.16 yields

$$a_{11} \begin{vmatrix} a_{22} & a_{23} \\ a_{32} & a_{33} \end{vmatrix} + a_{13} \begin{vmatrix} a_{21} & a_{22} \\ a_{31} & a_{32} \end{vmatrix} = a_{11}(a_{22}a_{33} - a_{23}a_{32}) + a_{13}(a_{21}a_{32} - a_{22}a_{31}). \tag{4.19}$$

Obviously, 4.18 and 4.19 coincide. Hence, it is clear that the expansions of the 3-determinant with respect to the first row and the second column, respectively, yield the same. Similarly, one may verify that the expansion with respect to other rows or columns leads to the same result. Hence, the expansion theorem for 3-determinants is seen to be valid.

We still evaluate the 3×3 determinant by expanding with respect to the second row, and subsequently with respect to the second column, for the example of the determinant

$$\det A = \begin{vmatrix} 4 & 5 & 22 \\ 8 & 11 & 44 \\ 3 & 7 & 1 \end{vmatrix}.$$

This yields

(a) Expansion with respect to the second row:

$$\det A = (-1)^{2+1} a_{21} S_{21} + (-1)^{2+2} a_{22} S_{22} + (-1)^{2+3} a_{23} S_{23}$$

$$= -a_{21} \begin{vmatrix} a_{12} & a_{13} \\ a_{32} & a_{33} \end{vmatrix} + a_{22} \begin{vmatrix} a_{11} & a_{13} \\ a_{31} & a_{33} \end{vmatrix} - a_{23} \begin{vmatrix} a_{11} & a_{12} \\ a_{31} & a_{32} \end{vmatrix}$$

$$= -8 \begin{vmatrix} 5 & 22 \\ 7 & 1 \end{vmatrix} + 11 \begin{vmatrix} 4 & 22 \\ 3 & 1 \end{vmatrix} - 44 \begin{vmatrix} 4 & 5 \\ 3 & 7 \end{vmatrix} = -62. \tag{4.20}$$

(b) Expansion with respect to the second column:

$$\det A = (-1)^{2+1} a_{12} S_{12} + (-1)^{2+2} a_{22} S_{22} + (-1)^{2+3} a_{32} S_{32}$$

$$= -a_{12} \begin{vmatrix} a_{21} & a_{23} \\ a_{31} & a_{33} \end{vmatrix} + a_{22} \begin{vmatrix} a_{11} & a_{13} \\ a_{31} & a_{33} \end{vmatrix} - a_{32} \begin{vmatrix} a_{11} & a_{13} \\ a_{21} & a_{23} \end{vmatrix}$$

$$= -5 \begin{vmatrix} 8 & 44 \\ 3 & 1 \end{vmatrix} + 11 \begin{vmatrix} 4 & 22 \\ 3 & 1 \end{vmatrix} - 7 \begin{vmatrix} 4 & 22 \\ 8 & 44 \end{vmatrix} = -62. \tag{4.21}$$

5 The Triple Scalar Product

Definition: The triple scalar product of the three vectors **a**, **b**, and **c** is defined as

$$\mathbf{a} \cdot (\mathbf{b} \times \mathbf{c}),$$

that is, a combination of a scalar and vector product. The triple scalar product is therefore also denoted as a mixed product. The triple scalar product is a scalar.

Triple scalar product in component representation:

$$\begin{aligned}
\mathbf{a} \cdot (\mathbf{b} \times \mathbf{c}) &= (a_1, a_2, a_3) \cdot [(b_1, b_2, b_3) \times (c_1, c_2, c_3)] \\
&= (a_1, a_2, a_3) \cdot \begin{vmatrix} \mathbf{e}_1 & \mathbf{e}_2 & \mathbf{e}_3 \\ b_1 & b_2 & b_3 \\ c_1 & c_2 & c_3 \end{vmatrix} \\
&= (a_1, a_2, a_3) \cdot (b_2c_3 - b_3c_2, -b_1c_3 + b_3c_1, b_1c_2 - b_2c_1) \\
&= a_1(b_2c_3 - b_3c_2) - a_2(b_1c_3 - b_3c_1) + a_3(b_1c_2 - b_2c_1).
\end{aligned}$$

The three terms may again be combined to a determinant, such that

$$\mathbf{a} \cdot (\mathbf{b} \times \mathbf{c}) = \begin{vmatrix} a_1 & a_2 & a_3 \\ b_1 & b_2 & b_3 \\ c_1 & c_2 & c_3 \end{vmatrix} = (\mathbf{a} \times \mathbf{b}) \cdot \mathbf{c}. \tag{5.1}$$

Cyclic permutability: The factors of the triple scalar product may be permuted cyclically. One has

$$\mathbf{a} \cdot (\mathbf{b} \times \mathbf{c}) = \mathbf{b} \cdot (\mathbf{c} \times \mathbf{a}) = \mathbf{c} \cdot (\mathbf{a} \times \mathbf{b}).$$

These rules may be confirmed easily by successive permutations of the rows in the determinant (5.1). The following simplified notation for the triple scalar product may be found occasionally in the literature:

$$\mathbf{a}\cdot(\mathbf{b}\times\mathbf{c}) = [\mathbf{a}\,\mathbf{b}\,\mathbf{c}] = [\mathbf{b}\,\mathbf{c}\,\mathbf{a}] = [\mathbf{c}\,\mathbf{a}\,\mathbf{b}].$$

Geometrically, the triple scalar product represents the volume

$$V = \mathbf{a}\cdot(\mathbf{b}\times\mathbf{c}) = a\cos\varphi\, bc\sin\gamma$$
$$= abc\cos\varphi\,\sin\gamma$$

of a parallelepipedon formed by the three vectors (see figure).

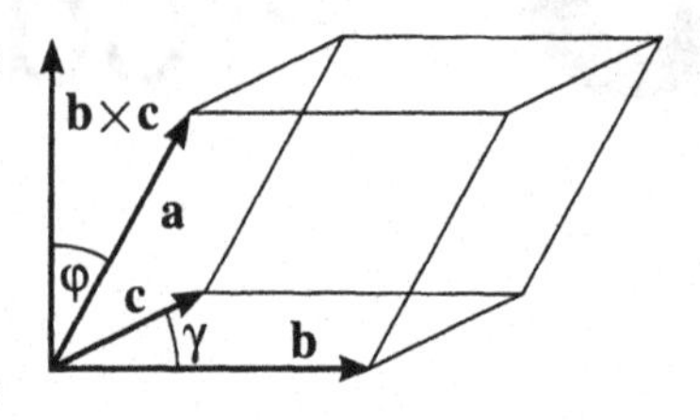

Illustration of the triple scalar product.

Note: The volume has a positive sign (+) if $\mathbf{a}$ lies on the side of $\mathbf{b}\times\mathbf{c}$, but a negative sign (−) if $\mathbf{a}$ lies on the side of $-\mathbf{b}\times\mathbf{c}$. Hence the volume might be associated with a sign. In general, however, this choice is not used, and a positive sign is always required. This is achieved by the definition $V = |\mathbf{a}\cdot(\mathbf{b}\times\mathbf{c})|$.

Properties of the triple scalar product: From

$$\mathbf{a}\cdot(\mathbf{b}\times\mathbf{c}) = 0 \quad \text{follows} \quad \varphi = \frac{\pi}{2} \quad \text{and/or} \quad \gamma = 0, \tag{5.2}$$

that is, the three vectors are coplanar or (and) two vectors lie on a straight line.

This is again a very clear proof of the theorems on determinants already mentioned above:

1. If two row vectors (or column vectors) are equal or proportional to each other, then the determinant equals zero.
2. When we permute two neighboring rows, the determinant changes by a factor (−1).

6 Application of Vector Calculus

Application in mathematics:

Problem 6.1: Distance vector

Calculate the length of the vector **a** that represents the distance vector between the points $\mathbf{r}_1$ and $\mathbf{r}_2$.

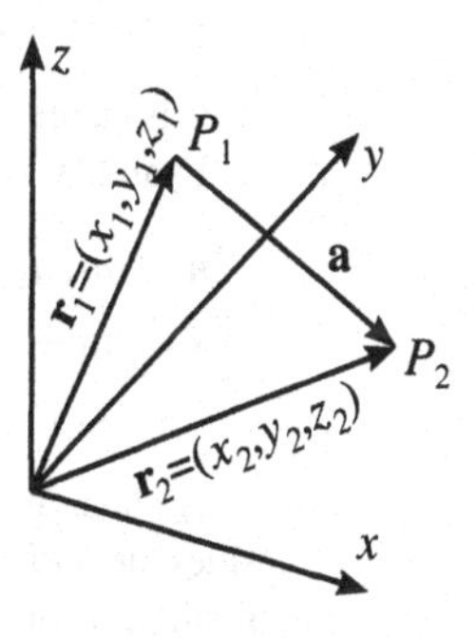

The distance vector between the points $\mathbf{r}_1$ and $\mathbf{r}_2$.

Solution

$$\begin{aligned}\mathbf{a} &= \mathbf{r}_2 - \mathbf{r}_1 \\ &= (x_2\mathbf{e}_1 + y_2\mathbf{e}_2 + z_2\mathbf{e}_3) - (x_1\mathbf{e}_1 + y_1\mathbf{e}_2 + z_1\mathbf{e}_3) \\ &= (x_2 - x_1)\mathbf{e}_1 + (y_2 - y_1)\mathbf{e}_2 + (z_2 - z_1)\mathbf{e}_3;\end{aligned}$$

hence **a** reads in row notation

$$\mathbf{a} = (x_2 - x_1, y_2 - y_1, z_2 - z_1),$$

and the magnitude of **a** is therefore

$$|\mathbf{a}| = \sqrt{(x_2 - x_1)^2 + (y_2 - y_1)^2 + (z_2 - z_1)^2}\,.$$

Problem 6.2: Projection of a vector onto another vector

Given

$$\mathbf{a} = (2, 1, 1),$$
$$\mathbf{b} = (1, -2, 2),$$
$$\mathbf{c} = (3, -4, 2),$$

what is the absolute value of the projection of the sum $(\mathbf{a} + \mathbf{b})$ onto the vector $\mathbf{c}$?

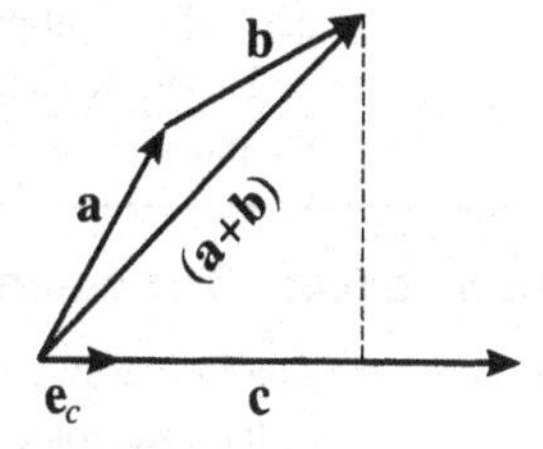

The projection of the sum $\mathbf{a} + \mathbf{b}$ onto the vector $\mathbf{c}$.

Solution This projection is given by the scalar product of $(\mathbf{a} + \mathbf{b})$ and the unit vector $\mathbf{e}_c$ along $\mathbf{c}$.

$$\mathbf{e}_c = \frac{\mathbf{c}}{|\mathbf{c}|} = \frac{(3, -4, 2)}{\sqrt{3^2 + 4^2 + 2^2}},$$

$$(\mathbf{a}+\mathbf{b}) = (2+1, 1-2, 1+2),$$

$$(\mathbf{a}+\mathbf{b}) \cdot \mathbf{e}_c = \frac{3 \cdot 3 + (-1) \cdot (-4) + 3 \cdot 2}{\sqrt{29}} = \frac{19}{\sqrt{29}}.$$

Problem 6.3: Equations of a straight line and of a plane

Let the points A and B be given by their position vectors $\mathbf{a}$ and $\mathbf{b}$. What is the equation of the straight line through A and B?

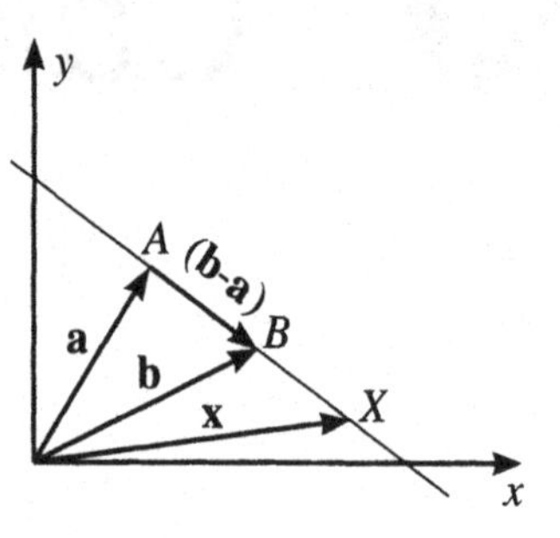

The point-direction form of a straight line.

Solution The straight line AB is parallel to $(\mathbf{b}-\mathbf{a})$. Moreover, it passes through point A. Hence, the equation determining any position vector $\mathbf{x}$ of a point X on the desired straight line reads

$$\mathbf{x} = \mathbf{a} + t(\mathbf{b}-\mathbf{a}),$$

with t being a real number (running parameter $-\infty < t < \infty$). If two points A and B are not given but one point A and a vector $\mathbf{u}$ specifying the orientation of the straight line are given, the equation of the straight line reads

$$\mathbf{x} = \mathbf{a} + t\mathbf{u}\,.$$

This is called the *point-direction form* of the equation of a straight line.

Example:

$$\mathbf{a} = (a_1, a_2, a_3), \qquad \mathbf{u} = (u_1, u_2, u_3),$$

$$\mathbf{x} = (a_1 + tu_1, a_2 + tu_2, a_3 + tu_3)$$

$$= (x, y, z).$$

A plane in space may be fixed by specifying besides the position vector $\mathbf{a}$ and the orientation vector $\mathbf{u}$ still a second orientation vector $\mathbf{v}$:

$$\mathbf{x}_E = \mathbf{a} + t\mathbf{u} + k\mathbf{v},$$

where $\mathbf{u} \upuparrows \mathbf{v}$ and also $\mathbf{u} \uparrow\downarrow \mathbf{v}$ and $k, t \in \mathrm{R}$. The notation $\upuparrows$ and $\uparrow\downarrow$ indicates that $\mathbf{u}$ and $\mathbf{v}$ are neither parallel nor antiparallel.

This is the *point-direction form* of the equation of the plane.

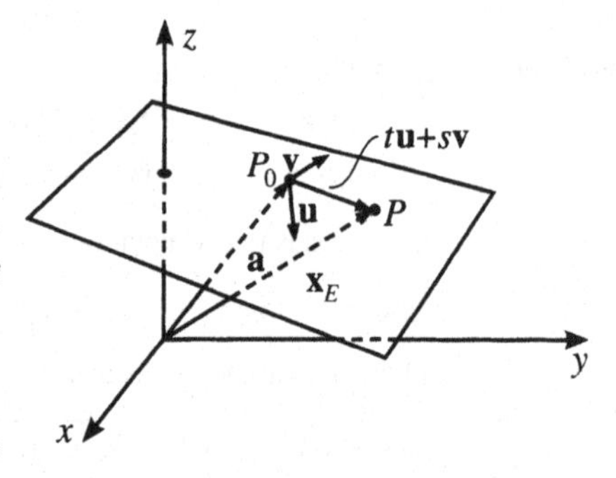

Representation of a plane in space spanned by the vectors $\mathbf{u}$ and $\mathbf{v}$ from point P_0.

Example 6.4: The cosine theorem

The cosine law of plane trigonometry is obtained by scalar multiplication of the equation $\mathbf{c} = \mathbf{a} - \mathbf{b}$ by itself:

$$\mathbf{c} \cdot \mathbf{c} = (\mathbf{a}-\mathbf{b}) \cdot (\mathbf{a}-\mathbf{b}) = \mathbf{a}^2 + \mathbf{b}^2 - 2\mathbf{a} \cdot \mathbf{b}$$

$$= a^2 + b^2 - 2ab\cos\gamma.$$

$$\Rightarrow \qquad c^2 = a^2 + b^2 - 2ab\cos\gamma.$$

For $\gamma = \pi/2$ there results the theorem of Pythagoras.

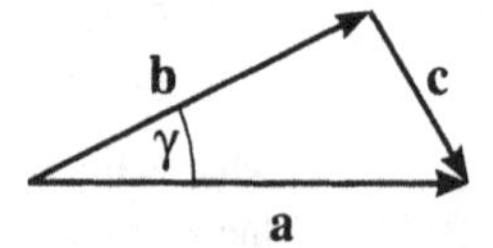

The vectors $\mathbf{a}$, $\mathbf{b}$, and $\mathbf{c}$ characterize the sides of the triangle.

Example 6.5: The theorem of Thales

In order to prove the theorem of Thales[1] we introduce the following vectors according to the sketch:

$$\overrightarrow{MA} = -\overrightarrow{MB} = \mathbf{a}, \qquad \overrightarrow{MC} = \mathbf{b}.$$

It holds that

$$|\mathbf{a}| = |\mathbf{b}|, \quad \overrightarrow{BC} = \mathbf{a}+\mathbf{b}, \quad \text{and} \quad \overrightarrow{AC} = \mathbf{b}-\mathbf{a}.$$

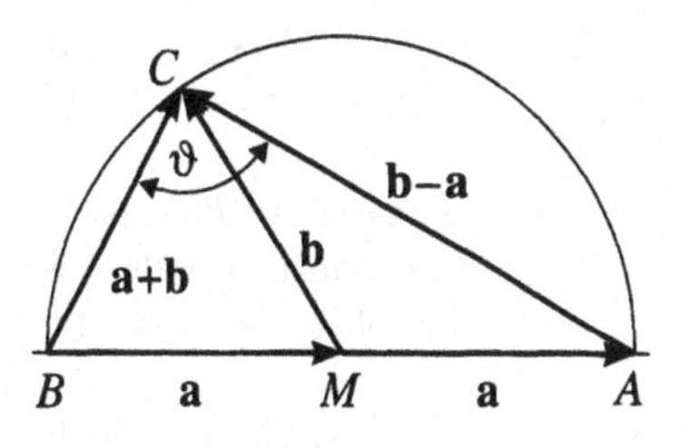

The theorem of Thales, demonstrated with the help of vectors.

The scalar product $(\mathbf{a}+\mathbf{b})\cdot(\mathbf{a}-\mathbf{b})$ has the value

$$(\mathbf{a}-\mathbf{b})\cdot(\mathbf{a}+\mathbf{b}) = \mathbf{a}^2 - \mathbf{b}^2 = |\mathbf{a}|^2 - |\mathbf{b}|^2 = 0.$$

For the angle enclosed by $(\mathbf{a}+\mathbf{b})$ and $(\mathbf{a}-\mathbf{b})$, it follows that $\vartheta = \pi/2$ or

$$(\mathbf{a}+\mathbf{b}) \perp (\mathbf{a}-\mathbf{b}) \quad \text{(theorem of Thales)}.$$

Example 6.6: The rotation matrix

The opposite figure shows into which vectors $\mathbf{e}_1'$ and $\mathbf{e}_2'$ the Cartesian unit vectors $\mathbf{e}_1$ and $\mathbf{e}_2$ are transformed under a rotation in the x, y-plane by the angle β around the z-axis:

$$\begin{aligned}
\mathbf{e}_1' &= \mathbf{e}_1 \cos\beta + \mathbf{e}_2 \sin\beta + \mathbf{e}_3 \cdot 0 \\
\mathbf{e}_2' &= \mathbf{e}_1(-\sin\beta) + \mathbf{e}_2\cos\beta + \mathbf{e}_3 \cdot 0 \\
\mathbf{e}_3' &= \mathbf{e}_1 \cdot 0 + \mathbf{e}_2 \cdot 0 + \mathbf{e}_3 \cdot 1\,.
\end{aligned} \tag{6.1}$$

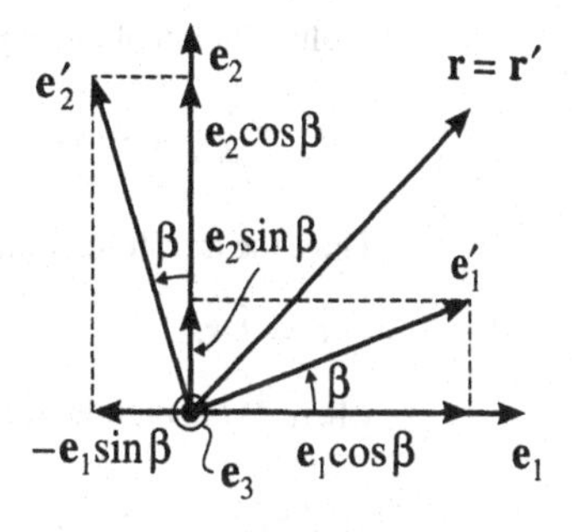

Case 1: vector **r** stays at rest; the coordinate system is rotated.

This system of equations may be written in matrix form (see equation 6.7):

$$\begin{pmatrix} \mathbf{e}_1' \\ \mathbf{e}_2' \\ \mathbf{e}_3' \end{pmatrix} = \begin{pmatrix} \cos\beta & \sin\beta & 0 \\ -\sin\beta & \cos\beta & 0 \\ 0 & 0 & 1 \end{pmatrix} \cdot \begin{pmatrix} \mathbf{e}_1 \\ \mathbf{e}_2 \\ \mathbf{e}_3 \end{pmatrix}$$

$$= \begin{pmatrix} d_{11}\mathbf{e}_1 + d_{12}\mathbf{e}_2 + d_{13}\mathbf{e}_3 \\ d_{21}\mathbf{e}_1 + d_{22}\mathbf{e}_2 + d_{23}\mathbf{e}_3 \\ d_{31}\mathbf{e}_1 + d_{32}\mathbf{e}_2 + d_{33}\mathbf{e}_3 \end{pmatrix} \tag{6.2}$$

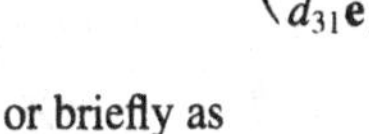

or briefly as

$$\mathbf{e}_\nu' = \sum_{\mu=1}^{3} d_{\nu\mu}\mathbf{e}_\mu,$$

[1] Named after *Thales of Milet*, b. about 624 BC—d. 546 BC. He is the first representative of the Ionic School. According to writings he did far travels (e.g., to Egypt) and was very active as a politician. The theorem named after him was for the first time strictly formulated by him.

where

$$d_{\nu\mu} = \mathbf{e}'_\nu \cdot \mathbf{e}_\mu \quad \text{or} \quad (d_{\nu\mu}) = \begin{pmatrix} \cos\beta & \sin\beta & 0 \\ -\sin\beta & \cos\beta & 0 \\ 0 & 0 & 1 \end{pmatrix}$$

represent the *direction cosines*. Note that $\sin\phi = \cos(\phi - \pi/2)$. The matrix $(d_{\nu\mu})$ describes the transformation of the base vectors. For a rotation in the three-dimensional space in the x, y-plane (i.e., about the z-axis), the rotation matrix reads

$$\widehat{D} = \begin{pmatrix} \cos\beta & \sin\beta & 0 \\ -\sin\beta & \cos\beta & 0 \\ 0 & 0 & 1 \end{pmatrix} \equiv (d_{\nu\mu}). \tag{6.3}$$

Case 1: $\mathbf{r} = \mathbf{r}'$ fixed in space. If $\mathbf{r} = \mathbf{r}'$ is fixed in space but the coordinate frame rotates, one has

$$\sum_\nu x_\nu \mathbf{e}_\nu = \sum_\mu x'_\mu \mathbf{e}'_\mu .$$

Multiplication of this equation with $\mathbf{e}'_\mu$ isolates x'_μ:

$$x'_\mu = \sum_\nu x_\nu(\mathbf{e}_\nu \cdot \mathbf{e}'_\mu) = \sum_\nu d_{\mu\nu} x_\nu .$$

Thus, the transformation of the components of a position vector that is kept fixed in space is given by

$$\mathbf{r}' = \widehat{D}\mathbf{r}, \tag{6.4}$$

where $\widehat{D}$ denotes the rotation matrix. Explicitly, this means because of $x'_1 = x'$, $x'_2 = y'$, $x'_3 = z'$:

$$\begin{pmatrix} x' \\ y' \\ z' \end{pmatrix}_{\text{new base}} = \begin{pmatrix} \cos\beta & \sin\beta & 0 \\ -\sin\beta & \cos\beta & 0 \\ 0 & 0 & 1 \end{pmatrix} \cdot \begin{pmatrix} x \\ y \\ z \end{pmatrix}_{\text{old base}} . \tag{6.5}$$

The addendum "new base" at the column tuple shall indicate that the components x', y', z' of the column tuple are to be interpreted as coefficients of the base vectors $\mathbf{e}'_1$, $\mathbf{e}'_2$, and $\mathbf{e}'_3$. Written explicitly, the vector in the new basis thus reads

$$\mathbf{r}' = x'\mathbf{e}'_1 + y'\mathbf{e}'_2 + z'\mathbf{e}'_3 .$$

Case 2: $\mathbf{r}$ is tightly fixed to the rotating coordinate frame. Thus, $\mathbf{r}$ rotates with the coordinate frame. This means

$$\begin{aligned} \sum_\nu x_\nu \mathbf{e}'_\nu &= \sum_\mu x'_\mu \mathbf{e}_\mu \\ \Rightarrow \quad x'_\mu &= \sum_\nu x_\nu(\mathbf{e}'_\nu \cdot \mathbf{e}_\mu) \\ &= \sum_\nu d_{\nu\mu} x_\nu \\ &= \sum_\nu \widetilde{d}_{\mu\nu} x_\nu . \end{aligned} \tag{6.6}$$

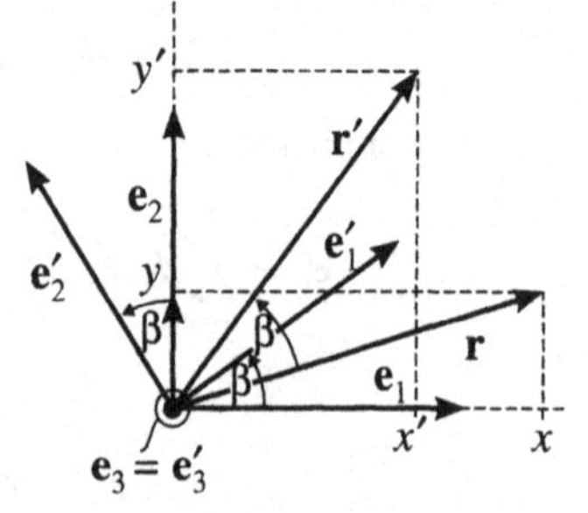

Case 2: vector **r** is rotated together with the coordinate system.

x'_μ are the new components of the rotated vector with respect to the fixed system $\mathbf{e}_\mu$: x_ν are old components of the vector with respect to the fixed system $\mathbf{e}_\mu$.

Note: Both x'_ν as well as x_ν in this case are defined in the old system (base $\mathbf{e}_\mu$). They denote the components of the new (rotated) and old (not rotated) vector, respectively!

In the preceding we have already used the matrix multiplication. It shall once again be clearly defined here.

Definition of the matrix product: The common element C_{ij} of the row i and the column j of the product matrix $\widehat{C} = \widehat{A} \cdot \widehat{B}$ is obtained by forming the sum

$$C_{ij} = \sum_k A_{ik} B_{kj}, \tag{6.7}$$

where $\widehat{A}$ and $\widehat{B}$ are the factor matrices.

Thus, the components of a vector $\mathbf{a} = (a_1, a_2, a_3)$ under rotations of the coordinate frame would change to

$$\begin{pmatrix} a'_1 \\ a'_2 \\ a'_3 \end{pmatrix}_{\text{new base}} = \mathbf{a}' = \begin{pmatrix} \cos\beta & \sin\beta & 0 \\ -\sin\beta & \cos\beta & 0 \\ 0 & 0 & 1 \end{pmatrix} \cdot \begin{pmatrix} a_1 \\ a_2 \\ a_3 \end{pmatrix}$$

$$= \begin{pmatrix} \cos\beta\, a_1 + \sin\beta\, a_2 \\ -\sin\beta\, a_1 + \cos\beta\, a_2 \\ a_3 \end{pmatrix},$$

$$a'_\mu = \sum_\nu d_{\mu\nu} a_\nu .$$

The vector itself remains fixed in space. Its components change, however, because the base was rotated (case 1). If the vector would rotate (case 2), then we would obtain according to 6.6

$$\begin{pmatrix} a'_1 \\ a'_2 \\ a'_3 \end{pmatrix}_{\text{new base}} = \begin{pmatrix} \cos\beta\, a_1 - \sin\beta\, a_2 \\ \sin\beta\, a_1 + \cos\beta\, a_2 \\ a_3 \end{pmatrix}; \qquad a'_\mu = \sum_\nu \widetilde{d}_{\mu\nu} a_\nu = \sum_\nu d_{\nu\mu} a_\nu,$$

where $\widetilde{d}_{\mu\nu} = d_{\nu\mu}$ is the *transposed* rotation matrix. The transposed of a matrix is simply the matrix reflected at the main diagonal (from the upper left to the lower right corner).

Application in physics:

Problem 6.7: Superposition of forces

Four coplanar forces are acting at the point 0, as shown in the sketch.

Calculate the net force acting at the point 0!

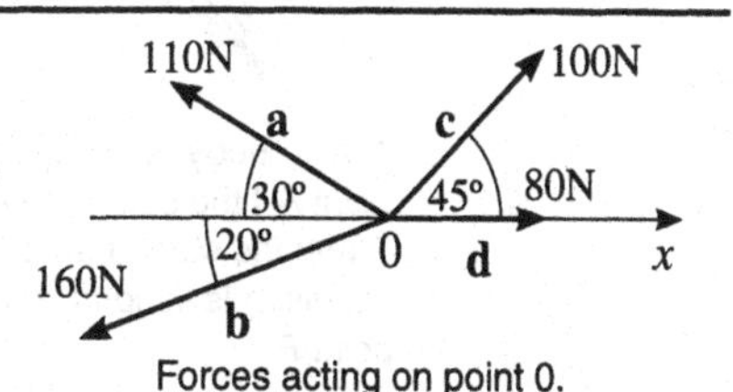

Forces acting on point 0.

Solution $\mathbf{a} = (-95.3, 55.0)\,\text{N}, \quad \mathbf{b} = (-150.4, -54.7)\,\text{N},$

$$\mathbf{c} = (70.7, 70.7)\,\text{N}, \quad \mathbf{d} = (80.0, 0.0)\,\text{N} \quad \left(\text{N} = \text{Newton} = 1\frac{\text{kg m}}{\text{s}^2}\right).$$

It holds that

$$\mathbf{F}_{\text{ges}} = \mathbf{a} + \mathbf{b} + \mathbf{c} + \mathbf{d} = (-95.0, 71.0)\,\text{N},$$
$$|\mathbf{F}_{\text{ges}}| = \sqrt{95.0^2 + 71.0^2}\,\text{N} = 118.6\,\text{N}.$$

We remember that

$$\mathbf{a} + \mathbf{b} = \sum_i a_i \mathbf{e}_i + \sum_i b_i \mathbf{e}_i = \sum_i (a_i + b_i)\mathbf{e}_i = (a_1 + b_1, a_2 + b_2, a_3 + b_3).$$

Graphical determination of the force: Representation by means of *polygon of forces*.

The angle β enclosed by $\mathbf{F}$ and the x-axis may be calculated easily. One has

$$\mathbf{F} = (-95.0, 71.0)\,\text{N}, \quad \frac{F_y}{F_x} = \tan\beta = -\frac{71.0}{95.0};$$

from there it follows that $\beta = 143°$.

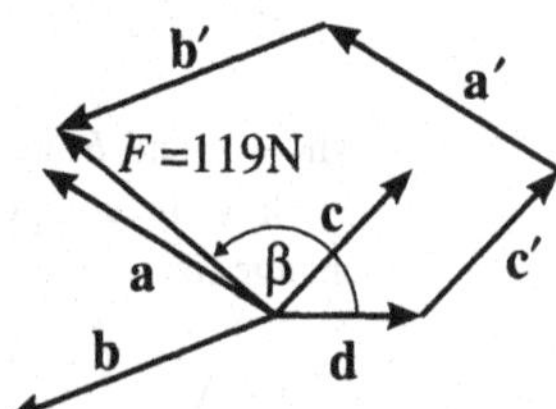

Graphical determination of the net force **F**.

Example 6.8: Equilibrium condition for a rigid body without fixed rotational axis

A rigid body is under the action of the forces $\mathbf{F}_i$ at the positions $\mathbf{r}_i$. We investigate the equilibrium at the point A (position vector $\mathbf{a}$) the body may rotate about. All forces $\mathbf{F}_i$ are now added and subtracted at A such that nothing is changed in total.

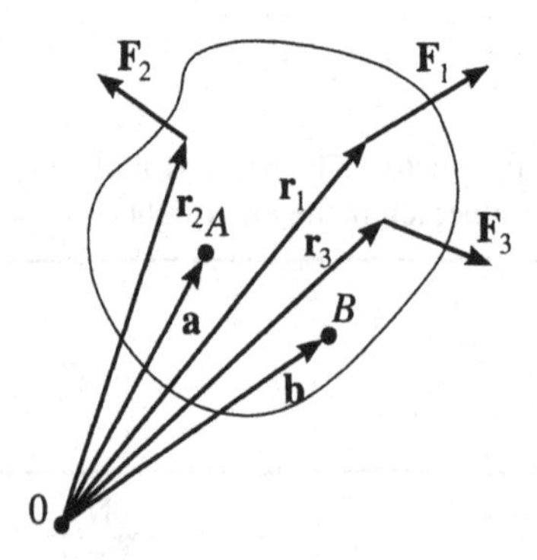

A rigid body is in equilibrium with respect to point A if the sum of all torques with respect to A and the sum of all forces in A vanish. If this condition is valid in A, it is also valid in every point B.

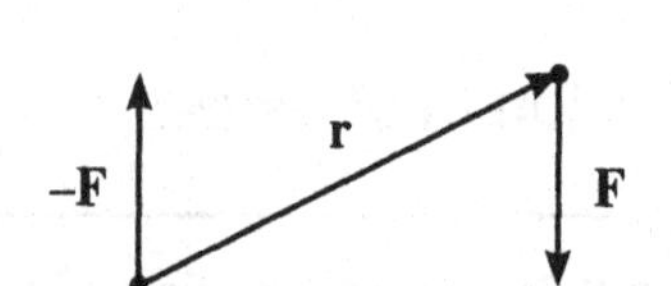

A pair of forces results in a torque $\mathbf{M} = \mathbf{r} \times \mathbf{F}$. These forces set the body on which they act into rotation.

The procedure is illustrated for the force $\mathbf{F}_1$ in the figure on the next page; for the other forces we proceed in the same way. Now the forces $\mathbf{F}_1$ at $\mathbf{r}_1$ and $-\mathbf{F}_1$ at $\mathbf{a}$ are forming a *pair of forces* that generates the *torque* (compare with the Problem 6.9)

$$\mathbf{M}_1(\mathbf{a}) = (\mathbf{r}_1 - \mathbf{a}) \times \mathbf{F}_1 \tag{6.8}$$

and will rotate the body. Similarly, all other forces $\mathbf{F}_i$ (at $\mathbf{r}_i$) and $-\mathbf{F}_i$ (at $\mathbf{a}$) are forming pairs of forces with the torques

$$\mathbf{M}_i(\mathbf{a}) = (\mathbf{r}_i - \mathbf{a}) \times \mathbf{F}_i. \tag{6.9}$$

The total force acting at the point A is therefore

$$\mathbf{F} = \sum_i \mathbf{F}_i, \tag{6.10}$$

and the total torque about A is

$$\mathbf{M}(\mathbf{a}) = \sum_i \mathbf{M}_i(\mathbf{a}) = \sum_i (\mathbf{r}_i - \mathbf{a}) \times \mathbf{F}_i. \tag{6.11}$$

At the point B (position vector $\mathbf{b}$) a similar construction would yield the total force

$$\mathbf{F} = \sum_i \mathbf{F}_i \tag{6.12}$$

and the total torque about B,

$$\mathbf{M}(\mathbf{b}) = \sum_i \mathbf{M}_i(\mathbf{b}) = \sum_i (\mathbf{r}_i - \mathbf{b}) \times \mathbf{F}_i. \tag{6.13}$$

The total force $\mathbf{F}$ tries to accelerate the body as a whole. The total torque tries to rotate the body. If there shall be equilibrium with respect to point A (position vector $\mathbf{a}$), then both the total force $\mathbf{F}$ and the total torque $\mathbf{M}(\mathbf{a})$ must vanish:

$$\mathbf{F} = 0, \tag{6.14}$$

$$\mathbf{M}(\mathbf{a}) = 0. \tag{6.15}$$

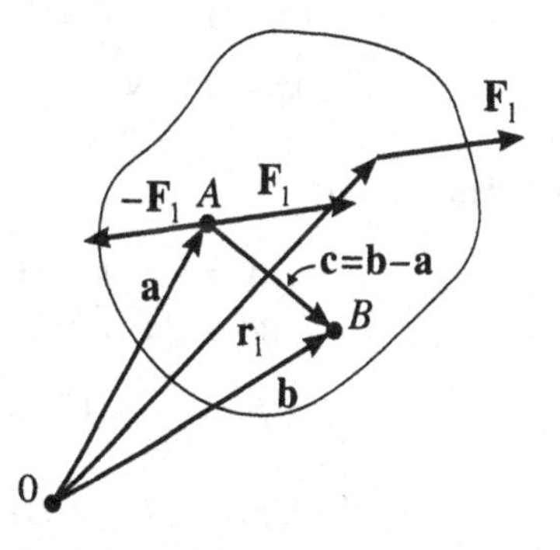

Equilibrium with respect to point A implies equilibrium with respect to point B.

The question arises of whether an equilibrium at point A also means an equilibrium at point B. To answer it, we recalculate the eqs. 6.14, 6.15 to the point B (eqs. 6.12 and 6.13). We realize: 6.14 is identical with 6.12. Further, it holds that

$$\begin{aligned}
\mathbf{M}(\mathbf{b}) &= \sum_i (\mathbf{r}_i - \mathbf{b}) \times \mathbf{F}_i = \sum_i (\mathbf{r}_i - (\mathbf{a} + \mathbf{c})) \times \mathbf{F}_i \\
&= \sum_i (\mathbf{r}_i - \mathbf{a}) \times \mathbf{F}_i - \sum_i \mathbf{c} \times \mathbf{F}_i \\
&= \mathbf{M}(\mathbf{a}) - \mathbf{c} \times \underbrace{\sum_i \mathbf{F}_i}_{=0 \text{ because of } 6.14} = \mathbf{M}(\mathbf{a}) = 0.
\end{aligned}$$

Therefore we may claim: If the equilibrium conditions 6.14,6.15 are fulfilled at a point A, then they also hold at any other point B.

Problem 6.9: Force and torque

The following external forces are acting on a body:

$$\mathbf{F}_1 = (10, 2, -1)\,\text{N} \quad \text{at point } P_1(2, 0, 0)\,\text{cm},$$
$$(\text{N} = \text{Newton} = 1\frac{\text{kg m}}{\text{s}^2})$$
$$\mathbf{F}_2 = (0, 0, 5)\,\text{N} \quad \text{at point } P_2(1, 3, 0)\,\text{cm}$$

and

$$\mathbf{F}_3 = (-6, 1, 8)\,\text{N} \quad \text{at point } P_3(6, 8, 1)\,\text{cm}.$$

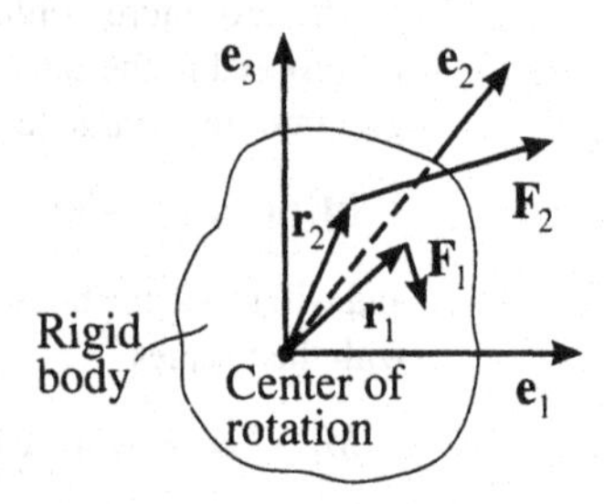

Illustration of the torque induced by two forces.

Find

(a) components, magnitude, and orientation of the resulting force $\mathbf{F}$,

(b) the torque with respect to P_2.

Remark (kp = kilopond). The kilopond is no longer a legal unit, all forces are measured in Newtons (N):

$$1\,\text{N} = 1\,\text{kg} \cdot \frac{\text{m}}{\text{s}^2} = 10^5\,\frac{\text{g cm}}{\text{s}^2} = 10^5\,\text{dyn},$$
$$1\,\text{kp} = 9.81\,\text{N}.$$

Solution (a)

$$\mathbf{F} = \mathbf{F}_1 + \mathbf{F}_2 + \mathbf{F}_3 = (4, 3, 12)\,\text{N},$$
$$|\mathbf{F}| = \sqrt{4^2 + 3^2 + 12^2}\,\text{N} = 13\,\text{N},$$
$$\cos\beta_1 = \frac{F_x}{|\mathbf{F}|} = 0.308, \quad \beta_1 = 72^\circ,$$
$$\cos\beta_2 = \frac{F_y}{|\mathbf{F}|} = 0.231, \quad \beta_2 = 77^\circ,$$
$$\cos\beta_3 = \frac{F_z}{|\mathbf{F}|} = 0.923, \quad \beta_3 = 23^\circ.$$

These are the direction cosines of the force. They describe the direction of force

$$\mathbf{n} = \frac{\mathbf{F}}{|\mathbf{F}|} = (\cos\beta_1, \cos\beta_2, \cos\beta_3) = (0,308;\ 0,231;\ 0,923).$$

(b) The torque of a force $\mathbf{F}_p$ acting at point $P(x, y, z)$, that is at the position $\mathbf{r} = (x, y, z)$, is defined with respect to the coordinate origin (center of rotation) as the vector

$$\mathbf{M} = \mathbf{r} \times \mathbf{F}_p.$$

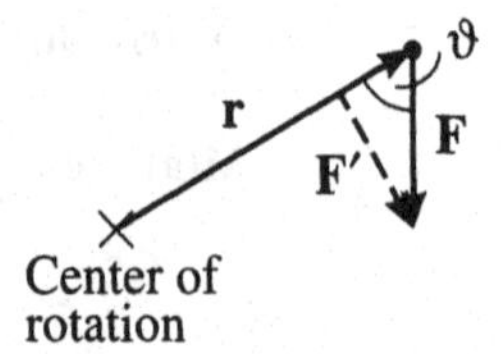

The absolute value of the torque results from the force component perpendicular to the distance vector.

Here $\mathbf{r}$ is the position vector from the center of rotation to the action point of the force $\mathbf{F}_p$. The magnitude of $\mathbf{M}$ is obviously given by

$M = r \cdot F'$, where $F' = F \sin\vartheta$ is the force component perpendicular to the position vector (compare in the figure). This may also be expressed as follows: M = distance from center of rotation to action point of the force *times* force component perpendicular to the distance vector.

This torque $\mathbf{M}$ is also caused by a pair of forces, as discussed in Example 6.8. If one adds at the center of rotation the forces $-\mathbf{F}$ and $\mathbf{F}$, in total $\mathbf{0}$ (compare to the figure), then the forces $-\mathbf{F}$ at the center of rotation and $\mathbf{F}$ at $\mathbf{r}$ form a pair of forces. The force $\mathbf{F}$ acting at the center of rotation presses onto the bearing of the rotation axis and is received there.

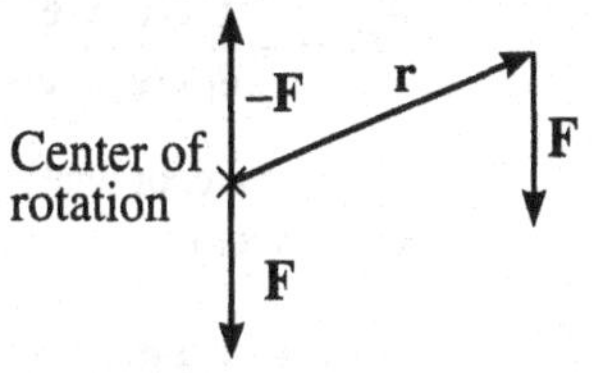

The pair of forces responsible for the torque.

If several forces $\mathbf{F}_\nu$ are acting on the rigid body at the points $\mathbf{r}_\nu$, the total torque is

$$\mathbf{M} = \sum_\nu \mathbf{M}_\nu = \sum_\nu \mathbf{r}_\nu \times \mathbf{F}_\nu.$$

In our example

$$\mathbf{M} = (\mathbf{r}_1 \times \mathbf{F}_1) + (\mathbf{r}_2 \times \mathbf{F}_2) + (\mathbf{r}_3 \times \mathbf{F}_3),$$
$$\mathbf{r}_1 = \mathbf{p}_1 - \mathbf{p}_2 = (1, -3, 0)\,\text{cm},$$
$$\mathbf{r}_2 = \mathbf{p}_2 - \mathbf{p}_2 = (0, 0, 0)\,\text{cm},$$
$$\mathbf{r}_3 = \mathbf{p}_3 - \mathbf{p}_2 = (5, 5, 1)\,\text{cm},$$

where $\mathbf{p}_1$, $\mathbf{p}_2$, and $\mathbf{p}_3$ are the position vectors of the points P_1, P_2, and P_3. Hence one obtains

$$\mathbf{M}_1 = \mathbf{r}_1 \times \mathbf{F}_1 = (3, 1, 32)\,\text{N cm},$$
$$\mathbf{M}_2 = \mathbf{r}_2 \times \mathbf{F}_2 = 0\,\text{N cm},$$
$$\mathbf{M}_3 = \mathbf{r}_3 \times \mathbf{F}_3 = (39, -46, 35)\,\text{N cm}.$$

The total torque is

$$\mathbf{M} = \mathbf{M}_1 + \mathbf{M}_2 + \mathbf{M}_3 = (42, -45, 67)\,\text{N cm},$$

and $|\mathbf{M}| = 91.0\,\text{N cm}$.

Problem 6.10: Forces in a three-leg stand

Find the rod forces in a three-leg stand that is movably linked at the points A, B, C to a vertical wall and loaded at the point D by the force $\mathbf{F}$.

Solution Only longitudinal forces may act in the rods, because of the movable links of the suspension rods (neglect of bending forces). The forces at the cut-out branching point D are to be considered as external forces and are obtained from the equilibrium condition

$$\mathbf{F}_1 + \mathbf{F}_2 + \mathbf{F}_3 + \mathbf{F} = 0. \qquad \textbf{(6.16)}$$

Using the unit vectors $\mathbf{e}_i$ $(i = 1, 2, 3)$ along the rod axes and the magnitudes F_i $(i = 1, 2, 3)$ of the rod forces, 6.16 may be written as

$$F_1\mathbf{e}_1 + F_2\mathbf{e}_2 + F_3\mathbf{e}_3 = -\mathbf{F}. \qquad \textbf{(6.17)}$$

To get the rod forces, 6.17 is scalar-multiplied successively by the vectors $\mathbf{e}_i \times \mathbf{e}_j$ $(i \neq j)$, where $(\mathbf{e}_i \times \mathbf{e}_j)$ by definition points perpendicular to $\mathbf{e}_i$, hence the scalar products $\mathbf{e}_i \cdot (\mathbf{e}_i \times \mathbf{e}_j)$ vanish.

Using the definition of the triple scalar product $\mathbf{A} \cdot (\mathbf{B} \times \mathbf{C})$, one then obtains from 6.17 for F_i $(i = 1, 2, 3)$

$$F_1 = -\frac{\mathbf{F} \cdot (\mathbf{e}_2 \times \mathbf{e}_3)}{\mathbf{e}_1 \cdot (\mathbf{e}_2 \times \mathbf{e}_3)}, \quad F_2 = -\frac{\mathbf{F} \cdot (\mathbf{e}_3 \times \mathbf{e}_1)}{\mathbf{e}_2 \cdot (\mathbf{e}_3 \times \mathbf{e}_1)}, \quad F_3 = -\frac{\mathbf{F} \cdot (\mathbf{e}_1 \times \mathbf{e}_2)}{\mathbf{e}_3 \cdot (\mathbf{e}_1 \times \mathbf{e}_2)}. \tag{6.18}$$

Putting a coordinate frame into the branching point D according to the above figure, one gets for the unit vectors

$$\begin{aligned} \mathbf{e}_1 &= (-\cos\alpha, \sin\alpha, 0), \\ \mathbf{e}_2 &= (\cos\alpha, \sin\alpha, 0), \\ \mathbf{e}_3 &= (0, \sin\beta, -\cos\beta). \end{aligned} \tag{6.19}$$

Insertion of equation 6.19 into equation 6.18 yields

$$\begin{aligned} \mathbf{F} \cdot (\mathbf{e}_2 \times \mathbf{e}_3) &= \begin{vmatrix} F_x & F_y & F_z \\ \cos\alpha & \sin\alpha & 0 \\ 0 & \sin\beta & -\cos\beta \end{vmatrix} \\ &= -F_x \sin\alpha\cos\beta + F_y \cos\alpha\cos\beta + F_z \cos\alpha\sin\beta \end{aligned} \tag{6.20}$$

and

$$\begin{aligned} \mathbf{e}_1 \cdot (\mathbf{e}_2 \times \mathbf{e}_3) &= \begin{vmatrix} -\cos\alpha & \sin\alpha & 0 \\ \cos\alpha & \sin\alpha & 0 \\ 0 & \sin\beta & -\cos\beta \end{vmatrix} \\ &= 2\sin\alpha\cos\alpha\cos\beta. \end{aligned} \tag{6.21}$$

From there one obtains for the component F_1

$$F_1 = \frac{1}{2}\left(\frac{F_x}{\cos\alpha} - \frac{F_y}{\sin\alpha} - \frac{F_z \tan\beta}{\sin\alpha}\right). \tag{6.22}$$

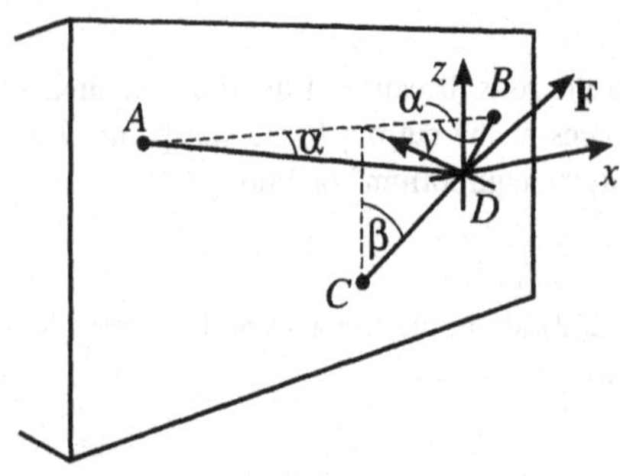

A three-leg stand fixed at a vertical wall.

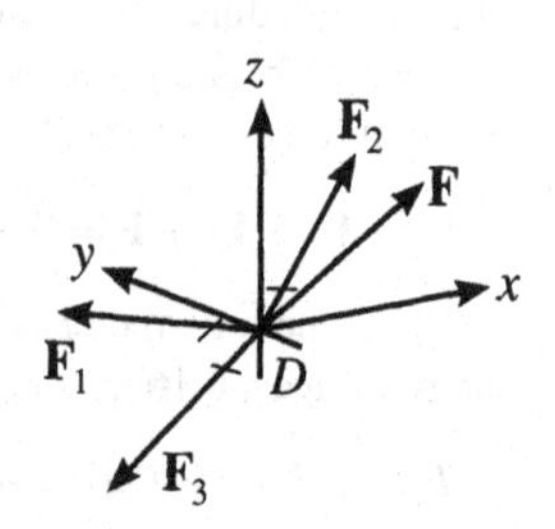

The forces acting at the three-leg stand.

The calculation of the scalar triple products for F_2 and F_3 from equation 6.18 runs in an analogous manner; one gets

$$F_2 = \frac{1}{2}\left(-\frac{F_x}{\cos\alpha} - \frac{F_y}{\sin\alpha} - \frac{F_z \tan\beta}{\sin\alpha}\right), \tag{6.23}$$
$$F_3 = \frac{F_z}{\cos\beta}.$$

Problem 6.11: Total force and torque

(a) Determine the components F_y, F_z of the force $\mathbf{F} = (2\,\mathrm{N}, F_y, F_z)$ acting at point $P_1(1, 2, 3)$ m such that it is perpendicular to the plane defined by the three points P_1, $P_2(2, 3, 4)$ m, and $P_3(2, 2, 1)$ m.

(b) What is the magnitude of the force $\mathbf{F}$ and which torque $\mathbf{M}$ does it apply with respect to the point $P_4(0, 1, 2)$ m?

(c) What is the component of the torque vector $\mathbf{M}$ that points perpendicular to the plane?

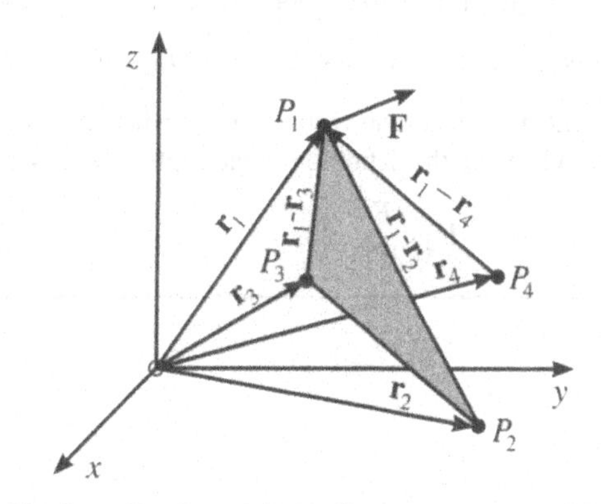

The force $\mathbf{F}$ acting at P_1 applies a torque around P_4.

Solution (a) Because the vectors $(\mathbf{r}_1 - \mathbf{r}_2)$ and $(\mathbf{r}_1 - \mathbf{r}_3)$ are within the represented plane, the vector product $\mathbf{R} = (\mathbf{r}_1 - \mathbf{r}_2) \times (\mathbf{r}_1 - \mathbf{r}_3)$ yields a normal vector $\mathbf{R}$ perpendicular to the plane. If the force $\mathbf{F}$ shall be perpendicular to the plane, that is, parallel to the vector $\mathbf{R}$, the following must hold:

$$\mathbf{R} = (R_x, R_y, R_z) = (\mathbf{r}_1 - \mathbf{r}_2) \times (\mathbf{r}_1 - \mathbf{r}_3) = \lambda\mathbf{F} = \lambda\,(2\,\mathrm{N}, F_y, F_z). \tag{6.24}$$

From there it follows that

$$\lambda = \frac{1}{2\,\mathrm{N}} R_x, \qquad F_y = \frac{1}{\lambda} R_y, \qquad F_z = \frac{1}{\lambda} R_z. \tag{6.25}$$

For the points $P_1(1, 2, 3)$ m, $P_2(2, 3, 4)$ m, $P_3(2, 2, 1)$ m, and $P_4(0, 1, 2)$ m specified in the problem, one easily gets the position vectors $\mathbf{r}_i$ $(i = 1, 2, 3, 4)$ as well as their differences $(\mathbf{r}_i - \mathbf{r}_j)$ $(i \neq j)$:

$$(\mathbf{r}_1 - \mathbf{r}_2) = (-1, -1, -1)\,\mathrm{m},$$
$$(\mathbf{r}_1 - \mathbf{r}_3) = (-1, 0, 2)\,\mathrm{m}, \tag{6.26}$$
$$(\mathbf{r}_1 - \mathbf{r}_4) = (1, 1, 1)\,\mathrm{m}.$$

For the vector product in equation 6.24, it then results that

$$\mathbf{R} = (\mathbf{r}_1 - \mathbf{r}_2) \times (\mathbf{r}_1 - \mathbf{r}_3) = (-2, 3, -1)\,\mathrm{m}^2 = (R_x, R_y, R_z).$$

Inserting these values in 6.25, one obtains

$$\lambda = \frac{1}{2\,\mathrm{N}} R_x = -1\,\mathrm{m}^2\mathrm{N}^{-1},$$
$$F_y = \frac{R_y}{\lambda} = -3\,\mathrm{N}, \tag{6.27}$$
$$F_z = \frac{R_z}{\lambda} = 1\,\mathrm{N}.$$

Thus, the components of the force **F** are

$$\mathbf{F} = (2, -3, 1)\,\mathrm{N}. \tag{6.28}$$

(b) The magnitude of the force **F** is obtained as

$$|\mathbf{F}| = (F_x^2 + F_y^2 + F_z^2)^{1/2} = (2^2 + 3^2 + 1)^{1/2}\,\mathrm{N} \approx 3.74\,\mathrm{N}. \tag{6.29}$$

The torque **M** with respect to the point P_4 results from the vector product

$$\mathbf{M} = (\mathbf{r}_1 - \mathbf{r}_4) \times \mathbf{F} = (1, 1, 1) \times (2, -3, 1)\mathrm{Nm} = (4, 1, -5)\,\mathrm{Nm}. \tag{6.30}$$

(c) The component of the torque vector **M** perpendicular to the plane, that is, along the orientation of the force **F**, results from the definition of the triple scalar product as

$$|\mathbf{M}_\mathbf{F}| = [(\mathbf{r}_1 - \mathbf{r}_4) \times \mathbf{F}] \cdot \frac{\mathbf{F}}{|\mathbf{F}|} = 0. \tag{6.31}$$

7 Differentiation and Integration of Vectors

Formation of the differential quotient: The vector $\mathbf{A}$ may occur as a function of a parameter. Let's consider, for example, the position vector $\mathbf{r}(t)$ that—as a function of the time t—describes the path of a mass point. If one decomposes $\mathbf{A}$ into its components with respect to fixed unit vectors, then these components are functions of the parameter. We write

$$\mathbf{A}(u) = A_x(u)\mathbf{e}_1 + A_y(u)\mathbf{e}_2 + A_z(u)\mathbf{e}_3. \tag{7.1}$$

The differential quotient of a vector is formed by differentiating its components separately, as corresponds to the differentiation rule for sums. Because the unit vectors are not variables, they are conserved under differentiation,

$$\begin{aligned}\frac{d\mathbf{A}(u)}{du} &= \lim_{\Delta u\to 0}\frac{\mathbf{A}(u+\Delta u)-\mathbf{A}(u)}{\Delta u}\\ &= \lim_{\Delta u\to 0}\left(\frac{A_x(u+\Delta u)-A_x(u)}{\Delta u}\mathbf{e}_1 + \frac{A_y(u+\Delta u)-A_y(u)}{\Delta u}\mathbf{e}_2\right.\\ &\qquad\left. + \frac{A_z(u+\Delta u)-A_z(u)}{\Delta u}\mathbf{e}_3\right).\end{aligned}$$

The limit of the sum is equal to the sum of the limits, that is, when passing to the limit, one obtains

$$\frac{d\mathbf{A}(u)}{du} = \frac{dA_x(u)}{du}\mathbf{e}_1 + \frac{dA_y(u)}{du}\mathbf{e}_2 + \frac{dA_z(u)}{du}\mathbf{e}_3. \tag{7.2}$$

By comparing (7.1) with (7.2), one notices that the differentiation of a vector in an arbitrary coordinate frame with fixed unit vectors amounts to the differentiation of the components of the vector. Generally, the rule for the n-fold differentiation of a vector reads

$$\frac{d^n \mathbf{A}(u)}{du^n} = \frac{d^n A_x(u)}{du^n}\mathbf{e}_1 + \frac{d^n A_y(u)}{du^n}\mathbf{e}_2 + \frac{d^n A_z(u)}{du^n}\mathbf{e}_3. \tag{7.3}$$

Example 7.1: Differentiation of a vector

$$\begin{aligned}
\mathbf{A}(u) &= \underbrace{(2u^2 - 3u)}_{A_x(u)}\mathbf{e}_1 + \underbrace{(5 \cdot \cos u)}_{A_y(u)}\mathbf{e}_2 - \underbrace{(3 \cdot \sin u)}_{A_z(u)}\mathbf{e}_3, \\
&= (2u^2 - 3u, 5 \cdot \cos u, -3 \cdot \sin u), \\
\frac{d\mathbf{A}(u)}{du} &= (4u - 3)\mathbf{e}_1 - (5 \cdot \sin u)\mathbf{e}_2 - (3 \cdot \cos u)\mathbf{e}_3 \\
&= (4u - 3, -5 \cdot \sin u, -3 \cdot \cos u), \\
\frac{d^2\mathbf{A}(u)}{du^2} &= 4\mathbf{e}_1 - 5 \cdot \cos u \cdot \mathbf{e}_2 + 3 \cdot \sin u \cdot \mathbf{e}_3 \\
&= (4, -5 \cdot \cos u, 3 \cdot \sin u).
\end{aligned}$$

For composite functions the usual differentiation rules apply. For example, for the product of a scalar function and a vector function, or for the scalar or vector product of two vector functions (parameter u), the product rule applies.

Differentiation of the product of a scalar and a vector:

$$\frac{d(\phi(u)\mathbf{A}(u))}{du} = \frac{d}{du}(\phi(u)A_x(u)\mathbf{e}_1 + \phi(u)A_y(u)\mathbf{e}_2 + \phi(u)A_z(u)\mathbf{e}_3).$$

Now

$$\frac{d(\phi A_x)}{du} = \frac{d\phi}{du}A_x + \phi\frac{dA_x}{du}$$

and analogously for the other components:

$$\frac{d}{du}(\phi A_i) = \frac{d}{du}(\phi)A_i + \frac{d}{du}(A_i)\phi \qquad (i = 1, 2, 3).$$

This yields

$$\frac{d(\phi(u) \cdot \mathbf{A}(u))}{du} = \frac{d\phi}{du}A_x\mathbf{e}_1 + \frac{d\phi}{du}A_y\mathbf{e}_2 + \frac{d\phi}{du}A_z\mathbf{e}_3 + \phi\frac{dA_x}{du}\mathbf{e}_1 + \phi\frac{dA_y}{du}\mathbf{e}_2 + \phi\frac{dA_z}{du}\mathbf{e}_3$$

or simply

$$\frac{d(\phi(u)\mathbf{A}(u))}{du} = \frac{d\phi}{du}\mathbf{A} + \phi\frac{d\mathbf{A}}{du}. \tag{7.4}$$

Differentiation of the scalar product: One has

$$\frac{d(\mathbf{A}(u)\cdot\mathbf{B}(u))}{du} = \frac{d}{du}\left(\sum_{i=1}^{3} A_i(u)B_i(u)\right) = \sum_{i=1}^{3}\frac{d}{du}(A_i(u)B_i(u))$$

$$= \sum_{i=1}^{3}\left(\frac{dA_i(u)}{du}B_i(u) + A_i(u)\frac{dB_i(u)}{du}\right)$$

and therefore,

$$\frac{d(\mathbf{A}(u)\cdot\mathbf{B}(u))}{du} = \frac{d\mathbf{A}}{du}\cdot\mathbf{B} + \mathbf{A}\cdot\frac{d\mathbf{B}}{du}. \tag{7.5}$$

Differentiation of the vector product: It is performed analogously to the differentiation of the scalar product. Because the vector product is not commutative, one has to take care of the ordering of the factors.

$$\frac{d}{du}(\mathbf{A}(u)\times\mathbf{B}(u)) = \frac{d\mathbf{A}(u)}{du}\times\mathbf{B}(u) + \mathbf{A}(u)\times\frac{d\mathbf{B}(u)}{du}. \tag{7.6}$$

This is easily proved by checking the individual components (e.g., the x-component) on both sides of the equation.

Example 7.2: Differentiation of the product of a scalar and a vector

For the scalar function $\varphi(x) = x + 5$ and the vector function $\mathbf{A}(x) = (x^2 + 2x + 1, 2x, x + 2)$ the second derivative of the products $\varphi\cdot\mathbf{A}$ is to be calculated.

The differentiation of the product yields

$$\frac{d^2(\varphi\mathbf{A})}{dx^2} = \frac{d}{dx}\left(\frac{d\varphi}{dx}\mathbf{A} + \varphi\frac{d\mathbf{A}}{dx}\right) = \frac{d^2\varphi}{dx^2}\mathbf{A} + 2\frac{d\varphi}{dx}\frac{d\mathbf{A}}{dx} + \varphi\frac{d^2\mathbf{A}}{dx^2}.$$

The derivatives of the individual functions read

$$\frac{d\varphi}{dx} = 1, \quad \frac{d^2\varphi}{dx^2} = 0, \quad \frac{d\mathbf{A}}{dx} = (2x + 2, 2, 1), \quad \frac{d^2\mathbf{A}}{dx^2} = (2, 0, 0).$$

From the above, it results that

$$\frac{d^2(\varphi\mathbf{A})}{dx^2} = (4x + 4, 4, 2) + (2x + 10, 0, 0) = (6x + 14, 4, 2).$$

Application: Position, velocity, and acceleration of a mass point on a defined trajectory may be represented as vectors. The position vector for the motion of the mass point on an arbitrary trajectory B is the vector from the origin of the coordinate frame to the mass point; the variation of the position of the mass point with the time may be represented as time variation of the position vector (compare with the figure).

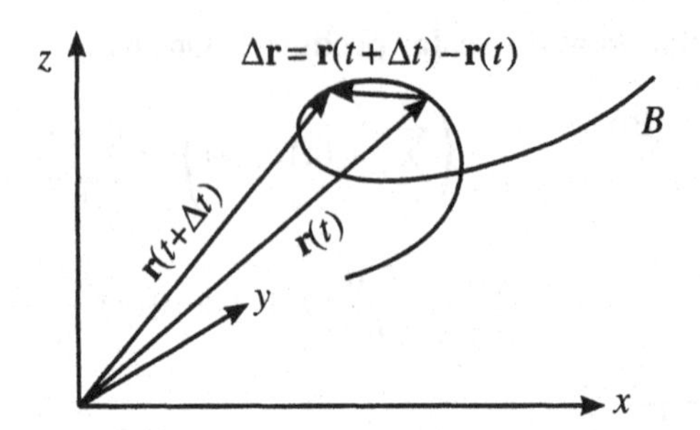

Definition of the orbital velocity: $\Delta\mathbf{r} = \mathbf{r}(t+\Delta t) - \mathbf{r}(t)$ is a secant vector to the orbit at the point $\mathbf{r}(t)$. The velocity is then given by $\mathbf{v} = \lim_{\Delta t\to 0} \Delta\mathbf{r}/\Delta t$.

The velocity is defined as the first derivative of the position vector $\mathbf{r}(t)$ of the orbital curve with respect to the time:

$$\mathbf{v} = \lim_{\Delta t\to 0} \frac{\Delta\mathbf{r}}{\Delta t} = \lim_{\Delta t\to 0} \frac{\mathbf{r}(t+\Delta t) - \mathbf{r}(t)}{\Delta t} = \frac{d\mathbf{r}}{dt}. \tag{7.7}$$

From equation (7.7) one notices that the vector of the velocity represents the limit position of the secant through the position vectors $\mathbf{r}(t+\Delta t)$ and $\mathbf{r}(t)$ divided by the time interval Δt in the limit $\Delta t \to 0$, that is, the velocity points along the tangent to the trajectory at the point $\mathbf{r}(t)$.

The acceleration is obtained as the first derivative of the velocity with respect to the time, or as the second derivative of the position vector with respect to the time:

$$\mathbf{a}(t) = \frac{d\mathbf{v}(t)}{dt} = \lim_{\Delta t\to 0} \frac{\Delta\mathbf{v}}{\Delta t} = \frac{d(d\mathbf{r}/dt)}{dt} = \frac{d^2\mathbf{r}(t)}{dt^2}. \tag{7.8}$$

Because the position vector is a vector, its derivatives with respect to the scalar time (t) are again vectors. Thus, the velocity and acceleration are vectors, too.

Problem 7.3: Velocity and acceleration on a space curve

Let the position vector be given by $\mathbf{r} = (t^3 + 2t, -3e^{-t}, t)$ m. Find the velocity and the acceleration as well as their magnitudes for the time points $t = 0$ s and $t = 1$ s.

Solution For the velocity and acceleration, we get

$$\mathbf{v}(t) = \dot{\mathbf{r}} = (3t^2 + 2, 3e^{-t}, 1)\,\frac{\text{m}}{\text{s}},$$

$$\mathbf{a}(t) = \ddot{\mathbf{r}} = (6t, -3e^{-t}, 0)\,\frac{\text{m}}{\text{s}^2}.$$

For the time $t = 0$, the results are

$$\mathbf{v}(0) = (2, 3, 1)\frac{\text{m}}{\text{s}}, \qquad \mathbf{a}(0) = (0, -3, 0)\,\frac{\text{m}}{\text{s}^2},$$

$$v(0) = \sqrt{14}\,\frac{\text{m}}{\text{s}}, \qquad a(0) = 3\,\frac{\text{m}}{\text{s}^2}.$$

For $t = 1$ s,

$$\mathbf{v}(1) = (5, \frac{3}{e}, 1)\frac{\text{m}}{\text{s}}, \qquad \mathbf{a}(1) = (6, -\frac{3}{e}, 0)\,\frac{\text{m}}{\text{s}^2},$$

$$v(1) = 5.22\,\frac{\text{m}}{\text{s}}, \qquad a(1) = 6.1\,\frac{\text{m}}{\text{s}^2}.$$

Example 7.4: Circular motion

The Cartesian components of a circular motion are given by

$$x(t) = R\cos\omega t,$$
$$y(t) = R\sin\omega t,$$
$$z(t) = 0.$$

ω is the so-called angular velocity or also angular frequency. It is related to the revolution period T via $\omega T = 2\pi$. The position vector is now

$$\mathbf{r}(t) = (x(t), y(t), z(t))$$
$$= x(t)\mathbf{e}_1 + y(t)\mathbf{e}_2 + z(t)\mathbf{e}_3,$$
$$\mathbf{r}(t) = (R \cdot \cos\omega t, R \cdot \sin\omega t, 0)$$
$$= R \cdot \cos\omega t\,\mathbf{e}_1 + R \cdot \sin\omega t\,\mathbf{e}_2 + 0\mathbf{e}_3.$$

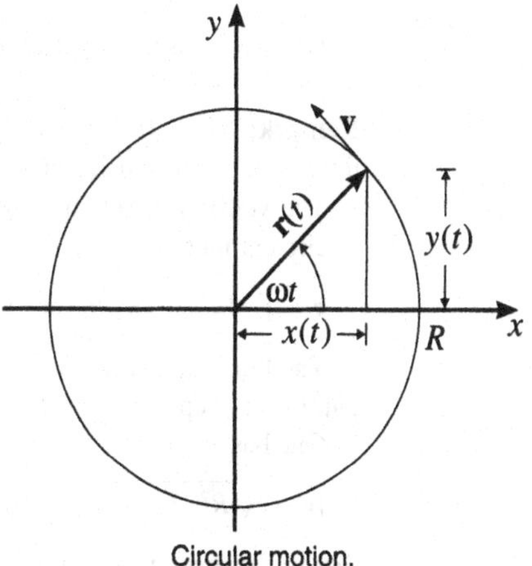

Circular motion.

For the velocity one gets

$$\mathbf{v} = \frac{d\mathbf{r}}{dt} = (-\omega R \cdot \sin\omega t, R\omega\cos\omega t, 0).$$

There holds

$$\mathbf{r}\cdot\mathbf{v} = \mathbf{r}\cdot\frac{d\mathbf{r}}{dt} = 0 \qquad \text{for any time point},$$
$$\Rightarrow \mathbf{v} \perp \mathbf{r},$$

which is immediately clear for a circular orbit.

For the magnitude of the velocity, one obtains

$$v = |\mathbf{v}| = \sqrt{\left(\frac{dx}{dt}\right)^2 + \left(\frac{dy}{dt}\right)^2 + \left(\frac{dz}{dt}\right)^2} = \sqrt{\omega^2 R^2 \sin^2\omega t + \omega^2 R^2 \cos^2\omega t + 0}$$
$$= \sqrt{\omega^2 R^2(\sin^2\omega t + \cos^2\omega t)} = \omega R = \frac{2\pi R}{T} = \frac{\text{circumference}}{\text{revolution period}}.$$

The acceleration is obtained as

$$\mathbf{b} = \frac{d\mathbf{v}}{dt} = \frac{d^2\mathbf{r}}{dt^2}$$
$$= (-\omega^2 R\cos\omega t, -\omega^2 R\sin\omega t, 0) = -\omega^2(R\cos\omega t, R\sin\omega t, 0),$$
$$= -\omega^2\mathbf{r}.$$

It turns out that the acceleration points opposite the orientation of the position vector (centripetal acceleration). The magnitude of the acceleration is given by

$$|\mathbf{b}| = \sqrt{\left(\frac{d^2x}{dt^2}\right)^2 + \left(\frac{d^2y}{dt^2}\right)^2 + \left(\frac{d^2z}{dt^2}\right)^2},$$

$$= \omega^2 R = \frac{v^2}{R}.$$

Example 7.5: The motion on a helix

The Cartesian coordinates of the helix read

$$x(t) = R\cos\omega t, \quad y(t) = R\sin\omega t, \quad z(t) = b\omega t.$$

The position vector is obtained by inserting in the relation

$$\mathbf{r}(t) = (x(t), y(t), z(t)),$$

that is, it holds that

$$\mathbf{r}(t) = (R\cos\omega t, R\sin\omega t, b\omega t).$$

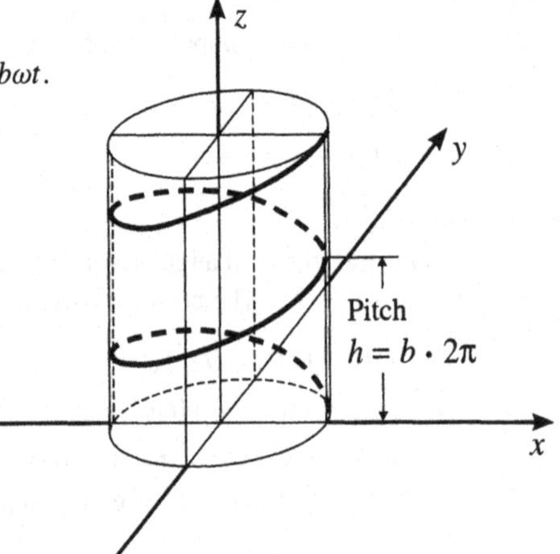

The helix and its pitch.

Remark: $b > 0$ means right-handed helix, $b < 0$ means left-handed helix.

The velocity results analogously to that of circular motion

$$\mathbf{v} = (-R\omega\sin\omega t, R\omega\cos\omega t, b\omega).$$

The third component $v_3 = b\omega$ implies a uniform (constant) upward velocity (z-direction) if the parameter t represents the time.

One has

$$|\mathbf{v}| = \sqrt{R^2\omega^2 + b^2\omega^2} = \omega\sqrt{R^2 + b^2},$$

that is, the magnitude of the velocity is constant.

The acceleration is the derivative of the velocity

$$\mathbf{b} = -\omega^2 \cdot (R\cos\omega t, R\sin\omega t, 0) = -\omega^2\mathbf{r}_\perp,$$

where

$$\mathbf{r}_\perp = (R\cos\omega t, R\sin\omega t, 0) = (\mathbf{r}\cdot\mathbf{e}_r)\mathbf{e}_r$$

and $\mathbf{e}_r = (\cos\omega t, \sin\omega t, 0)$ is the polar unit vector in the x, y-plane. We thus obtain the same acceleration as for the circular motion. For the magnitude, it holds that $|\mathbf{b}| = \omega^2 R$.

Integration of vectors: The integration rules may be applied also to vectors in the customary way. For a vector $\mathbf{A}$ depending on a parameter (e.g., u), it follows that

$$\int \mathbf{A}(u)\,du = \int \left(A_x(u)\mathbf{e}_1 + A_y(u)\mathbf{e}_2 + A_z(u)\mathbf{e}_3\right)du$$

$$= \int A_x(u)\,\mathbf{e}_1\,du + \int A_y(u)\,\mathbf{e}_2\,du + \int A_z(u)\,\mathbf{e}_3\,du.$$

If the unit vectors are constant, they may be pulled out before the integral symbol:

$$\int \mathbf{A}(u)\,du = \mathbf{e}_1 \int A_x(u)\,du + \mathbf{e}_2 \int A_y(u)\,du + \mathbf{e}_3 \int A_z(u)\,du.$$

Thus, we may formulate the following rule: A vector is integrated by integrating its components. This vector integration graphically means a summation of a large number of vectors according to the integral limits; for example, the sum of all forces acting on a body. More strictly speaking: $\mathbf{A}(u)$ is a vector density, and $d\mathbf{A} = \mathbf{A}(u)\,du$ is the vector associated with the interval du. These $d\mathbf{A}$ are summed to yield the integral. An example is the *impulse of force* $\mathbf{K}$, which is understood as the force $\mathbf{K}$ acting on a body over a time interval; thus $\mathbf{K} = \int_{\Delta t} \mathbf{F}(t')\,dt'$. The impulse of force is therefore the sum of the forces $\mathbf{F}(t')$ acting during the time interval. For more details, see Chapter 17, equations (17.14) and (17.15).

Example 7.6: Integration of a vector

$$\mathbf{A} = (2u^2 - 3u, 5\cos u, -3\sin u),$$

$$\int \mathbf{A}\,du = \left(\frac{2}{3}u^3 - \frac{3}{2}u^2 + C_1\right)\mathbf{e}_1 + (5\sin u + C_2)\mathbf{e}_2 + (3\cos u + C_3)\mathbf{e}_3$$

$$= \left(\frac{2}{3}u^3 - \frac{3}{2}u^2\right)\mathbf{e}_1 + (5\sin u)\mathbf{e}_2 + (3\cos u)\mathbf{e}_3 + C_1\mathbf{e}_1 + C_2\mathbf{e}_2 + C_3\mathbf{e}_3$$

$$= \left(\frac{2}{3}u^3 - \frac{3}{2}u^2, 5\sin u, 3\cos u\right) + \mathbf{C}.$$

The integration constants arising in the components are composed to the vector $\mathbf{C}$.

Problem 7.7: Integration of a vector

Calculate

$$\int_0^2 \mathbf{A}(n)\,dn \qquad \text{with} \quad \mathbf{A} = (3n^2 - 1, 2n - 3, 6n^2 - 4n).$$

Solution

$$\int_0^2 \mathbf{A}(n)\,dn = \int_0^2 (3n^2 - 1, 2n - 3, 6n^2 - 4n)\,dn$$

$$= \left[(n^3 - n, n^2 - 3n, 2n^3 - 2n^2)\right]_0^2$$

$$= (6, -2, 8).$$

Problem 7.8: Motion on a special space curve

(a) Which curve is passed by the vector

$$\mathbf{r}(t) = (x(t), y(t), z(t)) = (t\cos t, t\sin t, t)$$

when t is running from 0 to 2π?

(b) Calculate the velocity and acceleration of the point at the time t.

(c) What are the velocity and acceleration for $t = 0$ and $t = 2$?

(d) How do the magnitudes of radius vector, velocity, and acceleration vary for large time t?

Solution (a) We first consider the vector $\tilde{\mathbf{r}}(t)$ with $\tilde{z}(t) \equiv 0$ (projection onto the x, y-plane).

$$\tilde{\mathbf{r}}(t) = (t\cos t, t\sin t, 0).$$

Because

$$|\tilde{\mathbf{r}}(t)| = (t^2\cos^2 t + t^2\sin^2 t)^{1/2} = t,$$

there results a spiral line with a radius from 0 to 2π.

If $z(t) = t$ additionally runs from 0 to 2π, we obtain a spiral line on the surface of a cone of height 2π with the vortex at (0, 0, 0).

The figure at the top of the facing page illustrates this result.

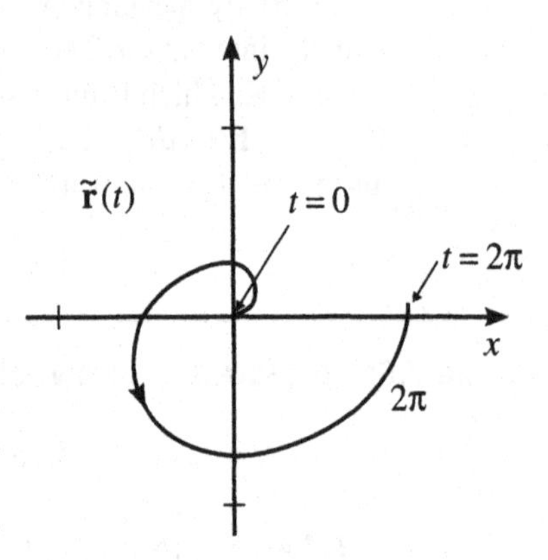

The resulting spiral line with a radius varying from 0 to 2π.

(b) For the velocity $\mathbf{v}(t)$ and acceleration $\mathbf{b}(t)$, it results that

$$\mathbf{v}(t) = \frac{d\mathbf{r}}{dt} = (\cos t - t\sin t, \sin t + t\cos t, 1),$$

$$\begin{aligned}\mathbf{b}(t) &= \frac{d\mathbf{v}}{dt} = \frac{d^2\mathbf{r}}{dt^2}\\ &= (-\sin t - \sin t - t\cos t, \cos t + \cos t - t\sin t, 0),\\ &= (-2\sin t - t\cos t, 2\cos t - t\sin t, 0),\end{aligned}$$

(c) One has

$$\begin{aligned}&\mathbf{v}(t=0) = (1, 0, 1); && |\mathbf{v}(t=0)| = \sqrt{2},\\ &\mathbf{v}(t=2) = (-2.23, 0.08, 1); && |\mathbf{v}(t=2)| = \sqrt{6},\\ &\mathbf{b}(t=0) = (0, 2, 0); && |\mathbf{b}(t=0)| = 2,\\ &\mathbf{b}(t=2) = (-0.99, -2.65, 0); && |\mathbf{b}(t=2)| = \sqrt{8}.\end{aligned}$$

(d)

$$|\mathbf{r}(t)| = \left(t^2\cos^2 t + t^2\sin^2 t + t^2\right)^{1/2} = \sqrt{2}\,|t|,$$

$$|\mathbf{v}(t)| = \left((\cos t - t\sin t)^2 + (\sin t + t\cos t)^2 + 1\right)^{1/2}$$

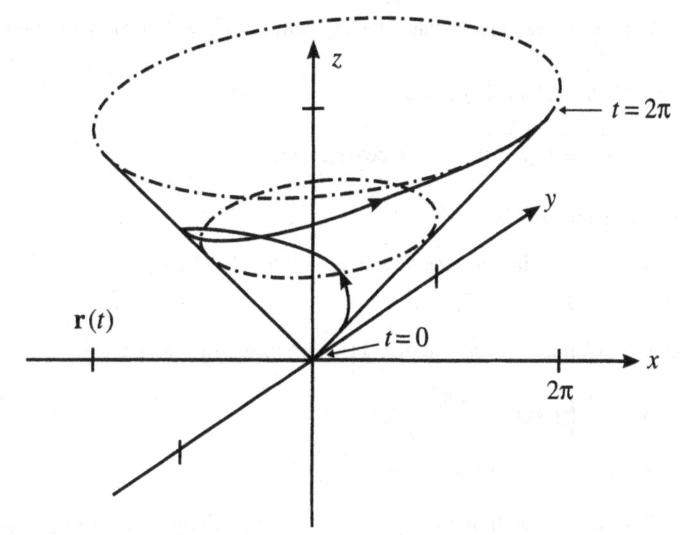

The spiral line on the surface of a cone.

$$= (2+t^2)^{1/2} = |t|\left(1+\frac{2}{t^2}\right)^{1/2}$$

$$= |t|\left\{1+\frac{1}{t^2} - O\left(\frac{1}{t^4}\right)\right\} \quad \xrightarrow{t\gg 1} \quad |t| \qquad \text{(by a series expansion of the square root),}$$

$$|\mathbf{b}(t)| = \left((2\sin t + t\cos t)^2 + (2\cos t - t\sin t)^2\right)^{1/2}$$

$$= (4+t^2)^{1/2} = |t|\left(1+\frac{4}{t^2}\right)^{1/2}$$

$$= |t|\left\{1+\frac{2}{t^2} - O\left(\frac{1}{t^4}\right)\right\} \quad \xrightarrow{t\gg 1} \quad |t| \qquad \text{(by a series expansion of the square root).}$$

Problem 7.9: Airplane landing along a special space curve

An airplane is landing. Thereby it is moving on the space curve

$$\mathbf{r}(t) = (x(t), y(t), z(t)) = (R\cos\omega t, R\sin\omega t, (H - b\omega t)),$$

with

$$R = 1000\,\text{m},$$
$$\omega = \frac{1}{7}\,\text{s}^{-1},$$
$$H = 400\,\text{m},$$
$$b = H/6\pi,$$
$$t \in [0, 42\pi]\,\text{s}.$$

What is the velocity of the plane on landing (at $t = 42\pi$ s)? Would you try landing this way?

Solution The velocity is calculated to be

$$\mathbf{v} = \frac{d\mathbf{r}}{dt} = (-\omega R \sin \omega t, \omega R \cos \omega t, -b\omega),$$

and its magnitude is

$$|\mathbf{v}| = (\omega^2 R^2 \sin^2 \omega t + \omega^2 R^2 \cos^2 \omega t + b^2 \omega^2)^{1/2}$$
$$= \omega (R^2 + b^2)^{1/2} .$$

Obviously, it is independent of t! Insertion of the values yields

$$|\mathbf{v}| = \frac{1}{7} \left(1000^2 + \frac{400^2}{(6\pi)^2} \right)^{1/2} \text{ m s}^{-1}$$
$$\approx 142.9 \text{ m s}^{-1} \quad \widehat{=} \quad 514.4 \text{ km h}^{-1} .$$

This kind of landing is certainly unsuitable; the approach velocity should better be reduced.

8 The Moving Trihedral (Accompanying Dreibein)—the Frenet Formulas

In some cases it may be simpler to express velocity and acceleration in *natural coordinates.* This means that the velocity and acceleration are not derived from the variation of the position vector with the time, but from its variation with the passed way s, the arc length, the starting point being arbitrary. Let the curve itself be given by the position vector $\mathbf{r}(t) = (x_1(t), x_2(t), x_3(t))$. For infinitesimally small segments, the increase of the arc length is $|d\mathbf{r}| = ds$.

The arc length s of the curve between the parameter values t_0 and t is then obtained by integration:

$$s(t) = \int_{t_0}^{t} ds = \int_{t_0}^{t} |d\mathbf{r}| = \int_{t_0}^{t} \frac{|d\mathbf{r}|}{dt}\, dt \tag{8.1}$$

$$= \int_{t_0}^{t} \sqrt{\left(\frac{dx_1}{dt}\right)^2 + \left(\frac{dx_2}{dt}\right)^2 + \left(\frac{dx_3}{dt}\right)^2}\, dt.$$

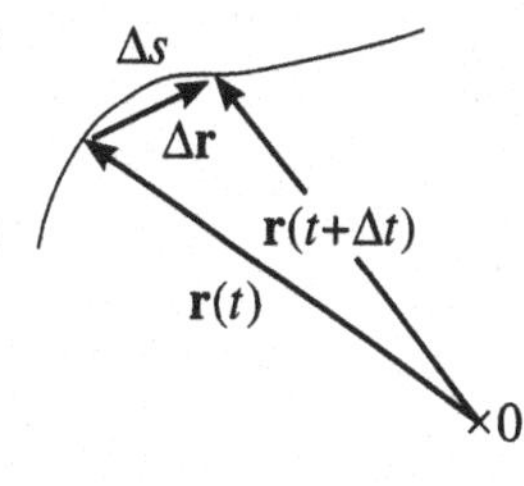

In the limit of small Δt, the absolute value of the secant vector $\Delta\mathbf{r}$ becomes the line element ds, that is, $|\Delta\mathbf{r}| \to ds$.

The magnitude of the velocity is

$$|\mathbf{v}| = \left|\frac{d\mathbf{r}}{dt}\right| = \frac{|d\mathbf{r}|}{dt} = \frac{ds}{dt}\,.$$

In order to become independent of the coordinate frame, a set of orthogonal unit vectors is put at the point of the trajectory of the mass point given by s. The set of unit vectors moves along with the mass point: it is therefore also called the *"moving trihedral"* or *"accompanying dreibein."* As unit vectors one uses

T tangent vector,

N principal normal vector,

B binormal vector.

Because the vectors form an orthonormalized set, it holds that $(\mathbf{N} \times \mathbf{B}) = \mathbf{T}$, cyclically permutable. In the following we give the precise definition of these three base vectors of the moving trihedral and show how they are calculated for a given space curve $\mathbf{r}(t)$.

The function $\mathbf{r}(t)$ describes a space curve depending on the time t as a parameter. To determine the moving trihedral, one has to convert the function $\mathbf{r}(t)$ into $\mathbf{r}(s)$; this is done by substituting the time $t = t(s)$ from $s = s(t)$ (compare with equation (8.1)).

The moving trihedral is determined from the local properties of the trajectory. $d\mathbf{r}/ds$ is a vector along the limit position of the secant, i.e., the tangent.

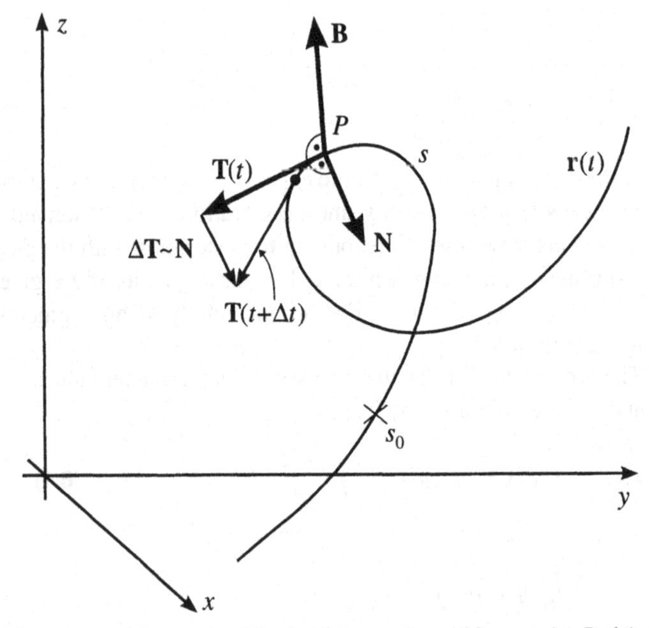

A curve in space and the moving trihedral (shown at an arbitrary point P of the curve).

The magnitude of this vector is $|\Delta\mathbf{r}|/\Delta s$. For infinitesimally small segments one has $|d\mathbf{r}| = ds$; thus $|d\mathbf{r}|/ds = 1$. Hence one has determined the tangent unit vector:

$$\mathbf{T} = \frac{d\mathbf{r}}{ds}.$$

Because $ds = |d\mathbf{r}|$, it also holds that

$$\mathbf{T} = \frac{d\mathbf{r}}{|d\mathbf{r}|} = \frac{d\mathbf{r}/dt}{|d\mathbf{r}/dt|} = \frac{\mathbf{v}}{|\mathbf{v}|}.$$

In order to determine the principal normal vector, one first forms

$$\mathbf{T}\cdot\mathbf{T} = 1.$$

By differentiating the scalar product of the tangent vector, one obtains

$$\frac{d}{ds}(\mathbf{T}\cdot\mathbf{T}) = \frac{d\mathbf{T}}{ds}\cdot\mathbf{T} + \mathbf{T}\cdot\frac{d\mathbf{T}}{ds} = 0.$$

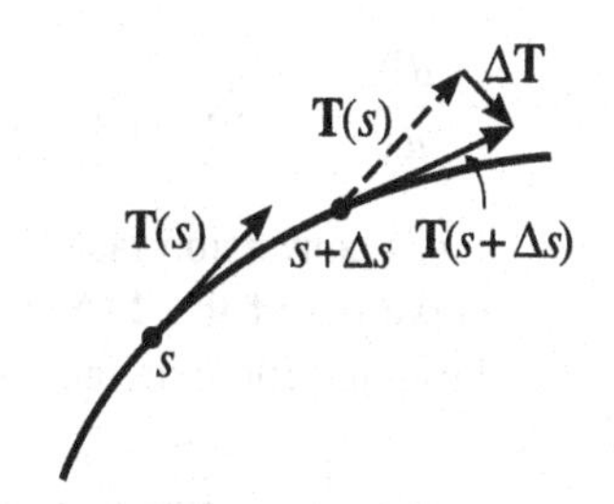

The difference vector $\Delta\mathbf{T}$ is a measure of the curvature of the curve. $\Delta\mathbf{T}$ points toward the "inner side" of the curve. A straight line has no curvature, and the normal vector **N** is not uniquely defined.

Because the commutative law holds for the scalar product, $\mathbf{T}\cdot d\mathbf{T}/ds$ is zero. This implies that $d\mathbf{T}/ds$ is perpendicular to $\mathbf{T}$.

The vector $d\mathbf{T}/ds$ gives the orientation of the principal normal vector. We characterize its position by constructing, besides the tangent defined by $\mathbf{T}(s)$, a second tangent $\mathbf{T}(s+\Delta s)$ (neighboring tangent) that differs from the first one only by an infinitesimal vector $\Delta\mathbf{T}$ (see figure). The principal normal vector lies in the plane spanned by the two tangents $\mathbf{T}(s)$ and $\mathbf{T}(s+\Delta s)$. Because the magnitude of $d\mathbf{T}/ds$ in general differs from unity, one still has to introduce a factor κ for normalization:

$$\kappa\cdot\mathbf{N} = \frac{d\mathbf{T}}{ds},$$

where $\kappa = |d\mathbf{T}/ds|$. This is the *first Frenet formula.* The value κ is always defined as positive. This is possible since the orientation of **N** may be chosen in an appropriate manner. The factor κ is called the *curvature* of the space curve.

The third unit vector, the binormal vector, is formed out of **T** and **N**:

$$\mathbf{B} = \mathbf{T}\times\mathbf{N}.$$

The *orientations* of all three unit vectors are functions of the arc length.

The vectors of the moving trihedral (accompanying dreibein) may be differentiated with respect to the arc length. The three differential quotients are called *Frenet's*[1] *formulas* and read

$$\frac{d\mathbf{T}}{ds} = \kappa\mathbf{N}, \tag{8.2}$$

$$\frac{d\mathbf{N}}{ds} = \tau\mathbf{B} - \kappa\mathbf{T}, \tag{8.3}$$

[1] *Jean Frédéric Frenet*, b. Feb. 7, 1816—d. June 12, 1900, Périgueux (Dordogne). In 1840 Frenet entered the École Normale in Paris as a scholar, was appointed as professor in 1848, and taught until 1868 at Lyon University. His research mainly concerned problems of *differential geometry,* and in 1847 he found the Frenet forms independently of Serret.

$$\frac{d\mathbf{B}}{ds} = -\tau\mathbf{N} \qquad \text{or} \qquad |\tau| = \left|\frac{d\mathbf{B}}{ds}\right|. \tag{8.4}$$

τ is a conversion factor and is called the *torsion*. The torsion describes the winding of the curve out of the $\mathbf{T}, \mathbf{N}$-plane. The quantity $d\mathbf{B}/ds$ is exactly a measure for this winding. From the curvature and torsion one gets

$$\varrho = \frac{1}{\kappa} \quad \text{curvature radius,} \qquad \sigma = \frac{1}{\tau} \quad \text{torsion radius.}$$

The curvature radius of a curve at a definite point equals the radius of the osculating circle having the same curvature as the curve at that point.

Formula (8.2) has already been introduced as a definition. For a transparent derivation of the remaining formulas, one utilizes the statement that any vector may be represented as a linear combination of the three unit vectors.

Derivation of the second Frenet formula: Because the moving trihedral spans the entire three-dimensional space, it holds that

$$\frac{d\mathbf{N}}{ds} = \alpha\mathbf{T} + \beta\mathbf{N} + \gamma\mathbf{B}\,,$$

where α, β, γ are to be determined. Because $\mathbf{N}$ is a unit vector, $\mathbf{N}\cdot\mathbf{N} = 1$.

By differentiating the scalar product $\mathbf{N}\cdot\mathbf{N}$, one obtains

$$\frac{d}{ds}(\mathbf{N}\cdot\mathbf{N}) = \frac{d}{ds}(1) = 0$$

or in other notation (product rule) $d\mathbf{N}/ds\cdot\mathbf{N}+\mathbf{N}\cdot d\mathbf{N}/ds = 0$, or, because the commutative law holds for the scalar product:

$$2\mathbf{N}\cdot\frac{d\mathbf{N}}{ds} = 0.$$

Because in the nontrivial case neither $\mathbf{N}$ nor $d\mathbf{N}/ds$ is equal to zero, this means that $d\mathbf{N}/ds$ is perpendicular to $\mathbf{N}$, i.e., there is no component of $d\mathbf{N}/ds$ along $\mathbf{N}$. Therefore,

$$\beta = 0; \qquad \text{i.e.,} \quad \frac{d\mathbf{N}}{ds} = \alpha\mathbf{T} + \gamma\mathbf{B}. \tag{8.5}$$

Moreover, according to the definition of the unit vectors, $\mathbf{T}\cdot\mathbf{N} = 0$. By forming the first derivative of this scalar product one has

$$\frac{d\mathbf{T}}{ds}\cdot\mathbf{N} + \mathbf{T}\cdot\frac{d\mathbf{N}}{ds} = 0. \tag{8.6}$$

Using the first Frenet formula, we find

$$\frac{d\mathbf{T}}{ds}\cdot\mathbf{N} = \kappa\mathbf{N}\cdot\mathbf{N} = \kappa. \tag{8.7}$$

By inserting (8.7) in (8.6), we obtain

$$\kappa + \mathbf{T} \cdot \frac{d\mathbf{N}}{ds} = 0, \quad \text{or} \quad \mathbf{T} \cdot \frac{d\mathbf{N}}{ds} = -\kappa.$$

Multiplication of equation (8.5) by **T** yields

$$\mathbf{T} \cdot \frac{d\mathbf{N}}{ds} = \alpha \mathbf{T} \cdot \mathbf{T} + \gamma \mathbf{B} \cdot \mathbf{T} = \alpha.$$

Because $\mathbf{T} \cdot d\mathbf{N}/ds = -\kappa$, it follows that $\alpha = -\kappa$. Hence

$$\frac{d\mathbf{N}}{ds} = -\kappa \mathbf{T} + \tau \mathbf{B},$$

where $\gamma = \tau$ is defined and inserted as conversion factor.

Derivation of the third Frenet formula: We first try with the preceding trick and start from $\mathbf{B} \cdot \mathbf{N} = 0$. If we differentiate the scalar product $\mathbf{B} \cdot \mathbf{N}$, the product rule yields

$$\frac{d\mathbf{B}}{ds} \cdot \mathbf{N} + \mathbf{B} \cdot \frac{d\mathbf{N}}{ds} = 0.$$

But this does not help immediately. We therefore simply start from the definition of **B**. Because $\mathbf{B} = \mathbf{T} \times \mathbf{N}$, it follows that

$$\frac{d\mathbf{B}}{ds} = \frac{d}{ds}(\mathbf{T} \times \mathbf{N}) = \frac{d\mathbf{T}}{ds} \times \mathbf{N} + \mathbf{T} \times \frac{d\mathbf{N}}{ds}. \tag{8.8}$$

The first term of the equation may be transformed as follows:

$$\frac{d\mathbf{T}}{ds} \times \mathbf{N} = \kappa \mathbf{N} \times \mathbf{N} = 0. \tag{8.9}$$

By inserting equation (8.9) in (8.8), it follows that

$$\frac{d\mathbf{B}}{ds} = \mathbf{T} \times \frac{d\mathbf{N}}{ds}.$$

With

$$\frac{d\mathbf{N}}{ds} = \tau \mathbf{B} - \kappa \mathbf{T},$$

it follows that

$$\frac{d\mathbf{B}}{ds} = \mathbf{T} \times (\tau \mathbf{B} - \kappa \mathbf{T}), \qquad \frac{d\mathbf{B}}{ds} = \tau(\mathbf{T} \times \mathbf{B}) - \kappa(\mathbf{T} \times \mathbf{T}).$$

Because

$$\mathbf{T} \times \mathbf{B} = -\mathbf{N} \quad \text{and} \quad \mathbf{T} \times \mathbf{T} = 0,$$

it follows that

$$\frac{d\mathbf{B}}{ds} = -\tau \mathbf{N}.$$

Darboux rotation vector: The Frenet formulas can be formulated in a very elegant way. To this end, we define a vector **D** as follows:

$$\mathbf{D} = \tau\mathbf{T} + \kappa\mathbf{B}.$$

This vector **D** is called the Darboux[2] rotation vector. We now consider

$$\begin{aligned}\mathbf{D} \times \mathbf{T} &= (\tau\mathbf{T} + \kappa\mathbf{B}) \times \mathbf{T} \\ &= \tau(\mathbf{T} \times \mathbf{T}) + \kappa(\mathbf{B} \times \mathbf{T}).\end{aligned}$$

Because $\mathbf{T} \times \mathbf{T} = 0$ and $\mathbf{B} \times \mathbf{T} = \mathbf{N}$, it follows that

$$\mathbf{D} \times \mathbf{T} = \kappa\mathbf{N}. \tag{8.10}$$

Correspondingly, one has

$$\begin{aligned}\mathbf{D} \times \mathbf{N} &= (\tau\mathbf{T} + \kappa\mathbf{B}) \times \mathbf{N} \\ &= \tau(\mathbf{T} \times \mathbf{N}) + \kappa(\mathbf{B} \times \mathbf{N}).\end{aligned}$$

Because $\mathbf{B} \times \mathbf{N} = -\mathbf{T}$ and $\mathbf{T} \times \mathbf{N} = \mathbf{B}$, it follows that

$$\mathbf{D} \times \mathbf{N} = \tau\mathbf{B} - \kappa\mathbf{T} \tag{8.11}$$

and there holds

$$\begin{aligned}\mathbf{D} \times \mathbf{B} &= (\tau\mathbf{T} + \kappa\mathbf{B}) \times \mathbf{B} \\ &= \tau(\mathbf{T} \times \mathbf{B}) + \kappa(\mathbf{B} \times \mathbf{B}).\end{aligned}$$

Because $\mathbf{B} \times \mathbf{B} = 0$ and $\mathbf{T} \times \mathbf{B} = -\mathbf{N}$, it follows that

$$\mathbf{D} \times \mathbf{B} = -\tau\mathbf{N}. \tag{8.12}$$

Using (8.10), (8.11), and (8.12), one may rewrite Frenet's formulas in the following, highly symmetric form:

$$\frac{d\mathbf{T}}{ds} = \mathbf{D} \times \mathbf{T}, \qquad \frac{d\mathbf{N}}{ds} = \mathbf{D} \times \mathbf{N}, \qquad \frac{d\mathbf{B}}{ds} = \mathbf{D} \times \mathbf{B}.$$

[2] *Jean Gaston Darboux*, b. Aug. 14, 1842, Nîmes—d. Feb. 23, 1917, Paris. Darboux came from modest relations. After graduating from École Polytechnique and École Normale in 1861, he decided for a teacher's profession at the École Normale. Supported by influential Parisian scientists, he got two teaching assignments after his doctorate in 1866. In 1881 he was appointed as professor. From 1880 on, he rendered merits as dean of the faculty of natural sciences on reorganizing the Sorbonne. From 1900 he served as permanent secretary of the Académie des Sciences. His main results concern the *theory of areas*. But he always aimed at joining to possibly all branches of mathematics, to penetrate them geometrically, and to work out the organic connection between mechanics, variational calculus, theory of partial differential equations, and theory of invariants.

Problem 8.1: Curvature and torsion

Prove the relation

$$\frac{d\mathbf{r}}{ds} \cdot \left(\frac{d^2\mathbf{r}}{ds^2} \times \frac{d^3\mathbf{r}}{ds^3}\right) = \frac{\tau}{\varrho^2}.$$

Solution By inserting Frenet's formulas and $\mathbf{T} = d\mathbf{r}/ds$, it follows that

$$\begin{aligned}
\frac{d\mathbf{r}}{ds} \cdot \left(\frac{d^2\mathbf{r}}{ds^2} \times \frac{d^3\mathbf{r}}{ds^3}\right) &= \mathbf{T} \cdot \left[\frac{d\mathbf{T}}{ds} \times \left(\kappa \frac{d\mathbf{N}}{ds} + \frac{d\kappa}{ds}\mathbf{N}\right)\right] \\
&= \mathbf{T} \cdot \left(\frac{d\mathbf{T}}{ds} \times \kappa \frac{d\mathbf{N}}{ds}\right) = \mathbf{T} \cdot (\kappa \mathbf{N} \times \kappa(\tau \mathbf{B} - \kappa \mathbf{T})) \\
&= \mathbf{T}\kappa^2 \cdot (\mathbf{N} \times (\tau \mathbf{B} - \kappa \mathbf{T})) = \kappa^2 \mathbf{T} \cdot ((\mathbf{N} \times \tau \mathbf{B}) - (\mathbf{N} \times \kappa \mathbf{T})) \\
&= \kappa^2 \mathbf{T} \cdot (\mathbf{N} \times \tau \mathbf{B}) = \kappa^2 \tau \mathbf{T} \cdot \mathbf{T} = \kappa^2 \tau = \frac{\tau}{\varrho^2}.
\end{aligned}$$

Examples on Frenet's formulas:

Example 8.2: Frenet's formulas for the circle

Given the position vector

$$\mathbf{r}(t) = (R\cos\omega t, R\sin\omega t, 0),$$

calculate the vectors of the moving trihedral.

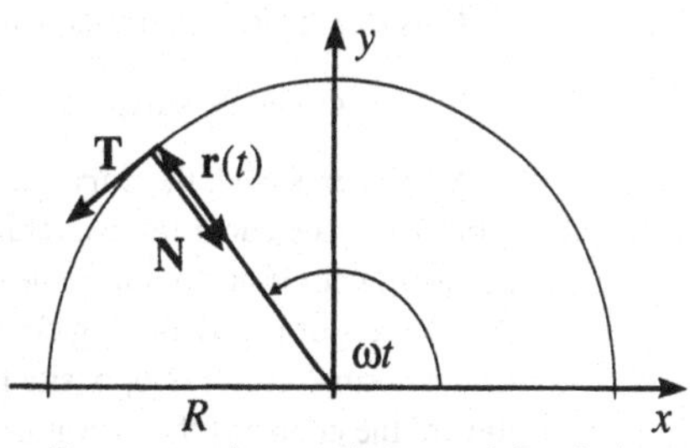

Tangent and normal vector of the circle.

Tangent vector: One has

$$\mathbf{T} = \frac{d\mathbf{r}}{ds},$$

and with $ds = |d\mathbf{r}|$, it follows that

$$\mathbf{T} = \frac{d\mathbf{r}}{|d\mathbf{r}|} = \frac{d\mathbf{r}/dt}{|d\mathbf{r}/dt|} = \frac{\mathbf{v}}{|\mathbf{v}|}.$$

The velocity is

$$\frac{d\mathbf{r}}{dt} = \mathbf{v} = (-R\omega\sin\omega t, R\omega\cos\omega t, 0) = R\omega(-\sin\omega t, \cos\omega t, 0),$$

$$\left|\frac{d\mathbf{r}}{dt}\right| = |\mathbf{v}| = \sqrt{R^2\omega^2\sin^2\omega t + R^2\omega^2\cos^2\omega t} = R\omega.$$

Hence, for the tangent vector one obtains

$$\mathbf{T} = \frac{R\omega(-\sin\omega t, \cos\omega t, 0)}{R\omega} = (-\sin\omega t, \cos\omega t, 0).$$

Normal vector: According to the first Frenet formula

$$\kappa \mathbf{N} = \frac{d\mathbf{T}}{ds} = \frac{d\mathbf{T}/dt}{ds/dt}.$$

We start from the time derivative:

$$\frac{d\mathbf{T}}{dt} = -\omega(\cos\omega t, \sin\omega t, 0),$$

$$\frac{ds}{dt} = \left|\frac{d\mathbf{r}}{dt}\right| = R\omega,$$

$$\frac{d\mathbf{T}}{ds} = \frac{-\omega(\cos\omega t, \sin\omega t, 0)}{R\omega}.$$

This means

$$\kappa \mathbf{N} = -\frac{1}{R}(\cos\omega t, \sin\omega t, 0) = -\frac{\mathbf{r}}{R^2}.$$

One has

$$|(\cos\omega t, \sin\omega t, 0)| = 1.$$

Because the curvature κ is defined as a positive quantity, thus always $\kappa > 0$, the following holds

$$\kappa \mathbf{N} = \frac{1}{R}(-\cos\omega t, -\sin\omega t, 0), \quad \text{thus} \quad \kappa = |\kappa|\,|\mathbf{N}| = \frac{1}{R}.$$

Thus $\kappa = 1/R$, and consequently one has

$$\mathbf{N} = (-\cos\omega t, -\sin\omega t, 0).$$

As was expected, the curvature radius $\varrho = 1/\kappa = R$ because R is the radius of the circle. For an arbitrary space curve the curvature radius in general varies continuously; it equals the radius of the osculating circle at a point of the curve. The geometric position of the centers of curvature of a curve (the centers of the osculating circles) is called the *evolute*. For the example of a circle the orientation of the normal vector is opposite to that of the position vector. The normal unit vector always points toward the center of the curvature circle. In this case the evolute is the center of the circle.

Binormal vector: The vector $\mathbf{B}$ is calculated from $\mathbf{B} = \mathbf{T} \times \mathbf{N}$.

$$\mathbf{B} = \begin{vmatrix} \mathbf{e}_1 & \mathbf{e}_2 & \mathbf{e}_3 \\ -\sin\omega t & \cos\omega t & 0 \\ -\cos\omega t & -\sin\omega t & 0 \end{vmatrix} = \mathbf{e}_3(\sin^2\omega t + \cos^2\omega t) = \mathbf{e}_3.$$

$$\mathbf{B} = (0, 0, 1).$$

$$\frac{d\mathbf{B}}{ds} = (0, 0, 0) = -\tau \mathbf{N}, \quad \Rightarrow \tau = 0.$$

The torsion (winding) equals zero because the curve lies within a plane. One easily realizes that the torsion vanishes for all plane curves because $\mathbf{T}$ and $\mathbf{N}$ are within the plane, therefore $\mathbf{B} = \mathbf{T} \times \mathbf{N} \perp$ to the plane and is therefore constant. Hence, from the third Frenet formula and from $d\mathbf{B}/ds = 0$, it follows that $\tau = 0$. The torsion specifies how fast the curve is running out (winding out) of the plane.

Example 8.3: Moving trihedral and helix

The moving trihedral of a helix is calculated analogously to the case of a circle. The position vector describing the helix in space reads

$$\mathbf{r}(t) = (R\cos\omega t, R\sin\omega t, b\omega t).$$

Tangent vector:

$$\mathbf{T} = \frac{\mathbf{v}}{|\mathbf{v}|} = \frac{d\mathbf{r}/dt}{ds/dt} = \frac{(-R\omega\sin\omega t, R\omega\cos\omega t, b\omega)}{\sqrt{R^2\omega^2(\sin^2\omega t + \cos^2\omega t) + b^2\omega^2}}$$
$$= \frac{(-R\sin\omega t, R\cos\omega t, b)}{\sqrt{R^2+b^2}}.$$

Normal vector:

$$\frac{d\mathbf{T}}{ds} = \frac{d\mathbf{T}/dt}{ds/dt} = \frac{-R\omega(\cos\omega t, \sin\omega t, 0)}{\sqrt{R^2+b^2}\cdot\omega\sqrt{R^2+b^2}} = |\kappa|\mathbf{N}.$$

The curvature κ is always defined as positive; correspondingly, the orientation of $\mathbf{N}$ is fixed (compare p. 51). One thus obtains

$$\mathbf{N} = (-\cos\omega t, -\sin\omega t, 0), \qquad |\kappa| = \frac{R}{R^2+b^2}.$$

The curvature of the helix is somewhat smaller than that of the circle, which is geometrically plausible.

Binormal vector: One forms the cross product

$$\mathbf{B} = \mathbf{T}\times\mathbf{N}.$$

In determinant notation:

$$\mathbf{B} = \frac{1}{\sqrt{R^2+b^2}}\begin{vmatrix} \mathbf{e}_1 & \mathbf{e}_2 & \mathbf{e}_3 \\ -R\sin\omega t & R\cos\omega t & b \\ -\cos\omega t & -\sin\omega t & 0 \end{vmatrix}$$
$$= \mathbf{e}_1\frac{b\sin\omega t}{\sqrt{R^2+b^2}} + \mathbf{e}_2\frac{-b\cos\omega t}{\sqrt{R^2+b^2}} + \mathbf{e}_3\frac{(R\sin^2\omega t + R\cos^2\omega t)}{\sqrt{R^2+b^2}}$$
$$= \frac{1}{\sqrt{R^2+b^2}}(b\sin\omega t, -b\cos\omega t, R).$$

For $b \to 0$, $\mathbf{B} = \overrightarrow{\text{constant}} = (0, 0, 1)$. To calculate the torsion, one forms

$$\frac{d\mathbf{B}}{ds} = \frac{d\mathbf{B}/dt}{ds/dt} = \frac{(1/\sqrt{R^2+b^2})(b\omega\cos\omega t, b\omega\sin\omega t, 0)}{\omega\sqrt{R^2+b^2}}$$
$$= \frac{b}{R^2+b^2}(\cos\omega t, \sin\omega t, 0),$$

$$\frac{d\mathbf{B}}{ds} = -\tau\mathbf{N}.$$

The vector **N** has already been calculated above: $\mathbf{N} = (-\cos\omega t, -\sin\omega t, 0)$. From there follows the torsion of the helix. It holds that

$$-\tau = \frac{-b}{R^2+b^2}, \qquad \tau = \frac{b}{R^2+b^2}.$$

The torsion radius: $\sigma = 1/\tau = (R^2+b^2)/b$.

For $b = 0$, it follows that $\tau = 0$. τ is a measure for the variation of **B**, i.e., for $d\mathbf{B}/ds$. In other words: τ is a measure of how the curve is winding out of the plane.

The three unit vectors **T**, **N**, and **B** define three planes that have particular names:

T and **N** span the *osculating plane,*

N and **B** span the *normal plane,*

B and **T** span the *rectifying plane.*

Remark: For a straight line $\mathbf{r}(t) = \mathbf{a} + t\mathbf{e}$ $\kappa = 0$ $(\varrho = \infty)$ and $\tau = 0$ $(\sigma = \infty)$. **N** and **B** may then be put arbitrarily $\perp$ to $\mathbf{T} = \mathbf{e}$. This is quite clear.

Velocity and acceleration of a mass point on an arbitrary space curve: For arbitrary space curves it is sometimes convenient to express the velocity and the acceleration by means of the new unit vectors. After introducing the vector **T** one has

$$\mathbf{T} = \frac{\mathbf{v}}{|\mathbf{v}|}, \qquad \mathbf{v} = |\mathbf{v}| \cdot \mathbf{T} = v\mathbf{T}.$$

This relation may be used to derive the acceleration.

$$\mathbf{b} = \frac{d^2\mathbf{r}}{dt^2} = \frac{d\mathbf{v}}{dt} = \frac{d}{dt}(v\,\mathbf{T}) = \frac{dv}{dt}\mathbf{T} + v\frac{d\mathbf{T}}{dt}.$$

By transforming the second term, one obtains for the acceleration

$$\frac{d\mathbf{T}}{dt} = \frac{d\mathbf{T}}{ds}\frac{ds}{dt} = \frac{d\mathbf{T}}{ds}v,$$

$$\mathbf{b} = \frac{dv}{dt}\mathbf{T} + v^2\frac{d\mathbf{T}}{ds} = \frac{dv}{dt}\mathbf{T} + v^2\kappa\mathbf{N} = \frac{dv}{dt}\mathbf{T} + \frac{v^2}{\varrho}\mathbf{N}.$$

The acceleration is composed of two components: the *tangential acceleration* $dv/dt\,\mathbf{T}$ pointing in the tangential direction, and the *centripetal acceleration* $v^2/\varrho\,\mathbf{N}$ pointing toward the center of the circle of curvature. For a uniform motion of a mass point on a circle (Example 7.4) there exists only the centripetal acceleration, because $dv/dt = 0$ due to the uniformity of motion.

The Evolute and the Evolvent: The evolute $\mathbf{E}(t)$ of a curve $\mathbf{r}(t)$ is the geometric position of the centers of curvature (centers of the osculating circles) of the curve $\mathbf{r}(t)$:

$$\mathbf{E}(t) = \mathbf{r}(t) + \varrho(t)\mathbf{N}(t) = \mathbf{r}(t) + \frac{1}{\kappa(t)}\mathbf{N}(t),$$

where $1/\kappa = \varrho$ is the curvature radius of the curve at point $\mathbf{r}$. For *plane curves*, ($\tau = 0$) this holds: The tangents of the evolutes are simultaneously normals of the initial curve, because

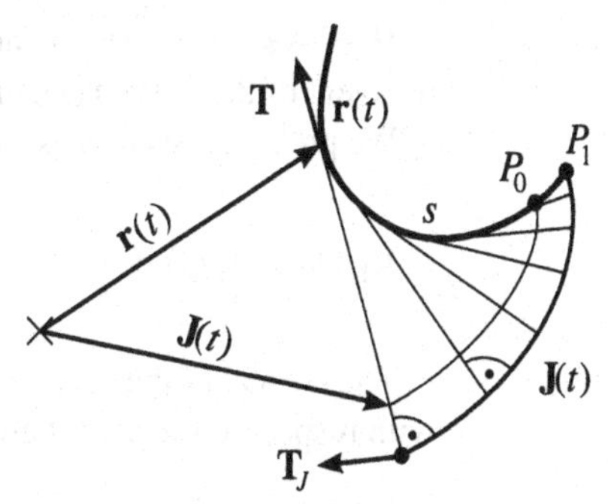

Parallel evolvents $\mathbf{J}(t)$ belonging to the curve $\mathbf{r}(t)$.

$$\begin{aligned}\frac{d\mathbf{E}}{dt} &= \frac{d\mathbf{r}}{dt} + \frac{d\varrho}{dt}\mathbf{N}(t) + \varrho\frac{d\mathbf{N}}{dt} \\ &= \mathbf{T}\frac{ds}{dt} + \frac{d\varrho}{dt}\mathbf{N}(t) + \varrho\frac{ds}{dt}(\tau\mathbf{B} - \kappa\mathbf{T}) \\ &= \mathbf{T}\left(\frac{ds}{dt} - \underbrace{\varrho\kappa}_{=1}\frac{ds}{dt}\right) + \frac{d\varrho}{dt}\mathbf{N} \qquad \text{(because } \tau = 0\text{)} \\ &= \frac{d\varrho}{dt}\mathbf{N}.\end{aligned}$$

The *evolvent* (or *involute* or *unwinding curve*) $\mathbf{J}(t)$ is the geometric position of the arc length s plotted along the tangents:

$$\mathbf{J}(t) = \mathbf{r}(t) - s(t)\cdot\mathbf{T}(t).$$

s is measured from an initial point P_0. Depending on the choice of the initial point P_0, one obtains a family of curves, whereby two evolvents in each point have a constant relative distance in normal direction. Such curves are called *parallel curves.* This is seen immediately by demonstrating that the tangent to the evolvent is perpendicular to the tangent of the initial curve, i.e., $\mathbf{T}\cdot\mathbf{T}_J = 0$. But this is evident because

$$\mathbf{T}_J \sim \frac{d\mathbf{J}}{dt} = \underbrace{\frac{ds}{dt}\mathbf{T} - \frac{ds}{dt}\mathbf{T}}_{=0} - s(t)\frac{d\mathbf{T}}{dt}; \quad \text{thus} \quad \mathbf{T}_J \sim -\mathbf{N}.$$

If one is dealing with *plane curves,* then the construction of the evolutes and of the evolvents are in some kind of inverse relation with respect to each other. One finds:

I. One of the evolvents of an evolute is the initial curve itself, symbolically written as

$$\mathbf{J}_{\mathbf{E}_\mathbf{r}}(s) = \mathbf{r}(s).$$

II. The evolute of each evolvent of a curve is the initial curve itself, that is,

$$\mathbf{E}_{\mathbf{J}_\mathbf{r}}(s) = \mathbf{r}(s).$$

Here we have written the corresponding initial curve as an index; thus $\mathbf{J}_\mathbf{r}(s)$ is the evolvent of the curve $\mathbf{r}(s)$, and $\mathbf{E}_{\mathbf{J}_\mathbf{r}}(s)$ is again the evolute of the evolvent $\mathbf{J}_\mathbf{r}(s)$.

We prove the second assertion: It reads

$$\mathbf{E}_{\mathbf{J}_\mathbf{r}}(s) = \mathbf{J}_\mathbf{r}(s) + \frac{1}{\kappa_J}\mathbf{N}_\mathbf{J} = (\mathbf{r}(s) - s\mathbf{T}_\mathbf{r}(s)) + \frac{1}{\kappa_J}\mathbf{N}_\mathbf{J}\,.$$

The normal of the curve $\mathbf{J}_\mathbf{r}$ is obtained by differentiation of the tangent vector $\mathbf{T}_\mathbf{J} = -\mathbf{N}_\mathbf{r}$ with respect to the arc length s_J (i.e. not with respect to $s \equiv s_r$!). Therefore,

$$\begin{aligned}\mathbf{N}_\mathbf{J} &= \frac{1}{\kappa_J}\frac{d\mathbf{T}_\mathbf{J}}{ds_J} = \frac{1}{\kappa_J}\frac{ds}{ds_J}\left(\frac{-d\mathbf{N}_\mathbf{r}}{ds}\right)\\ &= -\frac{1}{\kappa_J}\frac{ds}{ds_J}(\tau\mathbf{B}_\mathbf{r} - \kappa\mathbf{T}_\mathbf{r}) = \frac{\kappa}{\kappa_J}\frac{ds}{ds_J}\mathbf{T}_\mathbf{r}\end{aligned}$$

if the torsion τ vanishes (plane curve!).

The derivative of the arc length of the curve $\mathbf{r}$ with respect to the arc length of the evolvent $\mathbf{J}_\mathbf{r}$ is obtained because

$$\mathbf{T}_\mathbf{J} = \frac{d\mathbf{J}_\mathbf{r}}{ds_J} = \frac{d\mathbf{J}_\mathbf{r}}{ds}\frac{ds}{ds_J}; \quad \text{thus} \quad \frac{ds}{ds_J} = \frac{|\mathbf{T}_\mathbf{J}|}{|d\mathbf{J}_\mathbf{r}/ds|} = \frac{1}{s|d\mathbf{T}_\mathbf{r}/ds|} = \frac{1}{s\kappa}.$$

Because $\mathbf{N}_\mathbf{J}$ and $\mathbf{T}_\mathbf{r}$ are unit vectors, it must hold that

$$\frac{\kappa}{\kappa_J}\frac{ds}{ds_J} = 1 \quad \text{or} \quad \kappa_J = \kappa\frac{ds}{ds_J} = \kappa\frac{1}{s\kappa} = \frac{1}{s}.$$

We see that the curvature radius of the evolvent just equals the corresponding arc length s of the "unwound" curve, as is expected clearly.

For the evolute of the evolvent, we now obtain

$$\begin{aligned}\mathbf{E}_{\mathbf{J}_\mathbf{r}} &= \mathbf{r}(s) - s\mathbf{T}_\mathbf{r}(s) + \frac{1}{\kappa_J}\mathbf{N}_\mathbf{J}\\ &= \mathbf{r}(s) - s\mathbf{T}_\mathbf{r}(s) + \frac{1}{\kappa_J}\mathbf{T}_\mathbf{r}(s)\\ &= \mathbf{r}(s) - s\mathbf{T}_\mathbf{r}(s) + s\mathbf{T}_\mathbf{r}(s) = \mathbf{r}(s).\end{aligned}$$

Thus, assertion **II** is proved. Assertion **I** may be proved in a similar way.

Note: By adding a term pointing along the binormal direction, the definition of the evolute may be generalized in such a way that the assertion holds also for general space curves with torsion $\tau \neq 0$ (compare to Example 8.6).

Example 8.4: Evolvent of a circle

The evolvent of a circle is a spiral. The centers of curvature of this spiral are located on the circle, which therefore is the evolute of the spiral (compare the figure).

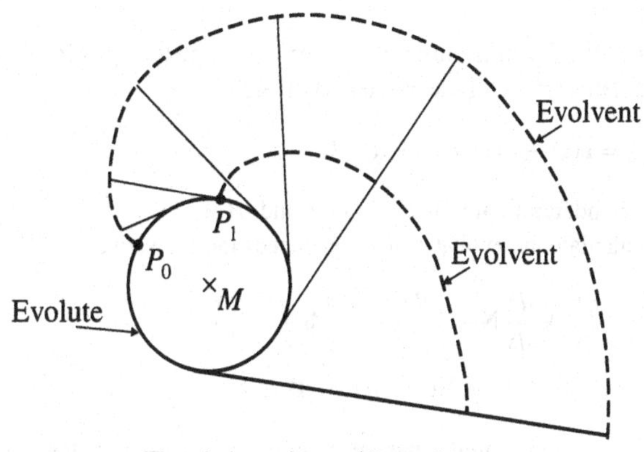

The circle and two of its evolvents.

Problem 8.5: Arc length

Calculate the arc length of the space curve given by

$$\mathbf{r}(t) = 3\cosh(2t)\,\mathbf{e}_x + 3\sinh(2t)\,\mathbf{e}_y + 6t\mathbf{e}_z$$

for the interval $0 \le t \le \pi$. Outline the curve!

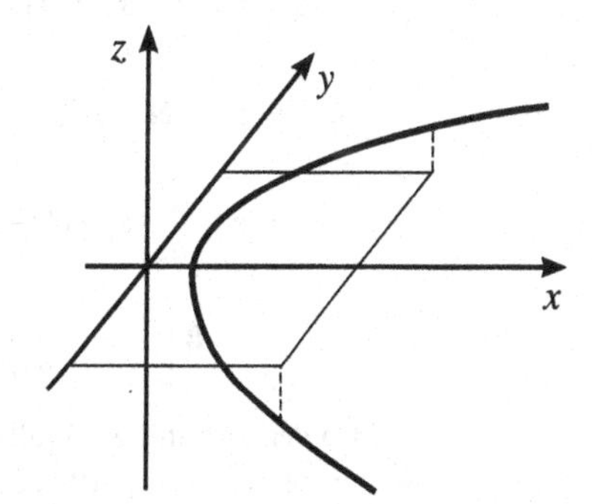

The space curve.

Solution One has

$$s = \int ds = \int \frac{ds}{dt}dt = \int \left|\frac{d\mathbf{r}}{dt}\right| dt,$$

because $ds = |d\mathbf{r}|$,

$$\frac{d\mathbf{r}}{dt} = 6\sinh(2t)\,\mathbf{e}_x + 6\cosh(2t)\,\mathbf{e}_y + 6\mathbf{e}_z,$$

$$\left|\frac{d\mathbf{r}}{dt}\right| = 6\sqrt{\sinh^2(2t) + \cosh^2(2t) + 1} = 6\sqrt{2\cosh^2(2t)},$$

because $\sinh^2 x = \cosh^2 x - 1 \rightarrow |d\mathbf{r}/dt| = 6\sqrt{2}\cosh(2t)$,

$$s = \int_0^{\pi} 6\sqrt{2}\cosh(2t)\,dt = \frac{1}{2}6\sqrt{2}\int_0^{2\pi} \cosh x\,dx = 3\sqrt{2}\sinh(2\pi).$$

The space curve comes from the first octant, intersects the x, y-plane at the point (3,0,0), and enters the eighth octant—twisted hyperbola: Consider the x, y-components: $x = \cosh 2t$, $y = \sinh 2t$, then form $x^2 - y^2 = \cosh^2 2t - \sinh^2 2t = 1 \rightarrow x^2 - y^2 = k$ (k = constant). This is the equation of a hyperbola, see Chapter 26.

Example 8.6: Generalization of the Evolute

The definition of the evolute may be extended to the case of nonplanar curves, i.e., curves with torsion $\tau(s) \neq 0$ in such a way that

$$\mathbf{J}_{\mathbf{E_r}}(s) = \mathbf{r}(s) \tag{8.13}$$

further holds. For this purpose we start from a general *ansatz* allowing that the evolute runs out of the osculating plane of the curve $\mathbf{r}(s)$, namely

$$\mathbf{E}(s) = \mathbf{r}(s) + \lambda(s)\mathbf{N}(s) + \mu(s)\mathbf{B}(s) \tag{8.14}$$

with two indeterminate functions $\lambda(s)$ and $\mu(s)$.

To calculate the evolvent of $\mathbf{E}$, one needs the derivative

$$\begin{aligned}\frac{d\mathbf{E}}{ds} &= \frac{d\mathbf{r}}{ds} + \frac{d\lambda}{ds}\mathbf{N} + \lambda\frac{d\mathbf{N}}{ds} + \frac{d\mu}{ds}\mathbf{B} + \mu\frac{d\mathbf{B}}{ds} \\ &= \mathbf{T}(1-\kappa\lambda) + \mathbf{N}(\dot{\lambda} - \mu\tau) + \mathbf{B}(\dot{\mu} + \tau\lambda),\end{aligned} \tag{8.15}$$

where the Frenet formulas have been utilized. The dot denotes differentiation with respect to s.

The evolvent of the evolute then has the form

$$\begin{aligned}\mathbf{J}_{\mathbf{E_r}}(s) &= \mathbf{E_r}(s) - s_E(s)\mathbf{T_E}(s) = \mathbf{E_r}(s) - s_E(s)\frac{d\mathbf{E}}{ds_E}(s) \\ &= \mathbf{r} + \lambda\mathbf{N} + \mu\mathbf{B} - s_E\frac{ds}{ds_E}\frac{d\mathbf{E}}{ds} \\ &= \mathbf{r} - \frac{ds}{ds_E}s_E\mathbf{T}(1-\kappa\lambda) + \mathbf{N}\left[\lambda - s_E\frac{ds}{ds_E}\left(\dot{\lambda} - \mu\tau\right)\right] \\ &\quad + \mathbf{B}\left[\mu - s_E\frac{ds}{ds_E}\left(\dot{\mu} + \tau\lambda\right)\right].\end{aligned} \tag{8.16}$$

In order to fulfill 8.13, all additional terms on the right side of 8.16 must vanish. Because the vectors of the moving trihedral are orthogonal, one is led to three independent equations:

$$1 - \kappa\lambda = 0; \tag{8.17}$$

$$\lambda - s_E\frac{ds}{ds_E}(\dot{\lambda} - \mu\tau) = 0; \tag{8.18}$$

$$\mu - s_E\frac{ds}{ds_E}(\dot{\mu} + \tau\lambda) = 0. \tag{8.19}$$

The first equation again yields the old result

$$\lambda(s) = \frac{1}{\kappa(s)}. \tag{8.20}$$

We now resolve equation 8.18:

$$s_E\frac{ds}{ds_E} = \frac{\lambda}{\dot{\lambda} - \mu\tau} = \frac{1/\kappa}{(-1/\kappa^2)\dot{\kappa} - \mu\tau} = \frac{-\kappa}{\dot{\kappa} + \mu\tau\kappa^2} \tag{8.21}$$

and insert this in equation 8.19:

$$\mu + \frac{\kappa}{\dot{\kappa} + \mu\tau\kappa^2}(\dot{\mu} + \tau\lambda) = 0. \tag{8.22}$$

This is a differential equation of first order for the function $\mu(s)$,

$$\dot{\mu} + \tau\kappa\mu^2 + \frac{\dot{\kappa}}{\kappa}\mu + \frac{\tau}{\kappa} = 0. \tag{8.23}$$

In order to solve 8.23, we multiply by κ,

$$(\kappa\dot{\mu} + \dot{\kappa}\mu) + \tau\kappa^2\mu^2 + \tau = 0. \tag{8.24}$$

We substitute $Y(s) = \kappa(s)\mu(s)$; hence

$$\frac{d}{ds}Y + \tau\left(Y^2 + 1\right) = 0. \tag{8.25}$$

This may be integrated by separation of variables,

$$-\int \frac{dY}{Y^2+1} = +\int ds\,\tau + C,$$

hence

$$+\text{arccot}Y = \int_0^s ds'\tau(s') + C$$

or

$$\mu(s) = \frac{1}{\kappa(s)}\cot\left(\int_0^s ds'\,\tau(s') + C\right). \tag{8.26}$$

The generalized definition of the evolute therefore reads

$$\mathbf{E}(s) = \mathbf{r}(s) + \frac{1}{\kappa(s)}\mathbf{N}(s) + \frac{1}{\kappa(s)}\cot\left(\int_0^s ds'\,\tau(s') + C\right)\mathbf{B}(s). \tag{8.27}$$

Because C is an arbitrary constant, there exists an entire set of evolutes.

9 Surfaces in Space

It may happen that the position vector is not a function of one parameter only but depends on two parameters u and v:

$$\mathbf{r}(u, v) = (x(u, v), y(u, v), z(u, v)).$$

The position vector then describes a surface in space. This shall be visualized: Let $\mathbf{r}$ be a function of two parameters u and v. We first choose a fixed value v_1 for v and let u vary continuously. $\mathbf{r}(u, v_1)$ then describes a space curve (compare to the figure).

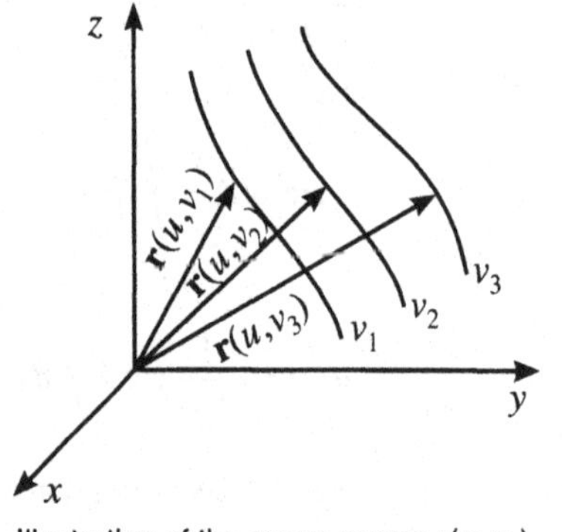

Illustration of the space curves $\mathbf{r}(u, v_n)$.

Now we choose another fixed value of v that is not widely spaced from v_1 and denote it by v_2. u is again varied continuously.

There results a space curve $\mathbf{r}(u, v_2)$ that does not differ too much from $\mathbf{r}(u, v_1)$. This procedure may be repeated many times, and one obtains many neighboring space curves (see figure overleaf).

Then, the same procedure may be performed in the opposite manner. By choosing a fixed value for u and varying v continuously, one obtains distinct neighboring lines $\mathbf{r}(u_n, v)$ for a fixed u_n (see next figure).

If the spacings between u and v become more and more dense, one obtains a surface in space. One may form the derivative along such a curve (e.g., fixed $u = u_2$ and varying v). The derivative in which one of the parameters is considered variable while the other parameters are considered fixed is called a *partial derivative* and is denoted by a round ∂ (spoken: "d partial" or "d partially derived with respect to").

$$u = u_i = \text{constant}: \quad \frac{d\mathbf{r}(u_i, v)}{dv} = \mathbf{r}_v = \frac{\partial \mathbf{r}(u, v)}{\partial v}.$$

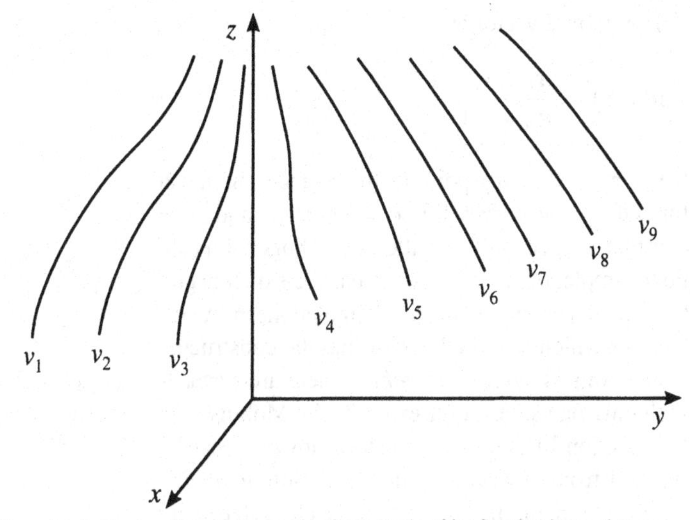

Coordinate lines with varying u are characterized by fixed values of $v_1, v_2, \ldots$.

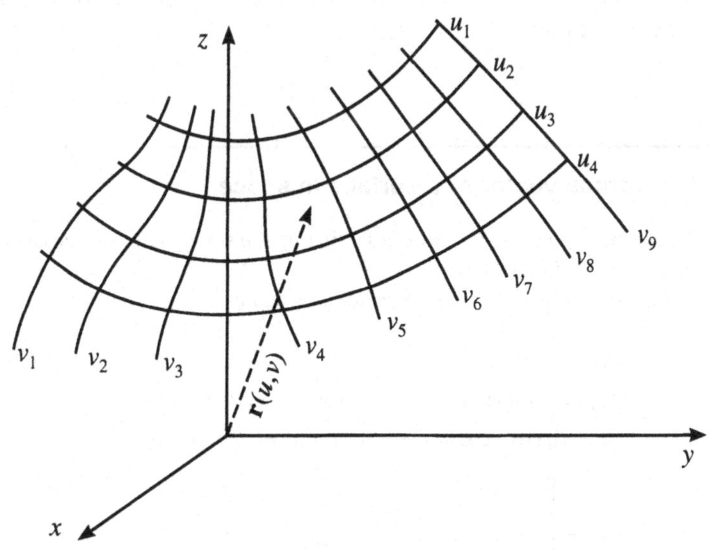

The net of coordinate lines.

In the same way one forms the tangent vector $\mathbf{r}_u$:

$$v = v_i = \text{constant} : \quad \frac{d\mathbf{r}(u, v_i)}{du} = \mathbf{r}_u = \frac{\partial \mathbf{r}(u, v)}{\partial u}.$$

The plane fixed by $\mathbf{r}_u$ and $\mathbf{r}_v$ is called the tangent plane of the surface. From $\mathbf{r}_u$ and $\mathbf{r}_v$, one easily forms the normal vector $\mathbf{n}$, which is perpendicular to the tangent plane.

The normal vector is

$$\mathbf{n}(u, v) = \frac{\mathbf{r}_u \times \mathbf{r}_v}{|\mathbf{r}_u \times \mathbf{r}_v|}.$$

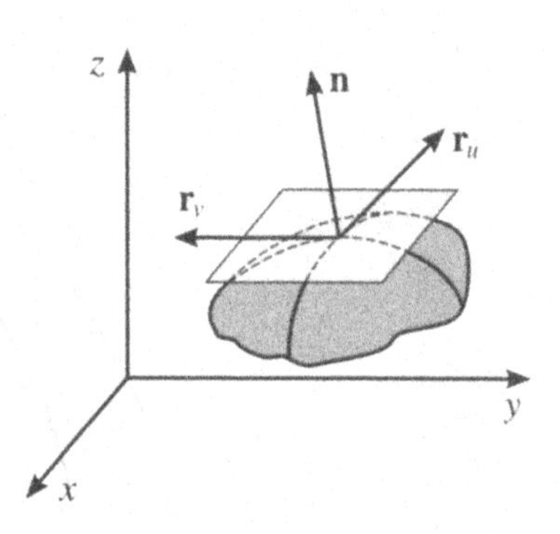

Illustration of a surface in space with tangent and normal vectors and a tangent plane in one point of the surface.

If $\mathbf{r}_u \cdot \mathbf{r}_v = 0$ at any point of the surface, the mesh formed by the curves for u = constant and v = constant, respectively, is called an orthogonal mesh. For example, the meridians and parallels of constant latitude of a sphere form an orthogonal mesh. A surface for which a normal vector may be constructed at any point is called *orientable*. There are surfaces with only one side, as for example the Möbius strip (see Section 14). On such a surface, any point can be reached from any other point by a *continuous displacement* of the normal vector. Such surfaces are *not orientable*. Orientable surfaces have inner and outer sides. By a continuous displacement of the normal vector, one stays always on the same side of a orientable surface. The normal $\mathbf{n}$ of an orientable surface is defined as positive for external (convex) surfaces, and negative for concave ones.

Example 9.1: Normal vector of a surface in space

The position vector $\mathbf{r}(u, v) = a \cos u \sin v\, \mathbf{e}_1 + a \sin u \sin v\, \mathbf{e}_2 + a \cos v\, \mathbf{e}_3$ with variable parameters describes a surface in space.

Find the normal vector as a function of u and v.

Solution

$$\mathbf{r}_u = -a \sin u \sin v\, \mathbf{e}_1 + a \cos u \sin v\, \mathbf{e}_2 + 0\, \mathbf{e}_3,$$
$$\mathbf{r}_v = a \cos u \cos v\, \mathbf{e}_1 + a \sin u \cos v\, \mathbf{e}_2 - a \sin v\, \mathbf{e}_3,$$

$$\mathbf{r}_u \times \mathbf{r}_v = \begin{vmatrix} \mathbf{e}_1 & \mathbf{e}_2 & \mathbf{e}_3 \\ -a \sin u \sin v & a \cos u \sin v & 0 \\ a \cos u \cos v & a \sin u \cos v & -a \sin v \end{vmatrix}$$

$$= -a^2 \cos u \sin^2 v\, \mathbf{e}_1 - a^2 \sin u \sin^2 v\, \mathbf{e}_2 - a^2 \sin v \cos v\, \mathbf{e}_3,$$

$$\begin{aligned} |\mathbf{r}_u \times \mathbf{r}_v| &= a^2 \sqrt{\cos^2 u \sin^4 v + \sin^2 u \sin^4 v + \sin^2 v \cos^2 v} \\ &= a^2 \sqrt{(\cos^2 u + \sin^2 u) \sin^4 v + \sin^2 v \cos^2 v} \\ &= a^2 \sqrt{\sin^2 v\, (\sin^2 v + \cos^2 v)} \\ &= a^2 |\sin v|, \end{aligned}$$

$$\mathbf{n} = (-\cos u \sin v, -\sin u \sin v, -\cos v) \quad \text{for } \sin v > 0.$$

The result means that the normal vector always points opposite to the position vector, which is the case for a sphere. One may easily prove that the function of the position vector represents a sphere, by calculating the magnitude of the position vector:

$$x = a \cos u \sin v,$$

$$y = a \sin u \sin v,$$

$$z = a \cos v\,.$$

The absolute value (the normalization) $|\mathbf{r}| = r$ of the position vector is calculated from

$$\begin{aligned} |\mathbf{r}|^2 = x^2 + y^2 + z^2 &= a^2(\cos^2 u\ \sin^2 v + \sin^2 u\ \sin^2 v + \cos^2 v) \\ &= a^2(\sin^2 v\,(\cos^2 u + \sin^2 u) + \cos^2 v) \\ &= a^2. \end{aligned}$$

From there it follows that $r = a =$ constant, i.e., the given position vector determines the surface of a sphere.

Because

$$\mathbf{r}_u \cdot \mathbf{r}_v = -a^2 \sin u\ \cos u\ \sin v\ \cos v + a^2 \sin u\ \cos u\ \sin v\ \cos v + 0\,(-a \sin v) = 0,$$

the mesh spanned by the u, v-lines represents orthogonal coordinates. One easily confirms that the u-v-lines are the meridians and parallels of equal latitude on a sphere.

10 Coordinate Frames

In an n-dimensional space one may always define n linearly independent *base vectors* out of which any arbitrary vector may be composed by a linear combination. For the sake of simplicity, vectors of magnitude unity are usually adopted as base vectors.

Corresponding to the number of base vectors, the position of an arbitrary point may be specified by n independent real numbers u_i, $i = 1, \ldots, n$. Each *coordinate frame* is characterized by a mutually unique assignment between the space points and these n numbers, the *coordinates*.

A vector in the n-dimensional space reads

$$\mathbf{r} = \sum_{i=1}^{n} u_i \mathbf{e}_i ,$$

where the n base vectors $\mathbf{e}_i$ again shall satisfy the orthonormality relation $\mathbf{e}_i \cdot \mathbf{e}_j = \delta_{i,j}$. The scalar product of two n-dimensional vectors $\mathbf{a} = \{a_i\}$ and $\mathbf{b} = \{b_i\}$ may be defined by $\mathbf{a} \cdot \mathbf{b} = \sum_{i=1}^{n} a_i b_i$, in analogy to the three-dimensional space.

The introduction of a coordinate frame implies that the coordinates of a space-fixed point change if the frame is displaced or rotated. From there it follows that for any special system a *reference point* and a definite *orientation* in space must be given.

Physically seen, both quantities may be fixed by tying the coordinate frame, for example, in a rigid body as a reference body; in a completely empty space it would make no sense to speak of the position of a point. Of course, a coordinate frame must not be "at rest" (e.g., all frames tied to the earth are accelerated frames due to the earth's rotation).

Special examples

1. The position of a point on an arbitrarily curved line ($n = 1$) is already specified by giving one number. In the simplest case one adopts the arc length s measured from a reference point in a defined direction of motion as a "natural parameter." This is a one-dimensional space.

A caterpillar on a blade of grass.

2. The surface of the earth, although being formed in a highly complicated manner (mountains, etc.), is an area with $n = 2$. Each point on it may thus be uniquely determined by two numbers. As is known this may be achieved by fixing two angular quantities: geographic length and latitude. Arbitrarily chosen reference quantities are the zero meridian through Greenwich (geographic length = 0) and the equator (geographic latitude = 0). This is a two-dimensional space.

An ant crawling on a sphere.

In order to change from one coordinate frame (q_1, q_2, q_3) to another one (here specifically the Cartesian frame: x, y, z), the following equations have to be set up:

Transformation equations:

$$\begin{aligned} q_1 &= q_1(x, y, z) & \qquad x &= x(q_1, q_2, q_3) \\ q_2 &= q_2(x, y, z) \quad \text{and their inversion} & y &= y(q_1, q_2, q_3) \\ q_3 &= q_3(x, y, z) & z &= z(q_1, q_2, q_3). \end{aligned} \tag{10.1}$$

Cartesian coordinates: Given are the three base vectors $\mathbf{e}_1, \mathbf{e}_2, \mathbf{e}_3$ along the directions of three mutually perpendicular axes. The coordinates x, y, z of a point P are the projections of the position vector $\mathbf{r} = \overrightarrow{OP}$ onto the axes,

$$\mathbf{r} = x\mathbf{e}_1 + y\mathbf{e}_2 + z\mathbf{e}_3, \qquad |\mathbf{e}_i| = 1.$$

By convention the three unit vectors form a right-handed frame. Because they are mutually perpendicular, they constitute an *orthogonal frame*. Moreover, the unit vectors are always parallel to the axes, that is, fully independent of the position of the point P in space.

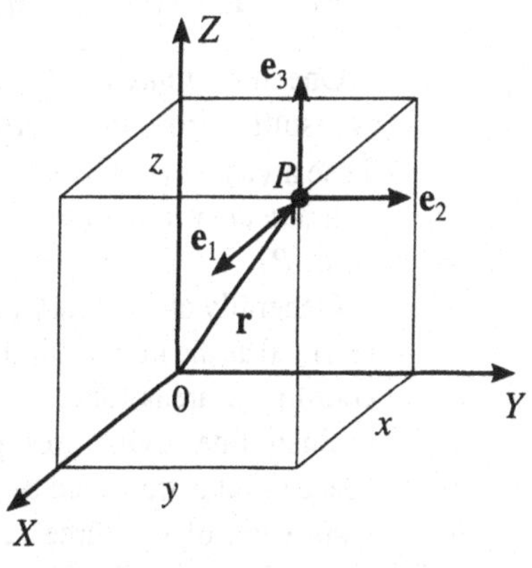

The definition of Cartesian coordinates.

This *constancy of direction of the unit vectors* combined with their orthogonality is the reason for preferred usage of Cartesian coordinates. For many special problems with particular symmetry, it turns out as convenient to use coordinate frames that are adapted to the geometric conditions and therefore simplify the calculations. For example, the motion of a plane pendulum may be described in terms of one angular coordinate, the motion of a spherical pendulum in terms of two angular quantities.

Curvilinear coordinate frames: To explain the denotation, we suppose the coordinates x, y, z of $\mathbf{r}$ to be expressed by q_1, q_2, q_3 according to (10.1). There results

$$\mathbf{r}(q_1, q_2, q_3) = \{x(q_1, q_2, q_3), y(q_1, q_2, q_3), z(q_1, q_2, q_3)\}.$$

Two of these three coordinates q_1, q_2, q_3 are now kept constant; let only the third one be variable. All points satisfying this condition are located on a space curve. There arise the three coordinate lines:

$$\begin{aligned} L_1: &\quad \mathbf{r} = \mathbf{r}(q_1 \qquad , q_2 = c_2, q_3 = c_3), \\ L_2: &\quad \mathbf{r} = \mathbf{r}(q_1 = c_1, q_2 \qquad , q_3 = c_3), \\ L_3: &\quad \mathbf{r} = \mathbf{r}(q_1 = c_1, q_2 = c_2, q_3 \qquad). \end{aligned} \tag{10.2}$$

As is immediately seen from the scheme, the three coordinate lines have exactly one common intersection point $P\ (c_1, c_2, c_3)$.

In the Cartesian frame these lines are straight lines parallel to the three axes. If, however, at least one of the lines is not straight, one speaks of curvilinear coordinates. One may still proceed one step further and keep only one of the three coordinates constant, while the other two remain variable. There arise two-dimensional (in general curved) areas in space.

Coordinate areas:

$$\begin{aligned} F_1: &\quad \mathbf{r} = \mathbf{r}(q_1 = c_1, q_2 \qquad , q_3), \\ F_2: &\quad \mathbf{r} = \mathbf{r}(q_1 \qquad , q_2 = c_2, q_3), \\ F_3: &\quad \mathbf{r} = \mathbf{r}(q_1 \qquad , q_2 \qquad , q_3 = c_3). \end{aligned} \tag{10.3}$$

One may imagine the coordinate lines as resulting from the intersection of two of these areas. In the Cartesian system the coordinate areas are planes with the common point P.

Generally an arbitrary point may be represented as the intersection point of its three coordinate areas (and, of course, also coordinate lines). One presupposes that each space point is traversed by exactly one area from each of the three sets of coordinate areas. The three fixed parameters of these areas are the coordinates of the point.

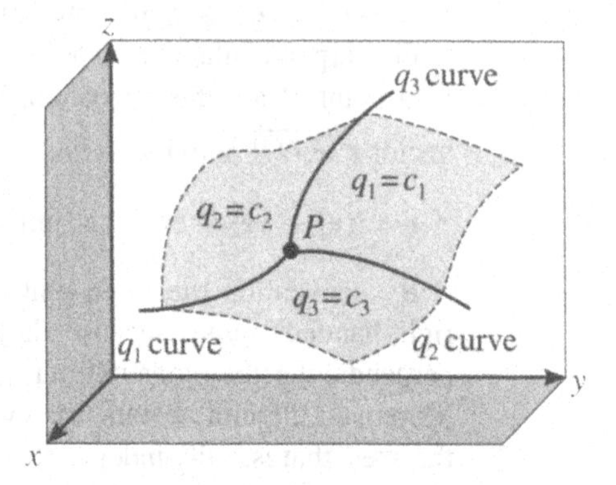

Illustration of coordinate surfaces.

The vector $\mathbf{r}(q_1, q_2, q_3)$ as a function of the three parameters q_1, q_2, q_3 describes a space region. Actually, if one of the coordinates is kept fixed, e.g., $q_3 = \overline{q}_3$, according to Chapter 9 we are dealing with an area in space. If q_3 changes to $q_3 = \overline{q}_3 + \Delta\overline{q}_3$, a neighboring area emerges. If q_3 is running continuously, there emerge more and more arbitrarily densely located areas in space that, in total, cover a space region.

General specification of base vectors: As normalized base vector (unit vector) $\mathbf{e}_{q_1}$ at point P, we choose a vector of magnitude 1 tangential to the coordinate line $q_2 = c_2, q_3 = c_3$ at P. Its orientation shall correspond to the direction of passage of the coordinate line with increasing value q_1.

This introduction of the unit vector corresponds exactly to the geometric meaning of the partial derivative; hence, $\mathbf{e}_{q_1}$ may be calculated by partial differentiation of the position vector with respect to q_1 and subsequent normalization:

$$\mathbf{e}_{q_1} = \frac{\partial \mathbf{r}/\partial q_1}{|\partial \mathbf{r}/\partial q_1|} \quad \text{or} \quad \frac{\partial \mathbf{r}}{\partial q_1} = h_1 \mathbf{e}_{q_1} \quad \text{or} \quad \frac{\partial \mathbf{r}}{\partial q_i} = h_i \mathbf{e}_{q_i}; \quad i = 1, 2, 3. \qquad \textbf{(10.4)}$$

Here h_i are *scaling factors,* namely $h_i = |\partial \mathbf{r}/\partial q_i|$. In curvilinear coordinate frames the direction of at least one of the coordinate lines changes by definition. Therefore, these frames are, contrary to the Cartesian frame, coordinate frames with *variable unit vectors.*

Cylinder coordinates: The coordinates used are

φ: angle between the projection of the position vector onto the x, y-plane and the x-axis,

ϱ: separation of the point from the z-axis,

z: length of the projection of the position vector onto the z-axis (as in the Cartesian frame).

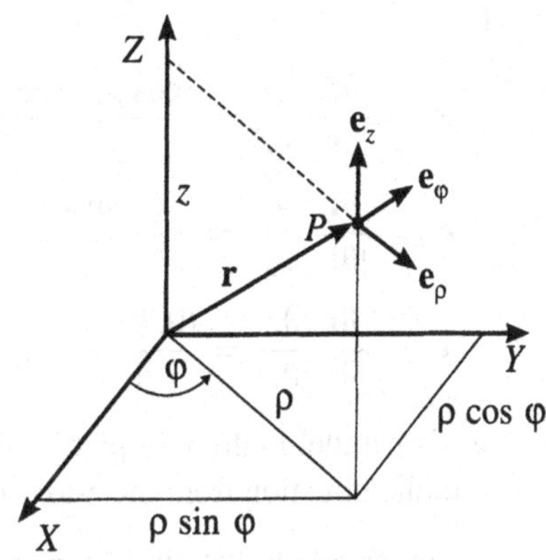

The definition of cylindrical coordinates.

The coordinate areas extend to infinity (see figure, showing limited sections) and are

$$\begin{aligned} \varrho &= \varrho_1: && \text{circular cylinders about the } z\text{-axis,} \\ \varphi &= \varphi_1: && \text{half-planes containing the } z\text{-axis,} \\ z &= z_1: && \text{planes parallel to the } x, y\text{-plane.} \end{aligned} \qquad \textbf{(10.5)}$$

Coordinate lines are two straight lines and a circle.

Transformation equations: From the figure one may directly read off the relations:

$$\mathbf{r} = (x_1, x_2, x_3) = (\varrho \cos\varphi, \varrho \sin\varphi, z)$$

or in detail:

$$\begin{aligned} x &= \varrho \cos\varphi, & \varrho &= \sqrt{x^2 + y^2}, \\ y &= \varrho \sin\varphi, & \varphi &= \arctan\frac{y}{x} = \arcsin\frac{y}{\varrho}, \\ z &= z, & z &= z. \end{aligned} \qquad \textbf{(10.6)}$$

To ensure that one point cannot be characterized by distinct combinations of coordinates, we agree on the following restrictions:

$$\varrho \geq 0; \qquad 0 \leq \varphi < 2\pi.$$

The representation is not completely unique since the angle remains indefinite for points with $\varrho = 0$. But inversely—and this is the more important requirement—to each triple ϱ, φ, z only one space point is associated.

Unit vectors: According to the geometric introduction as tangent vectors to the coordinate lines, the unit vectors $\mathbf{e}_\varrho, \mathbf{e}_\varphi, \mathbf{e}_z$ are given by

$$\mathbf{e}_\varrho = \frac{\partial \mathbf{r}/\partial \varrho}{|\partial \mathbf{r}/\partial \varrho|} = \frac{(\cos\varphi, \sin\varphi, 0)}{1},$$

$$\mathbf{e}_\varphi = \frac{\partial \mathbf{r}/\partial \varphi}{|\partial \mathbf{r}/\partial \varphi|} = \varrho\,\frac{(-\sin\varphi, \cos\varphi, 0)}{\varrho},$$

$$\mathbf{e}_z = \frac{\partial \mathbf{r}/\partial z}{|\partial \mathbf{r}/\partial z|} = \frac{(0,0,1)}{1}. \qquad \textbf{(10.7)}$$

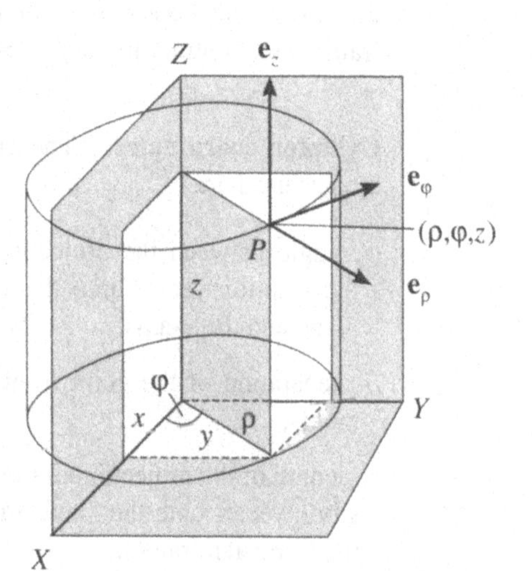

Illustration of cylindrical coordinates.

$\mathbf{e}_\varrho$ is parallel to the x, y-plane and points in radial direction from the z-axis outward.

$\mathbf{e}_\varphi$ is tangent to the circle $z = z_1$, $\varrho = \varrho_1$, that is, also parallel to the x, y-plane.

$\mathbf{e}_z$ corresponds to the Cartesian $\mathbf{e}_3$.

Thus, $\mathbf{e}_\varrho$ and $\mathbf{e}_\varphi$ may be projected onto the x, y-plane without any changes. One has

$$\mathbf{e}_\varrho = \cos\varphi\,\mathbf{e}_1 + \sin\varphi\,\mathbf{e}_2,$$

$$\mathbf{e}_\varphi = \cos\left(\varphi + \frac{\pi}{2}\right)\mathbf{e}_1 + \sin\left(\varphi + \frac{\pi}{2}\right)\mathbf{e}_2 = -\sin\varphi\,\mathbf{e}_1 + \cos\varphi\,\mathbf{e}_2,$$

$$\mathbf{e}_\varrho = (\cos\varphi, \sin\varphi, 0),$$
$$\mathbf{e}_\varphi = (-\sin\varphi, \cos\varphi, 0), \qquad \textbf{(10.8)}$$
$$\mathbf{e}_z = (0, 0, 1).$$

The same result follows by partial differentiation of $\mathbf{r}$ with respect to ϱ, φ, z and subsequent normalization (see equation (10.7)).

To check the unit vectors, we form the triple scalar product

$$\mathbf{e}_\varrho \cdot (\mathbf{e}_\varphi \times \mathbf{e}_z) = \begin{vmatrix} \cos\varphi & \sin\varphi & 0 \\ -\sin\varphi & \cos\varphi & 0 \\ 0 & 0 & 1 \end{vmatrix} = 1.$$

This is the unit volume spanned by the vectors $\mathbf{e}_\varrho$, $\mathbf{e}_\varphi$, $\mathbf{e}_z$. Thus, the cylindrical coordinates form an *orthogonal* frame with *variable* unit vectors. For solving kinematic problems it is important to know the *derivative of the unit vectors* with respect to time. Let the functions $\varrho(t), \varphi(t), z(t)$ be known. The generalization of the chain rule for a function of several variables then yields

$$\begin{aligned}
\frac{d\mathbf{e}_\varrho}{dt} &= \frac{\partial \mathbf{e}_\varrho}{\partial \varrho}\frac{d\varrho}{dt} + \frac{\partial \mathbf{e}_\varrho}{\partial \varphi}\frac{d\varphi}{dt} + \frac{\partial \mathbf{e}_\varrho}{\partial z}\frac{dz}{dt} \\
&= 0 + (-\sin\varphi, \cos\varphi, 0)\dot{\varphi} + 0 \quad = \dot{\varphi}\,\mathbf{e}_\varphi, \\
\frac{d\mathbf{e}_\varphi}{dt} &= (-\cos\varphi, -\sin\varphi, 0)\dot{\varphi} \quad = -\dot{\varphi}\,\mathbf{e}_\varrho, \\
\frac{d\mathbf{e}_z}{dt} &= 0.
\end{aligned} \tag{10.9}$$

The derivative of a vector $\mathbf{e}$ of constant magnitude has no component along the direction of $\mathbf{e}$ and hence must be perpendicular to it: $\mathbf{e}\cdot\mathbf{e} = \text{constant} \Rightarrow \mathbf{e}\cdot d\mathbf{e}/dt = 0$!

The equations given above fulfill this condition! We still note that from now on we shall frequently abbreviate the time derivative of a quantity by a dot above this quantity, as, for example, $d\varphi/dt \equiv \dot{\varphi}$ or $d\mathbf{e}_\varrho/dt \equiv \dot{\mathbf{e}}_\varrho$, etc.

Velocity and acceleration in cylindrical coordinates: Let a point move along a path described by the position vector $\mathbf{r}(t)$. One has

(a) the velocity $\mathbf{v}(t) = d\mathbf{r}/dt$,

(b) the acceleration $\mathbf{b}(t) = d^2\mathbf{r}/dt^2 = d\mathbf{v}/dt$.

In cylindrical coordinates let $\varrho(t), \varphi(t), z(t)$ be given. The position vector is

$$\mathbf{r} = \varrho\mathbf{e}_\varrho + z\mathbf{e}_z. \tag{10.10}$$

Note: These base vectors are now not fixed but are coordinate-dependent by themselves. One has to take care in component representation: For instance one cannot simply differentiate $\mathbf{r} = (\varrho, 0, z)$! In order to avoid errors, one has to write out the vector, as, for example,

(a)

$$\dot{\mathbf{r}} = \dot{\varrho}\mathbf{e}_\varrho + \varrho\dot{\mathbf{e}}_\varrho + \dot{z}\mathbf{e}_z + z\dot{\mathbf{e}}_z.$$

This yields the *velocity*:

$$\dot{\mathbf{r}} = \dot{\varrho}\mathbf{e}_\varrho + \varrho\dot{\varphi}\mathbf{e}_\varphi + \dot{z}\mathbf{e}_z. \tag{10.11}$$

(b)

$$\ddot{\mathbf{r}} = (\ddot{\varrho}\mathbf{e}_\varrho + \dot{\varrho}\dot{\mathbf{e}}_\varrho) + (\dot{\varrho}\dot{\varphi}\mathbf{e}_\varphi + \varrho\ddot{\varphi}\mathbf{e}_\varphi + \varrho\dot{\varphi}\dot{\mathbf{e}}_\varphi) + (\ddot{z}\mathbf{e}_z + \dot{z}\dot{\mathbf{e}}_z).$$

This yields the *acceleration*:

$$\ddot{\mathbf{r}} = (\ddot{\varrho} - \varrho\dot{\varphi}^2)\mathbf{e}_\varrho + (\varrho\ddot{\varphi} + 2\dot{\varrho}\dot{\varphi})\mathbf{e}_\varphi + \ddot{z}\mathbf{e}_z. \tag{10.12}$$

Hence, in the cylindric frame both the velocity and acceleration are composed of three components: a radial component, an azimuthal component, and a component in the z-direction.

Spherical coordinates According to the figure below, the coordinates are

r: length of the position vector,

ϑ: angle between the position vector and the z-axis (polar angle),

φ: azimuth (as in the cylindric frame).

The previous figure illustrates the various coordinate areas and lines. The point P is the intersection point of a circular cone about the z-axis with the vortex at the origin 0, a half-plane including the z-axis, and a sphere with the center at 0 that results by keeping the radius r constant and varying the two angles. The coordinate lines are two circles and a straight line: (1) $r =$ constant, $\varphi =$ constant, ϑ variable — meridian; (2) $r =$ constant, $\vartheta =$ constant, φ variable — parallel of constant latitude; (3) $\varphi =$ constant, $\vartheta =$ constant, r variable — radial ray.

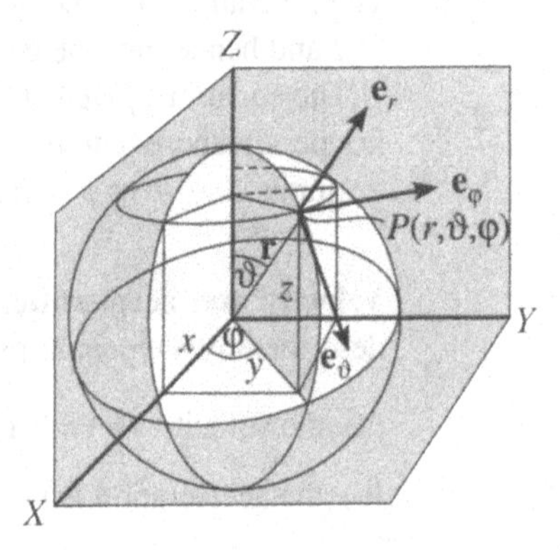

The definition of spherical coordinates.

The coordinate areas are a conical area ($\vartheta =$ constant), a half-plane ($\varphi =$ constant), and a spherical area ($r =$ constant).

Transformation equations

$$\mathbf{r} = x_1\mathbf{e}_1 + x_2\mathbf{e}_2 + x_3\mathbf{e}_3 = r\sin\vartheta\cos\varphi\,\mathbf{e}_1 + r\sin\vartheta\sin\varphi\,\mathbf{e}_2 + r\cos\vartheta\,\mathbf{e}_3.$$

When the equations are written in detail, we get

$$\begin{aligned} x &= r\sin\vartheta\cos\varphi, & r &= \sqrt{x^2+y^2+z^2}, \\ y &= r\sin\vartheta\sin\varphi, & \varphi &= \arctan\frac{y}{x}, \\ z &= r\cos\vartheta, & \vartheta &= \arctan\left(\sqrt{x^2+y^2}/z\right). \end{aligned} \tag{10.13}$$

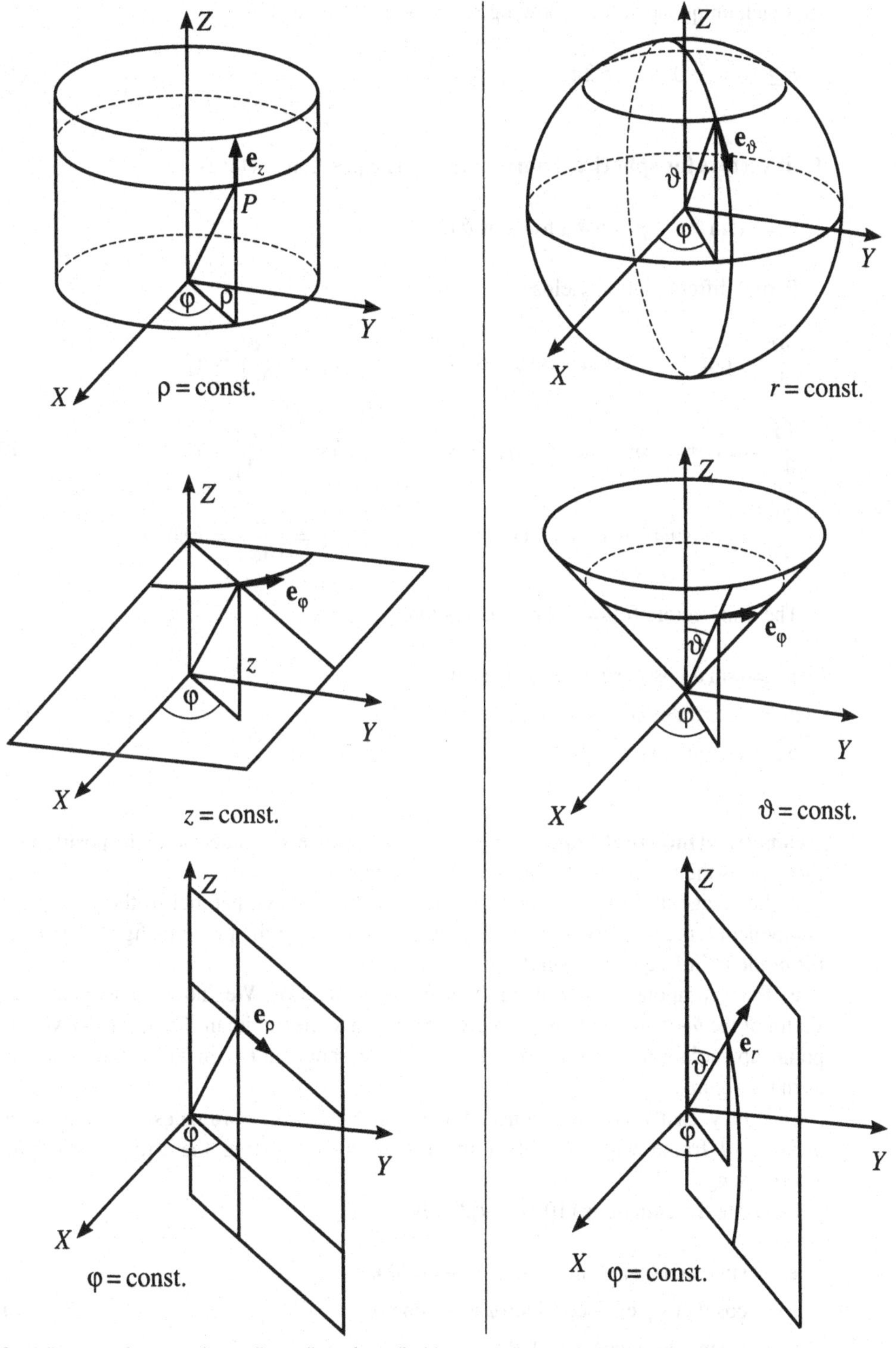

Coordinate surfaces and coordinate lines for cylindrical coordinates (left) and spherical coordinates (right).

To reach uniqueness, the following restrictions are agreed upon:

$$r \geq 0, \qquad 0 \leq \varphi < 2\pi, \qquad 0 \leq \vartheta < \pi. \tag{10.14}$$

Unit vectors for spherical coordinates: The position vector is

$$\mathbf{r} = r(\sin\vartheta\cos\varphi, \sin\vartheta\sin\varphi, \cos\vartheta).$$

Partial differentiation yields

$$\begin{aligned}
\frac{\partial \mathbf{r}}{\partial r} &= (\sin\vartheta\cos\varphi, \sin\vartheta\sin\varphi, \cos\vartheta), & h_r &= \left|\frac{\partial \mathbf{r}}{\partial r}\right| = 1, \\
\frac{\partial \mathbf{r}}{\partial \vartheta} &= r(\cos\vartheta\cos\varphi, \cos\vartheta\sin\varphi, -\sin\vartheta), & h_\vartheta &= \left|\frac{\partial \mathbf{r}}{\partial \vartheta}\right| = r, \\
\frac{\partial \mathbf{r}}{\partial \varphi} &= r(-\sin\vartheta\sin\varphi, \sin\vartheta\cos\varphi, 0), & h_\varphi &= \left|\frac{\partial \mathbf{r}}{\partial \varphi}\right| = r\sin\vartheta.
\end{aligned} \tag{10.15}$$

The unit vectors follow by normalization:

$$\begin{aligned}
\mathbf{e}_r &= (\sin\vartheta\cos\varphi, \sin\vartheta\sin\varphi, \cos\vartheta); \\
\mathbf{e}_\vartheta &= (\cos\vartheta\cos\varphi, \cos\vartheta\sin\varphi, -\sin\vartheta); \\
\mathbf{e}_\varphi &= (-\sin\varphi, \cos\varphi, 0).
\end{aligned} \tag{10.16}$$

Geometrical interpretation: One has $r\mathbf{e}_r = \mathbf{r}$; hence, $\mathbf{e}_r$ points along the position vector, that is, it is the normal to the surface of the sphere.

$\mathbf{e}_\varphi$ lies tangential to the circle $r = r_1$, $\vartheta = \vartheta_1$, namely, parallel to the x, y-plane. Its component representation may accordingly be seen from the previous figure when setting the circle radius equal to $r\sin\vartheta$.

$\mathbf{e}_\vartheta$ has a component $\sin\vartheta$ along the negative z-direction. We know that $\mathbf{e}_\vartheta$ is the tangent vector of the ϑ-coordinate line, namely the tangent to the meridian. The question whether $\mathbf{e}_\vartheta$ points upward or downward is decided by the z-component $(-\sin\vartheta)$: $\mathbf{e}_\vartheta$ points downward as in the figure.

One may easily convince oneself that the spherical coordinates also constitute an *orthogonal* frame with variable unit vectors, by evaluating the triple scalar product $\mathbf{e}_r \cdot (\mathbf{e}_\vartheta \times \mathbf{e}_\varphi) = 1$.

We write the equations (10.16) explicitly:

$$\begin{aligned}
\mathbf{e}_r &= \sin\vartheta\cos\varphi\,\mathbf{e}_1 + \sin\vartheta\sin\varphi\,\mathbf{e}_2 + \cos\vartheta\,\mathbf{e}_3; \\
\mathbf{e}_\vartheta &= \cos\vartheta\cos\varphi\,\mathbf{e}_1 + \cos\vartheta\sin\varphi\,\mathbf{e}_2 - \sin\vartheta\,\mathbf{e}_3; \\
\mathbf{e}_\varphi &= -\sin\varphi\,\mathbf{e}_1 + \cos\varphi\,\mathbf{e}_2 + 0\,\mathbf{e}_3.
\end{aligned} \tag{10.17}$$

and solve them for $\mathbf{e}_1, \mathbf{e}_2, \mathbf{e}_3$ according to Cramer's rule[1]. For example, for $\mathbf{e}_1$ one finds

$$\mathbf{e}_1 = \begin{vmatrix} \mathbf{e}_r & \sin\vartheta\sin\varphi & \cos\vartheta \\ \mathbf{e}_\vartheta & \cos\vartheta\sin\varphi & -\sin\vartheta \\ \mathbf{e}_\varphi & \cos\varphi & 0 \end{vmatrix} \Bigg/ \begin{vmatrix} \sin\vartheta\cos\varphi & \sin\vartheta\sin\varphi & \cos\vartheta \\ \cos\vartheta\cos\varphi & \cos\vartheta\sin\varphi & -\sin\vartheta \\ -\sin\varphi & \cos\varphi & 0 \end{vmatrix},$$

$$= \frac{\mathbf{e}_r \sin\vartheta\cos\varphi + \mathbf{e}_\vartheta \cos\vartheta\cos\varphi + \mathbf{e}_\varphi(-\sin\varphi)}{\sin^2\vartheta\cos^2\varphi + \cos^2\vartheta\cos^2\varphi + \sin^2\varphi},$$

$$\mathbf{e}_1 = \sin\vartheta\cos\varphi\,\mathbf{e}_r + \cos\vartheta\cos\varphi\,\mathbf{e}_\vartheta - \sin\varphi\,\mathbf{e}_\varphi, \tag{10.18}$$

and similarly for $\mathbf{e}_2$ and $\mathbf{e}_3$:

$$\begin{aligned} \mathbf{e}_2 &= \sin\vartheta\sin\varphi\,\mathbf{e}_r + \cos\vartheta\sin\varphi\,\mathbf{e}_\vartheta + \cos\varphi\,\mathbf{e}_\varphi; \\ \mathbf{e}_3 &= \cos\vartheta\,\mathbf{e}_r - \sin\vartheta\,\mathbf{e}_\vartheta\,. \end{aligned} \tag{10.19}$$

Velocity and acceleration in spherical coordinates: To calculate the velocity and acceleration in spherical coordinates, we still need the time derivatives $\dot{\mathbf{e}}_r, \dot{\mathbf{e}}_\vartheta, \dot{\mathbf{e}}_\varphi$. One finds

$$\begin{aligned} \dot{\mathbf{e}}_r &= \frac{\partial \mathbf{e}_r}{\partial\vartheta}\dot{\vartheta} + \frac{\partial \mathbf{e}_r}{\partial\varphi}\dot{\varphi} \\ &= (\cos\vartheta\cos\varphi, \cos\vartheta\sin\varphi, -\sin\vartheta)\dot{\vartheta} + (-\sin\vartheta\sin\varphi, \sin\vartheta\cos\varphi, 0)\dot{\varphi} \\ &= \dot{\vartheta}\mathbf{e}_\vartheta + \sin\vartheta\,\dot{\varphi}\mathbf{e}_\varphi, \end{aligned} \tag{10.20}$$

and similarly

$$\begin{aligned} \dot{\mathbf{e}}_\vartheta &= -\dot{\vartheta}\,\mathbf{e}_r + \cos\vartheta\,\dot{\varphi}\,\mathbf{e}_\varphi, \\ \dot{\mathbf{e}}_\varphi &= -\sin\vartheta\,\dot{\varphi}\,\mathbf{e}_r - \cos\vartheta\,\dot{\varphi}\,\mathbf{e}_\vartheta\,. \end{aligned} \tag{10.21}$$

Now we may calculate the velocity in spherical coordinates. The following hold:

$$\begin{aligned} \mathbf{r} &= r\mathbf{e}_r, \\ \dot{\mathbf{r}} &= \dot{r}\mathbf{e}_r + r\dot{\mathbf{e}}_r \\ &= \dot{r}\mathbf{e}_r + r\dot{\vartheta}\mathbf{e}_\vartheta + r\sin\vartheta\,\dot{\varphi}\,\mathbf{e}_\varphi, \end{aligned} \tag{10.22}$$

$$\begin{aligned} \ddot{\mathbf{r}} &= \ddot{r}\mathbf{e}_r + \dot{r}\dot{\mathbf{e}}_r + \dot{r}\dot{\vartheta}\mathbf{e}_\vartheta + r\ddot{\vartheta}\mathbf{e}_\vartheta + r\dot{\vartheta}\dot{\mathbf{e}}_\vartheta \\ &\quad + \dot{r}\sin\vartheta\,\dot{\varphi}\,\mathbf{e}_\varphi + r\cos\vartheta\,\dot{\vartheta}\,\dot{\varphi}\,\mathbf{e}_\varphi + r\sin\vartheta\,\ddot{\varphi}\,\mathbf{e}_\varphi + r\sin\vartheta\,\dot{\varphi}\,\dot{\mathbf{e}}_\varphi \end{aligned}$$

(after inserting (10.20) and (10.21))

[1] *Gabriel Cramer*, b. July 31, 1704, Geneva, as son of a physician—d. Jan. 4, 1752, Bagnols near Nîmes. After his studies at the university of Geneva, Cramer became appointed as professor for philosophy and mathematics. From 1727-1729 he made an informative trip through many European countries. After his return home Cramer held important municipal posts in Geneva. His rapidly decaying state of health led him to southern France where he soon died. His main work is the *Introduction à l'Analyse des Lignes Courbes Algébriques* (1750), where among other things the theory of solving systems of equations by means of determinants is outlined.

$$= \underbrace{(\ddot{r} - r\dot{\vartheta}^2 - r\sin^2\vartheta\,\dot{\varphi}^2)}_{b_r}\mathbf{e}_r + \underbrace{\left(\frac{1}{r}\frac{d}{dt}(r^2\dot{\vartheta}) - r\sin\vartheta\cos\vartheta\,\dot{\varphi}^2\right)}_{b_\vartheta}\mathbf{e}_\vartheta$$

$$+ \underbrace{\left(\frac{1}{r\sin\vartheta}\frac{d}{dt}\left(r^2\sin^2\vartheta\,\dot{\varphi}\right)\right)}_{b_\varphi}\mathbf{e}_\varphi$$

$$\equiv b_r\mathbf{e}_r + b_\vartheta\mathbf{e}_\vartheta + b_\varphi\mathbf{e}_\varphi. \tag{10.23}$$

If $\vartheta \equiv \pi/2$, that is, $\sin\vartheta = 1$, $\dot{\vartheta} = 0$, $\cos\vartheta = 0$, (10.22) and (10.23) turn into

$$\dot{\mathbf{r}} = \dot{r}\mathbf{e}_r + r\dot{\varphi}\mathbf{e}_\varphi$$

and

$$\ddot{\mathbf{r}} = (\ddot{r} - r\dot{\varphi}^2)\mathbf{e}_r + (2\dot{r}\dot{\varphi} + r\ddot{\varphi})\mathbf{e}_\varphi,$$

respectively. These expressions for velocity and acceleration in plane polar coordinates are already known from the discussion on cylinder coordinates.

Problem 10.1: Velocity and acceleration in cylindrical coordinates

A particle moves with constant velocity v along the heart curve or cardioid $r = k(1 + \cos\varphi)$ (Greek *kardia* = heart). Find the acceleration a, its magnitude, and the angular velocity. (Note that r denotes here the coordinate ϱ of the cylindrical coordinate frame.)

y
k
φ
2k
x
−k

The heart curve or cardioid.

Solution The differentiation of the path equation with respect to time yields

$$r = k(1 + \cos\varphi), \tag{10.24}$$

$$\dot{r} = -k\sin\varphi\,\dot{\varphi}, \tag{10.25}$$

$$\ddot{r} = -k(\dot{\varphi}^2\cos\varphi + \ddot{\varphi}\sin\varphi). \tag{10.26}$$

For the discussion below it is useful to conclude from 10.24 that

$$\cos\varphi = \frac{r}{k} - 1 \quad \text{and} \quad \sin^2\varphi = 1 - \left(\frac{r}{k} - 1\right)^2 = 2\frac{r}{k} - \frac{r^2}{k^2}. \tag{10.27}$$

According to 10.25 we obtain

$$\dot{r}^2 = k^2\sin^2\varphi\,\dot{\varphi}^2 = k^2\left(2\frac{r}{k} - \frac{r^2}{k^2}\right)\dot{\varphi}^2 = 2kr\dot{\varphi}^2 - r^2\dot{\varphi}^2. \tag{10.28}$$

Because we are dealing with plane polar coordinates, we write for the radius vector

$$\mathbf{r} = r\mathbf{e}_r, \tag{10.29}$$

$$\dot{\mathbf{r}} = \dot{r}\mathbf{e}_r + r\dot{\varphi}\mathbf{e}_\varphi, \tag{10.30}$$

$$\ddot{\mathbf{r}} = (\ddot{r} - r\dot{\varphi}^2)\mathbf{e}_r + (r\ddot{\varphi} + 2\dot{r}\dot{\varphi})\mathbf{e}_\varphi. \tag{10.31}$$

Because the velocity is given as constant, from 10.30 it follows that

$$v = \sqrt{\dot{r}^2 + r^2\dot{\varphi}^2}\,,$$

and with 10.28 it follows for the angular velocity that

$$\dot{\varphi} = \frac{v}{\sqrt{2kr}}\,,$$

because namely

$$v = \left(\sqrt{\left(2\frac{r}{k} - \frac{r^2}{k^2}\right)k^2 + r^2}\right)\dot{\varphi} = \sqrt{2kr}\dot{\varphi};$$

hence

$$\dot{\varphi} = \frac{v}{\sqrt{2kr}}\,. \qquad (10.32)$$

For $r \to 0$, obviously $\dot{\varphi} \to \infty$. This is due to the "turn-over" of the polar angle at $r = 0$ (compare the remark at the end of the problem). The $\mathbf{e}_r$-component of the acceleration is

$$a_r = \ddot{\mathbf{r}} \cdot \mathbf{e}_r = \ddot{r} - r\dot{\varphi}^2 = -k\left(\frac{v^2}{2kr}\cos\varphi + \sin\varphi\,\ddot{\varphi}\right) - \frac{v^2}{2k}\,. \qquad (10.33)$$

The angular acceleration $\ddot{\varphi}$ follows from 10.32, whereby $\dot{v} = 0$:

$$\ddot{\varphi} = -\frac{v\dot{r}}{2r\sqrt{2kr}} = \frac{v^2\sin\varphi}{4r^2}\,. \qquad (10.34)$$

Equation 10.34 inserted in 10.33 yields

$$\begin{aligned} a_r &= -k\frac{v^2}{4r^2}\left(r\frac{2}{k}\cos\varphi + \sin^2\varphi\right) - \frac{v^2}{2k} \\ &= -k\frac{v^2}{4k^2(1+\cos\varphi)^2}\left(1 + 2\cos\varphi + \cos^2\varphi\right) - \frac{v^2}{2k}\,, \\ a_r &= -\frac{3}{4}\frac{v^2}{k}, \qquad \text{radial acceleration.} \end{aligned} \qquad (10.35)$$

For the second component of the acceleration (azimuthal acceleration),

$$\begin{aligned} a_\varphi &= \ddot{\mathbf{r}} \cdot \mathbf{e}_\varphi = r\ddot{\varphi} + 2\dot{r}\dot{\varphi} \\ &= \frac{v^2\sin\varphi}{4r} - 2k\frac{v^2\sin\varphi}{2kr} = -\frac{3}{4}\frac{v^2\sin\varphi}{r} = -\frac{3}{4}\frac{v^2}{k}\cdot\frac{\sin\varphi}{1+\cos\varphi}\,. \end{aligned} \qquad (10.36)$$

Obviously, $a_\varphi \to -\infty$ for $\varphi \to 180°$ (the angle φ turns over — compare the remark at the end of the problem).

Because the acceleration components $a_r\mathbf{e}_r$ and $a_\varphi\mathbf{e}_\varphi$ are orthogonal, the magnitude of the acceleration is given by

$$a = \sqrt{a_r^2 + a_\varphi^2} = \frac{3}{4}\frac{v^2}{k}\sqrt{1 + \frac{\sin^2\varphi}{(1+\cos\varphi)^2}} = \frac{3}{4}\frac{v^2}{k}\sqrt{\frac{2}{1+\cos\varphi}}\,.$$

For the total acceleration, it also holds that $a \to \infty$ for $\varphi \to 180°$.

Remark: The angular velocity 10.32 and the angular acceleration 10.34 become infinite for $r = 0$. This singularity is implied by the choice of the coordinate frame and is independent of the motion along the cardioid. Consider, for example, the uniform motion of a particle on an arbitrary path in polar coordinates. Let the origin be located on a normal to the path. Because $\omega = v/d$, the angular velocity depends on the separation between origin and path: $\omega_1 < \omega_2 < \ldots < \omega_n$. In the limit with the origin located on the path, the angular velocity becomes infinite.

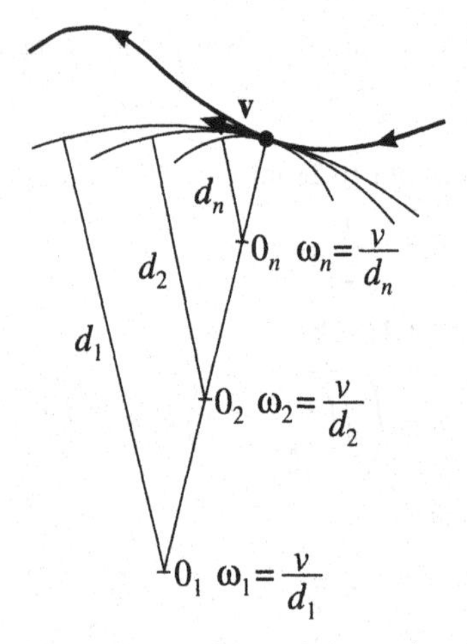

If the origin of the polar coordinates is located on the path, the angular velocity becomes infinite.

Problem 10.2: Representation of a vector in cylindrical coordinates

Write the vector $\mathbf{A} = z\mathbf{e}_1 + 2x\mathbf{e}_2 + y\mathbf{e}_3$ in cylindrical coordinates.

Solution For the solution, we make the *ansatz* $\mathbf{A} = A_\varrho \mathbf{e}_\varrho + A_\varphi \mathbf{e}_\varphi + A_z \mathbf{e}_z$. The unit vectors of the Cartesian frame have to be replaced by those of the cylinder system. Moreover, the components, namely, z, $2x$, and y, have to be expressed by cylindrical coordinates.

The system of equations

$$\mathbf{e}_\varrho = \mathbf{e}_1 \cos\varphi + \mathbf{e}_2 \sin\varphi,$$
$$\mathbf{e}_\varphi = -\mathbf{e}_1 \sin\varphi + \mathbf{e}_2 \cos\varphi$$

may be solved for $\mathbf{e}_1$, $\mathbf{e}_2$ and yields

$$\mathbf{e}_1 = \mathbf{e}_\varrho \cos\varphi - \mathbf{e}_\varphi \sin\varphi,$$
$$\mathbf{e}_2 = \mathbf{e}_\varrho \sin\varphi + \mathbf{e}_\varphi \cos\varphi.$$

It further holds that

$$x = \varrho\cos\varphi, \qquad y = \varrho\sin\varphi, \qquad z = z.$$

Insertion yields

$$\mathbf{A} = z(\mathbf{e}_\varrho \cos\varphi - \mathbf{e}_\varphi \sin\varphi) + 2\varrho\cos\varphi(\mathbf{e}_\varrho \sin\varphi + \mathbf{e}_\varphi \cos\varphi) + \varrho\sin\varphi\mathbf{e}_z.$$

Thus the components are

$$A_\varrho = z\cos\varphi + 2\varrho\cos\varphi\sin\varphi,$$
$$A_\varphi = 2\varrho\cos^2\varphi - z\sin\varphi,$$
$$A_z = \varrho\sin\varphi.$$

Problem 10.3: Angular velocity and radial acceleration

A rod rotates about P_1 in a plane with the angular velocity $\omega = ke^{\sin\varphi}$. At the time $t = 0$, let $\varphi = 0$. The straight line intersects a fixed circle of radius a at the point P_2.

(a) Find the angular acceleration of the rod.

(b) Find the velocity $\mathbf{v}_\varrho$ and the acceleration $\mathbf{b}_\varrho$ of the point P_2 along the rod.

(c) Find the velocity and the acceleration of the point P_2 with respect to the center of the circle.

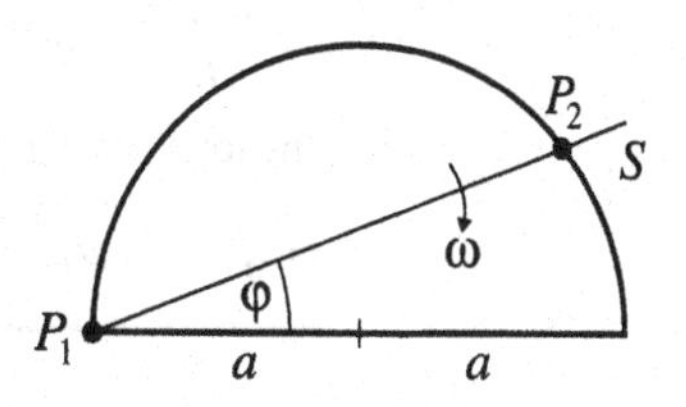

Motion around P_1 with angular velocity ω.

Solution (a) The angular velocity is

$$\omega = \dot{\varphi} = ke^{\sin\varphi}$$

$\Rightarrow$ for the angular acceleration

$$\begin{aligned}\dot{\omega} = \ddot{\varphi} &= k\omega\cos\varphi e^{\sin\varphi}\\ &= k^2e^{2\sin\varphi}\cos\varphi\,.\end{aligned}$$

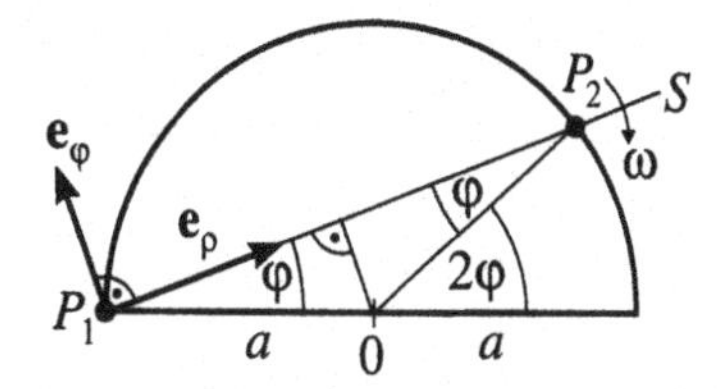

Motion around P_1 as seen from O.

(b) The position vector to the point P_2 on the rod is

$$\mathbf{r} = \varrho\mathbf{e}_\varrho\,, \qquad \text{where} \quad \varrho = 2a\cos\varphi \quad \Rightarrow \quad \mathbf{r} = 2a\cos\varphi\,\mathbf{e}_\varrho\,.$$

The velocity of P_2 is obtained from the relation

$$\dot{\mathbf{r}} = \dot{\varrho}\mathbf{e}_\varrho + \varrho\dot{\varphi}\mathbf{e}_\varphi\,,$$

and the acceleration is

$$\ddot{\mathbf{r}} = (\ddot{\varrho} - \varrho\dot{\varphi}^2)\mathbf{e}_\varrho + (\varrho\ddot{\varphi} + 2\dot{\varrho}\dot{\varphi})\mathbf{e}_\varphi\,.$$

Insertion yields

$$\begin{aligned}\mathbf{v} = \dot{\mathbf{r}} &= -2a\sin\varphi\,\dot{\varphi}\,\mathbf{e}_\varrho + 2a\cos\varphi\,\dot{\varphi}\,\mathbf{e}_\varphi\\ &= 2a\left(-\sin\varphi\,\dot{\varphi}\,\mathbf{e}_\varrho + \cos\varphi\dot{\varphi}\mathbf{e}_\varphi\right)\,,\end{aligned}$$

$$\mathbf{b} = \ddot{\mathbf{r}} = 2a\left[\left(-\ddot{\varphi}\sin\varphi - 2\dot{\varphi}^2\cos\varphi\right)\mathbf{e}_\varrho + \left(\ddot{\varphi}\cos\varphi - 2\dot{\varphi}^2\sin\varphi\right)\mathbf{e}_\varphi\right].$$

For the velocity and acceleration along the rod axis, that is, in the $\mathbf{e}_\varrho$-direction, one obtains

$$\mathbf{v}_\varrho = -2a\dot{\varphi}\sin\varphi\,\mathbf{e}_\varrho\,, \qquad \mathbf{b}_\varrho = -2a\left(\ddot{\varphi}\sin\varphi + 2\dot{\varphi}^2\cos\varphi\right)\mathbf{e}_\varrho.$$

The negative sign indicates that both $\mathbf{v}_\varrho$ as well as $\mathbf{b}_\varrho$ point toward the center of rotation P_1.

(c) The rotation angle of $\overline{OP_2}$ equals 2φ, and the velocity along the circle is

$$r_{P_2} = a\cdot 2\varphi \quad \Rightarrow \quad v_{P_2} = 2a\dot{\varphi} = 2ake^{\sin\varphi}\,.$$

The normal acceleration is

$$b_{P_2\mathbf{b}} = \frac{v_{P_2}^2}{a} = 4ak^2e^{2\sin\varphi}\,,$$

the tangential acceleration is

$$b_{P_2\mathrm{r}} = \frac{dv_{P_2}}{dt} = 2a\ddot{\varphi} = 2ak^2 \cos\varphi \, e^{2\sin\varphi} \,,$$

and the total acceleration is

$$b_{P_2} = \sqrt{b^2_{P_2\mathrm{b}} + b^2_{P_2\mathrm{r}}} = 2ak^2 e^{2\sin\varphi}\sqrt{4 + \cos^2\varphi} \,.$$

11 Vector Differential Operations

Scalar fields: The notion of scalar field means a function $\phi(x, y, z)$ that assigns a scalar, the value $\phi(x_1, y_1, z_1)$, to any space point $P(x_1, y_1, z_1)$. Examples are temperature fields $T(x, y, z)$ and density fields $\varrho(x, y, z)$ (e.g., mass density, charge density).

Vector fields: A vector field correspondingly means a function $\mathbf{A}(x, y, z)$ that assigns a vector $\mathbf{A}(x_1, y_1, z_1)$ to any space point $P(x_1, y_1, z_1)$.

Vector fields are, for instance, electric and magnetic fields, characterized by the field strength vectors **E** and **H**, or velocity fields $\boldsymbol{v}(x, y, z)$ in flowing liquids or gases.

The operations gradient, divergence, and curl (rotation)

Gradient: Given a scalar field $\phi(x, y, z)$, the gradient of the scalar field at a fixed position $P_0(x_0, y_0, z_0)$, denoted by grad $\phi(x_0, y_0, z_0)$, is a *vector pointing along the steepest ascent of* ϕ, the magnitude of which equals the change of ϕ per unit length of the path along the maximum ascent at the point $P_0(x_0, y_0, z_0)$.

In this way, any point of a scalar field can be associated with a gradient vector. The set of gradient vectors forms a vector field associated to the scalar field. Mathematically the so-defined vector field is given by the relation

$$\mathbf{A}(x, y, z) = \operatorname{grad}\phi = \mathbf{e}_1\frac{\partial}{\partial x}\phi + \mathbf{e}_2\frac{\partial}{\partial y}\phi + \mathbf{e}_3\frac{\partial}{\partial z}\phi. \tag{11.1}$$

To simplify the mathematical description, the following notation is used:

$$\operatorname{grad}\phi = \nabla\phi, \qquad \text{where} \quad \nabla = \mathbf{e}_1\frac{\partial}{\partial x} + \mathbf{e}_2\frac{\partial}{\partial y} + \mathbf{e}_3\frac{\partial}{\partial z}.$$

(∇: spoken "nabla" or "nabla operator".)

Definition of an operator: The nabla operator is a symbolic vector (vector operator) that, when applied to a function ϕ, generates the gradient of ϕ. Taken as such, the operator is meaningless; it has to operate on something, for example a scalar function $\phi(x, y, z)$.

We now demonstrate that the vector field $\nabla\phi$ has the properties quoted above. For this purpose we need the *total differential* of ϕ, namely

$$d\phi = \frac{\partial\phi}{\partial x}dx + \frac{\partial\phi}{\partial y}dy + \frac{\partial\phi}{\partial z}dz. \tag{11.2}$$

This quantity describes the main part of the total increase of the function ϕ if x changes by dx, y by dy, z by dz, that is,

$$\Delta\phi \approx \phi(x + dx, y + dy, z + dz) - \phi(x, y, z).$$

The Taylor expansion up to the first-order term yields

$$\begin{aligned}\phi(\mathbf{r} + d\mathbf{r}) &= \phi(x + dx, y + dy, z + dz)\\ &= \phi(x, y, z) + \frac{\partial\phi}{\partial x}dx + \frac{\partial\phi}{\partial y}dy + \frac{\partial\phi}{\partial z}dz + \cdots,\end{aligned}$$

and therefore

$$\begin{aligned}\Delta\phi = \phi(\mathbf{r} + d\mathbf{r}) - \phi(\mathbf{r}) &= \frac{\partial\phi}{\partial x}dx + \frac{\partial\phi}{\partial y}dy + \frac{\partial\phi}{\partial z}dz + \cdots\\ &= d\phi + \text{terms of higher order.}\end{aligned} \tag{11.3}$$

This explains the name *total differential* for the main part of the total increase of the function ϕ. We thereby have used the Taylor expansion of a function (up to the first terms in the small quantities dx, dy, dz). In Section 22 Taylor expansions will be outlined in detail and explained by numerous examples. We recommend that you have a look at this section now.

Using the infinitesimal position vector $d\mathbf{r} = (dx, dy, dz)$, we may also write the total differential as follows:

$$\begin{aligned}d\phi = \nabla\phi\cdot d\mathbf{r} &= \left(\frac{\partial\phi}{\partial x}, \frac{\partial\phi}{\partial y}, \frac{\partial\phi}{\partial z}\right)\cdot(dx, dy, dz)\\ &= \frac{\partial\phi}{\partial x}dx + \frac{\partial\phi}{\partial y}dy + \frac{\partial\phi}{\partial z}dz.\end{aligned} \tag{11.4}$$

Equipotential surfaces are surfaces on which the function ϕ takes a constant value, $\phi(x, y, z) = \text{constant}$.

As has been shown above, there is the relation

$$\nabla\phi\cdot d\mathbf{r} = d\phi, \qquad \text{with} \qquad d\mathbf{r} = (dx, dy, dz). \tag{11.5}$$

Because $d\phi$ represents the sum of the increases of ϕ in each direction $d\mathbf{r}$, $d\phi = 0$ means to stay on an equipotential surface. For this case, it holds that

$$0 = d\phi = \nabla\phi\cdot d\mathbf{r}_{ES}, \tag{11.6}$$

where $d\mathbf{r}_{ES}$ lies in the equipotential surface ES. The scalar product $\nabla\phi \cdot d\mathbf{r}_{ES}$ vanishes only then if the cosine of the enclosed angle vanishes (compare the opposite figure), provided that $\nabla\phi \neq 0$. This implies that $\nabla\phi$ and $d\mathbf{r}_{ES}$ are perpendicular to each other. Thus the gradient of ϕ is always perpendicular to the equipotential areas.

We now consider the increase $d\phi$ along the gradient vector at a fixed point of the scalar field: Here $d\mathbf{r}$ is parallel to $\nabla\phi$ and then $\nabla\phi\cdot d\mathbf{r}$ takes the maximum value. Therefore, the vector grad $\phi = \nabla\phi$ always points in the direction of the strongest increase of ϕ; see the opposite figure.

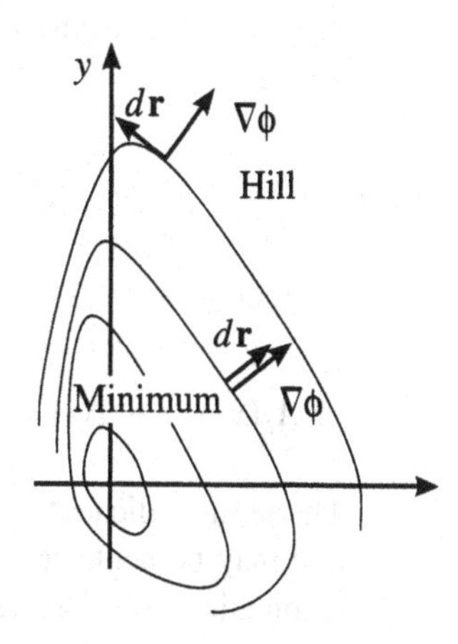

Equipotential lines and the direction of the gradient.

Divergence: Contrary to the gradient operation, the divergence is applied to vector fields. Given a vector field $\mathbf{A} = (A_x, A_y, A_z)$, we further imagine a cuboid-shaped *"control volume"* (rectangular box) with the edge lengths Δx, Δy, Δz.

The *"vector flow"* across an area represents the entity of vectors penetrating it perpendicularly, that is, the normal components of the vectors integrated over the entire area.

The lateral faces of the cuboid are denoted by $s_1, s_2, \ldots, s_6$.

We now calculate the vector flow across all lateral faces of the cuboid (rectangular box). The edge lengths Δx, Δy, Δz shall be chosen so small that the vector on the cuboid faces may be considered as nearly constant, such that the integration of the vector across the faces may be replaced by a simple summation. We shall count the vector flow as positive if it flows out of the volume, and negative if it flows into the volume.

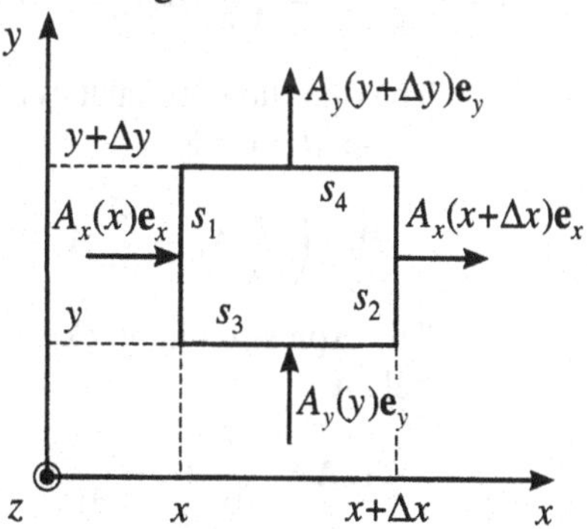

The flow across a cuboid. The extension in the z-direction (out of the paper plane) is not shown.

The vector flow through the faces is

$$\begin{aligned} s_1&: \quad -A_x(x)\Delta y\Delta z, \\ s_2&: \quad A_x(x+\Delta x)\Delta y\Delta z, \\ s_3&: \quad -A_y(y)\Delta x\Delta z, \\ s_4&: \quad A_y(y+\Delta y)\Delta x\Delta z, \end{aligned} \tag{11.7}$$

and in the third space direction is

$$\begin{aligned} s_5&: \quad -A_z(z)\Delta x\Delta y, \\ s_6&: \quad A_z(z+\Delta z)\Delta x\Delta y. \end{aligned} \tag{11.8}$$

A Taylor series expansion up to terms of first order, which is satisfied for small Δx, Δy, Δz, yields

$$\begin{aligned} A_x(x+\Delta x, y, z) &= A_x(x, y, z) + \frac{\partial}{\partial x} A_x(x, y, z)\Delta x + \cdots, \\ A_y(x, y+\Delta y, z) &= A_y(x, y, z) + \frac{\partial}{\partial y} A_y(x, y, z)\Delta y + \cdots, \\ A_z(x, y, z+\Delta z) &= A_z(x, y, z) + \frac{\partial}{\partial z} A_z(x, y, z)\Delta z + \cdots. \end{aligned} \tag{11.9}$$

The terms indicated by dots $\cdots$ are of higher order in the small increments Δx, Δy, Δz and may be neglected. The resulting vector flow through the control volume follows by summation over the lateral faces:

$$\begin{aligned} &\big(A_x(x+\Delta x, y, z) - A_x(x, y, z)\big)\Delta y \Delta z \\ &\quad + \big(A_y(x, y+\Delta y, z) - A_y(x, y, z)\big)\Delta x \Delta z \\ &\quad + \big(A_z(x, y, z+\Delta z) - A_z(x, y, z)\big)\Delta y \Delta x, \\ &\quad = \frac{\partial}{\partial x} A_x(x, y, z)\Delta x \Delta y \Delta z + \frac{\partial}{\partial y} A_y(x, y, z)\Delta x \Delta y \Delta z + \frac{\partial}{\partial z} A_z(x, y, z)\Delta x \Delta y \Delta z \\ &\quad = \left(\frac{\partial}{\partial x} A_x(x, y, z) + \frac{\partial}{\partial y} A_y(x, y, z) + \frac{\partial}{\partial z} A_z(x, y, z)\right)\Delta V. \end{aligned}$$

Thus the "flow" (total flow) through an infinitesimally small volume ($\Delta x \to dx$, $\Delta y \to dy$, $\Delta z \to dz$) reads

$$dV \cdot \left(\frac{\partial}{\partial x} A_x + \frac{\partial}{\partial y} A_y + \frac{\partial}{\partial z} A_z\right). \tag{11.10}$$

The expression in brackets is called divergence of the vector field $\mathbf{A}$:

$$\operatorname{div} \mathbf{A} = \frac{\partial}{\partial x} A_x + \frac{\partial}{\partial y} A_y + \frac{\partial}{\partial z} A_z. \tag{11.11}$$

Thus, the divergence represents the vector flow through a volume ΔV per unit volume. It may also be written in the form

$$\operatorname{div} \mathbf{A} = \nabla \cdot \mathbf{A}(x, y, z). \tag{11.12}$$

This last relation may be interpreted as analytic definition. As has been shown, it is identical with the geometric definition, namely:

Illustration of the divergence as flow of the vector field through a volume.

$$\operatorname{div} \mathbf{A} = \lim_{\Delta V \to 0} \frac{\text{flow of the vector field } \mathbf{A} \text{ through } \quad \Delta V}{\Delta V} = \lim_{\Delta V \to 0} \frac{\int_{\Delta F} \mathbf{A} \cdot \mathbf{n}\, dF}{\Delta V}. \tag{11.13}$$

While the argument of the gradient operation is a scalar, the divergence represents the scalar product of the operator ∇ and the vector $\mathbf{A}$. For a vanishing divergence, the total flow through an infinitesimal volume equals zero, that is, the in-flow just balances the out-flow. If at some point of the vector field div $\mathbf{A} > 0$, one says that the vector field there has a *source*; for div $\mathbf{A} < 0$, one speaks of a *sink* of the vector field. This is immediately clear from the definition of the divergence as net flow = out-flow − in-flow per unit volume.

Curl (Rotation):[1] The operation curl **A** assigns a vector field curl **A** to a given vector field **A**. The vector field curl **A** informs about possible "vortices" of the field **A** (a vortex exists if there is a closed curve in the vector field fulfilling the condition that the contour integral $\oint \mathbf{A} \cdot d\mathbf{s} \neq 0$—see theorem of Stokes). The mathematical formulation of curl **A** is given by

1. $\operatorname{curl} \mathbf{A} = \nabla \times \mathbf{A}$, or
2. $\mathbf{n} \cdot \operatorname{curl} \mathbf{A} = \lim_{\Delta F \to 0} \left(\oint \mathbf{A} \cdot d\mathbf{s} \right) / \Delta F$.

n is a unit normal vector on ΔF.

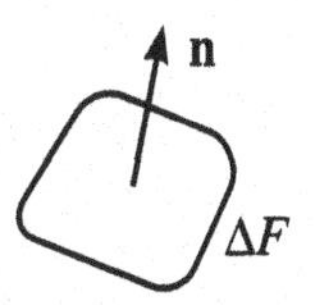

Oriented surface element.

The second definition states that the rotation may also be determined by forming the contour integral. The integration is performed over the vector field along a curve. More strictly speaking: One integrates over the projection of **A** onto $d\mathbf{s}$ along the tangent to the curve forming the border of ΔF. After division by ΔF, this yields the component of curl **A** along **n**.

The rotation is thus determined by two distinct definitions. The first of these reads in detail

$$\operatorname{curl} \mathbf{A} = \nabla \times \mathbf{A} = \begin{vmatrix} \mathbf{e}_1 & \mathbf{e}_2 & \mathbf{e}_3 \\ \partial/\partial x & \partial/\partial y & \partial/\partial z \\ A_x & A_y & A_z \end{vmatrix}$$

$$= \mathbf{e}_1 \left(\frac{\partial A_z}{\partial y} - \frac{\partial A_y}{\partial z} \right) + \mathbf{e}_2 \left(\frac{\partial A_x}{\partial z} - \frac{\partial A_z}{\partial x} \right) + \mathbf{e}_3 \left(\frac{\partial A_y}{\partial x} - \frac{\partial A_x}{\partial y} \right) \qquad (11.14)$$

One has to prove that both definitions are identical. Here we show the identity only for the x-component.

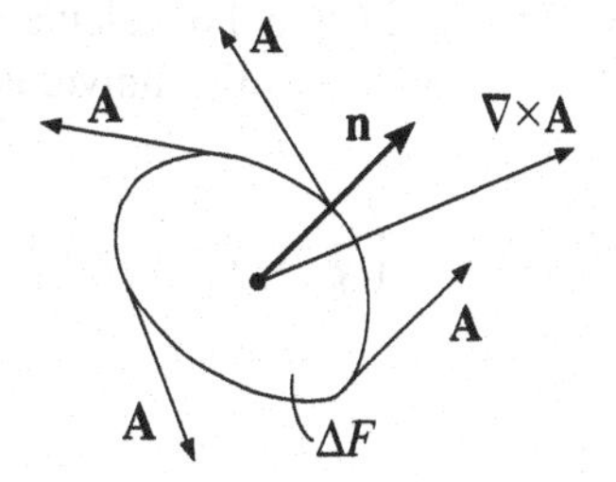

Illustration of a vector field **A** with vorticity on surface element ΔF with normal vector **n**.

***x*-component of the curl of A:** One may integrate about an area $\Delta F = 4 \Delta y \Delta z$ in the y, z-plane (see the lower figure). **n** then points along the x-axis, that is, $\mathbf{n} \cdot \operatorname{curl} \mathbf{A}$ just yields the x-component of curl **A**, namely $(\operatorname{curl} \mathbf{A})_x$.

[1] In German literature, curl is named rotation (rot), so curl **A** = rot **A**.

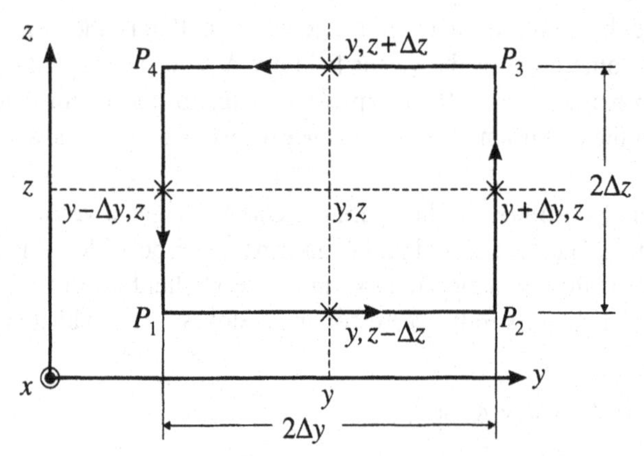

Calculating the x-component of the curl of **A**.

For the loop integral one has

$$\left(\oint \mathbf{A}\cdot d\mathbf{r}\right)_x = \underset{C}{\int\!\!\!\!\!\curvearrowleft} A_x dx + \underset{C}{\int\!\!\!\!\!\curvearrowleft} A_y dy + \underset{C}{\int\!\!\!\!\!\curvearrowleft} A_z dz, \tag{11.15}$$

$$= \underset{C}{\int\!\!\!\!\!\curvearrowleft}(A_y dy + A_z dz),$$

since for this orientation of the area concerned (see figure) $\Delta x = 0$ (i.e., Δx does not enter at all). In other words: Because x remains unchanged ($dx = 0$), $\int A_x dx$ drops out. For the exact definition of the contour or loop integral we refer to Chapter 12. It is recommended to study this section in brief right now.

Remark: $\oint$ shall indicate that the integration is performed over a closed curve (contour or loop integral) in the counter-clockwise direction. $\int\!\!\!\!\!\curvearrowleft$ means integration over a section of the curve. For calculating a contour integral, we employ the values of the functions in the *middle* of the individual sections (marked points).

$$\left(\oint \mathbf{A}\cdot d\mathbf{r}\right)_x = \int_{P_1}^{P_2} + \int_{P_2}^{P_3} + \int_{P_3}^{P_4} + \int_{P_4}^{P_1} (A_y dy + A_z dz)$$

$$\approx A_y(x,y,z-\Delta z)2\Delta y + A_z(x,y+\Delta y,z)2\Delta z$$

$$- A_y(x,y,z+\Delta z)2\Delta y - A_z(x,y-\Delta y,z)2\Delta z. \tag{11.16}$$

According to the Taylor expansion this yields

$$\approx \left[A_y - \frac{\partial A_y}{\partial z}\Delta z\right]2\Delta y + \left[A_z + \frac{\partial A_z}{\partial y}\Delta y\right]2\Delta z$$

$$-\left[A_y + \frac{\partial A_y}{\partial z}\Delta z\right] 2\Delta y - \left[A_z - \frac{\partial A_z}{\partial y}\Delta y\right] 2\Delta z$$

$$= 4\Delta y \Delta z \left[\frac{\partial A_z}{\partial y} - \frac{\partial A_y}{\partial z}\right].$$

The enclosed area is $\Delta F = 4\Delta y \Delta z$. From that, it follows that

$$\lim_{\Delta F \to 0} \left(\oint \frac{\mathbf{A} \cdot d\mathbf{s}}{\Delta F}\right)_x = \frac{\partial A_z}{\partial y} - \frac{\partial A_y}{\partial z}. \tag{11.17}$$

Hence, the x-components corresponding to the two definitions of curl **A** coincide. For the remaining two components, the equivalence of the definitions may be demonstrated in an analogous way (which will be skipped), q.e.d.

The second definition of curl **A** implies that the rotation at some point of the field **A** vanishes if the contour integral $\oint \mathbf{A} \cdot d\mathbf{s}$ (loop integral) enclosing this point equals zero — see the theorem of Stokes. From there originates the name "rotation." A finite value of the loop integral expresses a certain rotation, that is, vortex formation of the vector field (to be visualized as a flow field).

Multiple application of the vector operator nabla: Given a scalar field $f(\mathbf{r})$ and a vector field $\mathbf{g}(\mathbf{r})$, then

(a)

$$\nabla \cdot (\nabla f) = \frac{\partial^2 f}{\partial x^2} + \frac{\partial^2 f}{\partial y^2} + \frac{\partial^2 f}{\partial z^2} = \operatorname{div} \operatorname{grad} f(x, y, z) = \Delta f(x, y, z), \tag{11.18}$$

where Δ is introduced as a new operator:

$$\Delta = \frac{\partial^2}{\partial x^2} + \frac{\partial^2}{\partial y^2} + \frac{\partial^2}{\partial z^2} = \nabla \cdot \nabla$$

(Δ, spoken: delta, is called the Laplace operator[2]). $\nabla \cdot (\nabla f) = \operatorname{div} \nabla f$ is a scalar field.

[2] *Pierre Simon Laplace*, b. March 23, 1749, Beaumont-en-Auge—d. March 5, 1827, Paris. After his school education Laplace became a teacher in Beaumont and, by mediation of D'Alembert, became appointed as professor at the Military School of Paris. Because Laplace used to quickly modify his political convictions, he was swamped with honors both by Napoleon and by Louis XVIII. Among his works his *Analytic Theory of Probability* (1812) and the *Celestial Mechanics* (1799–1825) became significant. The theory of probability calculus contains, for example, the method of the *generating functions*, the *Laplace transformations* and the final formulation of the mechanical materialism. The *Celestial Mechanics* presents, for instance, the cosmologic hypothesis of Laplace, the theories of the earth's shape and of the moon's motion, the perturbation theory of planets, and the potential theory with the Laplace equation.

(b)

$$\nabla \times (\nabla f) = \text{curl grad } f = \begin{vmatrix} \mathbf{e}_1 & \mathbf{e}_2 & \mathbf{e}_3 \\ \frac{\partial}{\partial x} & \frac{\partial}{\partial y} & \frac{\partial}{\partial z} \\ \frac{\partial f}{\partial x} & \frac{\partial f}{\partial y} & \frac{\partial f}{\partial z} \end{vmatrix} \equiv 0.$$

Thereby it is of course required that f is twofold continuously differentiable. The physicist always presupposes functions that are sufficiently often continuously differentiable; this is also assumed below. Hence, a gradient field has no vortices!

(c)

$$\begin{aligned} \nabla(\nabla \cdot \mathbf{g}) &= \nabla \left(\frac{\partial g_x}{\partial x} + \frac{\partial g_y}{\partial y} + \frac{\partial g_z}{\partial z} \right) \\ &= \frac{\partial}{\partial x} \left(\frac{\partial g_x}{\partial x} + \frac{\partial g_y}{\partial y} + \frac{\partial g_z}{\partial z} \right) \mathbf{e}_1 + \frac{\partial}{\partial y} \left(\frac{\partial g_x}{\partial x} + \frac{\partial g_y}{\partial y} + \frac{\partial g_z}{\partial z} \right) \mathbf{e}_2 \\ &\quad + \frac{\partial}{\partial z} \left(\frac{\partial g_x}{\partial x} + \frac{\partial g_y}{\partial y} + \frac{\partial g_z}{\partial z} \right) \mathbf{e}_3 \\ &= \text{grad}(\text{div}\, \mathbf{g}) \quad \text{is a vector field.} \end{aligned}$$

(d)

$$\nabla \cdot (\nabla \times \mathbf{g}) = \text{div}(\text{curl}\, \mathbf{g}) = 0.$$

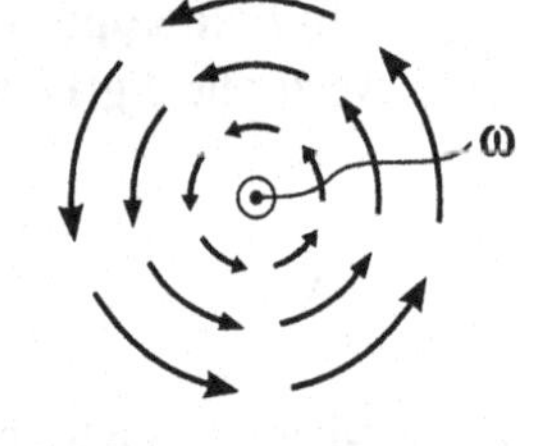

The velocity field of a rotating rigid body: $\mathbf{A} = \boldsymbol{\omega} \times \mathbf{r}$.

Hence, a rotation field has neither sources nor sinks, as is graphically clear: The vector field $\mathbf{A} = \boldsymbol{\omega} \times \mathbf{r}$ with $\overrightarrow{\boldsymbol{\omega} = \text{constant}}$ is so to speak an optimum vortex field (the velocity field of a rigid body rotating with the angular velocity $\boldsymbol{\omega}$).

For this maximum vortex field one has curl $\mathbf{A} = 2\boldsymbol{\omega}$, that is, it is a constant vector field that obviously is divergence-free. One should note the similarity of ∇ with a vector: The triple scalar product involving identical vectors vanishes.

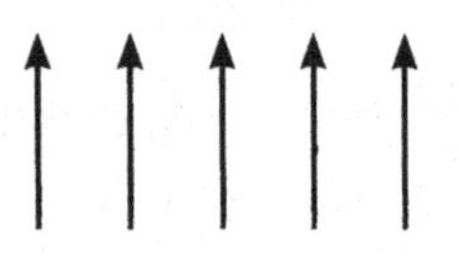
The rotation of the velocity field above: curl $\mathbf{A} =$ curl $\boldsymbol{\omega} \times \mathbf{r} = 2\boldsymbol{\omega}$.

(e)

$$\begin{aligned} \nabla \times (\nabla \times \mathbf{g}) &= \text{curl}(\text{curl}\, \mathbf{g}) \\ &= \nabla(\nabla \cdot \mathbf{g}) - (\nabla \cdot \nabla)\mathbf{g} \\ &= \text{grad}(\text{div}\, \mathbf{g}) - \Delta \mathbf{g} \end{aligned}$$

is a vector field.

The proof is simple, because according to the expansion theorem,

$$\mathbf{C} \times (\mathbf{B} \times \mathbf{A}) = \mathbf{B}(\mathbf{C} \cdot \mathbf{A}) - (\mathbf{C} \cdot \mathbf{B})\mathbf{A}.$$

This twofold application of the rotation operator physically and geometrically means that the vortices of the vortex field are calculated.

(f)

$$\operatorname{div}(\mathbf{B} \times \mathbf{C}) = \mathbf{C} \cdot (\operatorname{curl} \mathbf{B}) - \mathbf{B} \cdot (\operatorname{curl} \mathbf{C}).$$

(g)

$$\nabla \cdot (f\,\mathbf{g}) = \nabla f \cdot \mathbf{g} + f\,\nabla \cdot \mathbf{g}.$$

(h)

$$\nabla \times (f\,\mathbf{g}) = \nabla f \times \mathbf{g} + f\,\nabla \times \mathbf{g}.$$

Problem 11.1: Gradient of a scalar field

Given the scalar field $\varphi = x^2 + y^2 = r^2$, find the gradient of φ.

Solution

$$\nabla \varphi = 2(x\mathbf{e}_x + y\mathbf{e}_y) = 2\sqrt{x^2 + y^2}\mathbf{e}_r = 2r\mathbf{e}_r .$$

Problem 11.2: Determination of the scalar field from the associated gradient field

Let $\nabla \varphi = (1 + 2xy)\,\mathbf{e}_x + (x^2 + 3y^2)\,\mathbf{e}_y$. Find the associated scalar field.

Solution

$$\frac{\partial \varphi}{\partial x} = (1 + 2xy) \quad \Rightarrow \quad \varphi(x, y) = x + x^2 y + f_1(y),$$

$$\frac{\partial \varphi}{\partial y} = (x^2 + 3y^2) \quad \Rightarrow \quad \varphi(x, y) = x^2 y + y^3 + f_2(x).$$

By comparison:

$$f_1(y) = y^3 + C_1, \qquad f_2(x) = x + C_2;$$

thus,

$$\varphi(x, y) = x + x^2 y + y^3 + C\,.$$

Problem 11.3: Divergence of a vector field

Calculate the divergence of the field of the position vectors:

$$\mathbf{r} = x\,\mathbf{e}_1 + y\,\mathbf{e}_2 + z\,\mathbf{e}_3.$$

Solution

$$\operatorname{div} \mathbf{r} = \frac{\partial x}{\partial x} + \frac{\partial y}{\partial y} + \frac{\partial z}{\partial z} = 3.$$

Thus, the vector field $\mathbf{r}$ everywhere has a finite divergence (i.e., source density) of magnitude 3. To generate this field in practice by a flow, one would have to attach sources of intensity 3 to any space point.

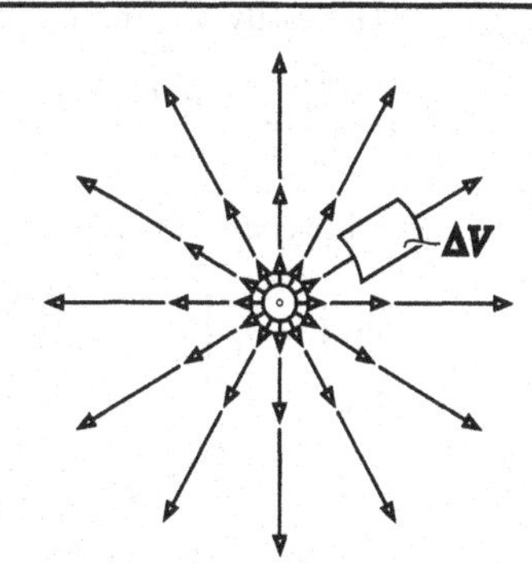

The field of the position vector $\mathbf{A}(x, y, z) = \mathbf{r}$. Flow out of the volume element ΔV is larger than flow into the volume element.

Problem 11.4: Rotation of a vector field

Calculate the rotation of the vector field

$$\mathbf{A} = 3x^2 y \mathbf{e}_1 + yz^2 \mathbf{e}_2 - xz\mathbf{e}_3.$$

Solution

$$\operatorname{curl}\mathbf{A} = \mathbf{e}_1\left(\frac{\partial(-xz)}{\partial y} - \frac{\partial yz^2}{\partial z}\right) + \mathbf{e}_2\left(\frac{\partial 3x^2 y}{\partial z} - \frac{\partial(-xz)}{\partial x}\right) + \mathbf{e}_3\left(\frac{\partial yz^2}{\partial x} - \frac{\partial 3x^2 y}{\partial y}\right)$$

$$= -2yz\,\mathbf{e}_1 + z\,\mathbf{e}_2 - 3x^2\,\mathbf{e}_3.$$

Problem 11.5: Electric field strength, electric potential

Let a positive electric charge of magnitude Q be localized at the origin of the coordinate frame. The field intensity $\mathbf{E}$ describing the electrostatic field is given by

$$\mathbf{E} = \frac{Q}{r^2}\mathbf{e}_r,$$

where r denotes the spatial distance from the coordinate origin, and $\mathbf{e}_r$ represents the corresponding unit vector in radial direction. Calculate the associated potential field (let U denote the potential field, then $\mathbf{E} = -\nabla U$) and show that it satisfies the Laplace equation $\Delta U = 0$, except for the origin.

Solution

$$\mathbf{E} = \frac{Q}{r^2}\mathbf{e}_r = \frac{Q}{r^2}\frac{\mathbf{r}}{r} \qquad \mathbf{E} = -\nabla U. \tag{11.19}$$

Because $\mathbf{E}$ points in the radial direction and the gradient means the derivative along this direction, one has

$$|\mathbf{E}| = -\frac{dU}{dr},$$

and because $\mathbf{E}$ is a function of r only, it follows that

$$U = -\int |\mathbf{E}| dr = -Q\int \frac{dr}{r^2} = Q\frac{1}{r} + C.$$

One easily confirms the relation 11.19 for this potential field, for example, for the x-component

$$-\frac{\partial U}{\partial x} = -\frac{\partial}{\partial x}\frac{Q}{r} = -Q\frac{\partial r}{\partial x}\frac{\partial}{\partial r}\left(\frac{1}{r}\right) = Q\frac{x}{r^3} = E_x,$$

etc. The constant C is usually set to zero, that is, the potential vanishes for $r \to \infty$.

$$\operatorname{div}\mathbf{E} = Q\left\{\frac{\partial}{\partial x}\frac{x}{(x^2+y^2+z^2)^{3/2}} + \frac{\partial}{\partial y}\frac{y}{(x^2+y^2+z^2)^{3/2}}\right. \tag{11.20}$$

$$\left. + \frac{\partial}{\partial z}\frac{z}{(x^2+y^2+z^2)^{3/2}}\right\} = 0, \tag{11.21}$$

$\operatorname{div}(\nabla U) = \Delta U = 0$ for $r \neq 0$. At $r = 0$, one has $\operatorname{div}\mathbf{E} = -\operatorname{div}\nabla U \neq 0$ (see below: the Gauss theorem).

Problem 11.6: Differential operations in spherical coordinates

Given a scalar field $\phi(r, \vartheta, \varphi)$ and a vector field $\mathbf{A}(r, \vartheta, \varphi)$, which are the relations for (a) $\nabla\phi$, (b) $\nabla \cdot \mathbf{A}$, (c) $\nabla \times \mathbf{A}$, (d) $\nabla^2\phi$ in spherical coordinates?

Spherical coordinates: A reminder.

Solution (a) **Gradient:** For the total differential, it holds that

$$d\phi = \nabla\phi \cdot d\mathbf{r}. \tag{11.22}$$

In spherical coordinates one has

$$d\phi = \frac{\partial\phi}{\partial r}dr + \frac{\partial\phi}{\partial\vartheta}d\vartheta + \frac{\partial\phi}{\partial\varphi}d\varphi, \tag{11.23}$$

$$\begin{aligned} d\mathbf{r} &= \frac{\partial\mathbf{r}}{\partial r}dr + \frac{\partial\mathbf{r}}{\partial\vartheta}d\vartheta + \frac{\partial\mathbf{r}}{\partial\varphi}d\varphi \\ &= \mathbf{e}_r dr + r\mathbf{e}_\vartheta\, d\vartheta + r\sin\vartheta\,\mathbf{e}_\varphi\, d\varphi, \end{aligned} \tag{11.24}$$

and

$$\nabla\phi = (\nabla\phi)_r\mathbf{e}_r + (\nabla\phi)_\vartheta\mathbf{e}_\vartheta + (\nabla\phi)_\varphi\mathbf{e}_\varphi .$$

The partial derivatives of the position vector have already been calculated in Chapter 10 when formulating the unit vectors:

$$\frac{\partial\mathbf{r}}{\partial r} = \mathbf{e}_r, \qquad \frac{\partial\mathbf{r}}{\partial\vartheta} = r\mathbf{e}_\vartheta, \qquad \frac{\partial\mathbf{r}}{\partial\varphi} = r\sin\vartheta\,\mathbf{e}_\varphi. \tag{11.25}$$

By insertion and comparison of coefficients, 11.22 immediately yields for the components of the gradient in spherical coordinates

$$(\nabla\phi)_r dr + (\nabla\phi)_\vartheta r\, d\vartheta + (\nabla\phi)_\varphi r\sin\vartheta\, d\varphi = \frac{\partial\phi}{\partial r}dr + \frac{\partial\phi}{\partial\vartheta}\cdot d\vartheta + \frac{\partial\phi}{\partial\varphi}d\varphi,$$

$$\begin{aligned} \nabla\phi &= \frac{\partial\phi}{\partial r}\mathbf{e}_r + \frac{1}{r}\frac{\partial\phi}{\partial\vartheta}\mathbf{e}_\vartheta + \frac{1}{r\sin\vartheta}\frac{\partial\phi}{\partial\varphi}\mathbf{e}_\varphi \\ &= (\nabla\phi)_r\mathbf{e}_r + (\nabla\phi)_\vartheta\mathbf{e}_\vartheta + (\nabla\phi)_\varphi\mathbf{e}_\varphi. \end{aligned} \tag{11.26}$$

(b) **Divergence:** The divergence may be expressed by the flow of the vector $\mathbf{A}$ across the surface of an infinitesimal volume element ΔV:

$$\operatorname{div}\mathbf{A} = \lim_{\Delta V\to 0}\frac{\int_{\Delta F}\mathbf{A}\cdot\mathbf{n}\,dF}{\Delta V}. \tag{11.27}$$

The figure shows the volume element with the magnitude

$$\Delta V = r^2\sin\vartheta\,\Delta r\,\Delta\vartheta\,\Delta\varphi. \tag{11.28}$$

Calculation of the flow components (to first approximation): The flow in the $\mathbf{e}_r$-direction across the area $ADHE$ is

$$\mathbf{A}(r,\vartheta,\varphi)\mathbf{e}_r\,\Delta F_r = A_r r^2\sin\vartheta\,\Delta\varphi\,\Delta\vartheta,$$

the flow across the back area $BCGF$ is

$$\mathbf{A}(r+\Delta r,\vartheta,\varphi)\mathbf{e}_r\Delta F_{r+\Delta r} = A_r r^2 \sin\vartheta\,\Delta\varphi\,\Delta\vartheta + \frac{\partial}{\partial r}\Big(r^2\sin\vartheta\,A_r\,\Delta\varphi\,\Delta\vartheta\Big)\Delta r.$$

The difference yields the contribution of the flow in the $\mathbf{e}_r$-direction to the surface integral in 11.27. The flow excess is

$$\sin\vartheta\frac{\partial}{\partial r}(r^2A_r)\Delta\varphi\,\Delta\vartheta\,\Delta r. \tag{11.29}$$

The flow excess in the $\mathbf{e}_\vartheta$-direction (areas $ABFE$ and $DCGH$) correspondingly follows as

$$r\frac{\partial}{\partial\vartheta}(\sin\vartheta\,A_\vartheta)\,\Delta\varphi\,\Delta r\,\Delta\vartheta. \tag{11.30}$$

The flow excess in the $\mathbf{e}_\varphi$-direction is

$$r\frac{\partial}{\partial\varphi}A_\varphi\,\Delta r\,\Delta\vartheta\,\Delta\varphi. \tag{11.31}$$

Summation of the contributions 11.29, 11.30, 11.31 yields the flow integral $\oint\mathbf{A}\cdot\mathbf{n}dF$. Then, 11.27 yields the expression for the divergence:

$$\nabla\cdot\mathbf{A} = \frac{1}{r^2}\frac{\partial}{\partial r}(r^2A_r) + \frac{1}{r\sin\vartheta}\frac{\partial}{\partial\vartheta}(\sin\vartheta\,A_\vartheta) + \frac{1}{r\sin\vartheta}\frac{\partial}{\partial\varphi}A_\varphi. \tag{11.32}$$

(c) **Curl (Rotation):** The geometric definition traces the rotation operation back to a contour integral:

$$\mathbf{n}\cdot\operatorname{curl}\mathbf{A} = \lim_{\Delta F\to 0}\frac{\oint\mathbf{A}\cdot d\mathbf{s}}{\Delta F}. \tag{11.33}$$

Component along $\mathbf{e}_r$:

The $\mathbf{e}_r$-component of the rotation is obtained when performing the contour integral along the curve $ADHEA$ ($\mathbf{n}=\mathbf{e}_r$). The enclosed area is then

$$\Delta F = r^2\sin\vartheta\,\Delta\vartheta\,\Delta\varphi \qquad \text{(compare fig.)}, \tag{11.34}$$

$$\oint_{ADHEA}\mathbf{A}\cdot d\mathbf{s} = \int_A^D + \int_D^H + \int_H^E + \int_E^A .$$

The partial integrals are

$$\int_A^D \mathbf{A}\cdot d\mathbf{s} = \mathbf{A}\cdot\mathbf{e}_\vartheta r\Delta\vartheta = A_\vartheta r\Delta\vartheta,$$

$$\int_E^A \mathbf{A}\cdot d\mathbf{s} = \mathbf{A}\cdot(-\mathbf{e}_\varphi)r\sin\vartheta\,\Delta\varphi = -A_\varphi r\sin\vartheta\,\Delta\varphi.$$

And to first approximation,

$$\int_H^E \mathbf{A} \cdot d\mathbf{s} = -\left(rA_\vartheta \Delta\vartheta + \frac{\partial}{\partial\varphi}(rA_\vartheta \Delta\vartheta)\Delta\varphi\right),$$

$$\int_D^H \mathbf{A} \cdot d\mathbf{s} = r \sin\vartheta\, A_\varphi \Delta\varphi + \frac{\partial}{\partial\vartheta}(r \sin\vartheta\, A_\varphi \Delta\varphi)\Delta\vartheta.$$

Then the contour integral along the closed curve is

$$\oint_{ADHEA} \mathbf{A} \cdot d\mathbf{s} = r\frac{\partial}{\partial\vartheta}(\sin\vartheta\, A_\varphi)\Delta\varphi\, \Delta\vartheta - r\frac{\partial}{\partial\varphi}(A_\vartheta)\Delta\vartheta\, \Delta\varphi. \tag{11.35}$$

From 11.33, 11.34, 11.35, it follows that the $\mathbf{e}_r$-component of the rotation is

$$\underset{r}{\operatorname{curl}} \mathbf{A} = \frac{1}{r \sin\vartheta}\left[\frac{\partial}{\partial\vartheta}(\sin\vartheta\, A_\varphi) - \frac{\partial}{\partial\varphi}A_\vartheta\right]. \tag{11.36}$$

Accordingly for the curve $AEFBA$ with $\Delta F = r \sin\vartheta\, \Delta r \Delta\varphi$,

$$\oint_{AEFBA} \mathbf{A} \cdot d\mathbf{s} = -\frac{\partial}{\partial r}(A_\varphi r \sin\vartheta\, \Delta\varphi)\Delta r + \frac{\partial}{\partial\varphi}(A_r \Delta r)\Delta\varphi$$

and because $\mathbf{n} = \mathbf{e}_\vartheta$, it follows that

$$\underset{\vartheta}{\operatorname{curl}} \mathbf{A} = \frac{1}{r \sin\vartheta}\left[\frac{\partial}{\partial\varphi}A_r - \sin\vartheta\frac{\partial}{\partial r}(rA_\varphi)\right]. \tag{11.37}$$

Investigation of the curve $ABCDA$ yields

$$\underset{\varphi}{\operatorname{curl}} \mathbf{A} = \frac{1}{r}\left(\frac{\partial}{\partial r}(rA_\vartheta) - \frac{\partial}{\partial\vartheta}A_r\right). \tag{11.38}$$

The results 11.36, 11.37, 11.38 may be combined into a determinant:

$$\nabla \times \mathbf{A} = \frac{1}{r^2 \sin\vartheta}\begin{vmatrix} \mathbf{e}_r & r\mathbf{e}_\vartheta & r\sin\vartheta\, \mathbf{e}_\varphi \\ \frac{\partial}{\partial r} & \frac{\partial}{\partial\vartheta} & \frac{\partial}{\partial\varphi} \\ A_r & rA_\vartheta & r\sin\vartheta\, A_\varphi \end{vmatrix}.$$

(d) **Laplace operator:** The Laplace operator is defined by

$$\nabla^2\phi = \operatorname{div} \nabla\phi. \tag{11.39}$$

Using the results 11.26 and 11.32, it follows that

$$\nabla^2\phi = \frac{1}{r^2}\frac{\partial}{\partial r}\left(r^2\frac{\partial\phi}{\partial r}\right) + \frac{1}{r^2 \sin\vartheta}\frac{\partial}{\partial\vartheta}\left(\sin\vartheta\frac{\partial\phi}{\partial\vartheta}\right) + \frac{1}{r^2 \sin^2\vartheta}\frac{\partial^2\phi}{\partial\varphi^2}. \tag{11.40}$$

Differential operators in arbitrary general (curvilinear) coordinates

In Chapter 10 we outlined curvilinear coordinates (e.g., spherical and cylindrical coordinates). In Problem 11.6, the differential operators ∇, div, and curl have been derived in spherical coordinates, basing on special considerations. Now we shall develop the general approaches for calculating differential operators in arbitrary curvilinear coordinates.

Brief repetition: Let $\mathbf{r}(x_\nu) = \sum_{\nu=1}^{3} x_\nu \mathbf{e}_\nu$ be the position vector in Cartesian coordinates x_ν ($\nu = 1, 2, 3$) that are related to the curvilinear coordinates q_σ ($\sigma = 1, 2, 3$) via $x_\nu = x_\nu$ (q_1, q_2, q_3). The x_ν may then be inserted in the position vector, which yields

$$\mathbf{r}(x_\nu) = \mathbf{r}(x_\nu(q_\sigma)) = \mathbf{r}(q_\sigma). \tag{11.41}$$

The new unit vectors $\mathbf{e}_{q_\sigma}$, which in general are characteristic for the point q_σ, may be defined at each point q_σ ($\sigma = 1, 2, 3$):

$$\mathbf{e}_{q_\sigma} = \frac{\partial \mathbf{r}(q_\mu)/\partial q_\sigma}{|\partial \mathbf{r}(q_\mu)/\partial q_\sigma|}, \qquad \sigma = 1, 2, 3, \tag{11.42}$$

or

$$\frac{\partial \mathbf{r}(q_\mu)}{\partial q_\sigma} = h_\sigma \mathbf{e}_{q_\sigma} \qquad \text{with} \quad h_\sigma = \left| \frac{\partial \mathbf{r}(q_\mu)}{\partial q_\sigma} \right|. \tag{11.43}$$

Here the h_σ ($\sigma = 1, 2, 3$) are *scaling factors*. The unit vectors $\mathbf{e}_{q_\sigma}$ point along the q_σ-coordinate line toward increasing q_σ.

The coordinate areas are obtained by solving the three equations $x_\nu = x_\nu(q_1, q_2, q_3)$ for q_σ:

$$q_\sigma = q_\sigma(x_1, x_2, x_3) = q_\sigma(x_\mu). \tag{11.44}$$

$q_\sigma = \text{constant} = c_\sigma$ ($\sigma = 1, 2, 3$) are the equations for the coordinate areas.

One may now construct other unit vectors $\mathbf{E}_{q_\sigma}$ at the point $P(x, y, z) = P(q_1, q_2, q_3)$ (see the figure), namely

$$\mathbf{E}_{q_\sigma} = \frac{\nabla q_\sigma}{|\nabla q_\sigma|}, \qquad \sigma = 1, 2, 3. \tag{11.45}$$

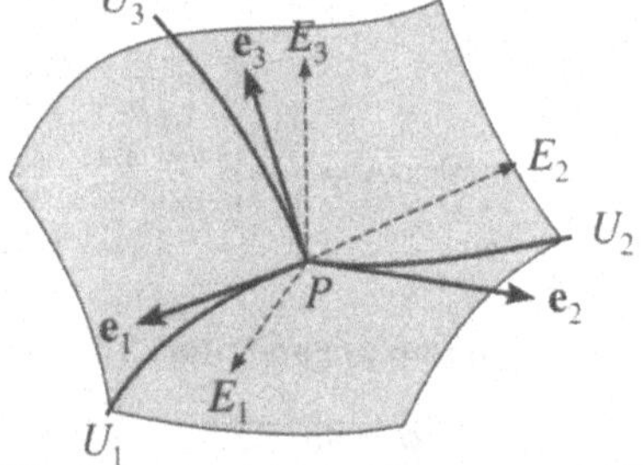

Different basis vectors at the same point P.

The $\mathbf{E}_{q_\sigma}$ are obviously perpendicular to the coordinate areas $q_\sigma = c_\sigma$. Thus, there are two sets of unit vectors at each point $P(q_\sigma)$, namely $\mathbf{e}_{q_\sigma}$ and $\mathbf{E}_{q_\sigma}$. In general, these sets are distinct. We shall demonstrate in the following that these two basic frames coincide only then if the curvilinear coordinates are orthogonal. One also has to take into account that both $\mathbf{e}_{q_\sigma}(q_1, q_2, q_3)$ as well as $\mathbf{E}_{q_\sigma}(q_1, q_2, q_3)$ depend on the point $P(q_1, q_2, q_3)$, that is, their orientations in general vary from point to point.

An arbitrary vector **A** may now be expressed in terms of both the base $\mathbf{e}_{q_\sigma}$ as well as the $\mathbf{E}_{q_\sigma}$,

$$\mathbf{A} = A_1\mathbf{e}_{q_1} + A_2\mathbf{e}_{q_2} + A_3\mathbf{e}_{q_3} = a_1\mathbf{E}_{q_1} + a_2\mathbf{E}_{q_2} + a_3\mathbf{E}_{q_3}. \tag{11.46}$$

The A_i and a_i, respectively, are the components of **A** in the bases concerned. Instead of the normalized base vectors $\mathbf{e}_{q_\sigma}$ or $\mathbf{E}_{q_\sigma}$ the nonnormalized vectors

$$\mathbf{b}_{q_\nu} = \frac{\partial \mathbf{r}\,(q_\sigma)}{\partial q_\nu} \qquad (\nu = 1, 2, 3) \tag{11.47}$$

and

$$\mathbf{B}_{q_\nu} = \nabla q_\nu \qquad (\nu = 1, 2, 3) \tag{11.48}$$

may also be used. They are called *unitary base vectors* and are in general not unit vectors. For an arbitrary vector **A**,

$$\mathbf{A} = C_1\frac{\partial \mathbf{r}}{\partial q_1} + C_2\frac{\partial \mathbf{r}}{\partial q_2} + C_3\frac{\partial \mathbf{r}}{\partial q_3} = C_1\mathbf{b}_{q_1} + C_2\mathbf{b}_{q_2} + C_3\mathbf{b}_{q_3} \tag{11.49}$$

and

$$\mathbf{A} = c_1\nabla q_1 + c_2\nabla q_2 + c_3\nabla q_3 = c_1\mathbf{B}_{q_1} + c_2\mathbf{B}_{q_2} + c_3\mathbf{B}_{q_3}.$$

The components C_ν $(\nu = 1, 2, 3)$ are called *contravariant* components and c_ν $(\nu = 1, 2, 3)$ *covariant* components of the vector **A**. They play an important role in the general theory of relativity where all coordinate frames are used on equal footing. In Cartesian coordinates the co- and contravariant components of a vector are equal to each other, as is immediately clear from their construction.

(a) **Arc length and volume element:** From $\mathbf{r} = \mathbf{r}(q_1, q_2, q_3)$, one obtains

$$d\mathbf{r} = \frac{\partial \mathbf{r}}{\partial q_1}dq_1 + \frac{\partial \mathbf{r}}{\partial q_2}dq_2 + \frac{\partial \mathbf{r}}{\partial q_3}dq_3 = h_1 dq_1\,\mathbf{e}_{q_1} + h_2 dq_2\,\mathbf{e}_{q_2} + h_3 dq_3\,\mathbf{e}_{q_3}. \tag{11.50}$$

Therefore, for the differential ds of the arc length, it results that

$$ds^2 = d\mathbf{r}\cdot d\mathbf{r}, \tag{11.51}$$

which for *orthogonal coordinates* $(\mathbf{e}_{q\mu}\cdot\mathbf{e}_{q\nu} = \delta_{\mu\nu})$ simplifies to

$$ds^2 = h_1^2\,dq_1^2 + h_2^2\,dq_2^2 + h_3^2\,dq_3^2. \tag{11.52}$$

For *nonorthogonal coordinates*, it holds that

$$\mathbf{b}_{q_\mu}\cdot\mathbf{b}_{q_\nu} = h_\mu h_\nu \mathbf{e}_{q_\mu}\cdot\mathbf{e}_{q_\nu} \equiv g_{\mu\nu} \not\equiv h_\mu h_\nu \delta_{\mu\nu}\,, \tag{11.53}$$

and therefore it follows from (11.50), (11.51), and (11.53) that

$$\begin{aligned}(ds)^2 &= d\mathbf{r}\cdot d\mathbf{r}\\ &= \left(\sum_\mu h_\mu dq_\mu \mathbf{e}_{q_\mu}\right)\cdot\left(\sum_\nu h_\nu dq_\nu \mathbf{e}_{q_\nu}\right)\\ &= \sum_{\mu,\nu} g_{\mu\nu}dq_\mu dq_\nu.\end{aligned} \tag{11.54}$$

This is the *fundamental quadratic* (or *metric*) form. The $g_{\mu\nu}$ are called *metric coefficients* (since they determine the measurement in the coordinates q_ν via the length element ds^2) or also *metric tensor* (briefly: metric). If $g_{\mu\nu} = 0$ for $\mu \neq \nu$, the coordinate frame is orthogonal. In this case $g_{11} = h_1^2$, $g_{22} = h_2^2$, $g_{33} = h_3^2$. The metric tensor is of basic importance in the general theory of relativity. It is determined there from the energy (mass-) distribution in space.

The equations enabling this are called Einstein equations.

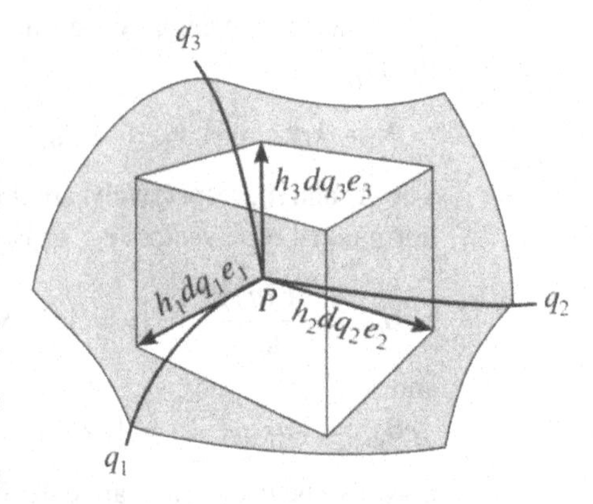

Illustration of the volume element.

The volume element dV may easily be calculated *for orthogonal coordinates* (see the figure):

$$\begin{aligned} dV &= |(h_1\, dq_1\, \mathbf{e}_{q_1}) \cdot [(h_2\, dq_2\, \mathbf{e}_{q_2}) \times (h_3\, dq_3\, \mathbf{e}_{q_3})]| \\ &= h_1 h_2 h_3\, dq_1\, dq_2\, dq_3 , \end{aligned} \tag{11.55}$$

because

$$\left|\mathbf{e}_{q_1} \cdot (\mathbf{e}_{q_2} \times \mathbf{e}_{q_3})\right| = 1 .$$

Problem 11.7: Reciprocal trihedral

Given are the three noncoplanar vectors $\mathbf{a}, \mathbf{b}, \mathbf{c}$ for which $\mathbf{a} \cdot (\mathbf{b} \times \mathbf{c}) \neq 0$. Show that the three *reciprocal* vectors

$$\mathbf{a}' = \frac{\mathbf{b} \times \mathbf{c}}{\mathbf{a} \cdot (\mathbf{b} \times \mathbf{c})}, \qquad \mathbf{b}' = \frac{\mathbf{c} \times \mathbf{a}}{\mathbf{a} \cdot (\mathbf{b} \times \mathbf{c})}, \qquad \mathbf{c}' = \frac{\mathbf{a} \times \mathbf{b}}{\mathbf{a} \cdot (\mathbf{b} \times \mathbf{c})} \tag{11.56}$$

are also noncoplanar and that

(a)

$$\mathbf{a}' \cdot \mathbf{a} = \mathbf{b}' \cdot \mathbf{b} = \mathbf{c}' \cdot \mathbf{c} = 1 ,$$

(b)

$$\begin{aligned} \mathbf{a}' \cdot \mathbf{b} &= \mathbf{a}' \cdot \mathbf{c} = 0 , \\ \mathbf{b}' \cdot \mathbf{a} &= \mathbf{b}' \cdot \mathbf{c} = 0 , \\ \mathbf{c}' \cdot \mathbf{a} &= \mathbf{c}' \cdot \mathbf{b} = 0 . \end{aligned} \tag{11.57}$$

(c) If $\mathbf{a} \cdot (\mathbf{b} \times \mathbf{c}) \equiv V$, then $\mathbf{a}' \cdot (\mathbf{b}' \times \mathbf{c}') = 1/V$.

Solution (a)

$$\mathbf{a}' \cdot \mathbf{a} = \frac{\mathbf{a} \cdot (\mathbf{b} \times \mathbf{c})}{\mathbf{a} \cdot (\mathbf{b} \times \mathbf{c})} = 1 .$$

In the same way one may conclude that

$$\mathbf{b}' \cdot \mathbf{b} = \mathbf{c}' \cdot \mathbf{c} = 1\,.$$

(b)

$$\mathbf{a}' \cdot \mathbf{b} = \mathbf{b} \cdot \mathbf{a}' = \cdot \frac{\mathbf{b} \cdot (\mathbf{b} \times \mathbf{c})}{\mathbf{a} \cdot (\mathbf{b} \times \mathbf{c})} = 0,$$

and similarly for the other cases.
(c) One has

$$\mathbf{a}' = \frac{\mathbf{b} \times \mathbf{c}}{V}, \qquad \mathbf{b}' = \frac{\mathbf{c} \times \mathbf{a}}{V}, \qquad \mathbf{c}' = \frac{\mathbf{a} \times \mathbf{b}}{V}.$$

Then it follows that

$$\begin{aligned}\mathbf{a}' \cdot (\mathbf{b}' \times \mathbf{c}') &= \frac{(\mathbf{b} \times \mathbf{c}) \cdot [(\mathbf{c} \times \mathbf{a}) \times (\mathbf{a} \times \mathbf{b})]}{V^3} \\ &= \frac{(\mathbf{a} \times \mathbf{b}) \cdot [(\mathbf{b} \times \mathbf{c}) \times (\mathbf{c} \times \mathbf{a})]}{V^3} \\ &= \frac{(\mathbf{a} \times \mathbf{b}) \cdot [\mathbf{c} \cdot ((\mathbf{b} \times \mathbf{c}) \cdot \mathbf{a}) - \mathbf{a} \cdot ((\mathbf{b} \times \mathbf{c}) \cdot \mathbf{c})]}{V^3} \\ &= \frac{[(\mathbf{a} \times \mathbf{b}) \cdot \mathbf{c}][(\mathbf{b} \times \mathbf{c}) \cdot \mathbf{a}]}{V^3} = \frac{[\mathbf{a} \cdot (\mathbf{b} \times \mathbf{c})]^2}{V^3} = \frac{V^2}{V^3} = \frac{1}{V}.\end{aligned}$$

From there it follows that $\mathbf{a}'$, $\mathbf{b}'$, $\mathbf{c}'$ are noncoplanar if $\mathbf{a}$, $\mathbf{b}$, $\mathbf{c}$ are noncoplanar.

Problem 11.8: Reciprocal coordinate frames

Let q_1, q_2, q_3 be general coordinates. Show that $\partial\mathbf{r}/\partial q_1$, $\partial\mathbf{r}/\partial q_2$, $\partial\mathbf{r}/\partial q_3$ and ∇q_1, ∇q_2, ∇q_3 form two reciprocal systems of vectors and that

$$\left\{\frac{\partial\mathbf{r}}{\partial q_1} \cdot \left(\frac{\partial\mathbf{r}}{\partial q_2} \times \frac{\partial\mathbf{r}}{\partial q_3}\right)\right\} \cdot \{\nabla q_1 \cdot (\nabla q_2 \times \nabla q_3)\} = 1\,.$$

Solution One has to show that

$$\frac{\partial\mathbf{r}}{\partial q_\nu} \cdot \nabla q_\mu = \begin{cases} 1 \text{ for } \nu = \mu, \\ 0 \text{ for } \nu \neq \mu, \end{cases} \tag{11.58}$$

where μ, ν may take any of the values 1,2,3. Now

$$d\mathbf{r} = \frac{\partial\mathbf{r}}{\partial q_1}dq_1 + \frac{\partial\mathbf{r}}{\partial q_2}dq_2 + \frac{\partial\mathbf{r}}{\partial q_3}dq_3.$$

and therefore after multiplication by ∇q_1

$$\nabla q_1 \cdot d\mathbf{r} = dq_1 = \left(\nabla q_1 \cdot \frac{\partial\mathbf{r}}{\partial q_1}\right)dq_1 + \left(\nabla q_1 \cdot \frac{\partial\mathbf{r}}{\partial q_2}\right)dq_2 + \left(\nabla q_1 \cdot \frac{\partial\mathbf{r}}{\partial q_3}\right)dq_3.$$

From there it follows that

$$\nabla q_1 \cdot \frac{\partial\mathbf{r}}{\partial q_1} = 1, \qquad \nabla q_1 \cdot \frac{\partial\mathbf{r}}{\partial q_2} = 0, \qquad \nabla q_1 \cdot \frac{\partial\mathbf{r}}{\partial q_3} = 0.$$

The other relations result in a similar way by forming $\nabla q_2 \cdot d\mathbf{r} = dq_2$ and $\nabla q_3 \cdot d\mathbf{r} = dq_3$. Thus, the reciprocity of the vector systems $\partial\mathbf{r}/\partial q_\nu$ and ∇q_ν is demonstrated.

From the preceding problem it then immediately follows that

$$\left\{\frac{\partial\mathbf{r}}{\partial q_1}\cdot\left(\frac{\partial\mathbf{r}}{\partial q_2}\times\frac{\partial\mathbf{r}}{\partial q_3}\right)\right\}\cdot\{\nabla q_1\cdot(\nabla q_2\times\nabla q_3)\} = 1.$$

This statement is equivalent to the following theorem on *Jacobi determinants:*

$$J\left(\frac{q_1, q_2, q_3}{x, y, z}\right) \overset{\text{def.}}{\equiv} \nabla q_1\cdot(\nabla q_2\times\nabla q_3) = \begin{vmatrix} \dfrac{\partial q_1}{\partial x} & \dfrac{\partial q_1}{\partial y} & \dfrac{\partial q_1}{\partial z} \\ \dfrac{\partial q_2}{\partial x} & \dfrac{\partial q_2}{\partial y} & \dfrac{\partial q_2}{\partial z} \\ \dfrac{\partial q_3}{\partial x} & \dfrac{\partial q_3}{\partial y} & \dfrac{\partial q_3}{\partial z} \end{vmatrix}, \tag{11.59}$$

which reads

$$J\left(\frac{x, y, z}{q_1, q_2, q_3}\right)\cdot J\left(\frac{q_1, q_2, q_3}{x, y, z}\right) = 1\,.$$

One can check easily that $\partial\mathbf{r}/\partial q_i$ and ∇q_i fulfill relation 11.56 from Problem 11.7.

(b) **Gradient in general orthogonal coordinates:** Let $\phi(q_1, q_2, q_3)$ be an arbitrary function. We look for the components f_1, f_2, f_3 of the gradient in the general base $\mathbf{e}_{q_\nu}$, that is,

$$\nabla\phi = f_1\mathbf{e}_{q_1} + f_2\mathbf{e}_{q_2} + f_3\mathbf{e}_{q_3}. \tag{11.60}$$

Because

$$\begin{aligned} d\mathbf{r} &= \frac{\partial\mathbf{r}}{\partial q_1}dq_1 + \frac{\partial\mathbf{r}}{\partial q_2}dq_2 + \frac{\partial\mathbf{r}}{\partial q_3}dq_3 \\ &= h_1\mathbf{e}_{q_1}dq_1 + h_2\mathbf{e}_{q_2}dq_2 + h_3\mathbf{e}_{q_3}dq_3\,, \end{aligned}$$

because of the presupposed orthogonality of the $\mathbf{e}_{q_\nu}$, it follows that

$$d\phi = \nabla\phi\cdot d\mathbf{r} = h_1 f_1\,dq_1 + h_2 f_2\,dq_2 + h_3 f_3\,dq_3.$$

But it also holds that

$$d\phi = \frac{\partial\phi}{\partial q_1}dq_1 + \frac{\partial\phi}{\partial q_2}dq_2 + \frac{\partial\phi}{\partial q_3}dq_3.$$

A comparison of the last two relations yields

$$\nabla\phi = \frac{\mathbf{e}_{q_1}}{h_1}\frac{\partial\phi}{\partial q_1} + \frac{\mathbf{e}_{q_2}}{h_2}\frac{\partial\phi}{\partial q_2} + \frac{\mathbf{e}_{q_3}}{h_3}\frac{\partial\phi}{\partial q_3}.$$

In operator notation this reads

$$\nabla = \mathbf{e}_{q_1}\left(\frac{1}{h_1}\frac{\partial}{\partial q_1}\right) + \mathbf{e}_{q_2}\left(\frac{1}{h_2}\frac{\partial}{\partial q_2}\right) + \mathbf{e}_{q_3}\left(\frac{1}{h_3}\frac{\partial}{\partial q_3}\right). \tag{11.61}$$

From there it follows especially for $\phi = q_1$ that

$$\nabla q_1 = \frac{\mathbf{e}_{q_1}}{h_1}, \tag{11.62}$$

and therefore $|\nabla q_1| = 1/h_1$ or generally $|\nabla q_\nu| = 1/h_\nu \quad (\nu = 1, 2, 3)$.

Because

$$\mathbf{E}_{q_\nu} = \frac{\nabla q_\nu}{|\nabla q_\nu|}$$

(compare to (11.45)), it results that

$$\mathbf{E}_{q_\nu} = \frac{\nabla q_\nu}{|\nabla q_\nu|} = h_\nu \nabla q_\nu = h_\nu \frac{\mathbf{e}_{q_\nu}}{h_\nu} = \mathbf{e}_{q_\nu} \qquad (\nu = 1, 2, 3).$$

This means that *for orthogonal coordinates the reciprocal base systems $\mathbf{E}_{q_\nu}$ and $\mathbf{e}_{q_\nu}$ coincide.* This happens, of course, in particular for Cartesian coordinates.

For the following, the relations

$$\begin{aligned}
\mathbf{e}_{q_1} &= h_2 h_3 \nabla q_2 \times \nabla q_3, \\
\mathbf{e}_{q_2} &= h_3 h_1 \nabla q_3 \times \nabla q_1, \\
\mathbf{e}_{q_3} &= h_1 h_2 \nabla q_1 \times \nabla q_2
\end{aligned} \tag{11.63}$$

are helpful. They may be checked quickly, for example,

$$h_2 h_3 \nabla q_2 \times \nabla q_3 = h_2 h_3 \left(\frac{\mathbf{e}_{q_2}}{h_2} \times \frac{\mathbf{e}_{q_3}}{h_3} \right) = \frac{h_2 h_3}{h_2 h_3} (\mathbf{e}_{q_2} \times \mathbf{e}_{q_3}) = \mathbf{e}_{q_1}. \tag{11.64}$$

(c) **Divergence in general orthogonal coordinates:** We shall now calculate

$$\operatorname{div} \mathbf{A} = \nabla \cdot (A_1 \mathbf{e}_{q_1} + A_2 \mathbf{e}_{q_2} + A_3 \mathbf{e}_{q_3})$$

in general coordinates. For this purpose we consider at first

$$\begin{aligned}
\nabla \cdot (A_1 \mathbf{e}_{q1}) &= \nabla \cdot (A_1 h_2 h_3 \nabla q_2 \times \nabla q_3) \\
&= (\nabla (A_1 h_2 h_3)) \cdot (\nabla q_2 \times \nabla q_3) + A_1 h_2 h_3 \nabla \cdot (\nabla q_2 \times \nabla q_3) \\
&= (\nabla (A_1 h_2 h_3)) \cdot \left(\frac{\mathbf{e}_{q_2}}{h_2} \times \frac{\mathbf{e}_{q_3}}{h_3} \right) + 0 \\
&= \nabla (A_1 h_2 h_3) \cdot \frac{\mathbf{e}_{q_1}}{h_2 h_3} \\
&= \left[\frac{\mathbf{e}_{q_1}}{h_1} \frac{\partial}{\partial q_1} (A_1 h_2 h_3) + \frac{\mathbf{e}_{q_2}}{h_2} \frac{\partial}{\partial q_2} (A_1 h_2 h_3) + \frac{\mathbf{e}_{q_3}}{h_3} \frac{\partial}{\partial q_3} (A_1 h_2 h_3) \right] \cdot \frac{\mathbf{e}_{q_1}}{h_2 h_3} \\
&= \frac{1}{h_1 h_2 h_3} \frac{\partial}{\partial q_1} (A_1 h_2 h_3).
\end{aligned}$$

Similarly it follows that

$$\nabla \cdot (A_2 \mathbf{e}_{q_2}) = \frac{1}{h_1 h_2 h_3} \frac{\partial}{\partial q_2}(A_2 h_1 h_3)$$

and

$$\nabla \cdot (A_3 \mathbf{e}_{q_3}) = \frac{1}{h_1 h_2 h_3} \frac{\partial}{\partial q_3}(A_3 h_1 h_2).$$

Therefore,

$$\begin{aligned} \operatorname{div} \mathbf{A} &= \nabla \cdot (A_1 \mathbf{e}_{q_1} + A_2 \mathbf{e}_{q_2} + A_3 \mathbf{e}_{q_3}) \\ &= \nabla \cdot A_1 \mathbf{e}_{q_1} + \nabla \cdot A_2 \mathbf{e}_{q_2} + \nabla \cdot A_3 \mathbf{e}_{q_3} , \end{aligned} \tag{11.65}$$

$$\operatorname{div} \mathbf{A} = \frac{1}{h_1 h_2 h_3} \left[\frac{\partial}{\partial q_1}(A_1 h_2 h_3) + \frac{\partial}{\partial q_2}(A_2 h_1 h_3) + \frac{\partial}{\partial q_3}(A_3 h_1 h_2) \right] . \tag{11.66}$$

(d) **Curl (Rotation) in general orthogonal coordinates:** We have to calculate

$$\begin{aligned} \nabla \times \mathbf{A} &= \nabla \times (A_1 \mathbf{e}_{q_1} + A_2 \mathbf{e}_{q_2} + A_3 \mathbf{e}_{q_3}) \\ &= \nabla \times (A_1 \mathbf{e}_{q_1}) + \nabla \times (A_2 \mathbf{e}_{q_2}) + \nabla \times (A_3 \mathbf{e}_{q_3}) . \end{aligned}$$

It suffices to consider, for example, the term $\nabla \times (A_1 \mathbf{e}_{q_1})$ in more detail. We obtain

$$\begin{aligned} \nabla \times (A_1 \mathbf{e}_{q_1}) &= \nabla \times (A_1 h_1 \nabla q_1) = \nabla(A_1 h_1) \times \nabla q_1 + A_1 h_1 \nabla \times \nabla q_1 \\ &= \nabla(A_1 h_1) \times \frac{\mathbf{e}_{q_1}}{h_1} + 0 \\ &= \left[\frac{\mathbf{e}_{q_1}}{h_1} \frac{\partial}{\partial q_1}(A_1 h_1) + \frac{\mathbf{e}_{q_2}}{h_2} \frac{\partial}{\partial q_2}(A_1 h_1) + \frac{\mathbf{e}_{q_3}}{h_3} \frac{\partial}{\partial q_3}(A_1 h_1) \right] \times \frac{\mathbf{e}_{q_1}}{h_1} \\ &= \frac{\mathbf{e}_{q_2}}{h_3 h_1} \frac{\partial}{\partial q_3}(A_1 h_1) - \frac{\mathbf{e}_{q_3}}{h_1 h_2} \frac{\partial}{\partial q_2}(A_1 h_1). \end{aligned}$$

Therefore,

$$\begin{aligned} \nabla \times \mathbf{A} = {} & \frac{\mathbf{e}_{q_1}}{h_2 h_3} \left[\frac{\partial}{\partial q_2}(A_3 h_3) - \frac{\partial}{\partial q_3}(A_2 h_2) \right] + \frac{\mathbf{e}_{q_2}}{h_3 h_1} \left[\frac{\partial}{\partial q_3}(A_1 h_1) - \frac{\partial}{\partial q_1}(A_3 h_3) \right] \\ & + \frac{\mathbf{e}_{q_3}}{h_1 h_2} \left[\frac{\partial}{\partial q_1}(A_2 h_2) - \frac{\partial}{\partial q_2}(A_1 h_1) \right] . \end{aligned}$$

In determinant notation this reads

$$\nabla \times \mathbf{A} = \frac{1}{h_1 h_2 h_3} \begin{vmatrix} h_1 \mathbf{e}_{q_1} & h_2 \mathbf{e}_{q_2} & h_3 \mathbf{e}_{q_3} \\ \dfrac{\partial}{\partial q_1} & \dfrac{\partial}{\partial q_2} & \dfrac{\partial}{\partial q_3} \\ A_1 h_1 & A_2 h_2 & A_3 h_3 \end{vmatrix} . \tag{11.67}$$

(e) **The delta (Laplace) operator in general (orthogonal) coordinates:** One has to calculate $\Delta\psi$ in orthogonal curvilinear coordinates. This does not provide any difficulties, because

$$\Delta\psi = \nabla\cdot\nabla\psi = \nabla\cdot\left(\frac{\mathbf{e}_{q_1}}{h_1}\frac{\partial}{\partial q_1} + \frac{\mathbf{e}_{q_2}}{h_2}\frac{\partial}{\partial q_2} + \frac{\mathbf{e}_{q_3}}{h_3}\frac{\partial}{\partial q_3}\right)\psi .$$

Using now equation (11.66) for the divergence, where obviously

$$A_\nu = \frac{1}{h_\nu}\frac{\partial}{\partial q_\nu} \qquad (\nu = 1, 2, 3),$$

one immediately finds that

$$\begin{aligned}\Delta\psi &= \nabla\cdot\nabla\psi \qquad (11.68)\\ &= \frac{1}{h_1h_2h_3}\left[\frac{\partial}{\partial q_1}\left(\frac{h_2h_3}{h_1}\frac{\partial\psi}{\partial q_1}\right) + \frac{\partial}{\partial q_2}\left(\frac{h_3h_1}{h_2}\frac{\partial\psi}{\partial q_2}\right) + \frac{\partial}{\partial q_3}\left(\frac{h_1h_2}{h_3}\frac{\partial\psi}{\partial q_3}\right)\right].\end{aligned}$$

(f) **Examples of special orthogonal coordinate frames**

1. Cylinder coordinates

$$\begin{aligned}\mathbf{r}(x, y, z) &= x\mathbf{e}_1 + y\mathbf{e}_2 + z\mathbf{e}_3 \qquad (11.69)\\ &= \varrho\cos\varphi\,\mathbf{e}_1 + \varrho\sin\varphi\,\mathbf{e}_2 + z\mathbf{e}_3 = \mathbf{r}(\varrho, \varphi, z).\end{aligned}$$

Here $\varrho \geq 0$, $0 \leq \varphi < 2\pi$, $-\infty < z < \infty$.

We identify $q_1 = \varrho$, $q_2 = \varphi$, $q_3 = z$. According to (11.42), it then follows that

$$\begin{aligned}\mathbf{e}_{q_1} &\equiv \mathbf{e}_\varrho = \cos\varphi\,\mathbf{e}_1 + \sin\varphi\,\mathbf{e}_2\,,\\ \mathbf{e}_{q_2} &\equiv \mathbf{e}_\varphi = -\sin\varphi\,\mathbf{e}_1 + \cos\varphi\,\mathbf{e}_2\,,\\ \mathbf{e}_{q_3} &\equiv \mathbf{e}_z = \mathbf{e}_3 . \qquad (11.70)\end{aligned}$$

Moreover,

$$h_1 \equiv h_\varrho = \left|\frac{\partial\mathbf{r}}{\partial\varrho}\right| = 1\,,$$

$$h_2 \equiv h_\varphi = \left|\frac{\partial\mathbf{r}}{\partial\varphi}\right| = \varrho\,,$$

$$h_3 \equiv h_z = \left|\frac{\partial\mathbf{r}}{\partial z}\right| = 1\,.$$

According to equation (11.61), it therefore follows that

$$\nabla\phi = \frac{\partial\phi}{\partial\varrho}\mathbf{e}_\varrho + \frac{1}{\varrho}\frac{\partial\phi}{\partial\varphi}\mathbf{e}_\varphi + \frac{\partial\phi}{\partial z}\mathbf{e}_z .$$

According to equation (11.65),

$$\operatorname{div}\mathbf{A} = \nabla \cdot \mathbf{A} = \frac{1}{\varrho}\left[\frac{\partial}{\partial \varrho}(\varrho A_\varrho) + \frac{\partial}{\partial \varphi}A_\varphi + \frac{\partial}{\partial z}(\varrho A_z)\right],$$

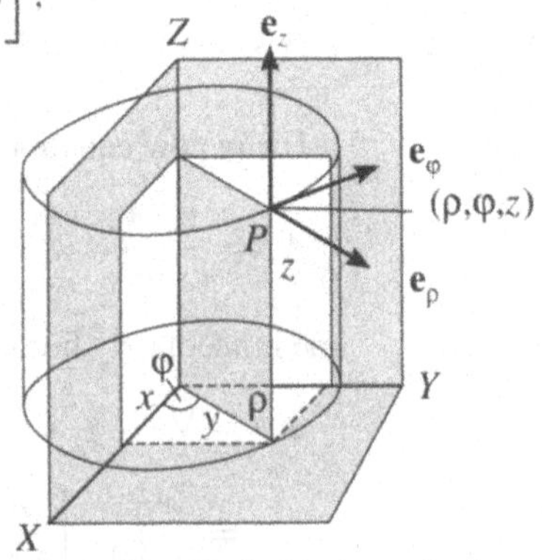

Illustration of cylindrical coordinates.

where

$$\mathbf{A} = A_\varrho \mathbf{e}_\varrho + A_\varphi \mathbf{e}_\varphi + A_z \mathbf{e}_z \equiv \sum_\nu A_\nu \mathbf{e}_{q_\nu}.$$

Moreover, according to equation (11.67),

$$\nabla \times \mathbf{A} = \frac{1}{\varrho}\begin{vmatrix} \mathbf{e}_\varrho & \varrho \mathbf{e}_\varphi & \mathbf{e}_z \\ \dfrac{\partial}{\partial \varrho} & \dfrac{\partial}{\partial \varphi} & \dfrac{\partial}{\partial z} \\ A_\varrho & \varrho A_\varphi & A_z \end{vmatrix}$$

$$= \frac{1}{\varrho}\left[\left(\frac{\partial A_z}{\partial \varphi} - \frac{\partial}{\partial z}(\varrho A_\varphi)\right)\mathbf{e}_\varrho + \left(\varrho\frac{\partial A_\varrho}{\partial z} - \varrho\frac{\partial A_z}{\partial \varrho}\right)\mathbf{e}_\varphi \right.$$
$$\left. + \left(\frac{\partial}{\partial \varrho}(\varrho A_\varphi) - \frac{\partial A_\varrho}{\partial_\varphi}\right)\mathbf{e}_z\right],$$

and according to equation (11.68),

$$\Delta\psi = \nabla^2\psi = \frac{1}{\varrho}\left[\frac{\partial}{\partial \varrho}\left(\varrho\frac{\partial\psi}{\partial \varrho}\right) + \frac{\partial}{\partial \varphi}\left(\frac{1}{\varrho}\frac{\partial\psi}{\partial \varphi}\right) + \frac{\partial}{\partial z}\left(\varrho\frac{\partial\psi}{\partial z}\right)\right]$$
$$= \frac{1}{\varrho}\frac{\partial}{\partial \varrho}\left(\varrho\frac{\partial\psi}{\partial \varrho}\right) + \frac{1}{\varrho^2}\frac{\partial^2\psi}{\partial \varphi^2} + \frac{\partial^2\psi}{\partial z^2}.$$

Cylinder coordinates are very useful when solving problems with axial symmetry.

2. Spherical coordinates

$$\begin{aligned}\mathbf{r}(x, y, z) &= x\mathbf{e}_1 + y\mathbf{e}_2 + z\mathbf{e}_3 \\ &= r\sin\vartheta\cos\varphi\,\mathbf{e}_1 + r\sin\vartheta\sin\varphi\,\mathbf{e}_2 + r\cos\vartheta\,\mathbf{e}_3 \\ &= \mathbf{r}(r, \vartheta, \varphi).\end{aligned} \tag{11.71}$$

Here $r \geq 0, 0 \leq \vartheta \leq \pi, 0 \leq \varphi < 2\pi$.

We choose $q_1 = r, q_2 = \vartheta, q_3 = \varphi$. According to equation (11.42),

$$\begin{aligned}\mathbf{e}_{q_1} &= \mathbf{e}_r = \sin\vartheta\cos\varphi\,\mathbf{e}_1 + \sin\vartheta\sin\varphi\,\mathbf{e}_2 + \cos\vartheta\,\mathbf{e}_3, \\ \mathbf{e}_{q_2} &= \mathbf{e}_\vartheta = \cos\vartheta\cos\varphi\,\mathbf{e}_1 + \cos\vartheta\sin\varphi\,\mathbf{e}_2 - \sin\vartheta\,\mathbf{e}_3, \\ \mathbf{e}_{q_3} &= \mathbf{e}_\varphi = -\sin\varphi\,\mathbf{e}_1 + \cos\varphi\,\mathbf{e}_2.\end{aligned} \tag{11.72}$$

Moreover,

$$h_1 = h_r = \left|\frac{\partial \mathbf{r}}{\partial r}\right| = 1,$$

$$h_2 = h_\vartheta = \left|\frac{\partial \mathbf{r}}{\partial \vartheta}\right| = r,$$

$$h_3 = h_\varphi = \left|\frac{\partial \mathbf{r}}{\partial \varphi}\right| = r\sin\vartheta.$$

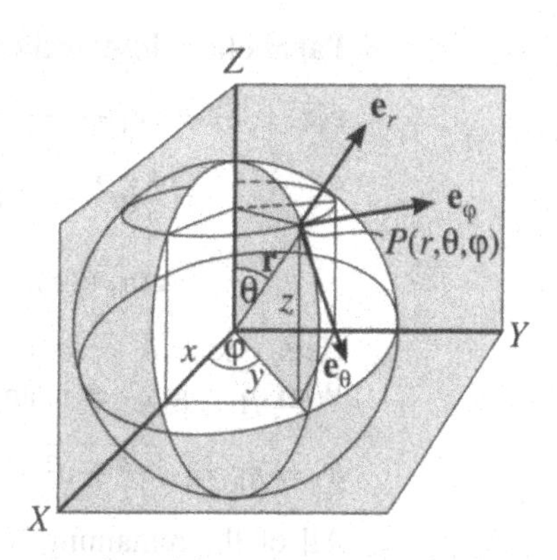

Illustration of spherical coordinates.

Therefore, according to equation (11.61),

$$\nabla\phi = \frac{\partial\phi}{\partial r}\mathbf{e}_r + \frac{1}{r}\frac{\partial\phi}{\partial\vartheta}\mathbf{e}_\vartheta + \frac{1}{r\sin\vartheta}\frac{\partial\phi}{\partial\varphi}\mathbf{e}_\varphi,$$

and according to equation (11.66),

$$\operatorname{div}\mathbf{A} = \nabla\cdot\mathbf{A} = \frac{1}{r^2}\frac{\partial}{\partial r}(r^2 A_r) + \frac{1}{r\sin\vartheta}\frac{\partial}{\partial\vartheta}(\sin\vartheta A_\vartheta) + \frac{1}{r\sin\vartheta}\frac{\partial A_\varphi}{\partial\varphi},$$

where $\mathbf{A} = A_r\mathbf{e}_r + A_\vartheta\mathbf{e}_\vartheta + A_\varphi\mathbf{e}_\varphi$.

Moreover, according to equation (11.67),

$$\nabla\times\mathbf{A} = \frac{1}{r\cdot r\sin\vartheta}\begin{vmatrix} \mathbf{e}_r & r\mathbf{e}_\vartheta & r\sin\vartheta\,\mathbf{e}_\varphi \\ \frac{\partial}{\partial r} & \frac{\partial}{\partial\vartheta} & \frac{\partial}{\partial\varphi} \\ A_r & rA_\vartheta & r\sin\vartheta A_\varphi \end{vmatrix}$$

$$= \frac{1}{r^2\sin\vartheta}\left\{\left[\frac{\partial}{\partial\vartheta}(r\sin\vartheta A_\varphi) - \frac{\partial}{\partial\varphi}(rA_\vartheta)\right]\mathbf{e}_r + \left[\frac{\partial A_r}{\partial\varphi} - \frac{\partial}{\partial r}(r\sin\vartheta A_\varphi)\right]r\mathbf{e}_\vartheta\right.$$

$$\left. + \left[\frac{\partial}{\partial r}(rA_\vartheta) - \frac{\partial A_r}{\partial\vartheta}\right]r\sin\vartheta\,\mathbf{e}_\varphi\right\},$$

and corresponding to equation (11.68),

$$\begin{aligned}\Delta\psi &= \nabla\cdot\nabla\psi \\ &= \frac{1}{r\cdot r\sin\vartheta}\left[\frac{\partial}{\partial r}\left(r\cdot r\sin\vartheta\frac{\partial\psi}{\partial r}\right) + \frac{\partial}{\partial\vartheta}\left(\frac{r\sin\vartheta}{r}\frac{\partial\psi}{\partial\vartheta}\right) + \frac{\partial}{\partial\varphi}\left(\frac{r}{r\sin\vartheta}\frac{\partial\psi}{\partial\varphi}\right)\right] \\ &= \frac{1}{r^2}\frac{\partial}{\partial r}\left(r^2\frac{\partial\psi}{\partial r}\right) + \frac{1}{r^2\sin\vartheta}\frac{\partial}{\partial\vartheta}\left(\sin\vartheta\frac{\partial\psi}{\partial\vartheta}\right) + \frac{1}{r^2\sin^2\vartheta}\frac{\partial^2\psi}{\partial\varphi^2}.\end{aligned}$$

Spherical coordinates are highly useful when solving problems with spherical symmetry.

3. Parabolic cylindrical coordinates

$$\begin{aligned}\mathbf{r}(x, y, z) &= x\mathbf{e}_1 + y\mathbf{e}_2 + z\mathbf{e}_3 \\ &= \frac{1}{2}(u^2 - v^2)\mathbf{e}_1 + uv\mathbf{e}_2 + z\mathbf{e}_3 \\ &= \mathbf{r}(u, v, z).\end{aligned} \tag{11.73}$$

Here $-\infty < u < \infty$, $v \geq 0$, $-\infty < z < \infty$.
With $q_1 = u$, $q_2 = v$, and $q_3 = z$, one easily evaluates

$$h_1 = h_u = \sqrt{u^2 + v^2}, \qquad h_2 = h_v = \sqrt{u^2 + v^2}, \qquad h_3 = h_z = 1.$$

All of the remaining follows according to the general methods (equations (11.61) – (11.68)) outlined. The figure illustrates these parabolic coordinates in the x, y-plane.

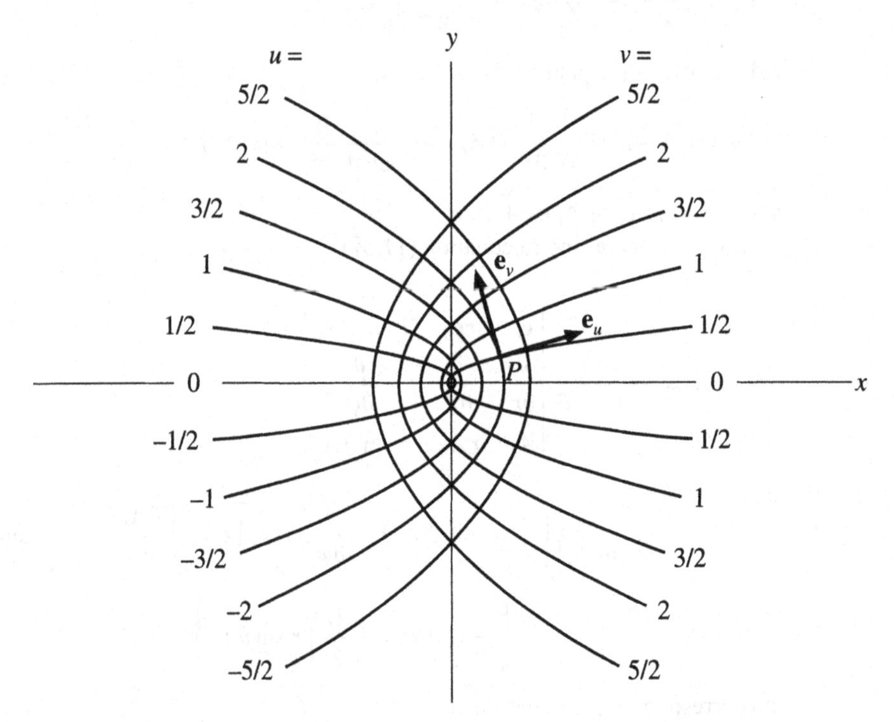

Projection of the coordinate surfaces of parabolic cylindrical coordinates into the x, y-plane. The z-coordinate of a point is identical to its Cartesian z-coordiante. The (variable) unit vectors $\mathbf{e}_u$ and $\mathbf{e}_v$ are shown in a point P.

4. Elliptic and hyperbolic cylindrical coordinates

$$\begin{aligned}\mathbf{r}(x, y, z) &= x\mathbf{e}_1 + y\mathbf{e}_2 + z\mathbf{e}_3 \\ &= a\cosh u \cos v\, \mathbf{e}_1 + a \sinh u \sin v\, \mathbf{e}_2 + z\mathbf{e}_3 \\ &= \mathbf{r}(u, v, z).\end{aligned} \tag{11.74}$$

Obviously,

$$x^2 = a^2 \cosh^2 u \cos^2 v,$$
$$y^2 = a^2 \sinh^2 u \sin^2 v,$$

and therefore

$$\frac{x^2}{a^2 \cosh^2 u} + \frac{y^2}{a^2 \sinh^2 u} = 1, \quad \text{and} \quad \frac{x^2}{a^2 \cos^2 v} - \frac{y^2}{a^2 \sin^2 v} = 1.$$

Here $u \geq 0, 0 \leq v < 2\pi, -\infty < z < \infty$.
With $q_1 = u, q_2 = v, q_3 = z$, it follows that

$$h_1 = h_u = a\sqrt{\sinh^2 u + \sin^2 v},$$
$$h_2 = h_v = a\sqrt{\sinh^2 u + \sin^2 v},$$
$$h_3 = h_z = 1.$$

All other operators follow according to the general equations (11.61) – (11.68). The projections of the coordinate areas u = constant, v = constant onto the x, y-plane are illustrated in the figure. They represent confocal ellipses or hyperbolas.

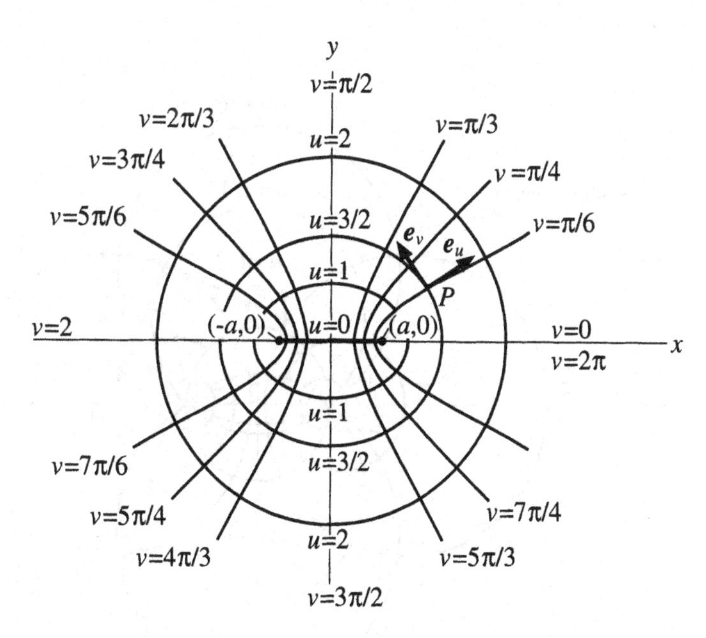

Projection of the coordinate surfaces of u = constant and v = constant in elliptic cylindrical coordinates into the x, y-plane.

5. Bipolar coordinates

$$\mathbf{r}(x, y, z) = x\mathbf{e}_1 + y\mathbf{e}_2 + z\mathbf{e}_3$$

$$= \frac{a \sinh v}{\cosh v - \cos u} \mathbf{e}_1 + \frac{a \sin u}{\cosh v - \cos u} \mathbf{e}_2 + z\mathbf{e}_3$$
$$= \mathbf{r}(u, v, z). \tag{11.75}$$

Here $0 \le u < 2\pi$, $-\infty < v < \infty$, $-\infty < z < \infty$.
With $q_1 = u$, $q_2 = v$, $q_3 = z$, one obtains

$$h_1 = h_u = \frac{a}{\cosh v - \cos u},$$
$$h_2 = h_v = \frac{a}{\cosh v - \cos u},$$
$$h_3 = h_z = 1.$$

The differential operators then follow according to the general rules (11.61)–(11.68).

For an easier identification of the coordinate areas u = constant and v = constant and their projection onto the x, y-plane, it is convenient to derive the following relations from (11.75):

$$x^2 + (y - a \cot u)^2 = a^2 \operatorname{cosec}^2 u, \qquad (x - a \coth v)^2 + y^2 = a^2 \operatorname{cosech}^2 v, \qquad z = z.$$

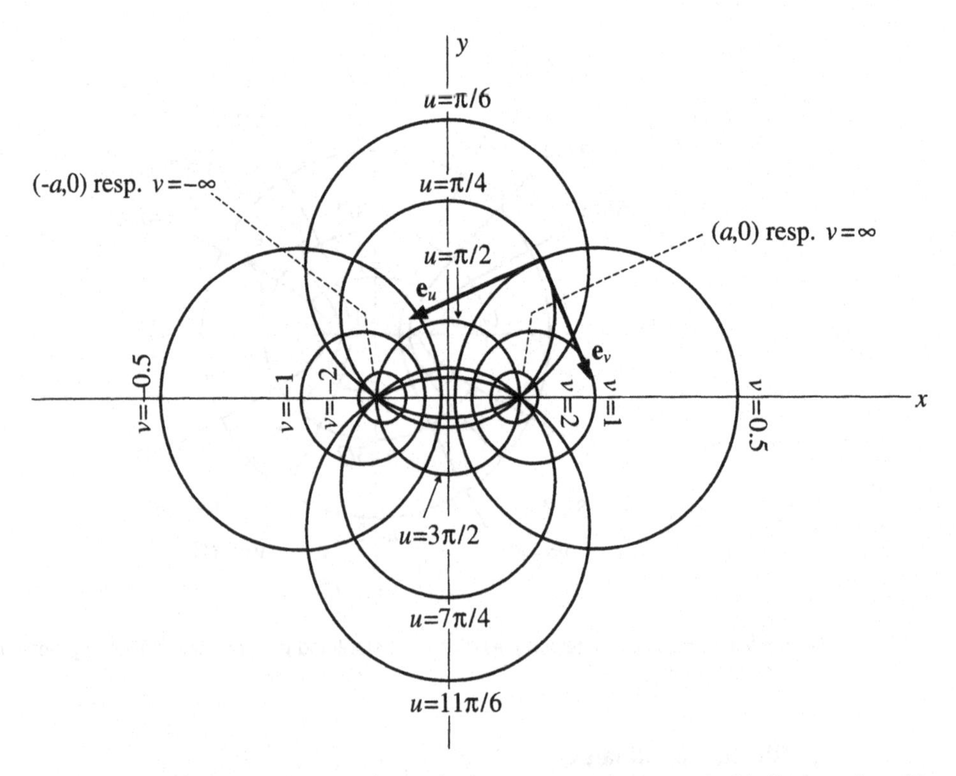

Projection of the bipolar coordinate surfaces of u = constant and v = constant in the x, y-plane. The (variable) unit vectors $\mathbf{e}_u$ and $\mathbf{e}_v$ are shown in an arbitrary point P.

12 Determination of Line Integrals

If $\mathbf{A}$ specifies a force field, the line integral (path integral) $\int_{P_1}^{P_2} \mathbf{A} \cdot d\mathbf{r}$ is the energy (work) that has to be supplied during a motion from P_1 to P_2, or is released, respectively. We shall make that clear now:

We ask for the work that is needed to move from the point P_1 along a space curve $\mathbf{r} = \mathbf{r}(t)$ in the force field (vector field) to the point P_2. We decompose the space curve into small path sections $\Delta\mathbf{r}$, calculate the expression $A \Delta r \cos(\mathbf{A}, \Delta\mathbf{r})$, which represents the wanted work on the section Δr, and sum up over all Δr. The work is then given by

The work integral (integral along a curve) along the curve C.

$$\sum_{\Delta r_i} A_i \Delta r_i \cos(\mathbf{A}_i, \Delta\mathbf{r}_i).$$

When changing to infinitesimally small path sections $d\mathbf{r}$, the work is then obtained as the line integral

$$\int_C \mathbf{A} \cdot d\mathbf{r} = \lim_{\Delta r_i \to 0} \sum_{\Delta r_i} A_i \Delta r_i \cos(\mathbf{A}_i, \Delta\mathbf{r}_i) = \lim_{\Delta r_i \to 0} \sum_{\Delta r_i} \mathbf{A}_i \cdot \Delta\mathbf{r}_i,$$

where $\mathbf{A} \cdot d\mathbf{r}$ is the scalar product of the field $\mathbf{A}$ and the vector $d\mathbf{r}$. C denotes the space curve $\mathbf{r}(t)$ between the initial point $\mathbf{r}(t_1)$ and the endpoint $\mathbf{r}(t_2)$.

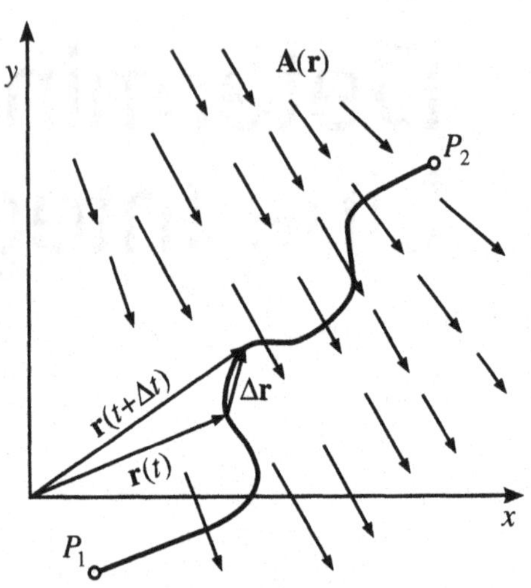

A path from P_1 to P_2 through the vector field $\mathbf{A}(\mathbf{r})$.

A line integral is calculated as follows:

We first form the indefinite integral. To this end we decompose the vector field **A** in its Cartesian components and insert this into the integral:

$$\int \mathbf{A} \cdot d\mathbf{r} = \int (A_x, A_y, A_z) \cdot d\mathbf{r}.$$

These Cartesian components still depend on the position, that is,

$$A_x = A_x(x, y, z), \qquad A_y = A_y(x, y, z), \qquad A_z = A_z(x, y, z).$$

The given space curve may also be written in components as

$$\mathbf{r}(t) = (x(t), y(t), z(t)).$$

To perform the integration, we need the components of the vector field **A** along the space curve depending on the parameter t. These are obtained by inserting the corresponding components of the space curve $\mathbf{r}(t)$ into A_x, A_y, and A_z:

$$A_x(t) = A_x(x(t), y(t), z(t));$$
$$A_y(t) = A_y(x(t), y(t), z(t));$$
$$A_z(t) = A_z(x(t), y(t), z(t)).$$

Because $\mathbf{r} = \mathbf{r}(t)$, we may form the total differential and write

$$d\mathbf{r} = \frac{d\mathbf{r}}{dt} dt.$$

Insertion yields the integral:

$$\int_c \mathbf{A} \cdot d\mathbf{r} = \int_c \big(A_x(x, y, z), A_y(x, y, z), A_z(x, y, z)\big) \cdot d\mathbf{r}$$

$$= \int_c \left[\left(A_x(x(t), y(t), z(t)), A_y(x(t), y(t), z(t)), A_z(x(t), y(t), z(t)) \right) \cdot \frac{d\mathbf{r}}{dt} \right] dt.$$

Because of

$$\frac{d\mathbf{r}}{dt} = \left(\frac{dx}{dt}, \frac{dy}{dt}, \frac{dz}{dt} \right),$$

one further has

$$\int_c \mathbf{A} \cdot d\mathbf{r} = \int \left[A_x(t) \frac{dx(t)}{dt} + A_y(t) \frac{dy(t)}{dt} + A_z(t) \frac{dz(t)}{dt} \right] dt.$$

This integral, as a rule, may be evaluated easily. Insertion of the limits after the integration yields the wanted line integral.

Example 12.1: Line integral over a vector field

The vector field $\mathbf{A}$ and the space curve $\mathbf{r} = \mathbf{r}(t)$ are given by

$$\mathbf{A} = (3x^2 - 6yz, 2y + 3xz, 1 - 4xyz^2),$$
$$\mathbf{r}(t) = (t, t^2, t^3).$$

The components of the space curve are

$$x = t \quad \Rightarrow \quad \dot{x} = 1,$$
$$y = t^2 \quad \Rightarrow \quad \dot{y} = 2t,$$
$$z = t^3 \quad \Rightarrow \quad \dot{z} = 3t^2.$$

We now insert

$$\int \mathbf{A} \cdot d\mathbf{r} = \int \left(A_x \frac{dx}{dt} + A_y \frac{dy}{dt} + A_z \frac{dz}{dt} \right) dt$$
$$= \int \left[(3t^2 - 6t^5) \cdot 1 + (2t^2 + 3t^4) \cdot 2t + (1 - 4t^9) \cdot 3t^2 \right] dt.$$

The integral with the limits $t_1 = 0$ and $t_2 = 2$ is then

$$\int_0^2 \mathbf{A} \cdot d\mathbf{r} = -4064.$$

---

13 The Integral Laws of Gauss and Stokes

Gauss Law:[1]

By means of the concept of divergence worked out in the preceding chapter, one may also

[1] *Carl Friedrich Gauss*, b. April 30, 1777, Brunswick—d. Feb. 23, 1855, Göttingen. Gauss was the son of a day laborer and attracted attention very early by his exceptional mathematical talent. The Duke of Brunswick sponsored the costs of his education as of 1791. Gauss studied from 1794–1798 in Göttingen and got his doctorate in 1799 in Helmstedt. As of 1807 Gauss was director of the observatory and professor at the university in Göttingen. He refused all offers to come, for example, to Berlin at the academy. Gauss started his scientific work in 1791 with investigations on the geometric-arithmetic mean, on the distribution of prime numbers, and in 1792 on the *foundations of geometry.* Already in 1794 he found the *least-squares method,* and from 1795 dates the intensive investigation of number theory, e.g., with the quadratic reciprocity law. In 1796 Gauss published his first paper containing the proof that, except for the known cases, regular n-gons may be constructed by means of circle and ruler if n is a Fermat prime number. In particular, this applies to the 17-gon. In his dissertation (1799) Gauss gave the first exact proof of the *fundamental law of algebra,* which was followed by further ones. From the unpublished works it is known that in the same year Gauss already had the foundations of the theory of elliptic and module functions. The first extensive work Gauss published in 1801 is his famous *Disquisitiones arithmeticae,* which are considered as the start of the more recent number theory. There one finds, for example, the theory of quadratic congruences and the first proof of the quadratic reciprocity theorem, the "theorema aureum," as well as the *theory of cyclotomy.*

Around 1801 Gauss became interested in astronomy. The results of these studies were as follows: In 1801 the orbit calculation of the planet Ceres; in 1809 and 1818 the investigations on secular perturbations; and in 1813 on the attraction by the general ellipsoid. In 1812 the treatise on the *hypergeometric series* was published; it contains the first correct and systematic study of convergence.

As of 1820 Gauss increasingly dealt with geodesy. The most important theoretical achievement is 1827's *theory of surfaces* with the "theorema egregium." Gauss also pursued practical geometry, he performed very extensive measurements in 1821–1825. Despite such costly work, in 1825 and 1831 his papers on *biquadratic remainders* appeared. The second of these treatises contains the representation of complex numbers in the plane and a new theory of prime numbers.

In his last years Gauss also became interested in physical problems. The most important results are 1833–1834's invention (together with W. Weber) of the electric telegraph and 1839–1840's *potential theory,* which became a new branch of mathematics.

Many important results of Gauss are only known from the diary and the letters. For example, already in 1816 Gauss had developed the *non-Euclidean geometry.* The reason for the attitude not to publish important results is to be seen in the extraordinarily high standard Gauss set also to the form of his works and in the attempt to avoid needless discussion.

calculate the excess of the outgoing over the incoming vector flow of a vector field **A** for an arbitrarily large volume V. For this purpose we decompose this volume into small volume elements dV, calculate the divergence for each volume element, and sum up over all volume elements, that is, the total flow is given by a volume integral:

$$\phi = \int_V \operatorname{div} \mathbf{A}\, dV.$$

Because the in- or outgoing vector flow of this volume has to pass across the surfaces F, it may also be represented by a surface integral

$$\phi = \int_F \mathbf{A} \cdot \mathbf{n}\, dF.$$

The combination of the surface integral with the integral over the volume yields the Gauss law:

$$\int_V \operatorname{div} \mathbf{A}\, dV = \int_F \mathbf{A} \cdot \mathbf{n}\, dF.$$

This relation clearly states: The sum of the partial flows out of each or into each volume element dV, respectively, equals the flow of the vector field **A** across the surface of this volume.

In the interior of the volume, the flows from one volume element into the next one mutually cancel. Hence, when integrating over the volume elements, there remains only the flow out of or into the total volume.

The proof of the Gauss theorem may be performed somewhat more formally by means of the definition of the divergence

$$\operatorname{div} \mathbf{A} = \lim_{\Delta V \to 0} \frac{\int_{\Delta F} \mathbf{A} \cdot \mathbf{n}\, dF}{\Delta V} = \lim_{\Delta V \to 0} \frac{\int_{\Delta F} \mathbf{A} \cdot d\mathbf{F}}{\Delta V}.$$

There is

$$\begin{aligned}
\int_V \operatorname{div} \mathbf{A}\, dV &= \lim_{\Delta V_i \to 0} \sum_i (\operatorname{div} \mathbf{A})_i \Delta V_i \\
&= \lim_{\Delta V_i \to 0} \sum_i \frac{1}{\Delta V_i} \int_{\Delta F_i} \mathbf{A} \cdot \mathbf{n}\, dF \Delta V_i \\
&= \lim_{\Delta V_i \to 0} \sum_i \int_{\Delta F_i} \mathbf{A} \cdot \mathbf{n}\, dF = \int_F \mathbf{A} \cdot \mathbf{n}\, dF.
\end{aligned}$$

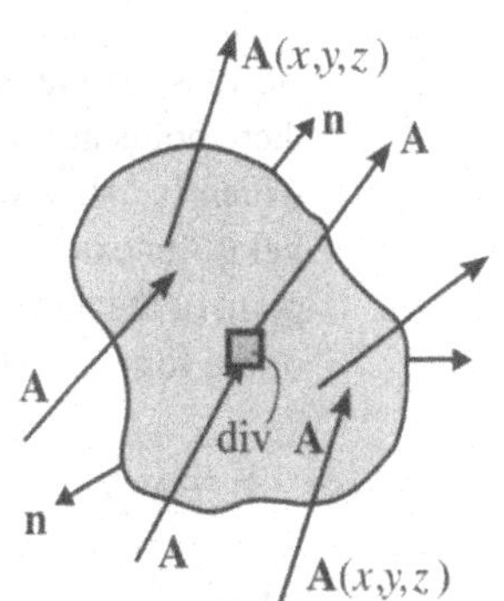

The divergence of the vector field **A** describes the sources and sinks of **A**.

The in-flows and out-flows at neighboring cells cancel each other, except for those on the surface.

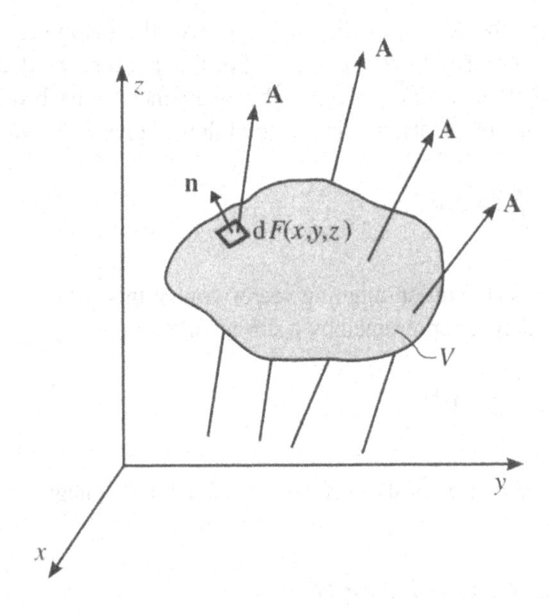

Illustration of the flow of the vector field **A** through the volume V. The flow through the surface equals the sum of the intensities of sources and sinks within the volume. This is the meaning of the Gauss theorem.

The Gauss theorem:

Besides the Gauss law we have just encountered, there holds also a theorem for special vector fields which is called the *Gauss theorem.* Central force fields, for example, the gravitational field of a mass point or the electrostatic field of a point charge, are of the form

$$\mathbf{K} = \kappa \frac{\mathbf{r}}{r^3}, \qquad (13.1)$$

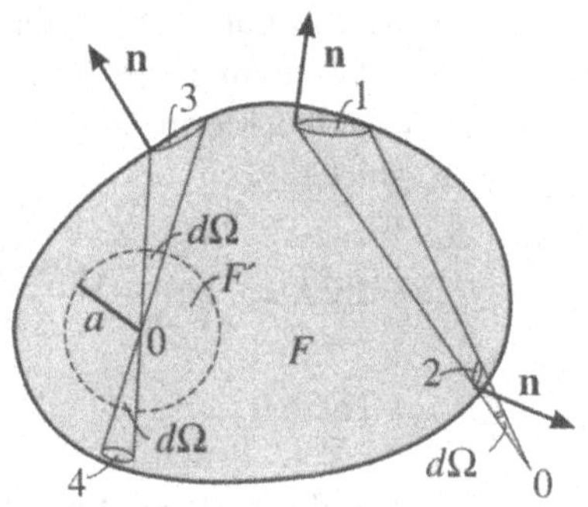

Proofing the Gauss theorem.

where κ is a coupling constant. For these fields the *Gauss theorem* holds:

Let F be a closed area and $\mathbf{r}$ the position vector of an arbitrary point (x, y, z) measured with respect to the origin O (center of force).

For the force flow through the area it holds that

$$\iint_F \mathbf{K}\cdot\mathbf{n}\,dF = \kappa \iint_F \frac{\mathbf{n}\cdot\mathbf{r}}{r^3}\,dF = \begin{cases} 4\pi\kappa & \text{if } O \text{ is inside } F, \\ 0 & \text{if } O \text{ is outside } F. \end{cases} \qquad (13.2)$$

The force flow of such a central force through a closed surface about the center of force O is therefore $4\pi\cdot$ intensity κ of the force field. This may be realized as follows: According to the *Gauss theorem*, it holds that

$$\iint_F \frac{\mathbf{n}\cdot\mathbf{r}}{r^3}\,dF = \iiint_V \nabla\cdot\frac{\mathbf{r}}{r^3}\,dV. \tag{13.3}$$

But now according to problem 11.5, $\operatorname{div}\mathbf{r}/r^3 = 0$ everywhere except for $r = 0$ (i.e., at the origin). Hence, the second case of equation (13.2) has been demonstrated: If O is outside F, then $\operatorname{div}\mathbf{r}/r^3 = 0$ holds everywhere inside the closed surface F.

But if O is within the surface, we form a spherical surface F' of radius a around O. For the closed volume limited by F and F', it then holds that

$$\iint_{F+F'} \frac{\mathbf{n}\cdot\mathbf{r}}{r^3}\,dF = \iint_F \frac{\mathbf{n}\cdot\mathbf{r}}{r^3}\,dF + \iint_{F'} \frac{\mathbf{n}\cdot\mathbf{r}}{r^3}\,dF = \iiint_{V-V'} \nabla\cdot\frac{\mathbf{r}}{r^3}\,dV = 0. \tag{13.4}$$

Here $V - V'$ is the volume enclosed by the surfaces F and F'. Within $V - V'$ again $\operatorname{div}\mathbf{r}/r^3 = 0$ everywhere, as the coordinate origin O lies outside this volume. From (13.4) it now follows that

$$\iint_F \frac{\mathbf{n}\cdot\mathbf{r}}{r^3}\,dF = -\iint_{F'} \frac{\mathbf{n}\cdot\mathbf{r}}{r^3}\,dF. \tag{13.5}$$

On the spherical surface F' holds: $\mathbf{n} = -\mathbf{r}/a$, where $|\mathbf{r}| = a$, such that

$$\frac{\mathbf{n}\cdot\mathbf{r}}{r^3} = -\frac{(\mathbf{r}/a)\cdot\mathbf{r}}{a^3} = -\frac{\mathbf{r}\cdot\mathbf{r}}{a^4} = -\frac{a^2}{a^4} = -\frac{1}{a^2}.$$

Therefore, for equation (13.5), it holds that

$$\iint_F \frac{\mathbf{n}\cdot\mathbf{r}}{r^3}\,dF = -\iint_{F'} \frac{\mathbf{n}\cdot\mathbf{r}}{r^3}\,dF = -\iint_{F'} \left(-\frac{1}{a^2}\right) dF = \frac{4\pi a^2}{a^2} = 4\pi. \tag{13.6}$$

This is the first statement of equation (13.2).

Geometric interpretation of the Gauss theorem:

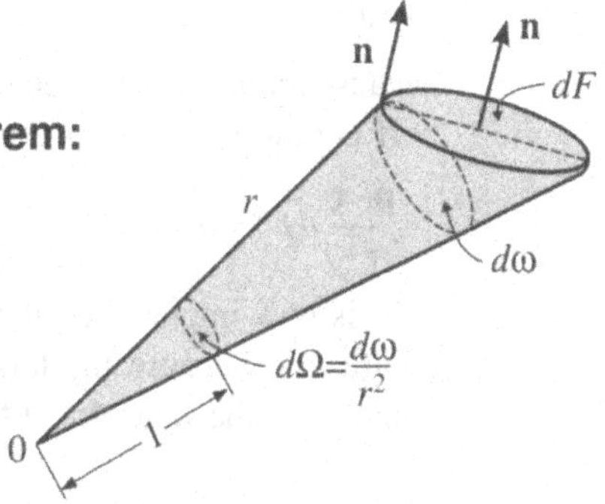

The area of the shadow under central projection of the surface dF on the unit sphere equals the solid angle.

Let dF be a surface element. If the border of this surface element is connected with O (see opposite sketch), there arises a cone.

$d\omega$ denotes that area cut out of a spherical surface with the center O and the radius r by this cone. The *solid angle* $d\Omega$ determined by the area dF and the point O is defined by

$$d\Omega = \frac{d\omega}{r^2} \tag{13.7}$$

and is numerically identical with the surface fraction cut out by the cone from a unit sphere of radius 1 centered about O.

The positive normal vector to the area dF is denoted by $\mathbf{n}$. If Θ is the angle between $\mathbf{n}$ and $\mathbf{r}$, there results the relation

$$\cos\Theta = \frac{\mathbf{n}\cdot\mathbf{r}}{r}. \tag{13.8}$$

From there follows the expression

$$d\omega = \pm dF\cos\Theta = \pm\frac{\mathbf{n}\cdot\mathbf{r}}{r}\,dF; \tag{13.9}$$

thus, one can write for $d\Omega$

$$d\Omega = \pm\frac{\mathbf{n}\cdot\mathbf{r}}{r^3}\,dF. \tag{13.10}$$

Depending on whether the vectors $\mathbf{n}$ and $\mathbf{r}$ enclose an acute or obtuse angle, the positive or negative sign in equations (13.9) and (13.10) is chosen.

Let F be the surface in the figure on page 114 that is characterized by the fact that any straight line may intersect it in at most two points. If O lies outside F, then, according to (13.10), for the area element 1 the following expression results:

$$\frac{\mathbf{n}\cdot\mathbf{r}}{r^3}\,dF = d\Omega. \tag{13.11}$$

Analogously, for the area element 2 it holds that

$$\frac{\mathbf{n}\cdot\mathbf{r}}{r^3}\,dF = -d\Omega. \tag{13.12}$$

Integration over these two regions yields the value zero, as their solid angle contributions mutually compensate. If the integration is now performed over the entire surface F, one immediately sees that the integral

$$\iint_F \frac{\mathbf{n}\cdot\mathbf{r}}{r^3}\,dF = 0, \tag{13.13}$$

because for any positive contribution there exists a corresponding negative contribution.

If O now lies within F, then for any of the area elements 3 and 4

$$\frac{\mathbf{n}\cdot\mathbf{r}}{r^3}\,dF = d\Omega. \tag{13.14}$$

This now implies that the contributions of the two regions to the surface integral are adding up. Because the total solid angle is identical to the surface of the unit sphere, namely has the value 4π, it follows that

$$\iint_F \frac{\mathbf{n}\cdot\mathbf{r}}{r^3}\,dF = 4\pi. \tag{13.15}$$

If the surface F has such a shape that a straight line may intersect it at more than two points (see figure), one may show that the considerations in the context of the figure on page 113 hold also in this case. If now O lies outside F, the cone with the apex at O cuts the surface F in an even number of positions. The contributions of these area elements to the surface integral compensate each other pairwise such that the surface integral over the area F equals zero. If, however, O lies within F, the cone cuts the surface in an odd number of positions.

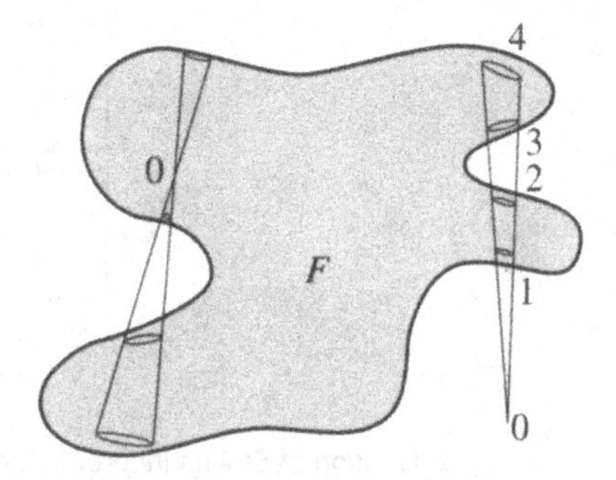

The Gauss theorem: The center of force lies within (left part), respectivly, out of (right part) the surface F.

Because the respective contributions to the surface integral cancel each other pairwise, the surface integration performed over the area F as shown in the figures on this page and on page 113 again yields the value 4π.

Stokes law:[3]

Given a vector field $\mathbf{A}$, we calculate the contour integral along a closed loop:

$$W = \oint_C \mathbf{A} \cdot d\mathbf{r}.$$

If we now interpret the closed loop s as the border of an arbitrary area, W may be thought as originating by summing up arbitrarily small partial contributions dW: These cancel out when integrating over the area elements, except for the path elements along the external free border representing the course of the borderline:

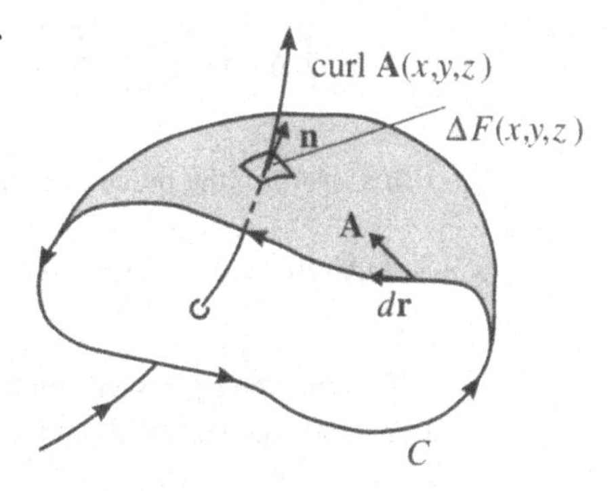

The Stokes theorem: The surface F is arbitrarily extended over the curve C, making C the boundary of F.

[3] *Sir George Gabriel Stokes*, b. Aug. 13, 1819, Skreen (Ireland)—d. Feb. 1, 1903, Cambridge. Since 1849 Stokes was professor of mathematics in Cambridge. Besides his contributions to analysis, such as the Stokes integral formula, he made important contributions to physics, for example, on fluorescence, and on the motion of viscous liquids. He also worked on geodesy.

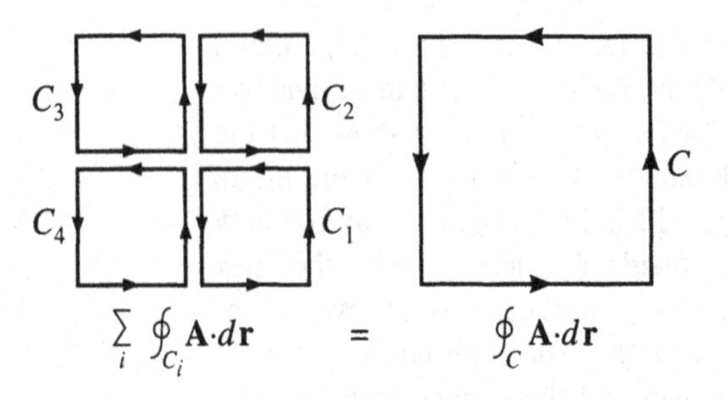

Explanation of Stokes theorem: The sum of the contour integrals over the curves C_i yields the contour integral over the border curve C.

The infinitesimal contributions dW may be represented by the flow of curl $\mathbf{A}$ through the area elements dF, as follows:

$$\Delta W_i = (\mathbf{n} \cdot \operatorname{curl} \mathbf{A})_i \Delta F_i = \oint_{C_{F_i}} \frac{\mathbf{A} \cdot d\mathbf{r}}{\Delta F_i} \cdot \Delta F_i = \oint_{C_{F_i}} \mathbf{A} \cdot d\mathbf{r}, \tag{13.16}$$

where $\mathbf{n}$ is the vector pointing perpendicular to the area element dF. We integrate (see also figure on this page) and obtain

$$W = \oint_C \mathbf{A} \cdot d\mathbf{r} = \sum_i \oint_{C_i} \mathbf{A} \cdot d\mathbf{r} = \sum_i (\mathbf{n} \cdot \operatorname{curl} \mathbf{A})_i \Delta F_i = \int_F \mathbf{n} \cdot \operatorname{curl} \mathbf{A}\, dF\,. \tag{13.17}$$

By inserting the preceding line into the contour integral, one obtains the Stokes law:

$$\oint_C \mathbf{A} \cdot d\mathbf{r} = \int_F \mathbf{n} \cdot \operatorname{curl} \mathbf{A}\, dF. \tag{13.18}$$

This may be expressed somewhat less precisely as follows: The sum of vortices over an area yields the vortex about the border of the area.

Problem 13.1: Path independence of a line integral

Show by means of the Stokes theorem that, assuming $\mathbf{A} = \nabla\phi$, the line integral from point P_1 to point P_2 is independent of the path.

Solution We first form the rotation of the vector field $\mathbf{A}$; because $\mathbf{A} = \nabla\phi$, we obtain

$$\operatorname{curl} \mathbf{A} = \operatorname{curl} \nabla\phi = \nabla \times \nabla\phi = 0.$$

By inserting this in the Stokes theorem, we obtain

$$\int_F \operatorname{curl} \mathbf{A} \cdot \mathbf{n}\, dF = \oint \mathbf{A} \cdot d\mathbf{r} = 0.$$

The above relation is fulfilled for *arbitrary but closed curves.* One has (compare figure):

$$\oint_{C_1+C_2} \mathbf{A}\cdot d\mathbf{r} = \int\limits_{P_{1_{C_1}}}^{P_2} \mathbf{A}\cdot d\mathbf{r} + \int\limits_{P_{2_{C_2}}}^{P_1} \mathbf{A}\cdot d\mathbf{r} = \int\limits_{P_{1_{C_1}}}^{P_2} \mathbf{A}\cdot d\mathbf{r} - \int\limits_{P_{1_{C_2}}}^{P_2} \mathbf{A}\cdot d\mathbf{r} = 0.$$

The line integral is path-independent since the path from P_1 to P_2 must not coincide with the path from P_2 to P_1, and nevertheless the relation

$$\int\limits_{P_{1_{C_1}}}^{P_2} \mathbf{A}\cdot d\mathbf{r} - \int\limits_{P_{1_{C_2}}}^{P_2} \mathbf{A}\cdot d\mathbf{r} = 0$$

is fulfilled. This may also be proved in an alternative way:

$$\int\limits_1^2 \nabla\phi\cdot d\mathbf{r} = \int\limits_1^2 d\phi = \phi(2) - \phi(1).$$

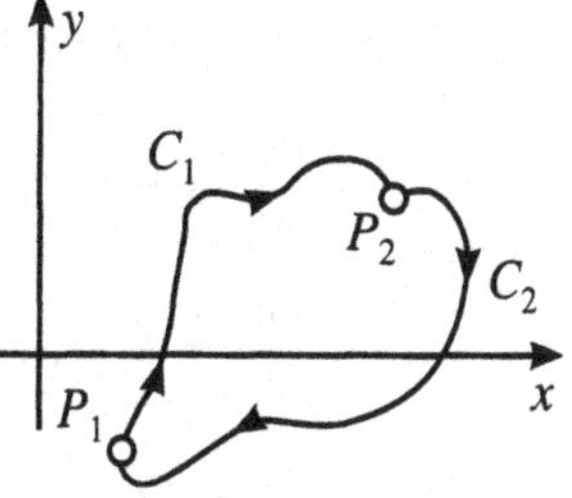

The line integral from P_1 to P_2 does not depend on the path for conservative vector fields (curl $\mathbf{A} = 0$). The line integral vanishes for closed paths.

Thus, the integral depends only on the function values ϕ at positions 1 and 2, but not on the special path of integration. This discovery is highly important because it allows us to understand for which force fields a potential exists.

Additional remark: If $\mathbf{A} = \operatorname{grad}\phi \;\Rightarrow\; \operatorname{curl}\mathbf{A} = 0$, because $\operatorname{curl}\operatorname{grad}\phi = 0$, and therefore, according to Stokes, $\oint \mathbf{A}\cdot d\mathbf{r} = 0$. Inversely, if for arbitrary closed paths $\oint \mathbf{A}\cdot d\mathbf{r} = 0$, then it follows from the definition of rotation $\mathbf{n}\cdot\operatorname{curl}\mathbf{A} = \lim_{\Delta F\to 0}(\oint \mathbf{A}\cdot d\mathbf{r}/\Delta F)$ that $\operatorname{curl}\mathbf{A} = 0$. From there in turn it follows that $\mathbf{A} = \nabla\phi$, where $\phi = \int_{\mathbf{r}_1}^{\mathbf{r}} \mathbf{A}(\mathbf{r}')d\mathbf{r}'$. The arising integral may then be taken along an arbitrary path from $\mathbf{r}_1$ to $\mathbf{r}$ (because of the path independence of the integral). This important statement shall now be proved:

Given

$$\mathbf{A}(\mathbf{r}) \quad \text{with} \quad \operatorname{curl}\mathbf{A}(\mathbf{r}) = 0,$$

if the line integral

$$\phi(\mathbf{r}) = \int\limits_{\mathbf{r}_1}^{\mathbf{r}} \mathbf{A}(\mathbf{r}')\cdot d\mathbf{r}' \tag{13.19}$$

is independent of the specially selected path, one has

$$\mathbf{A}(\mathbf{r}) = \nabla\phi(\mathbf{r})\,. \tag{13.20}$$

Proof: Because the integration contour may be chosen arbitrarily, we adopt especially

$$(x_1, y_1, z_1) \xrightarrow{x} (x, y_1, z_1) \xrightarrow{y} (x, y, z_1) \xrightarrow{z} (x, y, z) \quad \text{(see figure)},$$

$$\phi(x, y, z) = \int\limits_{\mathbf{r}_1}^{\mathbf{r}} \mathbf{A}(\mathbf{r})\cdot d\mathbf{r} = \int\limits_{\mathbf{r}_1}^{\mathbf{r}} [A_1(\mathbf{r})\,dx + A_2(\mathbf{r})\,dy + A_3(\mathbf{r})\,dz] \tag{13.21}$$

$$= \int\limits_{x_1}^{x} A_1\left(x', y_1, z_1\right) dx' + \int\limits_{y_1}^{y} A_2\left(x, y', z_1\right) dy' + \int\limits_{z_1}^{z} A_3\left(x, y, z'\right) dz'.$$

For the function ϕ constructed in this way it holds that

$$\frac{\partial \phi}{\partial z} = A_3(x, y, z),$$

$$\frac{\partial \phi}{\partial y} = A_2(x, y, z_1) + \int_{z_1}^{z} \frac{\partial A_3(x, y, z')}{\partial y} dz'$$

$$= A_2(x, y, z_1) + \int_{z_1}^{z} \frac{\partial A_2(x, y, z')}{\partial z'} dz'.$$

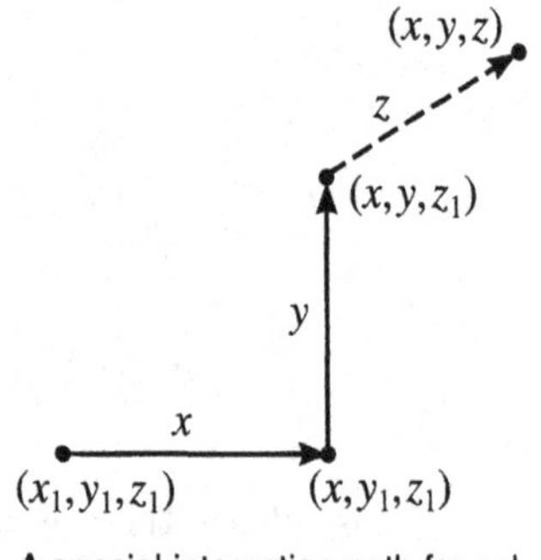

A special integration path for calculating $\phi(x, y, z)$.

Here we used $\operatorname{curl} \mathbf{A} = 0$, which means for the x-component $\partial A_3/\partial y = \partial A_2/\partial z$.

In the following we also employ the vanishing of the other components of $\operatorname{curl} \mathbf{A} = 0$:

$$\frac{\partial \phi}{\partial y} = A_2(x, y, z_1) + A_2(x, y, z')\Big|_{z_1}^{z} = A_2(x, y, z)$$

$$\frac{\partial \phi}{\partial x} = A_1(x, y_1, z_1) + \int_{y_1}^{y} \frac{\partial A_2(x, y', z_1)}{\partial x} dy' + \int_{z_1}^{z} \frac{\partial A_3(x, y, z')}{\partial x} dz'$$

$$= A_1(x, y_1, z_1) + \int_{y_1}^{y} \frac{\partial A_1(x, y', z_1)}{\partial y'} dy' + \int_{z_1}^{z} \frac{\partial A_1(x, y, z')}{\partial z'} dz'$$

(because $\operatorname{curl} \mathbf{A} = 0$)

$$= A_1(x, y_1, z_1) + A_1(x, y, z_1) - A_1(x, y_1, z_1) + A_1(x, y, z) - A_1(x, y, z_1).$$

The terms cancel each other out pairwise up to one summand, such that it finally remains that

$$\frac{\partial \phi}{\partial x} = A_1(x, y, z)\,.$$

In total we thus have demonstrated that the function $\phi(\mathbf{r})$ defined by the line integral $\phi(\mathbf{r}) = \int_{\mathbf{r}_1}^{\mathbf{r}} \mathbf{A} \cdot d\mathbf{r}$ satisfies the equation $\mathbf{A} = \nabla\phi$. Hence, for a given vector field $\mathbf{A}$ satisfying $\operatorname{curl} \mathbf{A} = 0$, one may always calculate the potential function $\phi(\mathbf{r})$ by a line integral. The function $-\phi(\mathbf{r})$ will later be called the *potential of the force field* $\mathbf{A}(\mathbf{r})$ (compare Problem 13.4 and Chapter 17).

This very detailed proof that, with definition 13.21 we have $\operatorname{grad} \phi(\mathbf{r}) = \mathbf{A}(\mathbf{r})$, can be given in a more succinct and elegant way: Because

$$\phi(\mathbf{r} + d\mathbf{r}) = \phi(\mathbf{r}) + \operatorname{grad} \phi \cdot d\mathbf{r},$$

we find

$$\operatorname{grad} \phi \cdot d\mathbf{r} = \phi(\mathbf{r} + d\mathbf{r}) - \phi(\mathbf{r}) = \int_{\mathbf{r}}^{\mathbf{r}+d\mathbf{r}} \mathbf{A}(\mathbf{r}') \cdot d\mathbf{r}' = \mathbf{A}(\mathbf{r}) \cdot d\mathbf{r}\,.$$

Because this holds for any arbitrary $d\mathbf{r}$, it follows that

$$\operatorname{grad}\phi(\mathbf{r}) = \mathbf{A}(\mathbf{r})\,.$$

Problem 13.2: Determination of the potential function

Show for the vector field

$$\mathbf{A} = (2xy + z^3, x^2 + 2y, 3xz^2 - 2)$$

that $\int \mathbf{A}\cdot d\mathbf{r}$ for a path from $(1,-1,1)$ to $(2,1,2)$ is independent of the path. Calculate the value of the integral. Find the potential function $\phi(x,y,z)$.

Solution One has $\operatorname{curl}\mathbf{A} = 0$. We check this, for example, for the x-component: $(\operatorname{curl}\mathbf{A})_x = \partial A_z/\partial y - \partial A_y/\partial z = 0 - 0 = 0$. The other components of curl **A** will be calculated similarly.

The integral $\phi(\mathbf{r}) = \int_{\mathbf{r}_1}^{\mathbf{r}} \mathbf{A}\cdot d\mathbf{r}$ is therefore path-independent, and $\mathbf{A} = \operatorname{grad}\phi = \nabla\phi$. According to (13.21), with the arbitrarily and hence effectively choosable $\mathbf{r}_1 = \{0,0,0\}$ (we sometimes adopt braces for notation of vectors), we then obtain

$$\begin{aligned}
\phi(x,y,z) &= \int_0^x A_1(x',y_1,z_1)\,dx' + \int_0^y A_2(x,y',z_1)\,dy' + \int_0^z A_3(x,y,z')\,dz' \\
&= 0 + \int_0^y (x^2+2y')\,dy' + \int_0^z (3xz'^2 - 2)\,dz' \\
&= x^2y' + y'^2\Big|_0^y + xz'^3 - 2z'\Big|_0^z \\
&= x^2y + y^2 + xz^3 - 2z.
\end{aligned}$$

Indeed we easily check

$$\nabla\phi = (2xy + z^3)\mathbf{e}_1 + (x^2+2y)\mathbf{e}_2 + (3z^2x - 2)\mathbf{e}_3.$$

Because $\mathbf{A} = \nabla\phi$, the line integral is path-independent. The value of the integral is determined as follows:

$$\begin{aligned}
\int_{(1,-1,1)}^{(2,1,2)} \mathbf{A}\cdot d\mathbf{r} &= \int_{(1,-1,1)}^{(2,1,2)} \nabla\phi\cdot d\mathbf{r} = \int_{(1,-1,1)}^{(2,1,2)} d\phi, \\
&= \phi(2,1,2) - \phi(1,-1,1), \\
&= (4+1+16-4) - (-1+1+1-2) = 18.
\end{aligned}$$

Of course, the line integral might be determined also in another way, by integrating, for instance, along an arbitrary contour (an arbitrary curve $\mathbf{r}(t)$ between the points $\{1,-1,1\}$ and $\{2,1,2\}$).

Problem 13.3: Vortex flow of a force field through a half-sphere

Let $\mathbf{A} = zx\mathbf{e}_x - (xy - 3z)\mathbf{e}_y + (4yz - x)\mathbf{e}_z$ be a given force field. Calculate the flow of curl **A** through the half-sphere above the x, y-plane. (Use spherical coordinates for the integration.)

There are

$$\begin{aligned}
\mathbf{A} &= xz\mathbf{e}_x - (xy - 3z)\mathbf{e}_y + (4yz - x)\mathbf{e}_z\,, \\
\operatorname{curl}\mathbf{A} &= (4z - 3)\mathbf{e}_x + (x+1)\mathbf{e}_y - y\mathbf{e}_z\,.
\end{aligned}$$

1. Solution: The upper half-sphere is parametrized by

$$\mathbf{r} = a \begin{pmatrix} \cos\varphi \sin\vartheta \\ \sin\varphi \sin\vartheta \\ \cos\vartheta \end{pmatrix} \quad \text{with} \quad 0 \le \varphi < 2\pi, \quad 0 \le \vartheta \le \frac{\pi}{2}.$$

The nonnormalized normal vector is

$$\mathbf{n} = \frac{\partial \mathbf{r}}{\partial \vartheta} \times \frac{\partial \mathbf{r}}{\partial \varphi} = \cdots = a \sin\vartheta\, \mathbf{r}.$$

The area element is given by $d\mathbf{F} = (\partial \mathbf{r}/\partial\vartheta) d\vartheta \times (\partial \mathbf{r}/\partial\varphi) d\varphi = \mathbf{n}\, d\vartheta\, d\varphi$. In the new coordinates curl $\mathbf{A}$ reads

$$\operatorname{curl}\mathbf{A} = \begin{pmatrix} 4a\cos\vartheta - 3 \\ a\cos\varphi\sin\vartheta + 1 \\ -a\sin\varphi\sin\vartheta \end{pmatrix}.$$

Thereby the integral becomes

$$\begin{aligned} I &= \iint \operatorname{curl}\mathbf{A}\cdot d\mathbf{F} \\ &= \int_0^{\pi/2} d\vartheta \int_0^{2\pi} d\varphi\, \mathbf{n}\cdot\operatorname{curl}\mathbf{A} \\ &= \int_0^{\pi/2} d\vartheta \int_0^{2\pi} d\varphi\, a^2 \sin\vartheta \big(4a\cos\varphi\sin\vartheta\cos\vartheta - 3\cos\varphi\sin\vartheta \\ &\qquad + a\sin\varphi\cos\varphi\sin^2\vartheta + \sin\varphi\sin\vartheta - a\sin\varphi\sin\vartheta\cos\vartheta\big) \\ &= 0, \end{aligned}$$

because

$$\int_0^{2\pi} \sin\varphi\, d\varphi = \int_0^{2\pi} \cos\varphi\, d\varphi = \int_0^{2\pi} \sin\varphi\cos\varphi\, d\varphi = 0.$$

2. Solution: According to the Stokes theorem,

$$I = \iint \operatorname{curl}\mathbf{A}\cdot d\mathbf{F} = \int_C \mathbf{A}\cdot d\mathbf{r},$$

where C is the border of the half-sphere,

$$\mathbf{r} = a \begin{pmatrix} \cos t \\ \sin t \\ 0 \end{pmatrix} \quad \text{with} \quad 0 \le t < 2\pi,$$

$$d\mathbf{r} = a \begin{pmatrix} -\sin t \\ \cos t \\ 0 \end{pmatrix} dt,$$

$$I = -a^3 \int_0^{2\pi} dt\, \sin t \cos^2 t = 0.$$

Problem 13.4: On the conservative force field

What is a conservative force field? Is the force field $\mathbf{F} = (3xz - y)\mathbf{e}_x - x\mathbf{e}_y + (3/2)x^2\mathbf{e}_z$ conservative? If yes, determine the potential V and the work A to be performed to move a particle from point (1, 1,1) to (2, 2, 2).

Solution One has $\mathbf{F} = (3xz - y)\mathbf{e}_x - x\mathbf{e}_y + (3/2)x^2\mathbf{e}_z$.

A force field $\mathbf{F}$ is conservative if it can be represented by $\mathbf{F} = -\nabla V$. Then it holds that $\operatorname{curl}\mathbf{F} = 0$, because $\operatorname{curl}(\nabla V) \equiv 0$.

One easily checks that $\nabla \times \mathbf{F} = \operatorname{curl}\mathbf{F} = 0$:

$$\operatorname{curl}\mathbf{F} = \begin{vmatrix} \mathbf{e}_x & \mathbf{e}_y & \mathbf{e}_z \\ \partial/\partial_x & \partial/\partial_y & \partial/\partial_z \\ 3xz - y & -x & \frac{3}{2}x^2 \end{vmatrix} = 0\,.$$

Thus, it holds that

$$\begin{aligned} \mathbf{F} = -\nabla V &= -\frac{\partial V}{\partial x}\mathbf{e}_x - \frac{\partial V}{\partial y}\mathbf{e}_y - \frac{\partial V}{\partial z}\mathbf{e}_z \\ &= (3xz - y)\mathbf{e}_x - x\mathbf{e}_y + \frac{3}{2}x^2\mathbf{e}_z. \end{aligned}$$

Comparison of coefficients yields:

$$(1)\quad \frac{\partial V}{\partial x} = -3xz + y; \qquad (2)\quad \frac{\partial V}{\partial y} = x; \qquad (3)\quad \frac{\partial V}{\partial z} = -\frac{3}{2}x^2.$$

By integration follows:

$$\begin{aligned} &(1)\quad V = -\frac{3}{2}x^2z + xy + f_1(y, z); \\ &(2)\quad V = \qquad\quad\; xy + f_2(x, z); \\ &(3)\quad V = -\frac{3}{2}x^2z \qquad\; + f_3(x, y). \end{aligned}$$

These equations coincide if one chooses

$$\begin{aligned} f_1(x, y) &= c, \\ f_2(x, z) &= -\frac{3}{2}x^2z + c, \\ f_3(x, y) &= xy + c\,. \end{aligned}$$

From that it follows that

$$V = -\frac{3}{2}x^2z + xy + c\,.$$

Because **F** is conservative, for the work A we have

$$A = \int \mathbf{F} \cdot d\mathbf{r} = -\int \nabla V d\mathbf{r} = -\int dV$$

$$\Rightarrow \quad A = -V(2,2,2) + V(1,1,1)$$

$$= -\left[-\frac{3}{2} x^2 z + xy + c \right]_{(1,1,1)}^{(2,2,2)} = 7\frac{1}{2} \, .$$

14 Calculation of Surface Integrals

Given an area F and a vector field $\mathbf{A}$, we look for the flow of the field through the area. For this purpose we subdivide the area into surface elements ΔF_i and calculate the product $\mathbf{A} \cdot \mathbf{n} \cdot \Delta F_i$, which represents the flow of the field $\mathbf{A}$ through the area element ΔF_i. Here $\mathbf{n}$ is the normal vector of magnitude 1 pointing perpendicularly to the area element ΔF_i. We now sum up these products over all i and, by changing to infinitesimal area elements, obtain the surface integral:

$$\int_F \mathbf{A} \cdot \mathbf{n} \, dF,$$

which represents the wanted flow. To calculate this integral, we convert ΔF_i to Cartesian coordinates. The area elements dF are always positive. We therefore set absolute bars:

$$|\mathbf{n} \cdot \mathbf{e}_3| dF = dx \, dy$$

or

$$dF = \frac{dx \, dy}{|\mathbf{n} \cdot \mathbf{e}_3|}.$$

We insert this expression into the surface integral and obtain

$$\int_F \mathbf{A} \cdot \mathbf{n} \, dF = \int_F \frac{\mathbf{A} \cdot \mathbf{n}}{|\mathbf{n} \cdot \mathbf{e}_3|} \, dx \, dy.$$

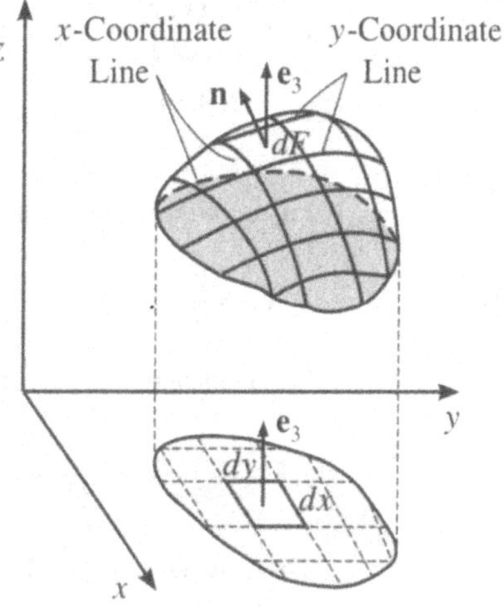

Example of a surface and its shadow area when calculating surface integrals.

The surface integral is thereby traced back to a double integral over the shadow area in the x-y-plane.

One now has to distinguish between two cases:

1. If $\mathbf{n}$ is parallel to z, then $dF = dx\,dy$, because $\mathbf{n} \cdot \mathbf{e}_3 = 1$, that is, the projection exactly corresponds to the prototype.
2. If $\mathbf{n}$ is inclined against $\mathbf{e}_3$, the projection is smaller than the prototype, that is, $dF > dx\,dy$. In this case $\mathbf{n} \cdot \mathbf{e}_3 < 1$, and the relation $|\mathbf{n} \cdot \mathbf{e}_3| \cdot dF = dx\,dy$ is fulfilled.

If the projection of the primordial area onto the x, y-plane (or any other plane) is not unique, such as for areas "hanging over," uniqueness may always be achieved after some appropriate subdivisions. In such cases the area integral turns into a sum of area integrals over partial areas.

Example 14.1: On the calculation of a surface integral

Given the surface $V \equiv 2x + 3y + 6z = 12$ (described by the position vector $\mathbf{r}(x, y) = \{x, y, (12 - 2x - 3y)/6\}$) and a vector field $\mathbf{A} = \{18z, -12, +3y\}$, find the flow of the field through the part of this area that is cut out by the three coordinate axes in the first octant.

For the calculation, the surface integral is traced back to an integral in the x, y-plane. The integral then takes the form

$$\int_F \frac{\mathbf{A} \cdot \mathbf{n}}{|\mathbf{n} \cdot \mathbf{e}_3|}\, dx\, dy.$$

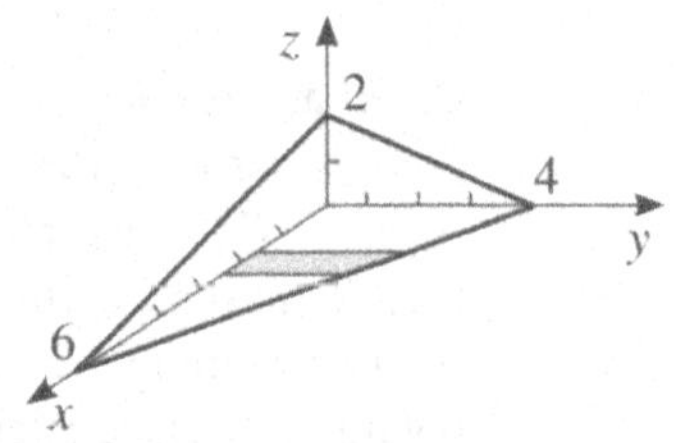

The y-integral runs from $y = 0$ to the intersection of the surface with the x, y-plane, i.e., to $y = 4 - (2/3)x$.

We evaluate the individual quantities separately:

$$\mathbf{n} = \frac{\nabla V(x, y, z)}{|\nabla V|} \qquad \text{(compare equation (11.6))}$$

$$= \frac{\{2, 3, 6\}}{7} = \overrightarrow{\text{constant}}.$$

For $\mathbf{n} \cdot \mathbf{e}_3$, it therefore results that

$$\mathbf{n} \cdot \mathbf{e}_3 = \frac{6}{7}.$$

To calculate $\mathbf{n}$, one may start also from the position vector $\mathbf{r}(x, y)$:

$$\mathbf{n} = \frac{\mathbf{r}_x \times \mathbf{r}_y}{|\mathbf{r}_x \times \mathbf{r}_y|} = \frac{(1, 0, -2/6) \times (0, 1, -3/6)}{|(1, 0, -3/6) \times (0, 1, -3/6)|} = \frac{(2/6, 1/2, 1)}{\sqrt{49/36}} = \frac{(2, 3, 6)}{7}.$$

For $\mathbf{A} \cdot \mathbf{n}$, it results that $\mathbf{A} \cdot \mathbf{n} = (36/7)z - (36/7) + (18/7)y$. From there one obtains for the surface integral

$$\int \mathbf{A} \cdot \mathbf{n}\, dF = \iint \left(\frac{36}{7}z - \frac{36}{7} + \frac{18}{7}y\right) \frac{7}{6}\, dx\, dy.$$

We replace z by $(12 - 2x - 3y)/6$ and multiply; then we find for the integral the following expression:

$$\int \mathbf{A} \cdot \mathbf{n}\, dF = \iint (12 - 2x - 3y + 3y - 6)\, dx\, dy$$
$$= \iint (6 - 2x)\, dx\, dy.$$

To get the limits of the integral, we consider the straight line along which the area V intersects the x, y-plane ($z = 0$):

$$2x + 3y = 12; \qquad y = 4 - \frac{2}{3}x\,.$$

From there it follows that the y-integration runs between the limits

$$y = 0 \quad \text{and} \quad y = 4 - \frac{2}{3}x\,.$$

The x-integration (integration) of all strips parallel to the y-axis (see figure) is performed between the limits $x = 0$ and $x = 6$.

Insertion of the calculated limits yields

$$\int \mathbf{A} \cdot \mathbf{n}\, dF = \int\limits_{x=0}^{6} \int\limits_{y=0}^{4-\frac{2}{3}x} (6 - 2x)\, dx\, dy$$
$$= \int\limits_{x=0}^{6} (6 - 2x) \left(\int\limits_{y=0}^{4-\frac{2}{3}x} dy \right) dx$$
$$= \int\limits_{x=0}^{6} (6 - 2x) \left(4 - \frac{2}{3}x \right) dx$$
$$= \int\limits_{x=0}^{6} \left(24 - 12x + \frac{4}{3}x^2 \right) dx$$
$$= 24\,.$$

Problem 14.2: Flow through a surface

Given the area $F \equiv x^2 + y^2 = 16$ and the vector field $\mathbf{A} = (z, x, -3y^2 z)$ between $z = 0$ and $z = 5$, find the flow of the field through the part of the area covering the first octant.

Solution Analogously to the first example, we evaluate $\mathbf{n} \cdot \mathbf{e}_2$ and $\mathbf{A} \cdot \mathbf{n}$; for this end we first determine the normal vector $\mathbf{n}$:

$$\mathbf{n} = \frac{\nabla F}{|\nabla F|} = \frac{(x, y, 0)}{4}\,.$$

For $\mathbf{A} \cdot \mathbf{n}$, we get

$$\mathbf{A}\cdot\mathbf{n} = \frac{zx}{4} + \frac{xy}{4}.$$

We obtain

$$\mathbf{n}\cdot\mathbf{e}_2 = \frac{y}{4}.$$

By inserting this into the surface integral, we get

$$\int \mathbf{A}\cdot\mathbf{n}\,dF = \frac{1}{4}\iint \frac{4(zx+xy)}{y}\,dx\,dz.$$

We replace $y = \sqrt{16-x^2}$ and integrate in the limits from $x = 0$ to $x = 4$ or from $z = 0$ to $z = 5$ (the shadow area in the x-z-plane):

$$\int \mathbf{A}\cdot\mathbf{n}\,dF = \int_{x=0}^{4}\int_{z=0}^{5}\left(\frac{zx}{\sqrt{16-x^2}} + x\right)dx\,dz\,.$$

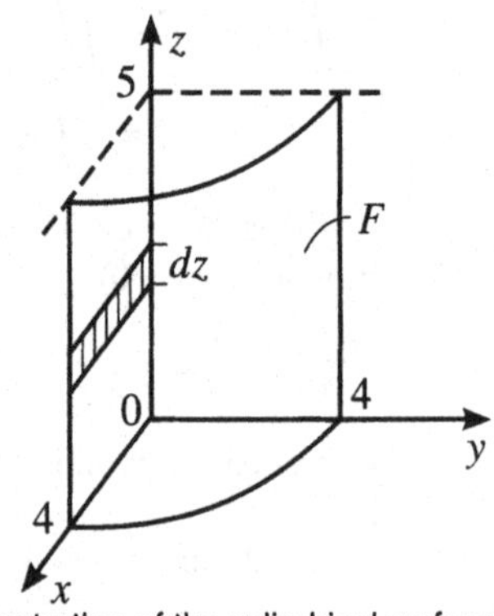

Illustration of the cylindrical surface F.

Integration over z yields

$$\begin{aligned}\int \mathbf{A}\cdot\mathbf{n}\,dF &= \int_{x=0}^{4}\left(\frac{z^2}{2}\frac{x}{\sqrt{16-x^2}} + zx\right)\Bigg|_0^5 dx\\ &= \int_{x=0}^{4}\left(\frac{1}{2}\frac{25x}{\sqrt{16-x^2}} + 5x\right)dx\\ &= -\frac{25}{2}\sqrt{16-x^2}\Bigg|_0^4 + \frac{5x^2}{2}\Bigg|_0^4 = 90.\end{aligned}$$

The Möbius strip: The areas in the examples treated so far were orientable, that is, for arbitrary travels over the area the normal vector of the area always remains on one side of the area. But there exist nonorientable areas; one example is the Möbius strip.[1]

In the case of the Möbius strip, there is no outer and inner side, that is, the Möbius strip has only one side. The vector flow through the Möbius strip vanishes; on the contrary, the vector flow through the represented orientable area in general does not vanish.

[1] *August Ferdinand Möbius*, b. Nov. 17, 1790, Schulpforta as son of a dance teacher—d. Sept. 26, 1868, Leipzig. Möbius attended the school in Schulpforta and then the university in Leipzig. A donation allowed him to go on a study trip, leading him among others to Gauss. In 1810 Möbius was appointed director of the observatory in Leipzig and later also served as a professor at the university. Möbius supported the development of geometry by his contributions to the extension of the traditional concept of coordinates, and to the (unconsciously) group-theoretical classification of geometry.

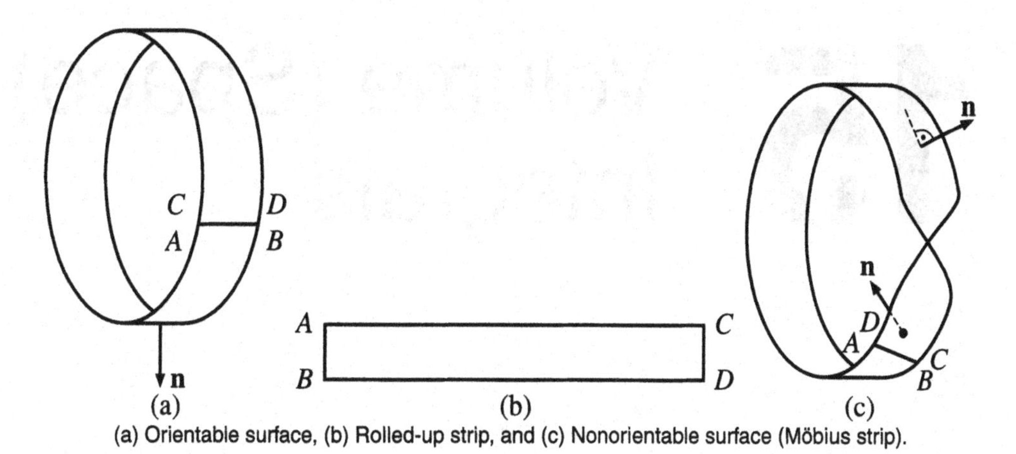

(a) Orientable surface, (b) Rolled-up strip, and (c) Nonorientable surface (Möbius strip).

15 Volume (Space) Integrals

Let $\varrho(x, y, z) = \varrho(\mathbf{r})$ be a scalar function of the position, for example, the mass density; the volume integral

$$\int_V \varrho \, dV \equiv \iiint_V \varrho(x, y, z) \, dx \, dy \, dz = \lim \sum_k \varrho(\mathbf{r}_k) \Delta V_k \tag{15.1}$$

then gives the total mass. ΔV_k thereby means small volume cells that in the limit turn over in $dxdydz$. Volume integrals may be performed also with a vector field $\mathbf{F}(\mathbf{r})$ (speaking more exactly: $\mathbf{F}(\mathbf{r})$ is a vector density, and $\mathbf{F}(\mathbf{r})dV$ is a vector):

$$\int_V \mathbf{F}(\mathbf{r}) \, dV = \int_V \mathbf{F}(x, y, z) \, dx \, dy \, dz = \lim \sum_k \mathbf{F}(\mathbf{r}_k) \Delta V_k \,. \tag{15.2}$$

This corresponds to the sum over all vectors of a vector field $\mathbf{F}$ in a volume V, for example, the sum over all forces acting on a rigid body. $\mathbf{F}(\mathbf{r})$ is then a force density, and $\mathbf{F}(\mathbf{r}) \, dV$ is the force acting on the volume dV. The mathematical evaluation of a volume integral is performed according to the following scheme: One constructs a grid consisting of planes parallel to the x, y-, y, z-, and x, z-planes; thus the volume V is subdivided into partial volumes (cuboids). In this case the triple integral over V may be written as an iterative integral of the form

$$\begin{aligned}
&\int_{x=a}^{b} \int_{y=g_1(x)}^{g_2(x)} \int_{z=f_1(x,y)}^{f_2(x,y)} F(x, y, z) \, dx \, dy \, dz \\
&\quad = \int_{x=a}^{b} \left\{ \int_{y=g_1(x)}^{g_2(x)} \left[\int_{z=f_1(x,y)}^{f_2(x,y)} F(x, y, z) \, dz \right] dy \right\} dx.
\end{aligned}$$

For the given subdivision the innermost integration has to be performed first. This innermost integration over z corresponds to integrating up columns of cross section $dx \, dy$ along the z-axis. The lower limit of the columns is given by the area $z = f_1(x, y)$, the upper limit by $z = f_2(x, y)$. The y-integration then corresponds to summing up these columns in strips parallel to the y-axis. The strips are limited by the function $g_1(x)$ and $g_2(x)$, respectively. The disks arising this way are integrated along the x-axis by means of the x-integration.

In general, one has to subdivide the volume into larger regions, such that the total triple integral may be calculated as sum over partial integrals. We still note that the integration may, of course, be performed in an arbitrary sequence. This will now be explained by the following examples.

Example 15.1: Calculation of a volume integral

Let $\varrho(\mathbf{r}) = 45x^2y$, and let the volume V be limited by the four planes $4x + 2y + z = 8$, $x = 0$, $y = 0$, $z = 0$. Calculate $\int \varrho(\mathbf{r})dV$ (see figure).

If ϱ means a mass density, then the integral represents the total mass of the volume V. One has

$$\int\limits_V \varrho(\mathbf{r})\,dV = \int\limits_{x=0}^{2}\int\limits_{y=0}^{4-2x}\int\limits_{z=0}^{8-4x-2y} (45x^2y)\,dz\,dy\,dx.$$

Here the integration is performed at first over z, then over y, and finally over x. The integration limits are determined as follows (see figure): z runs for fixed x and y from $z = 0$ up to the plane $z = 8 - 4x - 2y$. y runs from 0 up to the straight line $y = 4 - 2x$ in the x, y-plane (cut of the plane $4x + 2y + z = 8$ with the x, y-plane), and x runs from zero to 2 (intersection point of the straight line $y = 4 - 2x$ with the x-axis). The calculation now yields

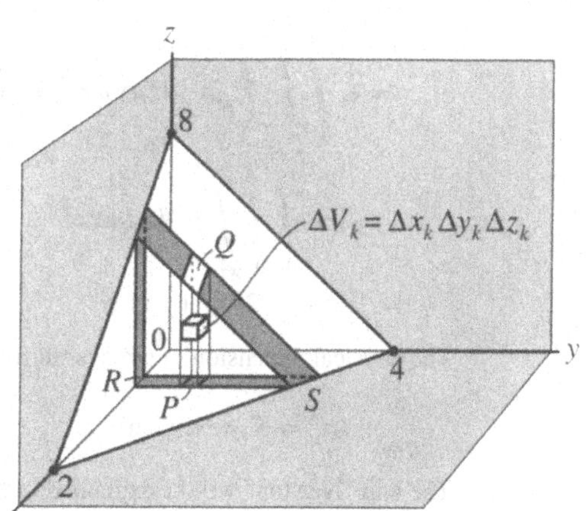

Illustration of the integration volume.

$$\int\limits_{x=0}^{2}\int\limits_{y=0}^{4-2x}\int\limits_{z=0}^{8-4x-2y} (45x^2y)dz\,dy\,dx$$

$$= 45\int\limits_{x=0}^{2}\int\limits_{y=0}^{4-2x} x^2y\left(z\Big|_0^{8-4x-2y}\right)dy\,dx$$

$$= 45\int\limits_{x=0}^{2}\int\limits_{y=0}^{4-2x} (x^2y)(8-4x-2y)\,dy\,dx$$

$$= 45\int\limits_{x=0}^{2}\left[x^2(8-4x)\left(\frac{y^2}{2}\Big|_0^{4-2x}\right) - 2x^2\left(\frac{y^3}{3}\Big|_0^{4-2x}\right)\right]dx$$

$$= 45\int\limits_{x=0}^{2} \frac{1}{3}x^2(4-2x)^3\,dx = 128.$$

Problem 15.2: Calculation of a total force from the force density

Integrate the force density $\mathbf{f} = (2xz, -x, y^2)$ N/cm³ over the volume V limited by the five areas $x = 0$, $y = 0$, $y = 6$ cm, $z = x^2$ cm, $z = 4$ cm (see figure).

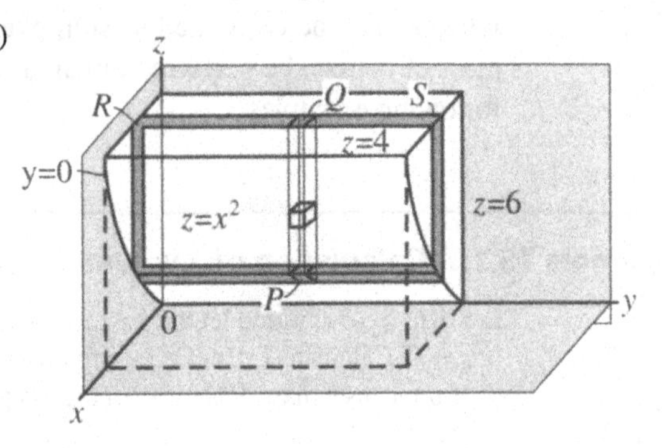

Illustration of the integration volume.

Solution The integral

$$\iiint_V \mathbf{f}(x, y, z)\, dV$$

obviously means the total force acting on the body with this volume. We obtain

$$\int_{x=0}^{2}\int_{y=0}^{6}\int_{z=x^2}^{4} (2xz\mathbf{e}_1 - x\mathbf{e}_2 + y^2\mathbf{e}_3)\, dz\, dy\, dx$$

$$= \mathbf{e}_1 \int_0^2\int_0^6\int_{z=x^2}^{4} 2xz\, dz\, dy\, dx + \mathbf{e}_2 \int_0^2\int_0^6\int_{x^2}^{4} (-x)\, dz\, dy\, dx$$

$$+ \mathbf{e}_3 \int_0^2\int_0^6\int_{x^2}^{4} y^2\, dz\, dy\, dx$$

$$= \quad 128\,\mathbf{e}_1 - 24\,\mathbf{e}_2 + 384\,\mathbf{e}_3\,.$$

The physical dimension of the overall force is, of course,

$$\frac{\mathrm{N}}{\mathrm{cm}^3} \cdot \mathrm{cm}^3 = \mathrm{N} = \text{Newton}.$$

The unit "Newton" will be explained in the Chapter 17.

PART II

NEWTONIAN MECHANICS

16 Newton's Axioms

The Newtonian[1] or classical mechanics is governed by three axioms, which are not independent of each other:

1. the law of inertia,
2. the fundamental equation of dynamics,
3. the interaction law,

and as a supplement: the theorems on independence concerning the superposition of forces and of motions.

Premises of Newtonian mechanics are as follows:

1. The absolute time; that means that the time is the same in all coordinate frames, that is, it is invariant: $t = t'$. One may determine in any coordinate frame whether events

[1]*Isaac Newton*, b. Jan. 4, 1643, Woolsthorpe (Lincolnshire)—d. March 31, 1727, London. Newton studied in 1660 at Trinity College in Cambridge, particularly with the eminent mathematician and theologian L. Barrow. After getting various academic degrees and making a series of essential discoveries, in 1669 Newton became successor of his teacher in Cambridge. In 1672 he was member and in 1703 president of the Royal Society. From 1688 to 1705, he was also member of Parliament, since 1696 attendant and since 1701 mint-master of the Royal mint. Newton's life's work comprises, besides theological, alchemistic, and chronological-historical writings, mainly works on optics and on pure and applied mathematics. In his investigations on optics he describes the light as a flow of corpuscles and by this way interprets the spectrum and the composition of light, as well as the Newton color rings, diffraction phenomena and double-refraction. His main opus *Philosophiae Naturalis Principia Mathematica* (printed in 1687) is fundamental for the evolution of exact sciences. It includes the definition of the most important basic concepts of physics, the three *axioms of mechanics* of macroscopic bodies, the principle of "actio et reactio," the *gravitational law,* the derivation of Kepler's laws, and the first publication on fluxion calculus. Newton also dealt with *potential theory* and with the equilibrium figures of rotating liquids. The ideas for the great work emerged mainly in 1665–1666 when Newton had left Cambridge because of the pestilence.

In mathematics Newton worked on the theory of series, for example, in 1669 on the binomial series, on interpolation theory, approximation methods, and the classification of cubic curves and conic sections. But Newton could not remove logical problems even with his fluxion calculus that was represented in 1704 in detail. His influence on the further development of mathematical sciences can hardly be judged, because Newton disliked publishing. When Newton made his fluxion calculus public, his kind of treatment of problems of analysis was already obsolete as compared to the calculus of Leibniz. The quarrel over whether Newton or Leibniz deserved priority for developing the infinitesimal calculus continued until the 20th century. Detailed studies have shown that they both obtained their results independently of each other [BR].

are simultaneous, because in classical physics one may imagine that signals are being exchanged with infinitely large velocity.

2. The absolute space; that means that a coordinate frame, being at absolute rest, which spans the full space, exists. This absolute space may be thought of as being represented by the world ether, which shall be at absolute rest and so to speak embodies the absolute space. Newton by himself did not believe in the ether; he could imagine the absolute space also as being empty. In most recent time the 2.7 Kelvin radiation has been discovered. This radiation is believed to originate from the Big Bang that presumably generated our universe. A coordinate frame in which this radiation is isotropic—of equal intensity in all directions—might also serve as such an absolute coordinate frame.
3. The mass being independent of the velocity.
4. The mass of a closed system of bodies (or mass points) is independent of the processes going on in this system, no matter what kind these processes are.

The concepts of absolute time and absolute space, as well as the velocity independence of the mass, are lost in the special theory of relativity. Finally, the fourth premise is no longer fulfilled in high-energy processes as, for example, $p + p \to p + p + \pi^+ + \pi^-$. Here new masses are generated.

Newton formulated his axioms essentially as follows:

Lex prima: Each body remains in its state of rest or uniform rectilinear motion as long as it is not forced by acting forces to change this state.

Lex secunda: The change of motion is proportional to the effect of the driving force and tends toward the direction of that straight line along which the force is acting.

Lex tertia: The action always equals the reaction, or the actions of two bodies onto each other are always of equal magnitude and of opposite direction.

Lex quarta: Supplement to the laws of motion: Rule of the parallelogram of forces, that is, forces add up like vectors. Thereby the superposition principle of the actions of forces is postulated (principle of unperturbed superposition).

Because we deal in the following only with *point mechanics*, we have to introduce the model representation of the mass point. Here one abstracts from shape, size, and rotational motions of a body and considers only its translational motion. Newton's axioms in modern form then read as follows:

Axiom 1: Any mass point remains in the state of rest or rectilinear uniform motion until this state is terminated by the action of other forces (i.e., by transfer of forces). This is a special case of the second axiom. Namely,

$$\mathbf{F} = 0, \quad \text{then} \quad m \cdot \mathbf{v} = \overrightarrow{\text{constant}}.$$

Because of the presupposed velocity independece of the mass, it then holds that

$$\mathbf{v} = \overrightarrow{\text{constant}}.$$

If the "quantity of motion" $\mathbf{p} = m \cdot \mathbf{v}$ is denoted as the *linear momentum* of the mass point, then the law of inertia is identical with the law of conservation of the linear momentum.

Axiom 2: The first time derivative of the linear momentum $\mathbf{p}$ of a mass point is equal to the force $\mathbf{F}$ acting on it:

$$\mathbf{F} = \frac{d(m \cdot \mathbf{v})}{dt} = \frac{d\mathbf{p}}{dt} = \dot{\mathbf{p}},$$

where

$$\mathbf{p} = m\mathbf{v}$$

is the linear momentum.[2].

Because in general the mass is a velocity-dependent quantity, that is, it is also time-dependent, it must not simply be pulled in front of the bracket. In the nonrelativistic Newtonian mechanics ($v \ll c;\ c = 3 \cdot 10^8\,\mathrm{m\,s^{-1}}$), the mass m is, however, treated as being independent of the time, and one thus obtains the dynamic fundamental equation:

$$\mathbf{F} = m\frac{d\mathbf{v}}{dt} = m\frac{d^2\mathbf{r}}{dt^2} = m\ddot{\mathbf{r}} = m\mathbf{a}.$$

That means that the acceleration $\mathbf{a}$ of a mass point is directly proportional to the force acting on it and coincides with the direction of the force.

If several forces are acting simultaneously onto a mass point, then the above relation according to the principle of superposition of forces reads

$$\frac{d\mathbf{p}}{dt} = \sum_{i=1}^{n} \mathbf{F}_i .$$

Axiom 3: The forces exerted by two mass points onto each other have equal magnitude and opposite directions; force = – counterforce:

$$\mathbf{F}_{ij} = -\mathbf{F}_{ji}, \quad \text{where } i \neq j.$$

Here $\mathbf{F}_{ij}$ is the force exerted by the jth point onto the ith point. $\mathbf{F}_{ji}$ is the force exerted by the ith onto the jth point.

Remark: The relation $\mathbf{F} = d(m\mathbf{v})/dt$ is *on the one hand a definition of the force, on the other hand a law.* The statutory aspect is that, for example, the first time derivative of the linear momentum occurs, but not the third or fourth or something else. Because the force

[2]The time derivatives are often abbreviated with a dot, for example $df/dt \equiv \dot{f}$, $\mathbf{v} = d\mathbf{r}/dt = \dot{\mathbf{r}}$, or $\mathbf{a} = d^2\mathbf{r}/dt^2 = \ddot{\mathbf{r}}$ We will use this notation wherever it seems appropriate.

is the derivative of a vector with respect to a scalar (the time), it is a vector itself. Hence, the addition of forces is governed, for instance, by the law of force parallelogram.

Problem 16.1: Single-rope pulley

A weight $W_1 = M_1 g$ hangs at the end of a rope. Here, $g = 9.81\ \mathrm{m/s^2}$ is the gravitational acceleration of all bodies at the surface of the earth. At the other end of the rope, hanging over a roller, a boy of weight $W_2 = M_2 g$ pulls himself upward. Let his acceleration relative to the tightly mounted roller be a. What is the acceleration of the weight W_1?

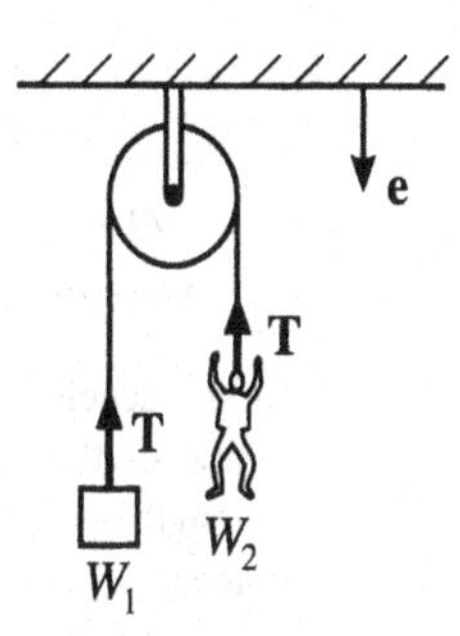

A boy and a weight hanging at the ends of a rope.

Solution Let b be the acceleration of W_1 and T the rope tension. The Newtonian equations of motion then read

(a) For the mass M_2 (boy):

$$-M_2 \cdot a\mathbf{e} = M_2 g\,\mathbf{e} - T\,\mathbf{e}; \tag{16.1}$$

(b) for the mass M_1 (weight W_1):

$$M_1 b\,\mathbf{e} = M_1 g\,\mathbf{e} - T\,\mathbf{e}. \tag{16.2}$$

These are two equations with two unknowns (T, b). Their solution may be given immediately:

$$T = M_2(a + g); \tag{16.3}$$

$$\begin{aligned} b &= g - \frac{T}{M_1} \\ &= g - \frac{M_2}{M_1}(a + g) \\ &= \frac{(M_1 - M_2)g - M_2 a}{M_1}. \end{aligned} \tag{16.4}$$

If $M_1 = M_2$, it follows that $b = -a$, as it should be. On the other hand, if $a = 0$, it follows that $b = \frac{(M_1 - M_2)}{M_1} g$ and vanishes for the case $M_1 = M_2$, as expected.

Problem 16.2: Double-rope pulley

A mass M_1 hangs at one end of a rope that is led over a roller A (compare the figure). The other end carries a second roller of mass M_2, which in turn carries a rope with the masses m_1 and m_2 fixed to its ends. The gravitational force is acting on all masses. Calculate the acceleration of the masses m_1 and m_2, as well as the tensions T_1 and T in the ropes.

Solution We introduce the unit vector $\mathbf{e} \perp$ pointing upward (see figure) and denote the string tensions by $\mathbf{T} = T\mathbf{e}$ and $\mathbf{T}_1 = T_1\mathbf{e}$, respectively (see figure). The individual masses are influenced by the string tension (i.e., the force in the rope) and by the gravitational force. We now write down the equations of motion for the individual masses according to Newton's fundamental law.

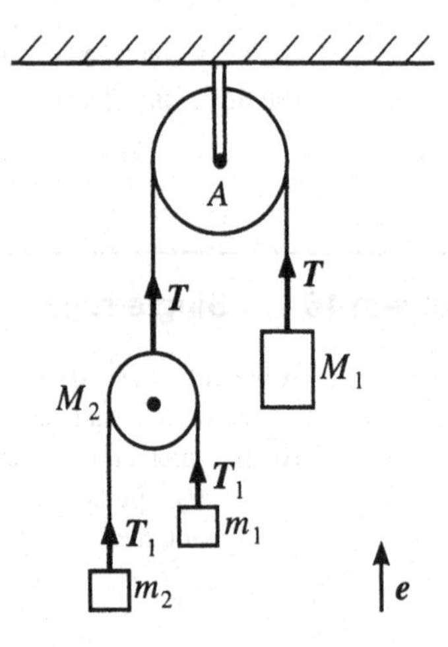

Masses and forces at the double-rope pulley.

$$\begin{aligned} M_1 a_1 \mathbf{e} &= -M_1 g\mathbf{e} + T\mathbf{e}\,, \\ -M_2 a_1 \mathbf{e} &= -M_2 g\mathbf{e} + T\mathbf{e} - 2T_1\mathbf{e}\,, \\ m_1(a_2 - a_1)\mathbf{e} &= -m_1 g\mathbf{e} + T_1\mathbf{e}\,, \\ m_2(-a_2 - a_1)\mathbf{e} &= -m_2 g\mathbf{e} + T_1\mathbf{e}\,. \end{aligned} \tag{16.5}$$

The acceleration of the mass M_1 has been denoted by $a_1\mathbf{e}$, that of the mass M_2 is then (because of the constant rope length) $-a_1\mathbf{e}$; the acceleration of the mass m_1 relative to the mass M_2 is $a_2\mathbf{e}$, that of the mass m_2 is $-a_2\mathbf{e}$. 16.5 represents a set of four equations with the four unknowns: a_1, a_2, T, T_1. Subtraction of the second equation from the first one yields

$$(M_1 + M_2)a_1 = -(M_1 - M_2)g + 2T_1\,. \tag{16.6}$$

The addition of the last two equations of 16.5 leads to

$$-(m_1 + m_2)a_1 + (m_1 - m_2)a_2 = -(m_1 + m_2)g + 2T_1\,. \tag{16.7}$$

The subtraction of 16.7 from 16.6 then yields a relation between a_1 and a_2:

$$(M_1 + M_2 + m_1 + m_2)a_1 - (m_1 - m_2)a_2 = (-M_1 + M_2 + m_1 + m_2)g\,. \tag{16.8}$$

A second relation of this kind is obtained by subtracting the last two equations 16.5 from each other, namely

$$-(m_1 - m_2)a_1 + (m_1 + m_2)a_2 = -(m_1 - m_2)g\,. \tag{16.9}$$

The accelerations a_1 and a_2 are now found from equations 16.8 and 16.9:

$$a_1 = \frac{-M_1(m_1 + m_2) + M_2(m_1 + m_2) + 4m_1 m_2}{(m_1 + m_2)(M_1 + M_2) + 4m_1 m_2} g; \tag{16.10}$$

$$a_2 = \frac{-2M_1(m_1 - m_2)}{(m_1 + m_2)(M_1 + M_2) + 4m_1 m_2} g, \tag{16.11}$$

such that the total acceleration of mass m_1 is obtained as

$$a_2 - a_1 = \frac{-M_1 m_1 + 3M_1 m_2 - M_2(m_1 + m_2) - 4m_1 m_2}{(m_1 + m_2)(M_1 + M_2) + 4m_1 m_2} g \tag{16.12}$$

and that of mass m_2 is

$$(-a_2 - a_1) = \frac{M_1(3m_1 - m_2) - M_2(m_1 + m_2) - 4m_1 m_2}{(m_1 + m_2)(M_1 + M_2) + 4m_1 m_2} g\,. \tag{16.13}$$

If all masses were identical ($M_1 = M_2 = m_1 = m_2$), then

$$a_2 - a_1 = -\frac{1}{2}g, \qquad a_2 = 0,$$

and

$$-a_2 - a_1 = -\frac{1}{2}g, \qquad a_1 = \frac{1}{2}g, \tag{16.14}$$

as one would expect. The string tension T_1 follows with 16.10 from equation 16.6 after a simple calculation as

$$\begin{aligned} T_1 &= \frac{1}{2}(M_1 + M_2)a_1 + \frac{1}{2}(M_1 - M_2)g \\ &= \frac{4m_1 m_2 M_1}{(m_1 + m_2)(M_1 + M_2) + 4m_1 m_2}g. \end{aligned} \tag{16.15}$$

The rope tension T is obtained from the first two equations 16.5, using 16.10 and 16.15, as

$$\begin{aligned} T &= \frac{(M_1 - M_2)a_1}{2} + \frac{(M_1 + M_2)g}{2} + T_1 \\ &= M_1 a_1 + M_1 g = M_1(a_1 + g) \\ &= \frac{2(m_1 + m_2)M_1 M_2 + 8m_1 m_2 M_1}{(m_1 + m_2)(M_1 + M_2) + 4m_1 m_2}g. \end{aligned} \tag{16.16}$$

According to 16.15 the rope tension T_1 vanishes if one of the masses m_1, m_2, M_1 vanishes. In this case the rope is rolling without tension, as is clearly expected. The rope tension T vanishes if either $M_1 = 0$, or M_2 and one of the masses m_1 or m_2 (or both) vanish. If $m_1 = m_2 = m = 0$ and if $M_1 \neq 0$, $M_2 \neq 0$, a limit $m \to 0$ can be taken:

$$T = \frac{2M_1 M_2}{M_1 + M_2}g.$$

This is the rope tension in the case of the single roller with the two masses M_1 and M_2 at the rope ends.

17 Basic Concepts of Mechanics

Inertial systems

We ask for the forces acting on a mass point P, as seen from two coordinate frames x, y, z and x', y', z' that are moving relative to each other, with correspondingly convected observers 0 and 0′, respectively. Let $\mathbf{r}$ and $\mathbf{r}'$ be the position vectors of P in x, y, z and in x', y', z', respectively. One then obtains the position vector from 0 to 0′ as the difference $\mathbf{r} - \mathbf{r}' = \mathbf{R}$.

According to Newton's basic equation:

$$\mathbf{F} = m\frac{d^2\mathbf{r}}{dt^2} \quad \text{and} \quad \mathbf{F}' = m\frac{d^2\mathbf{r}'}{dt^2}. \qquad \textbf{(17.1)}$$

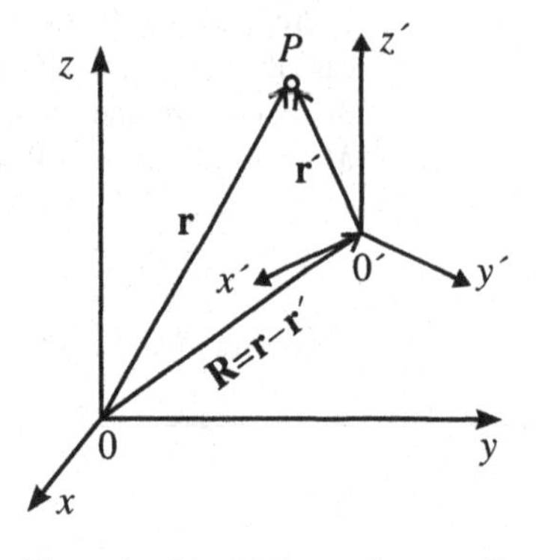

The point P in relation to the coordinate systems x, y, z and x', y', z'.

The difference of the observed forces is

$$\mathbf{F} - \mathbf{F}' = m\frac{d^2}{dt^2}(\mathbf{r} - \mathbf{r}') = m\frac{d^2\mathbf{R}}{dt^2}. \qquad \textbf{(17.2)}$$

Because $m \neq 0$, this difference vanishes then and only then if

$$\frac{d^2\mathbf{R}}{dt^2} = 0 \quad \text{or} \quad \frac{d\mathbf{R}}{dt} = \overrightarrow{\text{constant}} = \mathbf{v}_R. \qquad \textbf{(17.3)}$$

This means that the forces are then equal if the two coordinate frames are moving with constant velocity $\mathbf{v}_R$ relative to each other. Such systems are called *inertial systems* if one of them—and thus all of them—fulfills Newton's axioms. The statement that in such inertial systems the Newtonian equations (17.1) have the same form and the forces are also the same ($\mathbf{F} = \mathbf{F}'$) is called the *classical relativity principle*.

Measurement of masses

Masses are measured by comparison with an arbitrarily defined unit mass. If there are three distinct masses m_1, m_2, and m_3, with m_1 representing the unit mass, one may determine experimentally, for example, m_3, starting from the second and third Newton laws, as the quotient of the accelerations:

$$m_1 \frac{d\mathbf{v}_1}{dt} = -m_3 \frac{d\mathbf{v}_3}{dt}, \qquad m_1\,\mathbf{a}_1 = -m_3\,\mathbf{a}_3,$$

force = −counterforce.

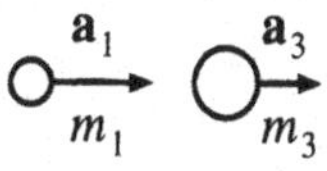

$\mathbf{a}_1$ $\mathbf{a}_3$ m_1 m_3

Central collision.

From there it follows that

$$m_3 = m_1 \frac{|\mathbf{a}_1|}{|\mathbf{a}_3|},$$

m_1 m_3

Noncentral collision.

where m_1 is the unit mass, and $\mathbf{a}_1$ and $\mathbf{a}_3$ may be determined. Thus, m_3 may be measured in units of m_1. In the process of measurement (collision) the basic laws (second and third Newtonian laws) are employed.

Correspondingly, it then also holds that

$$m_2 = m_1 \frac{|\mathbf{a}_1|}{|\mathbf{a}_2|}. \qquad \textbf{(17.4)}$$

Work

A force $\mathbf{F}$ causes a displacement of a mass point M by an infinitesimally small path element $d\mathbf{r}$ and thereby performs the work dW that is defined as follows:

$$dW = \mathbf{F} \cdot d\mathbf{r} = |\mathbf{F}||d\mathbf{r}| \cos(\mathbf{F}, d\mathbf{r}).$$

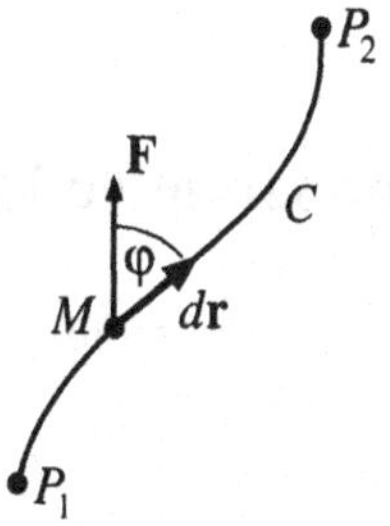

The unit of this scalar is therefore

$$\frac{\text{g cm}^2}{\text{s}^2} = 1 \text{ erg} \quad \text{or}$$

$$\frac{\text{kg m}^2}{\text{s}^2} = 1 \text{ Nm} \Rightarrow 1 \text{ erg} \mathrel{\widehat{=}} 10^{-7} \text{ Nm}.$$

P_2 $\mathbf{F}$ C φ M $d\mathbf{r}$ P_1

The work integral.

Here 1 Newton (N) = kg m/s^2 is the unit of force.

The total work W needed to move M along a curve C between the points P_1 and P_2 is given by the following line integral:

$$W = \int_C \mathbf{F} \cdot d\mathbf{r} = \int_{P_1}^{P_2} \mathbf{F} \cdot d\mathbf{r}. \qquad \textbf{(17.5)}$$

Power is work performed per unit time:

$$\frac{dW}{dt} = \mathbf{F} \cdot \frac{d\mathbf{r}}{dt} = \mathbf{F} \cdot \mathbf{v}. \tag{17.6}$$

The unit of power is $[\mathrm{g\,cm^2/s^3 = erg/s}]$ or $[\mathrm{kg\,m^2/s^3 = Nm/s}]$.

Kinetic energy

In order to accelerate a mass point and to bring it to a definite velocity, work must be performed. This work is then stored in the mass point in the form of *kinetic energy*. We therefore start from the integral of work:

$$\begin{aligned} W &= \int_{t_1}^{\mathbf{r}_2} \mathbf{F} \cdot d\mathbf{r} = \int_{t_1}^{\mathbf{r}_2} \mathbf{F} \cdot \mathbf{v}\, dt \\ &= \int_{t_1}^{t_2} m \frac{d\mathbf{v}}{dt} \cdot \mathbf{v}\, dt = \frac{1}{2} m \int_{\mathbf{v}_1}^{\mathbf{v}_2} d(\mathbf{v} \cdot \mathbf{v}) \\ &= \frac{1}{2} m (v_2^2 - v_1^2) = T_2 - T_1 , \end{aligned}$$

$$T = \frac{1}{2} m v^2 = \text{kin. energy}. \tag{17.7}$$

Conservative forces

A force is called conservative if the force field $\mathbf{F}$ may be represented by

$$\mathbf{F} = -\operatorname{grad} V(x, y, z) \qquad \text{(definition)}. \tag{17.8}$$

If this is true, then the work integrals are path-independent:

$$\begin{aligned} \int_{P_1}^{P_2} \mathbf{F} \cdot d\mathbf{r} = -\int_{P_1}^{P_2} \operatorname{grad} V \cdot d\mathbf{r} = -\int_{P_1}^{P_2} dV \qquad &\text{(see total differential; Chapter 11)} \\ = V(P_1) - V(P_2) \equiv V_1 - V_2 & \\ = -(V_2 - V_1). & \end{aligned} \tag{17.9}$$

Hence one has

$$W = V_1 - V_2 ,$$

where V is a scalar field that associates a numerical value to each space point. W is therefore path-independent. But this further means that for an integration along a closed curve the total work must vanish:

$$\oint_C \mathbf{F} \cdot d\mathbf{r} = 0 \tag{17.10}$$

for conservative forces. An equivalent requirement for conservative forces is

$$\operatorname{curl} \mathbf{F} = \nabla \times \mathbf{F} = 0\,.$$

Indeed we can also conclude this immediately from equation (17.8) since

$$\operatorname{curl} \operatorname{grad} V(\mathbf{r}) = 0.$$

Potential

If $\mathbf{F}(\mathbf{r}) = -\nabla V(\mathbf{r})$, then the scalar quantity $V(x, y, z)$ is called potential energy, scalar potential, or, briefly, potential:

$$V(x, y, z) = -\int_{(x_0,y_0,z_0)}^{(x,y,z)} \mathbf{F} \cdot d\mathbf{r}\,. \tag{17.11}$$

Example 17.1: Potential energy

Calculation of the potential energy between two points looks like

$$W = \int_{P_1}^{P_2} \mathbf{F} \cdot d\mathbf{r}$$

$$= \int_{P_1}^{(x_0,y_0,z_0)} \mathbf{F} \cdot d\mathbf{r} + \int_{(x_0,y_0,z_0)}^{P_2} \mathbf{F} \cdot d\mathbf{r}.$$

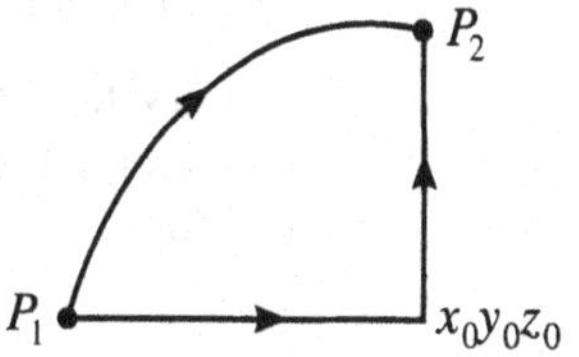

P_2

P_1 $x_0y_0z_0$

Calculation of the potential energy difference between in the points P_1 and P_2.

The presupposition is a conservative force field, and thus path-independence of the work integral.

$$W = -\int_{(x_0,y_0,z_0)}^{P_1} \mathbf{F} \cdot d\mathbf{r} + \int_{(x_0,y_0,z_0)}^{P_2} \mathbf{F} \cdot d\mathbf{r} = V(x_1y_1z_1) - V(x_2y_2z_2).$$

Therefore, the work represents a potential difference that is independent of the choice of the reference point. The potential itself is always defined relative to a reference point (x_0, y_0, z_0) and is therefore undetermined by an additive constant. The zero point of the potential may be set arbitrarily. This arbitrariness corresponds to the (arbitrary) additive constant in the potential.

Energy law

On deriving the kinetic energy, we found the following relation for the work:

$$W = T_2 - T_1.$$

For conservative fields there also holds the other relation between the same points P_1 and P_2:

$$W = V_1 - V_2.$$

This implies

$$T_2 + V_2 = T_1 + V_1. \tag{17.12}$$

This is the **energy conservation law** (briefly: energy law), where $T + V = E$ represents the total energy of the mass point.

Written out in detail, the energy conservation law reads

$$\frac{1}{2}m\mathbf{v}_2^2 + \left(-\int_{P_0}^{P_2} \mathbf{F}\cdot d\mathbf{r}\right) = \frac{1}{2}m\,\mathbf{v}_1^2 + \left(-\int_{P_0}^{P_1} \mathbf{F}\cdot d\mathbf{r}\right) \tag{17.13}$$

or

$$\frac{1}{2}m\,\mathbf{v}_2^2 + V_2 = \frac{1}{2}m\,\mathbf{v}_1^2 + V_1$$

or

$$E_2 = E_1\,.$$

The premises for this energy conservation law for the motion of a mass point are

1. The basic assumptions and basic laws of the Newtonian mechanics (e.g., nonrelativistic treatment of the mass);
2. conservative force fields, that is, the forces may be written as the negative gradient of a potential. For force fields that are constant in time, it then holds that $E = T + V =$ constant.

Equivalence of impulse of force and momentum change

If a mass point is affected by a force over a time interval $t = t_2 - t_1$, the time integral over this force is called *impulse of force:*

$$\int_{t_1}^{t_2} \mathbf{F}(t)\,dt = \text{impulse of force.} \tag{17.14}$$

The impulse of force is equivalent to the momentum change or momentum difference. This is seen as follows:

From the definition of the linear momentum $\mathbf{p} = m\mathbf{v}$ and from the second Newtonian fundamental equation, it follows that

$$\int_{t_1}^{t_2} \mathbf{F}\,dt = \int_{t_1}^{t_2} \frac{d}{dt}(m\,\mathbf{v})\,dt = \int_{t_1}^{t_2} d(m\,\mathbf{v}) = m\mathbf{v}_2 - m\mathbf{v}_1 = \mathbf{p}_2 - \mathbf{p}_1. \tag{17.15}$$

Thus a force acting on the mass causes a change of momentum: That is, only of its magnitude if $\mathbf{F}$ points along $\mathbf{p}_1$, and a change of both magnitude and direction if $\mathbf{F}$ points in an arbitrary angle relative to $\mathbf{p}_1$.

If the force $\mathbf{F}$ acts during the time Δt, the corresponding difference in momentum is $\mathbf{F}\Delta t = \mathbf{p}_2 - \mathbf{p}_1$. After the collision, the mass moves on a straight line with linear momentum $\mathbf{p}_2$.

Situation before (top) and after (bottom) the impulse of force.

Problem 17.2: Impulse of momentum by a time-dependent force field

A particle of mass $m = 2$ g moves in the time-dependent uniform force field:

$$\mathbf{F} = \left(24\frac{t^2}{\mathrm{s}^2}, 3\frac{t}{\mathrm{s}} - 16, -12\frac{t}{\mathrm{s}}\right) \text{ dyn.}$$

The initial conditions are

$$\mathbf{r}_{(t=0)} = \mathbf{r}_0 = (3, -1, 4) \text{ cm}$$

and

$$\mathbf{v}_{(t=0)} = \mathbf{v}_0 = (6, 15, -8)\,\frac{\text{cm}}{\text{s}}.$$

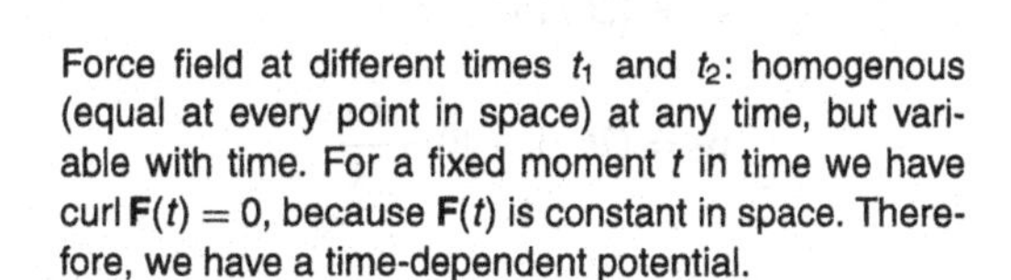

Force field at different times t_1 and t_2: homogenous (equal at every point in space) at any time, but variable with time. For a fixed moment t in time we have curl $\mathbf{F}(t) = 0$, because $\mathbf{F}(t)$ is constant in space. Therefore, we have a time-dependent potential.

Here we make use of the units of force

$$1 \text{ dyn} = \text{g}\frac{\text{cm}}{\text{s}^2} = 10^{-5}\text{ N} \quad \text{and} \quad 1\text{ N} = 1 \text{ Newton} = 1\text{kg}\frac{\text{m}}{\text{s}^2}.$$

Find the following quantities:

1. The kinetic energy at the time $t = 1$ s and $t = 2$ s.
2. The work performed by the field to move the particle from $\mathbf{r}_1 = \mathbf{r}_{(t=1\,\mathrm{s})}$ to $\mathbf{r}_2 = \mathbf{r}_{(t=2\,\mathrm{s})}$.
3. The linear momentum of the particle at $\mathbf{r}_1$ and $\mathbf{r}_2$.
4. The momentum transferred by the field to the particle over the time interval $t = 1$ s until $t = 2$ s.

Solution

To 1: $\mathbf{v}$ is obtained from $\mathbf{F} = m\mathbf{a} = m(d\mathbf{v}/dt)$ as

$$\mathbf{v} = \int d\mathbf{v} = \int \frac{\mathbf{F}}{m}\, dt + \mathbf{v}_0 .$$

Using the data of the problem, we get for $\mathbf{v}$

$$\mathbf{v}(t) = \left(4\frac{t^3}{\mathrm{s}^3}, \frac{3}{4}\frac{t^2}{\mathrm{s}^2} - 8\frac{t}{\mathrm{s}}, -3\frac{t^2}{\mathrm{s}^2}\right)\frac{\mathrm{cm}}{\mathrm{s}} + (6, 15, -8)\frac{\mathrm{cm}}{\mathrm{s}}$$

and

$$\mathbf{v}_{(t=1\,\mathrm{s})} = \left(10, 7\tfrac{3}{4}, -11\right)\frac{\mathrm{cm}}{\mathrm{s}},$$

$$\mathbf{v}_{(t=2\,\mathrm{s})} = (38, 2, -20)\frac{\mathrm{cm}}{\mathrm{s}}.$$

From there we get for the energy

$$T = \frac{1}{2}m\,\mathbf{v}^2 = \frac{1}{2}m\,v^2,$$

$$T_1 = 281\ \mathrm{erg}, \qquad T_2 = 1848\ \mathrm{erg}.$$

To 2: The work performed by the field equals the difference of the kinetic energies:

$$W = T_2 - T_1 = 1567\ \mathrm{erg}.$$

To 3: The momentum of the particle is $\mathbf{p} = m\mathbf{v}$:

$$\mathbf{p}_1 = \left(20, 15\tfrac{1}{2}, -22\right)\mathrm{g}\frac{\mathrm{cm}}{\mathrm{s}},$$

$$\mathbf{p}_2 = (76, 4, -40)\,\mathrm{g}\frac{\mathrm{cm}}{\mathrm{s}}.$$

To 4: The momentum transferred by the field is obtained from the difference of momenta $\mathbf{p}_2$ and $\mathbf{p}_1$:

$$\mathbf{p} = \mathbf{p}_2 - \mathbf{p}_1 = \left(56, -11\tfrac{1}{2}, -18\right)\mathrm{g}\frac{\mathrm{cm}}{\mathrm{s}}.$$

Problem 17.3: Impulse of force

A railway carriage of mass $m = 18000$ kg starts from a roll-off plateau of height 3 m. What is the change of momentum of the carriage, and which mean force is acting on it when colliding onto a buffer at the bottom of the hill, if the carriage within 0.2 s

(a) Comes to rest ?

(b) Is pushed back to a height of 0.5 m?

Discuss the momentum conservation.

Solution At the moment of impact the carriage has a momentum $\mathbf{p}_1$ that results from the potential energy at the start from the roll-off plateau:

$$\frac{1}{2} m\, v_1^2 = mgh \quad \Rightarrow \quad \mathbf{p}_1 = m\,\mathbf{v}_1 = m(2gh)^{1/2}\mathbf{e}_1 .$$

In case (a) the momentum $\mathbf{p}_2$ after the impact equals zero; hence

$$\begin{aligned}\Delta\mathbf{p} = \mathbf{p}_1 - \mathbf{p}_2 &= m(2gh)^{1/2}\mathbf{e}_1 \\ &= 138\,096.5 \text{ m kg s}^{-1} \cdot \mathbf{e}_1;\end{aligned}$$

the mean force acting over $\Delta t = 0.2$ s is then

$$\mathbf{F} = \frac{\Delta\mathbf{p}}{\Delta t} = 690\,482.4 \text{ N}.$$

In case (b) the momentum p_2 is given by

$$\mathbf{p}_2 = m\,\mathbf{v}_2 = -m(2gh')^{1/2}\mathbf{e}_1,$$

where h' is the height regained in the bouncing-back. The momentum change is then

$$\begin{aligned}\Delta\mathbf{p} = \mathbf{p}_1 - \mathbf{p}_2 &= m\mathbf{e}_1\left[(2gh)^{1/2} + (2gh')^{1/2}\right] \\ &= 194\,474.1 \text{ m kg s}^{-1}\,\mathbf{e}_1;\end{aligned}$$

for the mean force we obtain

$$\mathbf{F} = \frac{\Delta\mathbf{p}}{\Delta t} = 972\,370.7 \text{ N}.$$

The carriage alone does not represent a closed system: The reactive force imposed by the tightly mounted buffer is an external force; therefore, the momentum cannot be conserved.

Problem 17.4: The ballistic pendulum

The velocity of a bullet may be measured by means of the ballistic pendulum. This device consists of a string with negligible weight and a weight of mass m_G fixed to the string. The bullet (mass m_K, velocity v_K) is shot into the block, where it gets stuck. The arc length s covered by the center of the mass m_G is measured.

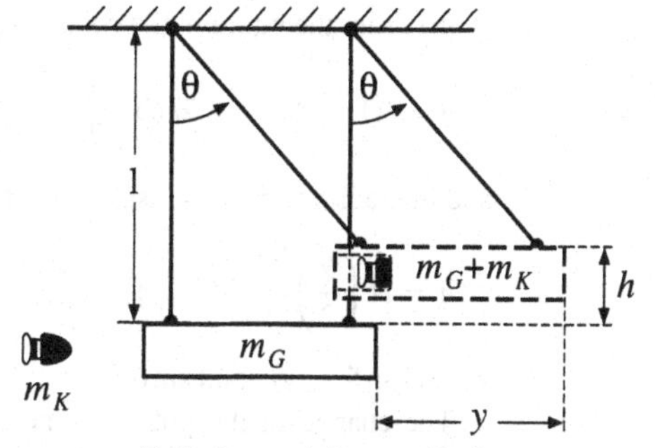

Ballistic pendulum and bullet.

(a) Determine the velocity of the block v_G after the collision, and

(b) Determine the velocity of the bullet v_K if the following quantities are specified: $m_G = 4$ kg, $l = 1.62$ m, $m_K = 0.055$ kg, $s = 6.5$ cm.

Solution (a) From the momentum conservation law, it follows that

$$m_K v_K = (m_G + m_K)v_G \tag{17.16}$$

and from there for the velocity v_G of the block just after the collision

$$v_G = \frac{m_K}{m_G + m_K} \cdot v_K . \tag{17.17}$$

For the kinetic energy it results immediately that

$$T = \frac{1}{2}(m_G + m_K) \cdot v_G^2 = \frac{m_K}{m_G + m_K}\left(\frac{1}{2}m_K v_K^2\right). \tag{17.18}$$

This energy coincides with the kinetic energy of the bullet reduced by the factor $m_K/(m_G + m_K)$. One may wonder why the kinetic energy of the block differs from the kinetic energy $\frac{1}{2}m_K v_K^2$ of the bullet? Where is the energy lost,

$$\Delta E = \frac{1}{2}m_K v_K^2 - \frac{m_K}{m_G + m_K}\left(\frac{1}{2}m_K v_K^2\right) = \frac{m_G}{m_G + m_K}\left(\frac{1}{2}m_K v_K^2\right)?$$

Obviously it must correspond to the heat released by the bullet getting stuck in the block. For $m_G \gg m_K$, almost the total energy of the bullet is converted to heat.

Another point is worth being mentioned: On calculating the velocity v_G of the block, we started from the momentum conservation law 17.16 but not, as might be thought first, from the energy conservation law ($\frac{1}{2}m_K v_K^2 = \frac{1}{2}(m_K + m_G)v_G^2$). Which of these two possibilities is now correct? The puzzle of two seemingly existing possibilities originates from the incomplete formulation of the problem. Actually, the percent fraction of the energy converted to heat ought to be specified in addition. Without any knowledge of this fraction we may, however, argue as follows: We know by experience that in the process of the bullet getting stuck, no small particles of the block (smallest pieces, molecules) are flying off, but rather the block moves as an entity. The block itself also becomes heated up by the friction of the bullet. In any case, the momentum conservation law must hold strictly, because the heat as a disordered molecular motion on the average does not carry off momentum, but for sure dissipates energy. In other words: Because the momentum conservation law 17.16 is strictly fulfilled, we actually may imagine that the energy loss ΔE has been converted to heat. If we had strictly required energy conservation without any production of heat, $\frac{1}{2}m_K v_K^2 = \frac{1}{2}(m_G + m_G)v_G^2$; this would imply a momentum loss, about which we would have no idea how it might evolve.

(b) From the figure in the context of the problem, we get for the height of lift of the block

$$h = l(1 - \cos\theta) = 2l\sin^2\frac{\theta}{2} \tag{17.19}$$

and in the limit of small displacements θ

$$h = 2l\left(\frac{\theta}{2}\right)^2 = 2l\left(\frac{y}{2l}\right)^2 = \frac{y^2}{2l}, \tag{17.20}$$

where $\sin\theta = y/l$ and $\sin\theta = \theta$.

The change of the potential energy of the block after being hit by the bullet is—at maximal elongation of the pendulum—according to the energy conservation law:

$$\Delta V = g(m_G + m_K)h = T = \frac{m_K}{(m_K + m_G)}\left(\frac{1}{2}m_K v_K^2\right). \tag{17.21}$$

From equations 17.20 and 17.21, we then obtain

$$gh = \frac{m_K^2}{2(m_G + m_K)^2}v_K^2 = g\frac{y^2}{2l}. \tag{17.22}$$

In the approximation $m_K + m_G \approx m_G$, the velocity v_K of the bullet is given by

$$v_K = \frac{m_G}{m_K} y \sqrt{\frac{g}{l}}. \tag{17.23}$$

Insertion of the data given in the formulation of the problem yields

$$v_K = \frac{4}{0.055} \cdot 6.5 \cdot 10^{-2} \sqrt{\frac{9.81}{1.62}} = 11.6 \, \frac{\text{m}}{\text{s}}.$$

Angular momentum and torque

Angular momentum and torque are always defined with respect to a fixed point, the pivot. If $\mathbf{r}$ is the vector from this point to the mass point, then the *angular momentum* is given by

$$\mathbf{L} = \mathbf{r} \times \mathbf{p}. \tag{17.24}$$

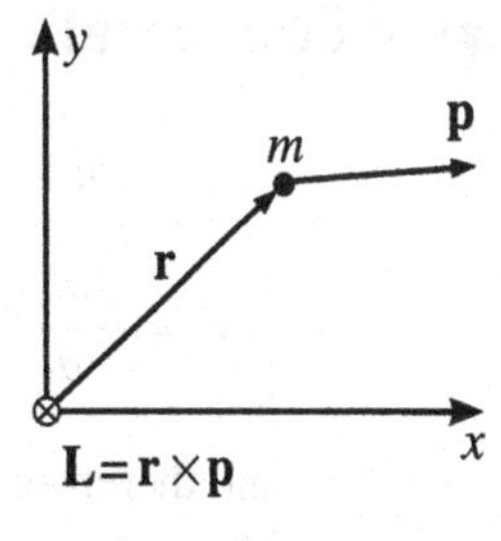

The definition of angular momentum: $\mathbf{L} = \mathbf{r} \times \mathbf{p}$.

If we put the coordinate frame into the reference point, then $\mathbf{r}$ is the position vector of the mass point, and $\mathbf{p}$ is its linear momentum.

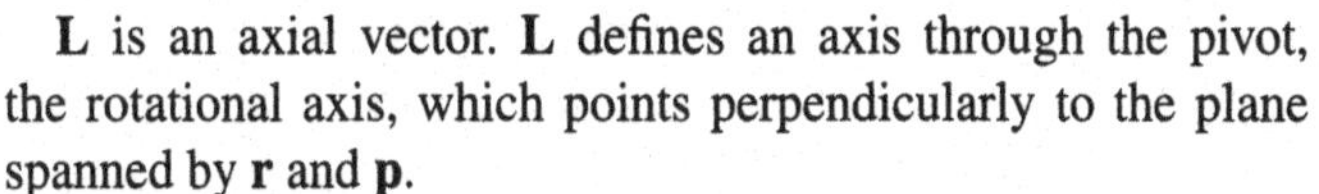

$\mathbf{L}$ is an axial vector. $\mathbf{L}$ defines an axis through the pivot, the rotational axis, which points perpendicularly to the plane spanned by $\mathbf{r}$ and $\mathbf{p}$.

The corresponding definition holds for the moment of the force, which is defined by

$$\mathbf{D} = \mathbf{r} \times \mathbf{F}, \tag{17.25}$$

and is also called *torque*. The time variation of the angular momentum is equal to the torque:

$$\dot{\mathbf{L}} = \mathbf{D},$$

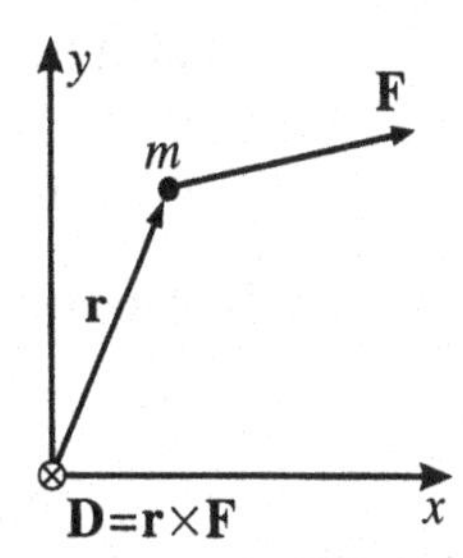

The definition of torque: $\mathbf{D} = \mathbf{r} \times \mathbf{F}$.

because

$$\dot{\mathbf{L}} = \frac{d\mathbf{L}}{dt} = \frac{d}{dt}(\mathbf{r} \times m\mathbf{v}) = \frac{d\mathbf{r}}{dt} \times m\mathbf{v} + \mathbf{r} \times \frac{d(m\mathbf{v})}{dt}$$

$$= \mathbf{v} \times m\mathbf{v} + \mathbf{r} \times \frac{d\mathbf{p}}{dt} = \mathbf{r} \times \mathbf{F}, \tag{17.26}$$

because $\mathbf{v} \times m\mathbf{v} = 0$.

The torque of the acting force $(\mathbf{r} \times \mathbf{F})$ is equal to the time variation of the angular momentum.

If, in particular, $\mathbf{D} = \mathbf{r} \times \mathbf{F} = 0 = \dot{\mathbf{L}}$, it then follows that $\mathbf{L} = \overrightarrow{\text{constant}}$. This is the *conservation law for the angular momentum.* But the quantity $\mathbf{r} \times \mathbf{F}$ vanishes only then (except for the trivial cases $\mathbf{r} = 0, \mathbf{F} = 0$) if $\mathbf{r}$ and $\mathbf{F}$ point along the same or the opposite direction. A force that acts exclusively parallel or antiparallel to the position vector is called the *central force.*

This implies that central forces obey the

Conservation law of angular momentum

$$\mathbf{L} = \overrightarrow{\text{constant}}\,, \qquad \text{because} \quad \mathbf{D} = 0.$$

Law of conservation of the linear momentum

As long as no forces are acting, the linear momentum $\mathbf{p}$ is a constant quantity. In general,

$$\mathbf{F} = \frac{d(m\mathbf{v})}{dt} = m\frac{d\mathbf{v}}{dt};$$

and therefore it follows for $\mathbf{F} = 0$ that

$$m\frac{d\mathbf{v}}{dt} = 0.$$

From there again we get

$$m\mathbf{v} = \mathbf{p} = \overrightarrow{\text{constant}}.$$

The momentum conservation law is identical to Newton's Lex prima.

Summary

The premises of the conservation laws concerning the energy, angular momentum, and linear momentum for a mass point in the Newtonian mechanics (compare to the introduction) are

(a) **Energy conservation:** If the forces acting on a mass point are *conservative* (gradient field: $\mathbf{F} = -\nabla V$), then the total energy $E = T + V$ of the mass point is conserved.

(b) **Angular momentum conservation:** The total angular momentum $\mathbf{L}$ is constant in time if the applied (external) torque vanishes, that is, if one is dealing with central force fields ($\mathbf{r} \times \mathbf{F} = 0$).

(c) **Momentum conservation:** If the total external force equals zero then the total linear momentum is conserved (equivalent to Newton's Lex prima).

The law of areas

(See also Chapter 26 on planetary motions; in particular, the Kepler laws.) The premises and contents of the three conservation laws (total energy, linear momentum, angular momentum) have been formulated already. The angular momentum conservation holds only in central force fields, as arise, for example, in planetary motion. Conservation of the angular momentum means constancy of its orientation as well as its magnitude.

Conservation of orientation

Conservation of $\mathbf{L} = \mathbf{r} \times \mathbf{p}$ means that the plane spanned by $\mathbf{r}$ and $\mathbf{p}$ remains fixed in spatial orientation; hence the motion proceeds in a plane.

Conservation of the *magnitude* of the angular momentum is often denoted as *law of areas*. The area covered by the "radius vector" $\mathbf{r}$ during the time element dt is

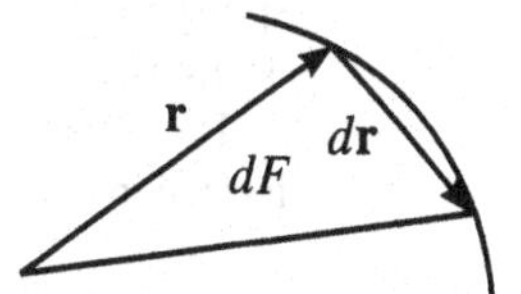

The area $dF = \frac{1}{2}|\mathbf{r} \times d\mathbf{r}|$ spanned by the vectors $\mathbf{r}$ and $d\mathbf{r}$.

$$dF = \frac{1}{2}\,|\mathbf{r} \times d\mathbf{r}\,| = \frac{1}{2}\,|\mathbf{r} \times \mathbf{v}\,|\,dt\,.$$

With $\mathbf{L} = \mathbf{r} \times \mathbf{p} = \mathbf{r} \times m\mathbf{v} = m(\mathbf{r} \times \mathbf{v})$, it holds that

$$dF = \frac{1}{2m}\,|\mathbf{L}|\,dt \qquad \text{or} \qquad \frac{dF}{dt} = \frac{1}{2m}\,|\mathbf{L}|\,,$$

where dF/dt is the area velocity of the radius vector (area covered per unit time). For the planetary motion, the law of areas is identical to the second Kepler law:

The radius vector of a planet covers equal areas in equal times.

The law of areas follows directly from the angular momentum conservation law and holds generally for arbitrary central fields, that is, also for the gravitational force, which is a central force with the sun as center. In the perihelion (closest distance to sun) the planet is moving faster than in the aphelion (largest distance to sun).

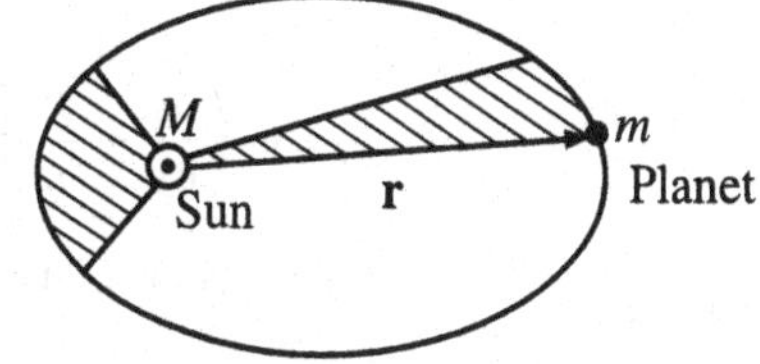

Illustration of the law of equal areas.

The situation in the example of Problem 17.5 is similar: The area velocity is constant, and hence the velocity v at $\mathbf{r} = \pm\mathbf{b}$ is maximum; at $\mathbf{r} = \pm\mathbf{a}$ it is minimum.

Example 17.5: Forces in the motion on an ellipse

We calculate the force to be applied to a mass point of constant mass to get it moving along the ellipse

$$\mathbf{r} = a\cos\omega t\,\mathbf{e}_1 + b\sin\omega t\,\mathbf{e}_2\,.$$

It is easy to verify from this parameterization the normal form of the equation of an ellipse (see also Chapter 26),

$$\frac{x^2}{a^2} + \frac{y^2}{b^2} = 1.$$

Starting from the second Newtonian axiom, the following *ansatz* results:

$$\begin{aligned}\mathbf{F} &= m\frac{d\mathbf{v}}{dt} = m\frac{d^2\mathbf{r}}{dt^2} = m\frac{d^2}{dt^2}(a\cos\omega t\,\mathbf{e}_1 + b\sin\omega t\,\mathbf{e}_2)\\ &= -m\omega^2\left[(a\cos\omega t)\,\mathbf{e}_1 + (b\sin\omega t)\,\mathbf{e}_2\right]\\ &= -m\omega^2\mathbf{r}.\end{aligned}$$

The force acts opposite to the direction of the position vector; it is an attractive central force. The force center lies at the midpoint of the ellipse.

Such forces that increase linearly with the distance play an important role, for example, for the spring (Hooke's law – see Section 18) and between the quarks, the primordial constituents of protons, neutrons, and mesons.

The planets also move around the sun along elliptic orbits. The sun as the center of attraction is located in one of the focal points of the ellipse. As we will see later in Chapter 26, the force acting is

$$\mathbf{F}_\mathrm{G} = -\gamma\frac{mM}{r^2}\frac{\mathbf{r}}{r},$$

that is, the gravitational force between the sun (mass M) and the planet (mass m).

We show that this force field is conservative. A necessary and sufficient condition for this property is the vanishing of the rotation of the force:

$$\operatorname{curl}\mathbf{F} = 0,$$

$\operatorname{curl}\mathbf{F} = -m\omega^2\operatorname{curl}\mathbf{r}$; hence it suffices to calculate the rotation of $\mathbf{r}$:

$$\begin{aligned}\operatorname{curl}\mathbf{r} &= \begin{vmatrix}\mathbf{e}_1 & \mathbf{e}_2 & \mathbf{e}_3\\ \dfrac{\partial}{\partial x} & \dfrac{\partial}{\partial y} & \dfrac{\partial}{\partial z}\\ x & y & z\end{vmatrix}\\ &= \mathbf{e}_1\left(\frac{\partial z}{\partial y} - \frac{\partial y}{\partial z}\right) + \mathbf{e}_2\left(\frac{\partial x}{\partial z} - \frac{\partial z}{\partial x}\right) + \mathbf{e}_3\left(\frac{\partial y}{\partial x} - \frac{\partial x}{\partial y}\right) = 0,\end{aligned}$$

namely, the rotation of the position vector vanishes.

Calculation of the potential at a point P (at the position r) with respect to the zero of the potential at point A (at the position a): We take a fixed point A on the ellipse (see sketch) and calculate the potential difference between A and the points of the trajectory given by $\mathbf{r}$.

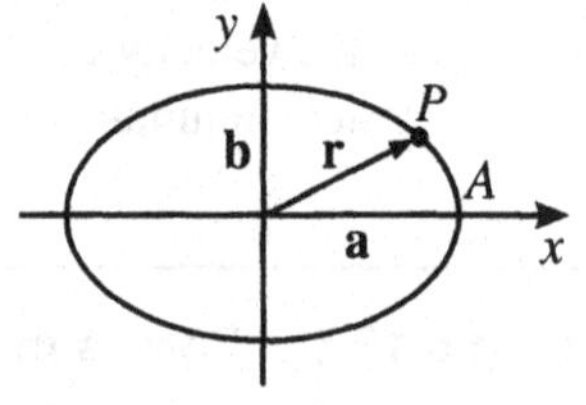

Illustration of elliptic motion.

$$V(x, y, z) = -\int_{\mathbf{a}}^{\mathbf{r}} \mathbf{F}\cdot d\mathbf{r} = m\omega^2\int_{\mathbf{a}}^{\mathbf{r}} \mathbf{r}\cdot d\mathbf{r}$$

$$= \frac{1}{2} m\omega^2 \int_{\mathbf{a}}^{\mathbf{r}} d(\mathbf{r}\cdot\mathbf{r})$$

$$= \frac{1}{2} m\omega^2 \mathbf{r}^2 \Big|_{\mathbf{a}}^{\mathbf{r}} = \frac{1}{2} m\omega^2 (\mathbf{r}^2 - \mathbf{a}^2).$$

With $\mathbf{r}^2 = r^2$, it results that

$$V(x, y, z) = \frac{1}{2} m\omega^2 (r^2 - a^2).$$

For $\mathbf{r} = \mathbf{a}$, one has $V(\mathbf{a}) = 0$, as it should be.

Calculation of the kinetic energy: The velocity is

$$\mathbf{v} = \dot{\mathbf{r}} = (-\omega a \sin\omega t)\mathbf{e}_1 + (\omega b \cos\omega t)\mathbf{e}_2 ,$$

$$|\mathbf{v}| = \sqrt{\omega^2 a^2 \sin^2\omega t + \omega^2 b^2 \cos^2\omega t} = v ,$$

$$T = \frac{1}{2} m v^2 = \frac{1}{2} m\omega^2 \left(a^2 \sin^2\omega t + b^2 \cos^2\omega t\right).$$

The kinetic energy is always positive and nonzero, as it must be in this case, to keep the mass point on the trajectory.

Calculation of the total energy: The total energy is the sum of $E = T + V$. By inserting the derived relations for T and V, we get

$$E = \frac{1}{2} m\omega^2 \left[a^2 \left(\sin^2\omega t + \cos^2\omega t\right) + b^2 \left(\cos^2\omega t + \sin^2\omega t\right) - a^2\right]$$

$$= \frac{1}{2} m\omega^2 \left(a^2 + b^2 - a^2\right)$$

$$= \frac{1}{2} m\omega^2 b^2 = \text{constant},$$

that is, the total energy is time-independent. The distinction of the half-axis b stems from our choice of referring the potential energy to the point $(x = a,\ y = 0)$. For $\mathbf{r} = \pm\mathbf{a}$ the total energy is exclusively kinetic energy; for $\mathbf{r} = \pm\mathbf{b}$ the kinetic energy is maximum, the potental energy is minimum, namely

$$V(\mathbf{b}) = \frac{1}{2} m\omega^2 (b^2 - a^2).$$

Problem 17.6: Calculation of angular momentum and torque

Find the torque $\mathbf{D}$ and the angular momentum $\mathbf{L}$ with respect to the origin for a mass point m moving on the trajectory $\mathbf{r} = (a\cos\omega t, b\sin\omega t)$.

Solution

$$\mathbf{r} = (a\cos\omega t, b\sin\omega t) \qquad \text{where} \quad a, b = \text{constant}$$

$$\mathbf{L} = \mathbf{r}\times\mathbf{p} = \mathbf{r}\times m\mathbf{v} = m(\mathbf{r}\times\mathbf{v}),$$

$$\mathbf{v} = \dot{\mathbf{r}} = (-a\omega \sin\omega t, b\omega\cos\omega t),$$

$$\mathbf{L} = \begin{vmatrix} \mathbf{e}_1 & \mathbf{e}_2 & \mathbf{e}_3 \\ a\cos\omega t & b\sin\omega t & 0 \\ -a\omega\sin\omega t & +b\omega\cos\omega t & 0 \end{vmatrix} \cdot m$$

$$= m\mathbf{e}_3(ab\,\omega\cos^2\omega t + ab\,\omega\sin^2\omega t)$$

$$= ab\,\omega\, m\mathbf{e}_3,$$

that is, $\mathbf{L}$ is time-independent, because $\mathbf{L} = \overrightarrow{\text{constant}}$. From there it follows that

$$\mathbf{D} = \dot{\mathbf{L}} = 0.$$

Hence, the force must be a central force. The mass point moves along an ellipse with the half-axes a and b, because

$$x = a\cos\omega t, \qquad y = b\sin\omega t,$$

and, therefore,

$$\frac{x^2}{a^2} + \frac{y^2}{b^2} = \cos^2\omega t + \sin^2\omega t = 1.$$

Problem 17.7: Show that the given force field is conservative

Show that the following force field is conservative:

$$\mathbf{F} = (y^2z^3 - 6xz^2)\mathbf{e}_1 + 2xyz^3\mathbf{e}_3 + (3xy^2z^2 - 6x^2z)\mathbf{e}_3\,.$$

Solution One has to show that $\operatorname{curl}\mathbf{F} = 0$:

$$\operatorname{curl}\mathbf{F} = \begin{vmatrix} \mathbf{e}_1 & \mathbf{e}_2 & \mathbf{e}_3 \\ \dfrac{\partial}{\partial x} & \dfrac{\partial}{\partial y} & \dfrac{\partial}{\partial z} \\ y^2z^3 - 6xz^2 & 2xyz^3 & 3xy^2z^2 - 6x^2z \end{vmatrix}$$

$$= \mathbf{e}_1\left[\frac{\partial}{\partial y}(3xy^2z^2 - 6x^2z) - \frac{\partial}{\partial z}(2xyz^3)\right]$$

$$+ \mathbf{e}_2\left[\frac{\partial}{\partial z}(y^2z^3 - 6xz^2) - \frac{\partial}{\partial x}(3xy^2z^2 - 6x^2z)\right]$$

$$+ \mathbf{e}_3\left[\frac{\partial}{\partial x}(2xyz^3) - \frac{\partial}{\partial y}(y^2z^3 - 6xz^2)\right]$$

$$= \mathbf{e}_1(6xyz^2 - 6xyz^2) + \mathbf{e}_2\left[(3y^2z^2 - 12xz) - (3y^2z^2 - 12xz)\right] + \mathbf{e}_3(2yz^3 - 2yz^3)$$

$$= 0,$$

that is, $\mathbf{F}$ is a conservative force field.

Problem 17.8: Force field, potential, total energy

(a) Show that $\mathbf{F} = \eta r^3 \mathbf{r}$ is conservative.

(b) Calculate the potential of a mass point in this field.

(c) What is the total energy of the mass point?

Solution (a)

$$\begin{aligned}
\operatorname{curl}\mathbf{F} &= \nabla \times \mathbf{F} \\
&= -\eta\Big[\mathbf{e}_1\left[3zy\left(x^2+y^2+z^2\right)^{1/2} - 3zy\left(x^2+y^2+z^2\right)^{1/2}\right] \\
&\quad + \mathbf{e}_2\left[3xz\left(x^2+y^2+z^2\right)^{1/2} - 3xz\left(x^2+y^2+z^2\right)^{1/2}\right] \\
&\quad + \mathbf{e}_3\left[3xy\left(x^2+y^2+z^2\right)^{1/2} - 3xy\left(x^2+y^2+z^2\right)^{1/2}\right]\Big] \\
&= 0,
\end{aligned}$$

where $|\mathbf{r}| = \sqrt{x^2+y^2+z^2}$ and $|\mathbf{r}|^3 = r^3 = (x^2+y^2+z^2)^{3/2}$ have been used.

(b) Potential:

$$V = \int_{\mathbf{r}_0=0}^{\mathbf{r}} \mathbf{F}\cdot d\mathbf{r} = \eta \int_{r_0=0}^{r} r^3\mathbf{r}\cdot d\mathbf{r} = \eta \int_{r_0=0}^{r} r^4 dr = \eta\frac{r^5}{5},$$

with $\mathbf{r}\cdot d\mathbf{r} = \frac{1}{2}d(\mathbf{r}\cdot\mathbf{r}) = \frac{1}{2}d(\mathbf{r}^2) = \frac{1}{2}d(r^2) = r\,dr$.

(c) Because the force field is conservative, the energy law $E = T + V = \text{constant}$ holds:

$$T = \frac{1}{2}m\dot{\mathbf{r}}^2; \qquad V(\mathbf{r}) = \frac{1}{5}\eta r^5.$$

It then follows that

$$E = \frac{1}{2}m\dot{\mathbf{r}}^2 + \frac{1}{5}\eta r^5.$$

Problem 17.9: Momentum and force at a ram pile

A crane lifts a mass of weight 1000 kg by 8.5 m upward. Afterward the weight falls onto a ram pile.

(a) Determine the transferred momentum.

(b) Determine the force acting onto the pile if the time of the impulse is 1/100 s.

Solution (a) After the crane releases the weight, it falls under the action of gravity with the velocity

$$v = gt.$$

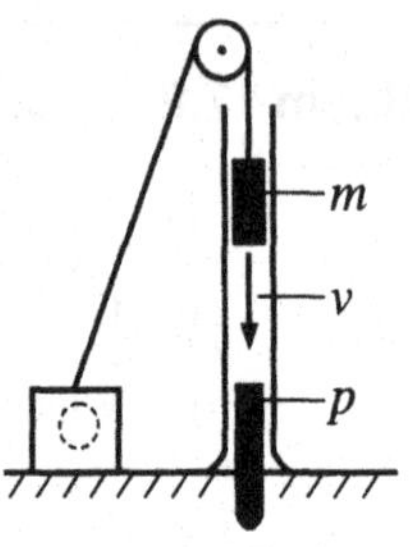

Forces at a ram pile.

From the considerations on the free fall we know that

$$s = \frac{1}{2}gt^2 \quad \text{and} \quad t = \sqrt{\frac{2s}{g}}$$

and thus obtain for the velocity of the falling mass

$$v = \sqrt{2sg} = \sqrt{2 \cdot 8.5 \cdot 9.81 \,\frac{\text{m}^2}{\text{s}^2}} = 12.9\frac{\text{m}}{\text{s}}$$

and for the momentum

$$p_1 = mv_1 = 1.29 \cdot 10^4 \,\frac{\text{kg m}}{\text{s}}.$$

After the impact onto the pile the momentum practically equals zero, that is,

$$p_2 \approx 0,$$

and the momentum change is

$$\Delta p = p_1 - p_2 \approx p_1.$$

Hence, the momentum transferred to the pile is

$$P = \Delta p = 1.29 \cdot 10^4 \,\frac{\text{kg m}}{\text{s}}.$$

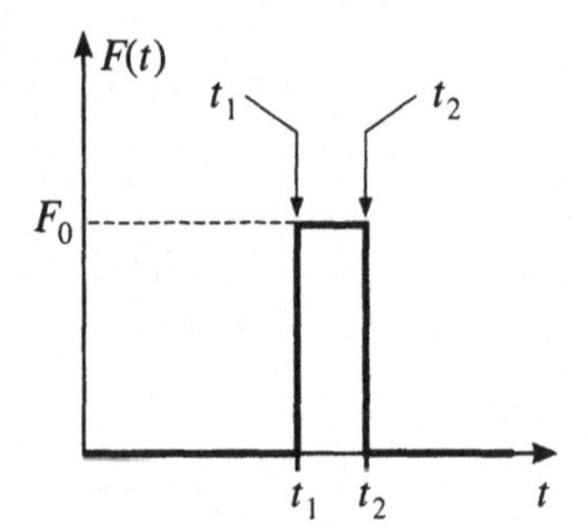

The force at the ram pile at a function of time.

(b) Assuming that the impulse is transferred within $1/100$ s and the force is constant over this time interval, one obtains for the acting force (see the figure)

$$F_0 = \frac{\Delta p}{\Delta t} = \frac{1.29 \cdot 10^4 \text{ kg m/s}}{10^{-2} \text{ s}} = 1.29 \cdot 10^6 \text{ N}.$$

Example 17.10: Elementary considerations on fictitious forces

A manned satellite was launched into an orbit about earth. We assume that gravity is absent everywhere within the satellite. We discuss the correctness of this assertion.

As is known, the only force acting on the satellite is the gravitational force of earth (see figure on next page). The acceleration

$$a_R = \frac{GM}{R^2} \tag{17.27}$$

points toward the center of earth; therefore, the satellite moves on a closed elliptical or spherical orbit.

If we consider the earth as being at rest, a fictitious force acts on each mass m in the frame of the satellite,

$$F_s = -ma_R,$$

which points away from the earth's center. In the satellite frame any body is under the action of the gravitation F_g and of the centrifugal fictitious force F_s,

$$F = F_g + F_s = m\frac{GM}{R^2} - ma_R.$$

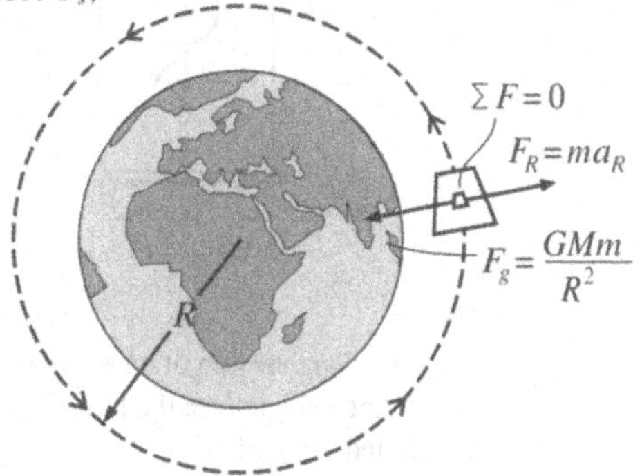

A manned satellite in a circular orbit around earth. Due to the gravitaional attraction of the earth, the same acceleration is acting on the satellite and every object within the satellite. In the system of the satellite, no forces act on these objects.

In view of equation 17.27 it is seen immediately that the resulting force on all bodies vanishes and that these bodies seemingly move acceleration-free within the satellite.

If we consider the problem in the earth-fixed system, then both the satellite as well as its objects are affected by the same acceleration and therefore follow the same path. Nothing is any more gravitation-free, and the objects in the satellite fall toward earth with the acceleration a_R (compare equation 17.27). But the satellite also falls with the same acceleration a_R, such that the relative acceleration between the objects and the satellite itself vanishes. Please note that this coordinate system is not an inertial system! In this example the centrifugal force just compensates the earth's attraction force. In other cases, for example, if an aircraft performs a loop, the centrifugal force may exceed the attractive force.

Another typical example for the appearance of fictitious forces is the acceleration meter. Let us consider a closed railway carriage in which a mass m is suspended at the ceiling by a string, allowing for free vibrations. If the carriage is accelerated, an observer sitting inside may notice that the pendulum is deflected by the angle θ against the vertical. The mass feels the fictitious force $F_s = -ma$, with a being the acceleration of the carriage. Because the resulting force must point away from the suspension point, the pendulum is deflected by the angle θ, because

$$\tan\theta = \frac{F_s}{F_g} = \frac{a}{g}.$$

If the carriage is at rest or in uniform motion, the string of the pendulum hangs vertically downward. (a) If the carriage is accelerated, then a fictitious force drives the mass in the opposite direction. Because both the fictitious force as well as the gravitational force are acting, the angle θ results.

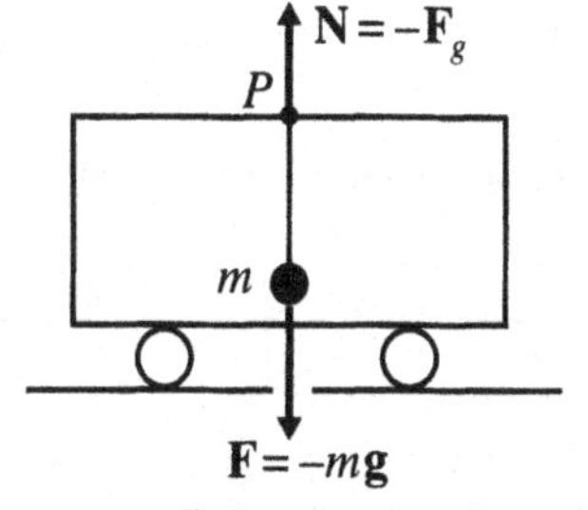

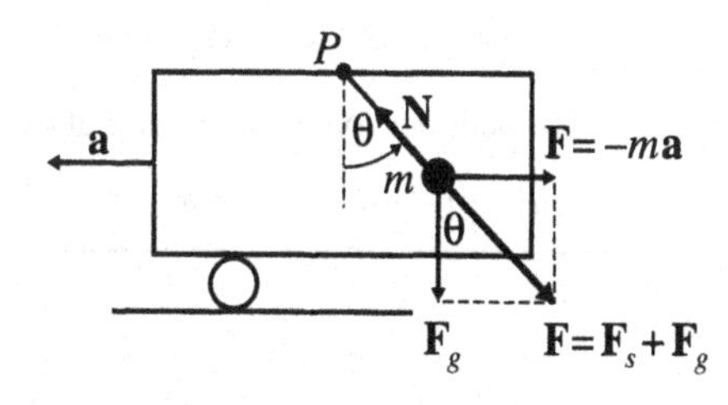

Railway carriage in uniform motion or at rest (left) or accelerated (right).

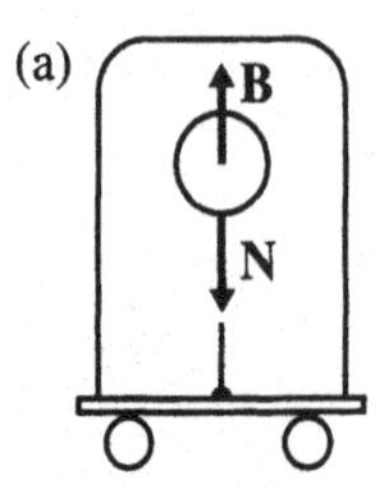

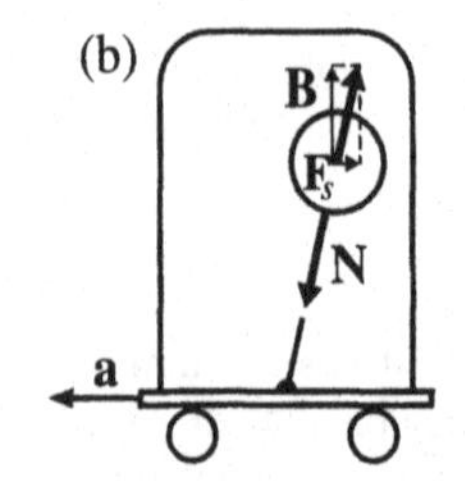

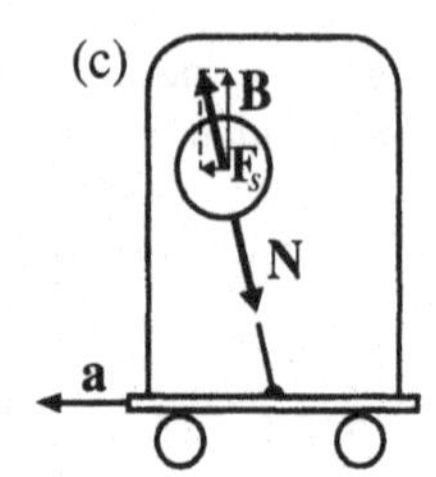

The balloon experiment.

An amusing version of an acceleration meter is a vehicle with a helium balloon below a glass bell fixed to the vehicle (see the figure below). In which direction will the balloon move if the vehicle is accelerated forward?

An acceleration meter is realized by a helium balloon below a glass bell fixed to a vehicle. If the vehicle is at rest or moves uniformly, the balloon stands vertically (a). If the vehicle is accelerated, the balloon is deflected in the same direction (c). Case (b) is wrong.

One might imagine intuitively that the balloon moves to the back since the sum of fictitious force F_s and buoyancy force B would point to the back (see figure — case (b). But this is wrong; the balloon moves forward (case (c)).

This may be explained as follows. Why does the balloon fly? It flies because the "pressure" below the balloon is higher than above it. This is due to the attractive force onto the air molecules. The difference of pressure causes a force that exceeds the attractive force onto the helium inside the balloon by the buoyancy force B. If the vehicle is accelerated, then the fictitious force acts on the air molecules; these drift to the backward side and create an overpressure that drives the balloon forward.

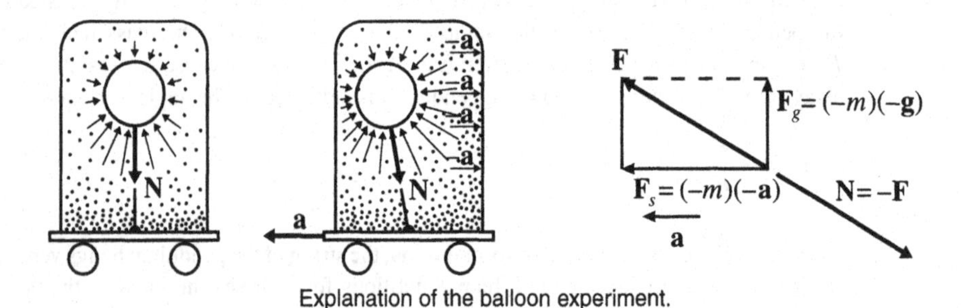

Explanation of the balloon experiment.

An ingenious theoretical trick may be applied to this example. Because the balloon is pushed upward against the gravitational force, we consider the balloon as an object of negative mass, $-m$. The gravitational force is then

$$F_g = (-m)(-g) = mg = B.$$

The fictitious force points parallel to the direction of acceleration, because one has

$$F_s = (-m)(-a) = +ma.$$

18 The General Linear Motion

We consider a linear (one-dimensional) motion of a mass point in the potential

$$V = V(x) = -\int_0^x F(x')\,dx'. \tag{18.1}$$

A potential always exists in this case because

$$\operatorname{curl}\mathbf{F}(x) = \begin{vmatrix} \mathbf{e}_1 & \mathbf{e}_2 & \mathbf{e}_3 \\ \dfrac{\partial}{\partial x} & \dfrac{\partial}{\partial y} & \dfrac{\partial}{\partial z} \\ F(x) & 0 & 0 \end{vmatrix} = 0. \tag{18.2}$$

This result is rather plausible, because rotation cannot develop in only one dimension.

In a conservative force field, the energy law holds:

$$E = T + V = \frac{1}{2}mv^2 + V(x) = \frac{1}{2}m\left(\frac{dx}{dt}\right)^2 + V(x). \tag{18.3}$$

In the one-dimensional problem this law always applies, provided that the forces are only position-dependent. Velocity-dependent forces (e.g., friction forces) in general may not be represented by a potential and hence are not conservative.

Equation (18.3) is a *differential equation of first order*; its solution yields the dependence of the position on time, that is $x(t)$.

Differential equations are equations for unknown functions (in our case $x(t)$) that also involve the derivatives of these functions (in our case $\dot{x}(t)$). If $d^n x/dt^n$ occurs as the highest derivative in the equation, the differential equation is called to be "of nth order."

Equation (18.3) is solved by "separation of variables" and subsequent definite integration:

$$\frac{1}{2}m\left(\frac{dx}{dt}\right)^2 = E - V(x). \tag{18.4}$$

The transformation is performed in such a way that all terms containing x stand on one side, and terms depending on t stand on the other side:

$$\pm\frac{dx}{\sqrt{(2/m)\big(E-V(x)\big)}}=dt. \tag{18.5}$$

The integration may then be performed and yields

$$\int\limits_{x_1}^{x}\frac{dx}{\sqrt{(2/m)\big(E-V(x)\big)}}=\pm\int\limits_{t_1}^{t}dt,$$

$$t=t(x)=t_1\pm\int\limits_{x_1}^{x}\frac{dx}{\sqrt{(2/m)\big(E-V(x)\big)}}. \tag{18.6}$$

The wanted solution is obtained by forming the inverse function $x=x(t)$ of the function $t=t(x)$.

As an application of the general linear motion, we shall investigate a motion in the *oscillator potential* (parabola potential):

$$V(x)=\frac{1}{2}k\,x^2$$

$$(k>0);\qquad \mathbf{r}=(x,0,0).$$

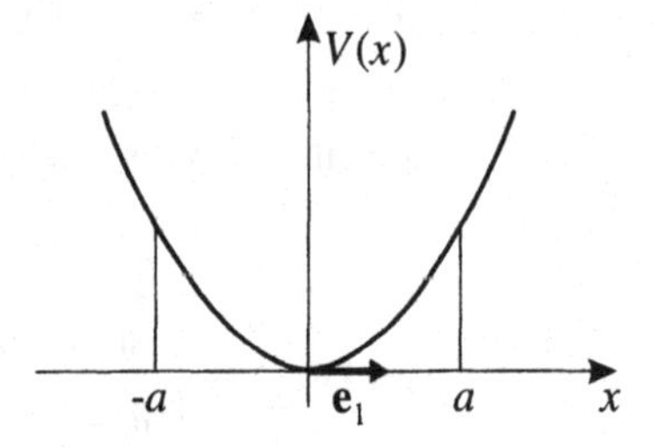

The potential of the linear harmonic oscillator.

The force $\mathbf{F}(x)$ results from the potential

$$\mathbf{F}(x)=-\operatorname{grad}V=-\nabla V=-\frac{\partial V}{\partial x}\mathbf{e}_1,$$

that is,

$$\mathbf{F}(x)=-kx\,\mathbf{e}_1. \tag{18.7}$$

Therefore, k is also called a *force constant.*

At the point $x=0$ no force is acting; here the body moves force-free. If $x>0$, the force is negative, $\mathbf{F}\uparrow\uparrow-\mathbf{e}_1$; if $x<0$, then $\mathbf{F}\uparrow\uparrow\mathbf{e}_1$. That means the force is backdriving and tries to counteract any displacement. One may expect that this type of motion is a vibrational one.

Let the following *initial conditions* be given: At the time $t=0$, let $x=a$ and $\dot{x}=v=0$, that is, the mass point is at rest at the position $x=a$ and is released at the time $t=0$. The total energy of the system is then

$$E=\frac{1}{2}k\,a^2. \tag{18.8}$$

Hence: $T + V = E$, or explicitly,

$$\frac{1}{2}m\left(\frac{dx}{dt}\right)^2 = \frac{1}{2}k\,a^2 - \frac{1}{2}k\,x^2.$$

From this it follows that

$$\frac{dx}{dt} = \pm\sqrt{\frac{k}{m}(a^2 - x^2)}, \qquad \frac{dx}{\sqrt{a^2 - x^2}} = \pm dt\sqrt{\frac{k}{m}}. \tag{18.9}$$

In the last step we have separated the variables x and t: To the left, dx together with the x-dependent factor $1/\sqrt{a^2 - x^2}$; to the right, dt multiplied by $\sqrt{k/m}$. From this equation, we get by indefinite integration

$$\int \frac{dx}{\sqrt{a^2 - x^2}} = \pm \int \sqrt{\frac{k}{m}}\,dt \tag{18.10}$$

or

$$\arcsin\left(\frac{x}{a}\right) = \sqrt{\frac{k}{m}}\,t + \text{constant} \tag{18.11}$$

We have adopted the positive sign in (18.9). One may easily check that the same result is obtained when using the negative sign. The function $\arcsin x$ is the inverse function of $\sin x$. The result of the integration becomes clear by differentiation: If $y = \arcsin x$, then $x = \sin y$, and further:

$$\frac{dy}{dx} = \frac{1}{dx/dy} = \frac{1}{\cos y} = (1 - \sin^2 y)^{-\frac{1}{2}} = (1 - x^2)^{-\frac{1}{2}}. \tag{18.12}$$

The integration constant is now determined from the initial conditions. At the time $t = 0$, there is $x = a$, and therefore

$$\text{constant} = \arcsin\left(\frac{a}{a}\right) = \frac{\pi}{2}.$$

The function obtained therefore reads

$$\arcsin\left(\frac{x}{a}\right) = \sqrt{\frac{k}{m}}\,t + \frac{\pi}{2}. \tag{18.13}$$

From there the inverse function $x = x(t)$ is obtained as

$$\frac{x}{a} = \sin\left(\sqrt{\frac{k}{m}}\,t + \frac{\pi}{2}\right) \quad \text{or}$$

$$x = a\sin\left(\sqrt{\frac{k}{m}}\,t + \frac{\pi}{2}\right). \tag{18.14}$$

We introduce the abbreviation $\omega = \sqrt{k/m}$; ω is called the *angular frequency*. $\omega = 2\pi\nu$; ν is the *frequency*. We thus obtain the function in the form

$$x = a \sin\left(\omega t + \frac{\pi}{2}\right) = a \cos \omega t. \tag{18.15}$$

For $t = T = 2\pi/\omega = 1/\nu$, the particle is again back at the starting point. T is called the *period of vibration* or the *vibration time*. For $t = T/2 = \pi/\omega$, we have $x = -a$ and $\dot{x} = 0$.

Here we are dealing with *harmonic vibrations*. For $x = \pm a$ the potential equals the total energy of the system, for $x = 0$ the kinetic energy. An example of a motion in a potential of the form $V(x) = \frac{1}{2}k x^2$ is the spring vibration for not too large displacements a (see figure).

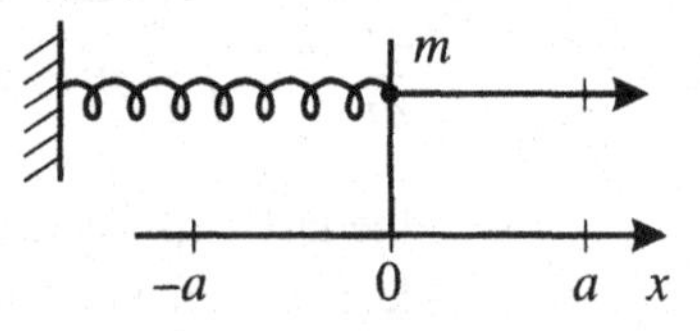

The vibrating spring: Its position at rest is $x = 0$. There, the spring is in equilibrium. Its elongation may be positive ($x > 0$—the spring is stretched) or negative ($x < 0$—the spring is compressed).

The spring vibration obeys the **Hooke[1] law**:

The force is proportional to the displacement. Hence there holds a linear force law:

$$\mathbf{F}(x) = -kx\mathbf{e}_1.$$

[1] *Robert Hooke*, British naturalist, b. July 18, 1635, Freshwater (Isle Wight)—d. March 3, 1703, London. Hooke was at first assistant with R. Boyle, became in 1662 Curator of Experiments of the Royal Society, in 1665 professor of geometry at Gresham College in London, and from 1677 to 1682 secretary of the Royal Society. Hooke improved already known methods and instruments, for example, the air pump and the composed microscope (described in his *Micrographia* in 1664). He frequently became involved in disputes on priority, for example, with Ch. Huygens, J. Hevelius, and in particular with I. Newton, with whom he was hostile. Hooke proposed, among other things, the melting point of ice as zero point of the thermometric scale (1664); he recognized the constancy of the melting point and boiling point of substances (1668); and he first observed the black spots on soap bubbles. He gave a conceptionally good definition of elasticity, and in 1679 he invented what we now call Hooke's law [BR].

19 The Free Fall

We consider the motion of a body under the influence of gravity. To make the problem tractable, we make a number of simplifications. We assume that the attraction by earth is constant, that is, the distance traversed in the fall shall be very small as compared with the earth's radius. Except for gravitation, no other forces shall act. This means: We neglect the air friction and consider the earth as an inertial frame. These simplifications will be dropped gradually to get a complete description of the problem.

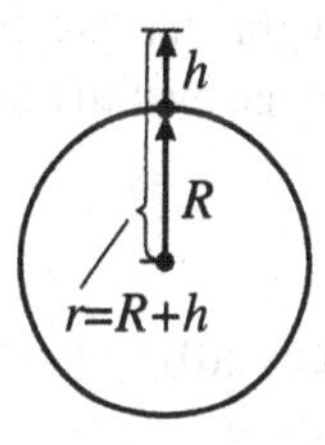

The earth with radius R. The height above the surface of the earth is z. z is supposed to be small compared to R.

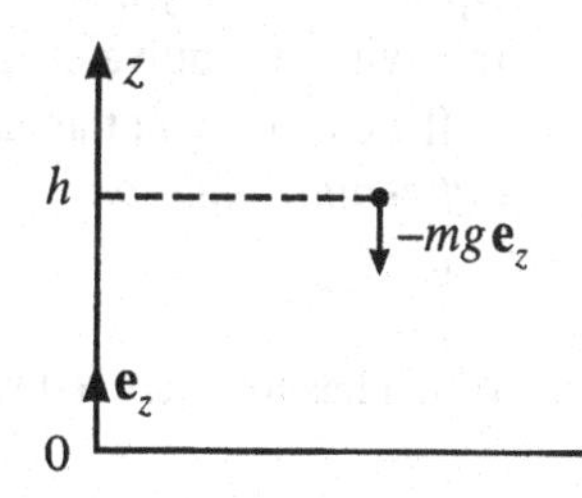

A useful coordinate system for motion near the earth's surface has its z-axis ($\mathbf{e}_z$) showing radially away outward from the center of the earth.

The gravitational force of a point mass M_s on another point mass m_s at the mutual distance $\mathbf{r}$ is

$$\mathbf{F}_{m_s} = -\gamma \frac{M_s m_s}{r^2} \frac{\mathbf{r}}{r}.$$

The distance vector $\mathbf{r}$ between two masses m_s and M_s.

This force law is fundamental for the classical (not generally relativistic) theory of gravitation. Here γ is the gravitational constant, which is given by (compare Example 26.1)

$$\gamma = 6.67 \cdot 10^{-11} \frac{\mathrm{N\,m^2}}{\mathrm{kg^2}}.$$

The earth, although being extended, is assumed to have its total mass M_s united at the earth's center. In the vicinity of the earth's surface, this force may be simplified:

$$\begin{aligned}\mathbf{F} &= -\gamma\frac{M_s m_s}{(R+z)^2}\mathbf{e}_r\\ &\approx -\gamma\frac{M_s m_s}{R^2}\left(1-2\frac{z}{R}\right)\mathbf{e}_r\\ &= -g\,m_s\left(1-2\frac{z}{R}\right)\mathbf{e}_r\\ &\approx -g\,m_s\mathbf{e}_r \quad \text{for } z \ll R.\end{aligned}$$

Here g is the gravitational acceleration

$$g = \gamma\cdot\frac{M_s}{R^2} = 9.81\,\frac{\mathrm{m}}{\mathrm{s}^2}.$$

According to the second Newtonian axiom, we therefore write for the free fall

$$m_t\ddot{z}\,\mathbf{e}_3 = \mathbf{F} = -m_s g\,\mathbf{e}_3. \tag{19.1}$$

The indices shall point out that the concept of "mass" denotes two basically distinct properties of the body. The inert mass m_t is a property exhibited by the body under changes of its state of motion (acceleration). The heavy mass m_s is the origin of gravitation. The equality of heavy and inert mass is therefore not at all trivial. Only in the general theory of relativity the equivalence of inertial forces and gravitational forces is shown.

If we cancel out the masses in (19.1) and change over to scalar notation, there results the *differential equation*

$$\ddot{z} = -g,$$

which has to be solved with the initial conditions $z(0) = h$, $\dot{z}(0) = 0$. We obtain

$$\frac{d\dot{z}}{dt} = -g,$$

from which it follows by integration that

$$\dot{z}(t) = -gt + C = -gt.$$

Because for $t = 0$ $\dot{z}(0) = 0$, there must be $C = 0$. A further integration yields

$$z(t) = h - \frac{1}{2}g\,t^2.$$

Vertical throw

If we solve the differential equations (19.1) with the initial conditions $z(0) = 0$ and $\dot{z}(0) = v_0$, we describe a vertical throw upward. The solution is

$$\dot{z}(t) = v_0 - gt, \tag{19.2}$$

$$z(t) = v_0 t - \frac{1}{2} g t^2. \tag{19.3}$$

The *time of ascent* $t = T$ may be determined as follows: At the reversal point we have $\dot{z}(T) = 0$, and by inserting in (19.2) we get

$$T = \frac{v_0}{g}.$$

If we now insert T into equation (19.3), we obtain the *maximum height of ascent*:

$$z(T) = -\frac{g v_0^2}{2g^2} + \frac{v_0^2}{g} = h; \qquad h = \frac{v_0^2}{2g}. \tag{19.4}$$

By means of (19.2) and (19.3), the velocity v may be given as function of the height of ascent z:

$$z = v_0 t - \frac{g}{2} t^2,$$

$$v(t) = \dot{z} = v_0 - gt;$$

hence

$$t = \frac{v_0 - v}{g}.$$

Now z is obtained as a function of v:

$$z(t) = -\frac{g(v_0 - v)^2}{2g^2} - \frac{2(v_0 v - v_0^2)}{2g} = \frac{v_0^2 - v^2}{2g} \qquad \left(\text{where } h = \frac{v_0^2}{2g}\right);$$

$$z(t) = h - \frac{v^2}{2g}.$$

Solving for v yields the wanted function

$$v(z) = \sqrt{2g(h - z)}.$$

$v(z)$ may also be determined via the energy conservation law, which must hold because $\operatorname{curl} \mathbf{F} = \operatorname{curl}(-m_s g \mathbf{e}_3) = 0$.

The potential is

$$V(\mathbf{r}) = -\int_0^z \mathbf{F} \cdot d\mathbf{r}$$

$$= -\int_0^z (0, 0, -mg) \cdot (dx, dy, dz)$$

$$= \int_0^z mg\, dz = mgz.$$

Thus, the energy law reads

$$E = \frac{m}{2}v^2 + mgz.$$

For $z = 0$, $v = v_0$, and $t = 0$, the total energy is

$$E = \frac{m}{2}v_0^2 + 0.$$

From there it follows that

$$E = \frac{m}{2}v_0^2 = \frac{m}{2}v^2 + mgz,$$

and with $v_0^2 = 2gh$, it further follows that

$$mgh - mgz = \frac{m}{2}v^2,$$

and therefore

$$v(z) = \sqrt{2g(h-z)}\,.$$

Inclined throw

We assume the same simplifications to apply as for the free fall. The initial velocity now has two components (in $\mathbf{e}_2$- and $\mathbf{e}_3$- directions).

Initial conditions: Let at the moment $t = 0$

$$\mathbf{r} = 0$$

and

$$\dot{\mathbf{r}} = \mathbf{v}_0 = v_0(\cos\alpha\,\mathbf{e}_2 + \sin\alpha\,\mathbf{e}_3);$$

α is the throw angle (see figure).

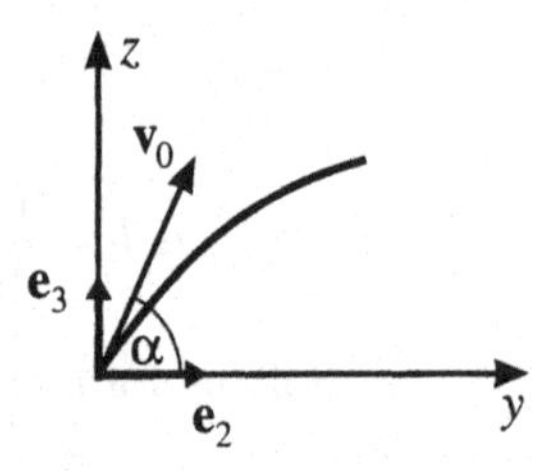

Initial condition of the inclined throw.

According to Newton's law, it again holds that

$$m\frac{d^2\mathbf{r}}{dt^2} = -mg\mathbf{e}_3$$

or

$$\frac{d\mathbf{v}}{dt} = -g\mathbf{e}_3\,.$$

After separation of variables and integration, we get

$$\mathbf{v}(t) = -gt\mathbf{e}_3 + \mathbf{c}_1.$$

From the initial conditions we obtain for $\mathbf{c}_1$

$$\mathbf{c}_1 = \mathbf{v}_0 = v_0(\cos\alpha\,\mathbf{e}_2 + \sin\alpha\,\mathbf{e}_3);$$

hence

$$\mathbf{v}(t) = (v_0\sin\alpha - gt)\mathbf{e}_3 + v_0\cos\alpha\,\mathbf{e}_2. \tag{19.5}$$

The *time of ascent* T is characterized by the vanishing of the $\mathbf{e}_3$-component of the velocity, $\mathbf{e}_3 \cdot \mathbf{v}(T) = 0$:

$$v_0 \sin\alpha - gT = 0,$$

and we obtain

$$T = \frac{v_0 \sin\alpha}{g}.$$

The position as a function of the time is obtained by integrating equation (19.5):

$$\mathbf{r}(t) = \left(v_0 t \sin\alpha - \frac{g}{2}t^2\right)\mathbf{e}_3 + v_0 t \cos\alpha\, \mathbf{e}_2. \tag{19.6}$$

Because $\mathbf{r}(t) = 0$ for $t = 0$, the integration constant also must be zero. The shape of the curve of motion is obtained by splitting (19.6) into components and eliminating the time. We have

$$y = t v_0 \cos\alpha; \qquad \text{thus} \quad t = \frac{y}{v_0 \cos\alpha}.$$

Furthermore, we have for the $\mathbf{e}_3$-component z:

$$z = -\frac{g}{2}t^2 + v_0 t \sin\alpha.$$

Inserting t yields

$$z = -\frac{g}{2}\left(\frac{y}{v_0 \cos\alpha}\right)^2 + y \tan\alpha. \tag{19.7}$$

This equation is a parabola equation of the form $-Ay^2 + By = z$, that is, a parabola downward open in the y, z-plane (see figure).

The *time of throw* t_0 that passes until the body again reaches the ground is obtained from the condition $z(t) = 0$ for $t \neq 0$. We then have

$$v_0 t_0 \sin\alpha - \frac{g}{2}t_0^2 = 0$$

and therefore

$$t_0 = \frac{2v_0 \sin\alpha}{g} = 2T.$$

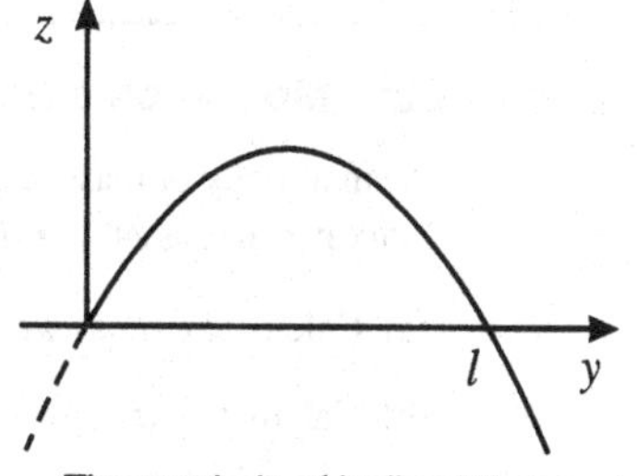

The parabola of inclined throw.

The time of throw is twice the time of ascent; thus, the curve of the throw motion is *symmetric*. The range of throw l is obtained by inserting the throw time $2T$ into (19.6):

$$l = 2T v_0 \cos\alpha = \frac{2v_0^2 \sin\alpha \cos\alpha}{g},$$

and converted:

$$l = \frac{v_0^2 \sin 2\alpha}{g}.$$

We immediately see that for the constant v_0 there is a *maximum range of throw* for the throw angle $\alpha = 45°$, because for $\sin 2\alpha = 1$, we have $\alpha = 45°$.

Problem 19.1: Motion of a mass in a constant force field

A mass point of mass m moves along a straight line under the action of the constant force $\mathbf{F}$. Its initial velocity at the time $t = 0$ is v_0.

Determine $v(t)$ and $x(t)$.

Solution The equation of motion holds:

$$F = mb = m\frac{dv}{dt}.$$

Separation of the variables and subsequent integration yield

$$\frac{F}{m}dt = dv,$$

$$v(t) - v(0) = \frac{F}{m}t, \quad \text{or}$$

$$v(t) = v_0 + \frac{F}{m}t.$$

The initial velocity v_0 corresponds to the position $x = x_0$; therefore, a further integration yields

$$x(t) = x_0 + v_0 t + \frac{F}{2m}t^2.$$

Problem 19.2: Motion on a helix in the gravitational field

A small body of mass m glides by its own weight $\mathbf{G} = \{0, 0, -mg\}$ frictionless downward along the helix $\mathbf{r} = \{a\cos\varphi(t), a\sin\varphi(t), c\varphi(t)\}$.

(a) Calculate $\varphi(t)$ as well as the path velocity and the guiding pressure as a function of the time.

(b) Calculate $\varphi(t)$ again by means of the energy law. The numerical data are $m = 1$ kg, $a = 2$ m, $c = 0.5$ m.

Solution (a) The motion of the mass m on the given helix is characterized by the following forces:

- the net weight $\mathbf{G} = \{0, 0, -mg\}$,
- the guiding pressure $\mathbf{F} = F_N\mathbf{N} + F_B\mathbf{B}$ normally to the path.

Thus, we need the tangent, normal, and binormal vector for further considerations.

We have

$$\mathbf{r} = \mathbf{r}(\varphi) = \{a\cos\varphi, a\sin\varphi, c\varphi\},$$

$$\mathbf{r}' = \frac{d\mathbf{r}}{d\varphi} = \{-a\sin\varphi, a\cos\varphi, c\},$$

$$\mathbf{r}'' = \frac{d^2\mathbf{r}}{d\varphi^2} = \{-a\cos\varphi, -a\sin\varphi, 0\}.$$

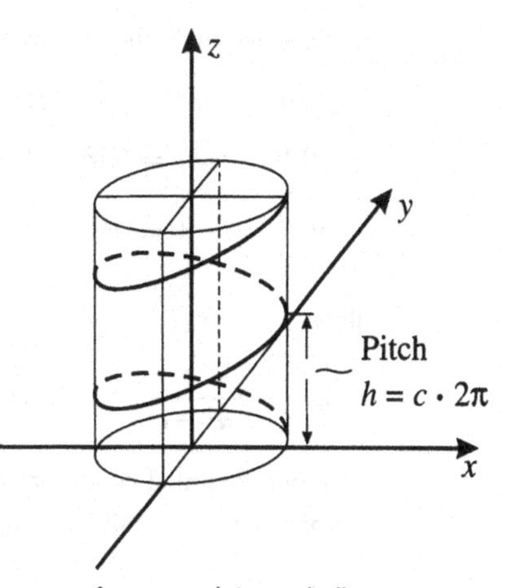

A mass point on a helix.

Thus, the vectors of the moving trihedral are

$$\mathbf{T} = \frac{\mathbf{r}'}{|\mathbf{r}'|} = \frac{1}{\sqrt{a^2+c^2}}\{-a\sin\varphi, a\cos\varphi, c\},$$

$$\mathbf{T}' = \frac{1}{\sqrt{a^2+c^2}}\{-a\cos\varphi, -a\sin\varphi, 0\},$$

$$\mathbf{N} = \frac{\mathbf{T}'}{|\mathbf{T}'|} = \frac{(a^2+c^2)^{-1/2}(-a\cos\varphi, -a\sin\varphi, 0)}{a(a^2+c^2)^{-1/2}}$$

$$= \{-\cos\varphi, -\sin\varphi, 0\},$$

$$\mathbf{B} = \mathbf{T}\times\mathbf{N}$$

$$= \frac{1}{\sqrt{a^2+c^2}}\begin{vmatrix} \mathbf{e}_x & \mathbf{e}_y & \mathbf{e}_z \\ -a\sin\varphi & a\cos\varphi & c \\ -\cos\varphi & -\sin\varphi & 0 \end{vmatrix}$$

$$= \frac{1}{\sqrt{a^2+c^2}}\{c\sin\varphi, -c\cos\varphi, a\}.$$

For the equation of motion we get

$$m\ddot{\mathbf{r}} = \mathbf{G} + \mathbf{F} = \mathbf{G} + F_N\mathbf{N} + F_B\mathbf{B},$$

and after scalar multiplication by **T**, **N**, and **B**:

$$\text{multiplication by } \mathbf{T}: \quad (m\ddot{\mathbf{r}} - \mathbf{G})\cdot\mathbf{T} = 0, \tag{19.8}$$

$$\text{multiplication by } \mathbf{N}: \quad (m\ddot{\mathbf{r}} - \mathbf{G})\cdot\mathbf{N} = F_N, \tag{19.9}$$

$$\text{multiplication by } \mathbf{B}: \quad (m\ddot{\mathbf{r}} - \mathbf{G})\cdot\mathbf{B} = F_B. \tag{19.10}$$

The time derivatives of **r** are

$$\dot{\mathbf{r}} = \frac{d}{dt}\mathbf{r} = \frac{d\mathbf{r}}{d\varphi}\frac{d\varphi}{dt} = \mathbf{r}'\dot{\varphi},$$

$$\ddot{\mathbf{r}} = \frac{d}{dt}(\dot{\mathbf{r}}) = \frac{d}{dt}(\mathbf{r}'\dot{\varphi}) = \mathbf{r}''\dot{\varphi}^2 + \mathbf{r}'\ddot{\varphi}.$$

After inserting the equation of the space curve, we find

$$\ddot{\mathbf{r}} = \{-a\cos\varphi, -a\sin\varphi, 0\}\dot{\varphi}^2 + \{-a\sin\varphi, a\cos\varphi, c\}\ddot{\varphi}.$$

Thereby the term $(m\ddot{\mathbf{r}} - G)$ in equations 19.8, 19.9, 19.10 turns into

$$m\ddot{\mathbf{r}} - \mathbf{G} = m\{-a\dot{\varphi}^2\cos\varphi - a\ddot{\varphi}\sin\varphi, -a\dot{\varphi}^2\sin\varphi + a\ddot{\varphi}\cos\varphi, c\ddot{\varphi} + g\}.$$

After scalar multiplication by the vectors of the moving trihedral, it follows for equation

19.9: $$F_N = (m\ddot{\mathbf{r}} - \mathbf{G}) \cdot \mathbf{N} = ma\dot{\varphi}^2, \qquad (19.11)$$

19.10: $$F_B = (m\ddot{\mathbf{r}} - \mathbf{G}) \cdot \mathbf{B} = mga\left(a^2 + c^2\right)^{-1/2}, \qquad (19.12)$$

19.8: $$0 = (m\ddot{\mathbf{r}} - \mathbf{G}) \cdot \mathbf{T} = \frac{m}{\sqrt{a^2 + c^2}}\left[\left(a^2 + c^2\right)\ddot{\varphi} + cg\right]; \qquad (19.13)$$

thus

$$\ddot{\varphi} = -g\frac{c}{a^2 + c^2}; \qquad \text{hence} \quad \varphi = C_2 + C_1 t - \frac{g}{2}t^2\frac{c}{a^2 + c^2}.$$

From the initial conditions $\varphi(t = 0) = \varphi_0$ and $\dot{\varphi}(t = 0) = 0$, it follows for the two integration constants: $C_1 = 0$ and $C_2 = \varphi_0$. Finally,

$$\varphi = \varphi(t) = \varphi_0 - \frac{c}{a^2 + c^2}\frac{g}{2}t^2,$$

$$\dot{\varphi}(t) = -cgt\left(a^2 + c^2\right)^{-1}.$$

For F_N we get according to equation 19.11:

$$F_N = \frac{mg^2ac^2t^2}{(a^2 + c^2)^2},$$

and for the resulting guiding pressure

$$F = \sqrt{F_N^2 + F_B^2} = mga(a^2 + c^2)^{-1/2} \cdot \sqrt{1 + \frac{g^2c^4}{(a^2 + c^2)^3}t^4} = F(t).$$

The path velocity is

$$\mathbf{v}(t) = \dot{\mathbf{r}} = \mathbf{r}'\dot{\varphi} = |\mathbf{r}'|\dot{\varphi}\mathbf{T} = \dot{\varphi}\sqrt{a^2 + c^2} \cdot \mathbf{T}.$$

The magnitude of the velocity—along the tangent—is

$$v(t) = \dot{\varphi}\sqrt{a^2 + c^2} = -gt\frac{c}{\sqrt{a^2 + c^2}}.$$

The negative sign characterizes the "downward motion" (for $c > 0$).

(b) To determine $\varphi(t)$ by means of the energy law, we compare the initial position $z_0(v(z_0) = 0)$ with an arbitrary intermediate position z. The result is

$$mgz_0 = mgz + \frac{m}{2}v^2$$

or, rewritten,

$$2g(z_0 - z) = v^2 = \dot{\mathbf{r}}^2 = \mathbf{r}'^2\dot{\varphi}^2.$$

Using $\mathbf{r}'^2 = a^2 + c^2$ and $z = c\varphi$ with $z_0 = c\varphi_0$, we obtain the following differential equation for $\varphi(t)$:

$$\dot{\varphi}^2 + \frac{2gc}{a^2 + c^2}(\varphi - \varphi_0) = 0,$$

or with the substitution $\psi = \varphi - \varphi_0$:

$$\dot{\psi}^2 + \frac{2gc}{a^2+c^2}\psi = 0 \qquad \text{or} \qquad \dot{\psi} = i\sqrt{\frac{2gc}{a^2+c^2}\cdot\psi} = \frac{d\psi}{dt}.$$

Separation of the variables leads to

$$i\sqrt{\frac{2gc}{a^2+c^2}}dt = \frac{d\psi}{\sqrt{\psi}},$$

and integration yields

$$i\sqrt{\frac{2gc}{a^2+c^2}}\int dt = \int\frac{d\psi}{\sqrt{\psi}} \qquad \text{or} \qquad i\sqrt{\frac{2gc}{a^2+c^2}}t = 2\sqrt{\psi},$$

and after forming the square:

$$\psi = -\frac{g}{2}t^2\frac{c}{a^2+c^2}.$$

Resubstitution finally yields

$$\varphi = \psi + \varphi_0 = \varphi(t) = \varphi_0 - \frac{c}{a^2+c^2}\frac{g}{2}t^2.$$

Problem 19.3: Spaceship orbits around the earth

A spaceship orbits around the earth at the height h above ground. Calculate (a) the orbital velocity, and (b) the orbital period such that zero gravity occurs in the spaceship. (c) Discuss these results for the case $h \ll R$.

Solution (a) Zero gravity $\Leftrightarrow$ earth attraction = centrifugal force.

$$\frac{mv^2}{R+h} = \frac{\gamma Mm}{(R+h)^2} = \frac{gR^2m}{(R+h)^2}, \qquad \text{because} \quad \frac{\gamma Mm}{R^2} = mg \quad \text{for } h = 0$$

$$\Rightarrow \quad v = \frac{R}{R+h}\sqrt{(R+h)g} \qquad \text{“orbital velocity.”}$$

(b)

$$v = \frac{\text{path length}}{\text{period}} = \frac{2\pi(R+h)}{T} \quad \Rightarrow \quad T = 2\pi\left(\frac{R+h}{R}\right)\sqrt{\frac{R+h}{g}}.$$

(c) For $h \ll R$ it follows that $v \approx \sqrt{Rg}$ and $T \approx 2\pi\sqrt{R/g}$.
The orbital velocity for $R = 6371$ km and $g = 9.81\text{m/s}^2$ is then

$$v \approx 7.9\,\frac{\text{km}}{\text{s}},$$

and the orbital period $T \approx 84$ min.

---

20 Friction

In general, any moving body undergoes a deceleration due to the interaction with its environment being at rest. The occurring friction forces are always directed opposite to the direction of motion; they are *not* conservative (the contour integral along a closed path is nonzero).

If we consider only the mechanical process, the energy conservation law does not apply: Kinetic energy is converted to heat.

Friction phenomena in a viscous medium

The friction of a body in gases and liquids is governed by the general *ansatz*

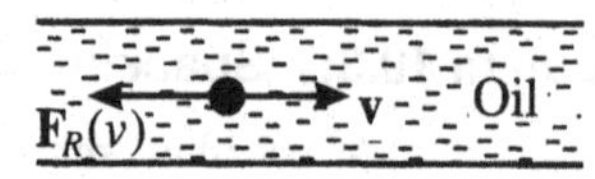

Illustration of friction in a viscous medium.

$$\mathbf{F}_R = -F(v)\frac{\mathbf{v}}{v}.$$

It always acts against the velocity $\mathbf{v}$. The function $F(v)$ is in general not simple and must be determined empirically.

As an *approximation* two approaches prove successful.

Stokes' friction $\mathbf{F}_R = -\beta \mathbf{v}$, $\beta = \text{constant} > 0$ (holds, e.g., for rapidly moving missiles or for the motion in viscous liquids).

Newtonian friction $\mathbf{F}_R = -\gamma v \mathbf{v}$, $\gamma = \text{constant} > 0$ (holds, e.g., for slowly moving missiles).

Friction phenomena between solid bodies: A solid body presses onto its support with the force $\mathbf{F}^{\perp}$. One may realize two distinct types of friction.

Friction of a solid body on a support.

(a) **Dynamic friction** ($v \neq 0$)

The effective friction force is over a wide range independent of the area of support and the velocity and is proportional to the force $F^{\perp}$ pressing the body onto the area (support load). Thus we may adopt the empirical *ansatz*:

$$\mathbf{F}_R = -\mu_g F^{\perp}\frac{\mathbf{v}}{v} \quad \textbf{(Coulomb)},$$

where μ_g is called the dynamic friction coefficient.

(b) **Static friction** ($v = 0$)

If the body is at rest, tractive forces **F** acting parallel to the support area are just compensated by static friction. This applies as long as the acting force remains below a maximum value that is proportional to the support load. Only if $F^{\parallel}$ becomes larger than a certain value $\mu_h F^{\perp}$ does the body begin to move. It is vividly clear that this "limit force" is proportional to the support load $F^{\perp}$.

Thus, the body remains at rest as long as

$$F^{\parallel} < \mu_h F^{\perp},$$

where μ_h is the static friction coefficient.

Thus, static friction obeys a similar law as dynamic friction does, although with another friction coefficient.

Empirically, we obtain the relation for the coefficients

$$0 < \mu_g < \mu_h.$$

Their magnitude depends sensitively on the surface properties.

The decomposition of the gravitational force in components normal and parallel to the inclination of the plane.

Example 20.1: Free fall with friction according to Stokes

As an example we consider the motion of a body (e.g., parachute) with the initial velocity $v = v_0$ at the time $t = 0$. The motion is one-dimensional; the equation of motion reads

$$m\ddot{z} = -mg - \beta\dot{z},$$

or

$$m\frac{dv}{dt} = (-mg - \beta v). \tag{20.1}$$

The gravitational force acts along the $-z$-direction; the friction force points opposite to the velocity.

After separation of the variables, we have

$$\frac{m\,dv}{mg + \beta v} = -dt,$$

$$m\int\limits_{v_0}^{v}\frac{dv}{mg + \beta v} = -\int\limits_{0}^{t} dt = -t.$$

The integral to the left is solved by substituting $mg + \beta v = u$ and $dv = du/\beta$:

$$m\int\limits_{v_0}^{v}\frac{dv}{mg + \beta v} = +\frac{m}{\beta}\int\limits_{mg+\beta v_0}^{mg+\beta v}\frac{du}{u} = \frac{m}{\beta}\ln\frac{mg + \beta v}{mg + \beta v_0}.$$

k is the unit vector in the negative z-direction, i.e., $\mathbf{k} = -\mathbf{e}_z$.

Therefore,

$$t = \frac{m}{\beta} \ln \left(\frac{mg + v_0 \beta}{mg + \beta v} \right).$$

Exponentiation of both sides of the equation yields

$$e^{\frac{\beta}{m}t} = \frac{mg + \beta v_0}{mg + \beta v};$$

and rewritten this reads

$$mg + \beta v_0 = (mg + \beta v) e^{\frac{\beta}{m}t}.$$

Solving for v leads to

$$v(t) = -\frac{mg}{\beta} + \left(\frac{mg}{\beta} + v_0 \right) e^{-\frac{\beta}{m}t}. \tag{20.2}$$

One easily sees from this velocity-time function that for increasing t the velocity $v(t)$ approaches a limit value, that is, for large times $v(t)$ becomes constant. Let us denote the limit velocity for large times by v_∞. According to 20.2,

$$v_\infty = \lim_{t \to \infty} v(t) = -\frac{mg}{\beta}. \tag{20.3}$$

This may already be concluded from the dynamic equation 20.1 for the case of a vanishing acceleration $\ddot{z} = 0$. In 20.2 we will approximate the exponential function by the first two terms of the corresponding Taylor expansion for small friction forces $(\beta/m)t \ll 1$:

$$v(t) = -\frac{mg}{\beta} + \left(v_0 + \frac{mg}{\beta} \right) \left(1 - \frac{\beta t}{m} + \cdots \right).$$

Investigation of the limit for $\beta \to 0$ yields

$$\lim_{\beta \to 0} v(t) = v_0 - gt,$$

that is the expected result for the case without friction.

We still determine $z(t)$ and its limit for $t \to \infty$: From 20.2 it follows by integration $(dz/dt = v(t))$ that

$$z(t) = -\frac{mgt}{\beta} - \frac{m}{\beta} \left(v_0 + \frac{mg}{\beta} \right) e^{-\frac{\beta}{m}t} + c_2,$$

where, because $z = 0$ for $t = 0$, the integration constant is

$$c_2 = \frac{m}{\beta} \left(v_0 + \frac{mg}{\beta} \right)$$

and therefore $z(t)$ finally reads

$$z(t) = -\frac{mgt}{\beta} + \frac{m}{\beta} \left(v_0 + \frac{mg}{\beta} \right) \left(1 - e^{-\frac{\beta}{m}t} \right),$$

$$\lim_{t \to \infty} z(t) = v_\infty t + \frac{m}{\beta} (v_0 - v_\infty).$$

That means that for large times z increases linearly with the time. From $z(t)$ one calculates the acceleration $a(t)$ as

$$\ddot{z}(t) = a(t) = \frac{-\beta}{m}\left(v_0 + \frac{mg}{\beta}\right)e^{-\frac{\beta}{m}t}.$$

It vanishes for large times. Then the gravitational force and the friction force just balance each other.

Example 20.2: The inclined throw with friction according to Stokes

Adopted initial conditions: At the time $t = 0$, let

$$\begin{aligned}\mathbf{r}(0) &= 0,\\ \mathbf{v}(0) &= \mathbf{v}_0\\ &= v_0\cos\alpha\,\mathbf{e}_2 + v_0\sin\alpha\,\mathbf{e}_3.\end{aligned}$$

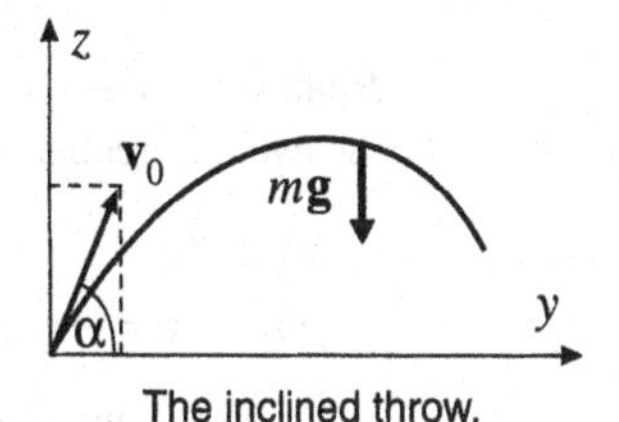

The inclined throw.

Equation of motion:

$$m\ddot{\mathbf{r}} = -\beta\dot{\mathbf{r}} - mg\mathbf{e}_3 \qquad \text{or} \qquad \dot{\mathbf{v}} + \frac{\beta}{m}\mathbf{v} = -g\mathbf{e}_3.$$

To integrate this vectorial differential equation, we multiply by $e^{\frac{\beta}{m}t}$:

$$\dot{\mathbf{v}}e^{\frac{\beta}{m}t} + \left(\frac{\beta}{m}\right)\mathbf{v}\,e^{\frac{\beta}{m}t} = -g\mathbf{e}_3 e^{\frac{\beta}{m}t}.$$

The left side of this equation is just the time derivative of $\mathbf{v}\,e^{\frac{\beta}{m}t}$ according to the product rule, so that it may be integrated right now:

$$\mathbf{v}\,e^{\frac{\beta}{m}t} = -\int g\,e^{\frac{\beta}{m}t}\mathbf{e}_3\,dt = -g\frac{m}{\beta}e^{\frac{\beta}{m}t}\mathbf{e}_3 + \mathbf{c}_1.$$

Because $\mathbf{v}(0) = \mathbf{v}_0$, $\mathbf{c}_1 = \mathbf{v}_0 + g\frac{m}{\beta}\mathbf{e}_3$. Ordered by components, the velocity is

$$\mathbf{v} = v_0\cos\alpha\,e^{-\frac{\beta}{m}t}\,\mathbf{e}_2 + \left[-\frac{mg}{\beta} + \left(v_0\sin\alpha + \frac{mg}{\beta}\right)e^{-\frac{\beta}{m}t}\right]\mathbf{e}_3$$

or

$$\mathbf{v} = -g\frac{m}{\beta}\left(1 - e^{-\frac{\beta}{m}t}\right)\mathbf{e}_3 + \mathbf{v}_0 e^{-\frac{\beta}{m}t}.$$

The position $\mathbf{r}(t)$ of the missile may be found by integration of the velocity:

$$\mathbf{r} = -\frac{m}{\beta}v_0\cos\alpha\,e^{-\frac{\beta}{m}t}\mathbf{e}_2 + \left[\frac{-mg}{\beta}t - \frac{m}{\beta}\left(v_0\sin\alpha + \frac{mg}{\beta}\right)e^{-\frac{\beta}{m}t}\right]\mathbf{e}_3 + \mathbf{c}_2$$

or

$$\mathbf{r} = -g\frac{m}{\beta}\left(t + \frac{m}{\beta}e^{-\frac{\beta}{m}t}\right)\mathbf{e}_3 - \mathbf{v}_0\frac{m}{\beta}e^{-\frac{\beta}{m}t} + \mathbf{c}_2\,.$$

Because $\mathbf{r}(0) = 0$, the integration constant is

$$\mathbf{c}_2 = \frac{m}{\beta} v_0 \cos\alpha \, \mathbf{e}_2 + \frac{m}{\beta}\left(v_0 \sin\alpha + \frac{mg}{\beta}\right)\mathbf{e}_3.$$

Inserting this integration constant, for the position we get

$$\mathbf{r} = \frac{m}{\beta} v_0 \cos\alpha \left(1 - e^{-\frac{\beta}{m}t}\right)\mathbf{e}_2 + \left[-\frac{mg}{\beta}t + \frac{m}{\beta}\left(v_0 \sin\alpha + \frac{mg}{\beta}\right)\left(1 - e^{-\frac{\beta}{m}t}\right)\right]\mathbf{e}_3.$$

Remark: The same results for $\mathbf{r}(t)$ and $\mathbf{v}(t)$ would have been found by separate considerations of the two differential equations

$$m\ddot{y} + \beta\dot{y} = 0,$$
$$m\ddot{z} + \beta\dot{z} = -mg.$$

If one adopts the *ansatz* of Newtonian friction, the equation of motion of the problem is no longer separable, because $m\ddot{\mathbf{r}} = -\beta|\dot{\mathbf{r}}|\dot{\mathbf{r}} - mg\mathbf{e}_3$ decays into

$$m\ddot{y} + \beta\sqrt{\dot{y}^2 + \dot{z}^2}\,\dot{y} = 0,$$
$$m\ddot{z} + \beta\sqrt{\dot{y}^2 + \dot{z}^2}\,\dot{z} = -mg,$$

that is, in a set of coupled nonlinear differential equations. In most cases such equations may not be solved analytically. The linearity and nonlinearity of differential equations is discussed in Chapters 23 and 25.

Discussion of the motion

For large times ($t \gg m/\beta$) the exponential factor $\exp(-\frac{\beta}{m}t)$ tends to zero. That means

(a) $\lim_{t\to\infty} \mathbf{v}(t) = -(mg/\beta)\mathbf{e}_3$. The motion turns over to the vertical fall with constant limit velocity. The horizontal velocity component vanishes for large times, namely, the motion in y-direction comes to rest.

(b) $\lim_{t\to\infty} y(t) = (m/\beta)v_0 \cos\alpha \equiv y_0$. With increasing time the motion in the horizontal direction tends asymptotically against the maximum distance y_0.

The path equation may be found explicitly by eliminating the time parameter from the equations for $\mathbf{r}\cdot\mathbf{e}_2$ and $\mathbf{r}\cdot\mathbf{e}_3$. We get

$$z(y) = \frac{m^2}{\beta^2} g \ln\left(1 - \frac{\beta y}{m v_0 \cos\alpha}\right) + \left(v_0 \sin\alpha + \frac{mg}{\beta}\right)\frac{y}{v_0 \cos\alpha}.$$

To investigate the trend of the trajectory for very low friction, we may employ the Taylor expansion of the logarithm:

$$\ln(1+x) = x - \frac{x^2}{2} + \frac{x^3}{3} - + \cdots.$$

Then

$$z(y) = -\frac{m^2}{\beta^2} g \left[\frac{\beta y}{m v_0 \cos\alpha} + \frac{1}{2}\left(\frac{\beta y}{m v_0 \cos\alpha}\right)^2\right] + \left(v_0 \sin\alpha + \frac{mg}{\beta}\right)\frac{y}{v_0 \cos\alpha} + \cdots$$

$$= -\frac{g}{2v_0^2 \cos^2\alpha} y^2 + y \tan\alpha + R.$$

Here

$$R = -\frac{1}{3}\frac{\beta}{m} g \left(\frac{y}{v_0 \cos\alpha}\right)^3 - \cdots$$

is a remainder term. Note that the first two terms are the same as in the formula of the parabola of inclined throw, equation (19.7).

From this relation we may realize that

(a) For vanishing friction the result approaches the throw parabola.

(b) If friction is present, the trajectory runs below the throw parabola; for small y, it osculates the parabola (osculation of second order).

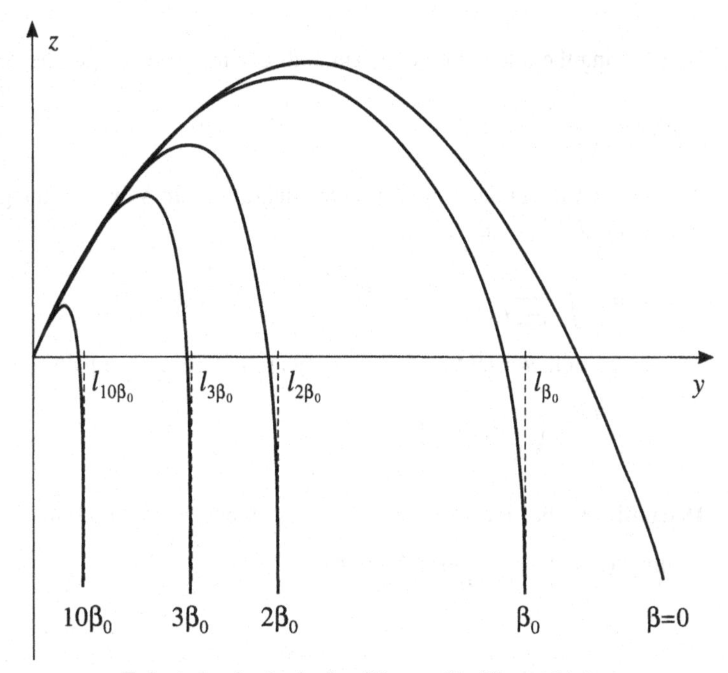

Trajectories for the inclined throw with different friction.

Motion in a viscous medium with Newtonian friction

We consider the motion of a body affected *only* by a velocity-dependent friction force. The case of Stokes friction has already been treated in the preceding example. We therefore now consider Newtonian friction.

Let the (necessarily rectilinear) motion proceed along the x-direction; the unit vector $\mathbf{e}_1$ is therefore omitted. We choose the initial conditions $v(t = 0) = v_0$, $x(t = 0) = 0$.

The equation of motion reads

$$m\ddot{x} = -\gamma\dot{x}^2,$$

and the separation of variables yields

$$m\frac{dv}{v^2} = -\gamma dt.$$

The integration leads to

$$-m\frac{1}{v} = -\gamma t + c_1.$$

From the initial conditions it follows that

$$c_1 = -\frac{m}{v_0}.$$

By inserting the integration constant and solving for v, we get the velocity

$$v(t) = \frac{m}{\gamma v_0 t + m} v_0.$$

The position is obtained by a further integration. To solve the integral we substitute $z = \gamma v_0 t + m, dz = \gamma v_0 dt$.

$$x = m v_0 \int \frac{dt}{\gamma v_0 t + m} = \frac{m}{\gamma}\int \frac{dz}{z} = \frac{m}{\gamma}\ln(\gamma v_0 t + m) + c_2.$$

The integration constant is $c_2 = -(m/\gamma)\ln m$. Hence, the path is

$$x(t) = \frac{m}{\gamma}\ln\left(\frac{\gamma}{m}v_0 t + 1\right).$$

Discussion: For increasing time $t \to \infty$, we have the two limits

$$\lim_{t\to\infty} v(t) = 0\,, \qquad \lim_{t\to\infty} x(t) = \infty.$$

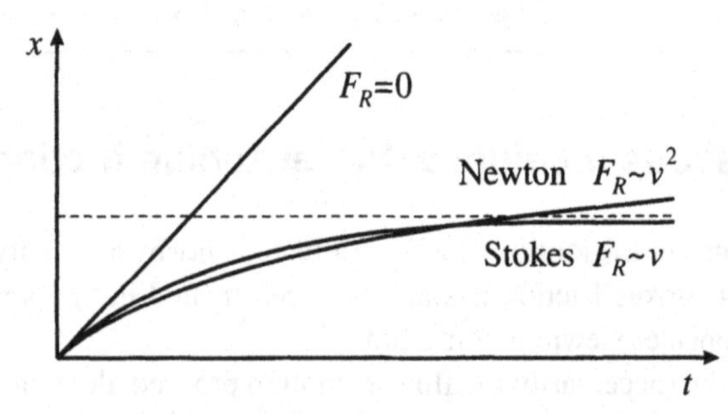

Illustration of the path $x(t)$ for different cases of friction.

Although the velocity becomes smaller and smaller, the body moves arbitrarily far under the influence of Newtonian friction.

Generalized *ansatz* for friction:

In the following we adopt a more general *ansatz* for the velocity-dependent friction force, which is of particular interest for low velocities, namely

$$\mathbf{F}_R = -\varrho v^n \frac{\mathbf{v}}{v}.$$

Here $n \geq 0$ because the friction shall decrease when v decreases. The *equation of motion* then reads

$$m\ddot{x} = -\varrho \dot{x}^n$$

or

$$\frac{dv}{v^n} = -\frac{\varrho}{m} dt \qquad \text{for } n \neq 1$$

($n = 1$ corresponds to Stokes friction). Integration yields

$$\frac{v^{-n+1}}{-n+1} = -\frac{\varrho}{m} t + C_1, \qquad C_1 = \frac{v_0^{1-n}}{1-n},$$

with $v(t=0) = v_0$. From there it follows for the velocity that

$$v(t) = \left(v_0^{1-n} - (1-n)\frac{\varrho}{m} t \right)^{1/(1-n)}.$$

Here one may distinguish two cases:

- $0 \leq n < 1$:
  The expression in the brackets may vanish. Therefore, the body comes to rest after some finite time t_0;

  $$t_0 = \frac{m}{\varrho} \frac{v_0^{1-n}}{1-n}.$$

  As soon as $t \geq t_0$, the derived formula no longer holds; the body remains at rest.

- $n > 1$:
  The body does not come to rest completely, but its velocity becomes arbitrarily small because

  $$\lim_{t\to\infty} v(t) = \lim_{t\to\infty} \frac{1}{(v_0^{-\alpha} + \alpha \frac{\varrho}{m} t)^{1/\alpha}} = 0 \quad \text{with } \alpha = n - 1 > 0.$$

The necessary second integration becomes simpler by considering $v(x(t))$. According to the chain rule,

$$\frac{dv}{dt} = \frac{dv}{dx} \cdot \frac{dx}{dt} = \frac{dv}{dx} v.$$

Insertion into the equation of motion yields

$$dv \cdot v^{1-n} = -\frac{\varrho}{m} dx.$$

Integration yields

$$\frac{1}{2-n} v^{2-n} = -\frac{\varrho}{m} x + C_2.$$

With

$$v(x=0) = v(t=0) = v_0,$$

we get

$$C_2 = \frac{1}{2-n} v_0^{2-n}.$$

Hence

$$x(v) = \frac{m}{\varrho} \frac{1}{n-2} (v^{2-n} - v_0^{2-n}).$$

Or, by inserting $v(t)$, the path is obtained as a function of the time as

$$x(t) = \frac{m}{\varrho} \frac{1}{n-2} \left[\left(v_0^{1-n} - (1-n) \frac{\varrho}{m} t \right)^{\frac{n-2}{n-1}} - v_0^{2-n} \right].$$

Here two distinct cases also exist, which may most simply be extracted from the function $x(v)$, namely

$$0 \le n < 2: \quad \lim_{v \to 0} x(v) = \frac{m}{\varrho} \frac{1}{2-n} v_0^{2-n} = l,$$

$$n > 2: \quad \lim_{v \to 0} x(v) = \frac{m}{\varrho \beta} \lim_{v \to 0} \left(\frac{1}{v^\beta} - \frac{1}{v_0^\beta} \right) = \infty$$

(with $\beta = n - 2 > 0$—for $n = 2$; see Newtonian friction).

In total, we thus may distinguish between three types of motion, namely

(a) $0 \le n < 1$:
The motion comes to *rest* at the time t_0 at the distance l.

(b) $1 \le n < 2$:
The velocity tends to zero while the body approaches a finite *limit point* at the distance l.

(c) $n \geq 2$:
The velocity asymptotically approaches zero while the distance *increases beyond any limit.*

The already treated cases are limit cases of the types of motion (b) (Stokes, $n = 1$) and (c) (Newton, $n = 2$). The figure illustrates the distinct trends. For very low velocities the friction force—independent of the coefficient—tends to zero the faster the larger the exponent n is. On the other hand, for small n the deceleration decreases so slowly (i.e., the decelerating force is for small n so strong) that the motion even comes to rest.

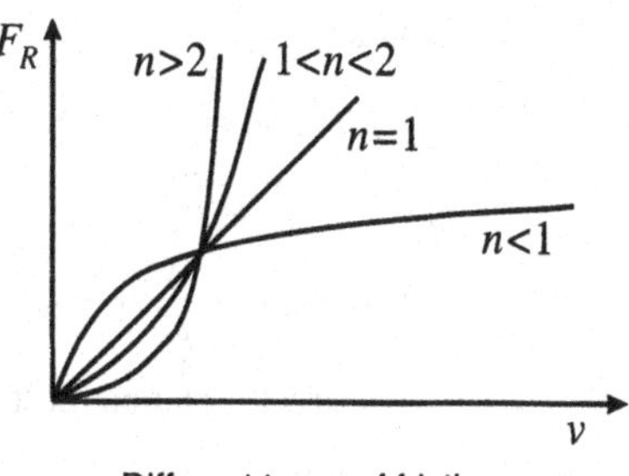

Different types of friction.

Problem 20.3: Free fall with Newtonian friction

A body begins to fall at the time $t = 0$ at the point $z = 0$ with the initial velocity v_0. Determine the fall velocity $v(t)$ and the path $z(t)$, assuming Newtonian friction. Which approximations hold for small times if $v_0 = 0$?

Solution The equation of motion reads

$$m\ddot{z} = -mg - \gamma\dot{z}|\dot{z}| \quad \text{or} \quad \dot{v} = -g\left(1 + \frac{\gamma}{mg}v|v|\right) = -g\left(1 + \frac{v|v|}{v_\infty^2}\right) .$$

With the abbreviation

$$v_\infty = \sqrt{\frac{mg}{\gamma}} \qquad \text{follows} \qquad v_\infty^2 \int\limits_{v_0}^{v} \frac{dv}{v_\infty^2 + v|v|} = -g \int\limits_0^t dt.$$

Integration of the equation of motion yields

$$-gt = \begin{cases} \left[v_\infty \arctan \dfrac{v}{v_\infty} \right]_{v_0}^{v} & \text{for } v \geq 0, \\ \left[v_\infty \operatorname{Artanh} \dfrac{v}{v_\infty} \right]_{v_0}^{v} & \text{for } -v_\infty < v < 0, \\ \left[v_\infty \operatorname{Arcoth} \dfrac{v}{v_\infty} \right]_{v_0}^{v} & \text{for } v < -v_\infty. \end{cases}$$

One therefore has to distinguish three cases, depending on the magnitude of the initial velocity.

1. $v_0 \geq 0$. Initial velocity and gravitational force are opposite to each other. After integration it follows that

$$-gt = v_\infty \arctan \frac{v}{v_\infty} - v_\infty \arctan \frac{v_0}{v_\infty}.$$

The meaning of the constant term becomes obvious if $v = 0$:

$$gt_0 = gt(v = 0) = v_\infty \arctan \frac{v_0}{v_\infty},$$

and thus

$$-g(t - t_0) = v_\infty \arctan \frac{v}{v_\infty}$$

or

$$v = -v_\infty \tan \frac{g}{v_\infty}(t - t_0).$$

Because the motion must be continuous also for $t = t_0$, the integration between t_0 and t yields

$$-g(t - t_0) = v_\infty \operatorname{Artanh} \frac{v}{v_\infty} - 0,$$

hence

$$v = -v_\infty \tanh \frac{g}{v_\infty}(t - t_0); \qquad t \geq t_0.$$

The body moves upward against the gravitational force, comes to rest at t_0, and falls downward.

2. $-v_\infty < v_0 \leq 0$. Integration yields

$$-gt = v_\infty \operatorname{Artanh} \frac{v}{v_\infty} - v_\infty \operatorname{Artanh} \frac{v_0}{v_\infty}.$$

If we imagine the velocity function as continued for negative times $t < 0$, the constant term again has a clear meaning:

$$t_0 = t(v = 0) = \frac{v_\infty}{g} \operatorname{Artanh} \frac{v_0}{v_\infty}.$$

The velocity function may then be expressed by

$$v = -v_\infty \tanh \frac{g}{v_\infty}(t - t_0).$$

3. $v_0 < -v_\infty$.

$$-gt = v_\infty \operatorname{Arcoth} \frac{v}{v_\infty} - v_\infty \operatorname{Arcoth} \frac{v_0}{v_\infty}.$$

Similar to the other cases we abbreviate

$$t(v = -\infty) = t_- = \frac{v_\infty}{g} \operatorname{Arcoth} \frac{v_0}{v_\infty}$$

and thus

$$v = -v_\infty \coth \frac{g}{v_\infty}(t - t_-).$$

In all three cases the velocity asymptotically approaches the limit velocity

$$-v_\infty = -\sqrt{\frac{mg}{\gamma}}.$$

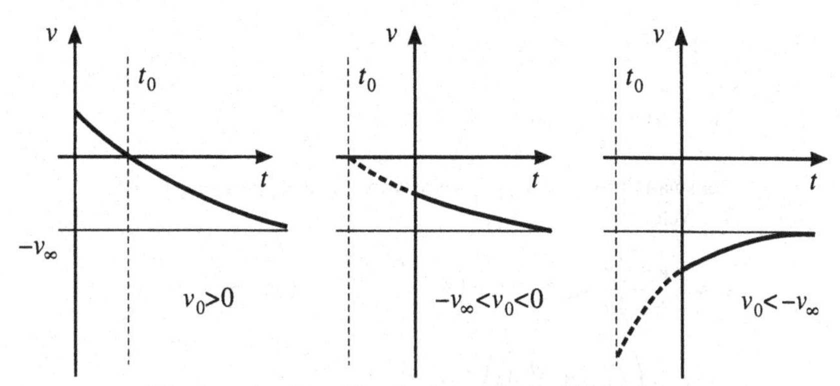

Velocity as function of time for three different initial velocities.

The function $z(t)$ may be calculated straightforward from $v(t)$.

1.

$$z = v_\infty \int \tan \frac{g}{v_\infty}(t_0 - t)\, dt = +\frac{m}{\gamma} \ln \cos \frac{g}{v_\infty}(t_0 - t) + K_1, \qquad t \le t_0;$$

$$z = -v_\infty \int \tanh \frac{g}{v_\infty}(t - t_0)\, dt = -\frac{m}{\gamma} \ln \cosh \frac{g}{v_\infty}(t - t_0) + K_1, \qquad t \ge t_0;$$

where

$$K_1 = -\frac{m}{\gamma} \ln \cos \frac{g}{v_\infty} t_0.$$

2.

$$z = -\frac{m}{\gamma} \ln \cosh \frac{g}{v_\infty}(t - t_0) + K_2\,,$$

$$K_2 = \frac{m}{\gamma} \ln \cosh \left(-\frac{g}{v_\infty} t_0 \right).$$

3.

$$z = -v_\infty \int \coth \frac{g}{v_\infty}(t - t_-)\, dt = -\frac{m}{\gamma} \ln \sinh \frac{g}{v_\infty}(t - t_-) + K_3\,,$$

$$K_3 = \frac{m}{\gamma} \ln \sinh \left(-\frac{g t_-}{v_\infty} \right).$$

In particular, for $v_0 = 0$ one has $t_0 = 0$, and therefore

$$v = -v_\infty \tanh \frac{gt}{v_\infty},$$

$$z = -\frac{v_\infty^2}{g} \ln \cosh \frac{gt}{v_\infty}\,.$$

There hold the series expansions

$$\sinh u = u + \frac{u^3}{3!} + \cdots,$$

$$\cosh u = 1 + \frac{u^2}{2!} + \frac{u^4}{4!} + \cdots,$$

$$\ln(1+u) = u - \frac{u^2}{2} + - \cdots.$$

For small times ($t \ll v_\infty/g$) we may then write approximately

1. Velocity:

$$v \approx -\frac{u + \frac{1}{6}u^3}{1 + \frac{1}{2}u^2} v_\infty \approx -v_\infty u \left(1 - \frac{1}{3}u^2\right) \qquad \text{with } u = \frac{gt}{v_\infty},$$

$$\approx -gt \left(1 - \frac{1}{3}\left(\frac{gt}{v_\infty}\right)^2\right).$$

2. Path:

$$z \approx -\frac{v_\infty^2}{g} \ln\left(1 + \frac{u^2}{2} + \frac{u^4}{24}\right)$$

$$\approx -\frac{v_\infty^2}{g}\left[\frac{u^2}{2} + \frac{u^4}{24} - \frac{1}{2}\left(\frac{u^4}{4} + \cdots\right)\right],$$

$$\approx -\frac{1}{2}gt^2 \left(1 - \frac{1}{6}\left(\frac{gt}{v_\infty}\right)^2\right).$$

Problem 20.4: Motion of an engine with friction

An engine of mass m moves without driving force but under the influence of the friction force $f(v) = \alpha + \beta v^2$ on horizontal rails. Let the initial velocity be v_0.

(a) After which time does the engine come to rest? What is the maximum deceleration time ($v_0 \to \infty$)?

(b) What distance has then been covered?

Solution The equation of motion reads

$$m\ddot{x} = -f(v) = -\alpha - \beta\dot{x}^2$$

and will be integrated after separation of the variables:

(a)

$$m\frac{dv}{dt} = -(\alpha + \beta v^2);$$

$$\frac{m}{\beta} \int_{v=v_0}^{0} \frac{dv}{\alpha/\beta + v^2} = -\int_{t=0}^{t_0} dt;$$

$$\frac{m}{\beta}\sqrt{\frac{\beta}{\alpha}}\left(\arctan 0 - \arctan\sqrt{\frac{\beta}{\alpha}}v_0\right) = -t_0;$$

$$t_0 = \frac{m}{\sqrt{\alpha\beta}} \arctan\sqrt{\frac{\beta}{\alpha}}v_0; \qquad \lim_{v_0\to\infty} t_0 = \frac{m}{\sqrt{\alpha\beta}}\frac{\pi}{2}.$$

(b) If we use x and v as variables, the equation of motion reads

$$m\frac{dv}{dx}\frac{dx}{dt} = -(\alpha + \beta v^2).$$

This is transformed and integrated:

$$\frac{m}{2\beta}\int\limits_{v=v_0}^{0}\frac{2\beta v}{\alpha+\beta v^2}dv = -\int\limits_{x=0}^{x_0}dx;$$

$$\frac{m}{2\beta}\left[\ln\alpha - \ln\left(\alpha+\beta v_0^2\right)\right] = -x_0.$$

The total path covered is therefore

$$x_0 = \frac{m}{2\beta}\ln\left(1+\frac{\beta}{\alpha}v_0^2\right).$$

For an infinite initial velocity v_0 the engine covers an infinite distance until the rest, although the deceleration to velocity 0 takes only the finite time $(m/\sqrt{\alpha\beta})(\pi/2)$.

Example 20.5: The inclined plane

So far we have considered the motion of a free massive body under the action of external forces. If its freedom of motion is, however, restricted to a defined area or line by certain constraints, one speaks of a *bound motion.* A *constraining force* must then act on the body that keeps it on the prescribed path.

In a motion on a solid area or rail the body undergoes a reactive force by the support that just balances the normal component of the force acting on it. When taking into account this constraining force, the equation of motion may be formulated according to the second Newtonian axiom.

The simplest example is the motion on the inclined plane.

(a) Without friction

We introduce the following denotations (compare figure):

$\mathbf{F}_s$: gravitational force;

$\mathbf{F}_s^{\parallel}, \mathbf{F}_s^{\perp}$: parallel and normal components of the weight force;

$\mathbf{F}_R$: reactive force;

s: covered path.

The following relations exist between the forces:

$$\mathbf{F}_s^{\perp} = -\mathbf{F}_R$$

(according to the third Newtonian axiom);

$$\mathbf{F}_s = \mathbf{F}_s^{\perp} + \mathbf{F}_s^{\parallel} = -mg\mathbf{e}_3;$$

$$\mathbf{F}_s^{\parallel} = mg\sin\alpha\,\mathbf{e} \qquad \text{(see figure).}$$

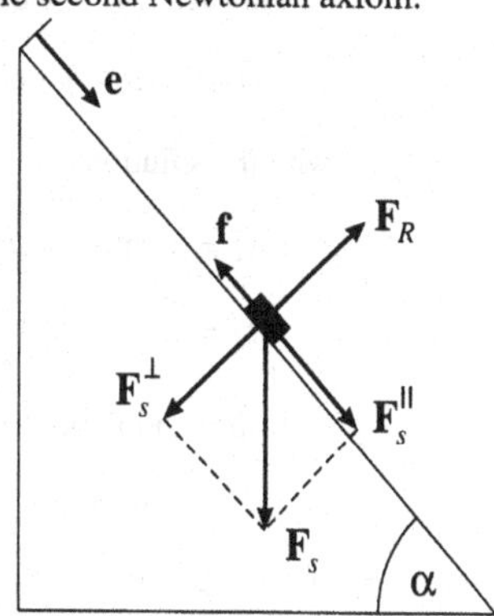

Forces acting on a mass at the inclined plane

The equation of motion is then

$$m\ddot{s}\mathbf{e} = \sum_i \mathbf{F}_i = \mathbf{F}_s + \mathbf{F}_R = \mathbf{F}_s^{\parallel} = mg \sin\alpha\, \mathbf{e}.$$

Only the parallel component of the weight force causes the acceleration (slope drift):

$$\ddot{s} = g \sin\alpha \equiv g'.$$

This is exactly the differential equation of the free fall with an earth acceleration reduced by the factor $\sin\alpha$. Twofold integration leads again to the solutions

$$v(t) = g \sin\alpha t + v_0,$$

$$s(t) = \frac{1}{2} g \sin\alpha t^2 + v_0 t + s_0.$$

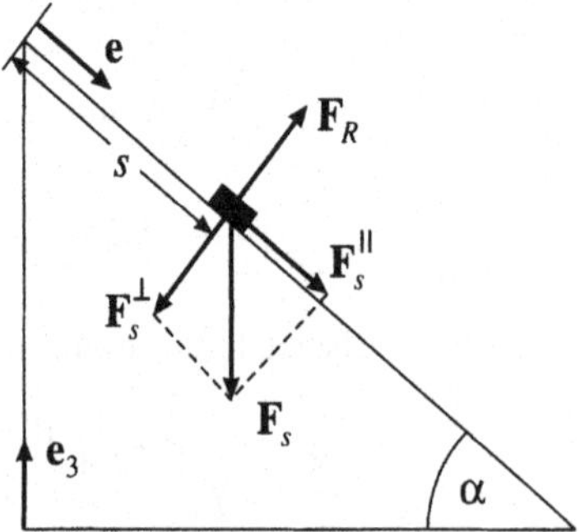

Decomposition of the forces on the inclined plane.

(b) With friction

Besides the constraining force $\mathbf{F}_R$ along the area normal, there also acts a parallel component $\mathbf{f}$ on the body that always points opposite to the friction force. According to the figure, the support force is

$$F_s^{\perp} = \cos\alpha\, mg,$$

and therefore the dynamic friction force is

$$\mathbf{f} = \mp\mu_g mg \cos\alpha\, \mathbf{e} \qquad \text{if } v \gtrless 0.$$

Hence, the equation of motion reads

$$m\ddot{s}\,\mathbf{e} = \mathbf{F}_s + \mathbf{F}_R + \mathbf{f} = \mathbf{F}_s^{\parallel} + \mathbf{f} = mg(\sin\alpha \mp \mu_g \cos\alpha)\mathbf{e}$$

if $v \gtrless 0$.

This again yields the differential equation

$$\ddot{s} = g(\sin\alpha \mp \mu_g \cos\alpha) \equiv g^{\parallel},$$

with the solutions

$$v(t) = g(\sin\alpha \mp \mu_g \cos\alpha)t + v_0, \qquad \text{for } v \gtrless 0,$$

$$s(t) = \frac{1}{2} g(\sin\alpha \mp \mu_g \cos\alpha)t^2 + v_0 t + s_0\,.$$

If the motion points downward ($v > 0$), we may distinguish among three distinct cases:

(a) $g^{\parallel} > 0$: that is, $\tan\alpha > \mu_g$ or $\alpha > \alpha_g = \arctan\mu_g$. The body is positively accelerated.

(b) $g^{\parallel} = 0$: $\tan\alpha = \mu_g$, $\alpha = \alpha_g$. The body moves uniformly, the gravitational force component and the friction mutually cancel.

(c) $g^{\parallel} < 0$: $\tan\alpha < \mu_g$. The body is decelerated and comes to rest after the time

$$t = \frac{v_0}{g(\mu_g \cos\alpha - \sin\alpha)}\,.$$

If $v < 0$, namely the direction of motion is upward, then $g^{\parallel} = g(\sin\alpha + \mu_g \cos\alpha) > 0$; the body comes to rest in any case. It depends on the magnitude of the coefficient μ_g of the now-acting static friction whether the body begins to move out of the state of rest.

The inclined plane allows us to determine the two friction coefficients by varying the slope angle α:

$\mu_g = \tan\alpha$ if the body uniformly moves ($v > 0$),

$\mu_h = \tan\alpha$ if the body just starts sliding.

Problem 20.6: Two masses on inclined planes

Two masses m_1 and m_2 are lying each on one of two joined planes that enclose the angles α and β with the horizontal (see figure). The two masses are connected by a massless and nonductile rope running over a roller fixed at point A.

Determine the acceleration a of the masses m_1 and m_2, taking the friction into account.

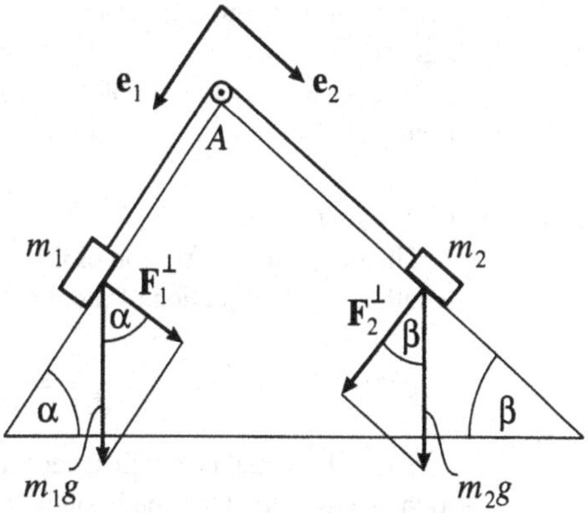

Two masses on inclined planes.

Solution The friction mentioned in the problem is dynamic $\left(\mathbf{F}_R = -\mu_g F^{\perp}\mathbf{v}/v\right)$. Because the velocity $\mathbf{v}$ points along $\mathbf{e}_1$ or $\mathbf{e}_2$, respectively, the quantity $\mathbf{v}/v$ just equals $\mathbf{e}_1$ or $\mathbf{e}_2$!

Hence

$$m_1 a\mathbf{e}_1 = m_1 g \sin\alpha\, \mathbf{e}_1 - T\mathbf{e}_1 - \mu_g F_1^{\perp}\mathbf{e}_1 \tag{20.4}$$

and

$$-m_2 a\mathbf{e}_2 = m_2 g \sin\beta\, \mathbf{e}_2 - T\mathbf{e}_2 + \mu_g F_2^{\perp}\mathbf{e}_2. \tag{20.5}$$

T is the string tension. The signs in front of the two last terms on the right-hand side (the friction terms) are valid only for positive acceleration, $a > 0$. We have to check this at the end of the calculation. We now have to calculate $F_1^{\perp}$ and $F_2^{\perp}$. From the sketch we see that

$$F_1^{\perp} = m_1 g \cos\alpha \quad \text{and} \quad F_2^{\perp} = m_2 g \cos\beta\,.$$

$F_1^{\perp}$ and $F_2^{\perp}$ inserted in 20.4 and 20.5 yield

$$m_1 a\mathbf{e}_1 = m_1 g \sin\alpha\, \mathbf{e}_1 - T\mathbf{e}_1 - \mu_g m_1 g \cos\alpha\, \mathbf{e}_1, \tag{20.6}$$

$$-m_2 a\mathbf{e}_2 = m_2 g \sin\beta\, \mathbf{e}_2 - T\mathbf{e}_2 + \mu_g m_2 g \cos\beta\, \mathbf{e}_2. \tag{20.7}$$

From 20.6 it follows that $T = m_1 g \sin\alpha - m_1 a - \mu_g m_1 g \cos\alpha$. T is now inserted in 20.7:

$$-a(m_1 + m_2) = m_2 g \sin\beta - m_1 g \sin\alpha + \mu_g m_1 g \cos\alpha + \mu_g m_2 g \cos\beta$$

$$\Leftrightarrow \quad a = \frac{m_1 \sin\alpha - m_2 \sin\beta - \mu_g m_1 \cos\alpha - \mu_g m_2 \cos\beta}{m_1 + m_2} g.$$

Thus, the acceleration has been determined. Finally, we consider two special cases:

(1) $\mu_g = 0$, that is, there is no friction; therefore,

$$a = \frac{m_1 \sin\alpha - m_2 \sin\beta}{m_1 + m_2} g\,.$$

(2) $\alpha = \beta = 90°$; the acceleration then becomes

$$a = \frac{m_1 - m_2}{m_1 + m_2} g.$$

Problem 20.7: A chain slides down from a table

A uniform chain of total length a hangs with a piece of length b ($0 \le b \le a$) over the edge of a plane table. Calculate the time in which the chain slides from the table under the influence of gravity but without friction. Let the initial velocity be 0 (see figure).

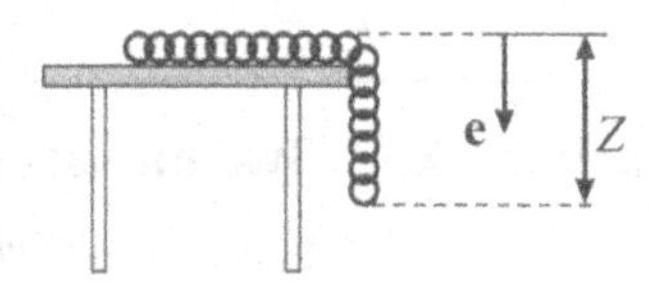

A chain sliding down from a table

Investigate the same problem, assuming now a dynamic friction μ_g.

Solution (a) Without friction

The length of the fraction of the chain hanging vertically down is denoted by z; the mass per unit length is ϱ. The equation of motion then reads

$$\varrho a\ddot{z} = \varrho z g \qquad \Leftrightarrow \qquad \frac{d^2 z}{dt^2} = \frac{g}{a} z. \tag{20.8}$$

This differential equation states that $z(t)$ differentiated twice with respect to t reproduces itself up to the factor g/a. This condition is fulfilled by one of the two independent exponential functions

$$e^{\sqrt{g/a}\,t} \quad \text{and} \quad e^{-\sqrt{g/a}\,t}$$

such that the general solution reads

$$z(t) = Ae^{\sqrt{g/a}\,t} + Be^{-\sqrt{g/a}\,t}. \tag{20.9}$$

A and B are integration constants, which are determined from the initial conditions

$$z(0) = b = A + B,$$

$$\dot{z}(0) = 0 = A\sqrt{\frac{g}{a}} - B\sqrt{\frac{g}{a}} \qquad \Rightarrow \qquad A - B = 0. \tag{20.10}$$

This yields $A = b/2$, $B = b/2$, and therefore for 20.9

$$z(t) = \frac{b}{2}\left(e^{\sqrt{g/a}\,t} + e^{-\sqrt{g/a}\,t}\right) = b\cosh\sqrt{\frac{g}{a}}\,t. \tag{20.11}$$

The time T of sliding follows from the condition

$$z(T) = a = \frac{b}{2}\left(e^{\sqrt{g/a}T} + e^{-\sqrt{g/a}T}\right). \tag{20.12}$$

From there it follows with $x = e^{\sqrt{g/a}T}$ that

$$\frac{2a}{b} = x + \frac{1}{x} \qquad \Leftrightarrow \qquad x^2 - \frac{2a}{b}x = -1$$

$$\Rightarrow \quad x_{1,2} = \frac{a}{b} \pm \sqrt{-1 + \frac{a^2}{b^2}} = \frac{a}{b} \pm \frac{1}{b}\sqrt{a^2 - b^2} = \frac{1}{b}(a \pm \sqrt{a^2 - b^2}). \tag{20.13}$$

One finds

$$T = \sqrt{\frac{a}{g}} \ln\left(\frac{a + \sqrt{a^2 - b^2}}{b}\right). \tag{20.14}$$

The negative root in 20.13 has to be ruled out because it leads to negative times, which is physically senseless. This we may realize as follows: To get positive times from 20.13 the argument of the logarithm must be ≥ 1. But the negative root is always ≤ 1 because

$$\frac{a}{b} - \sqrt{\frac{a^2}{b^2} - 1} \leq 1 \quad \Leftrightarrow \quad \frac{a}{b} \leq 1 + \sqrt{\frac{a^2}{b^2} - 1}$$

$$\Leftrightarrow \quad \frac{a^2}{b^2} \leq 1 + 2\sqrt{\frac{a^2}{b^2} - 1} + \frac{a^2}{b^2} - 1 \quad \Leftrightarrow \quad 0 \leq \sqrt{\frac{a^2}{b^2} - 1}.$$

Because $a/b \geq 1$, this last inequality is always fulfilled, and the first one obviously, too. One may easily check that for $b \to 0$, $T \to \infty$, as it should be.

(b) With friction

In this case the equation of motion 20.8 reads

$$\varrho a \ddot{z} = \varrho z g - \mu_g F_\perp = \varrho z g - \mu_g \varrho (a - z) g \tag{20.15}$$

$$\Rightarrow \quad \ddot{z} = \frac{g}{a} z - \frac{\mu_g g}{a}(a - z) = \frac{g}{a}(1 + \mu_g) z - \mu_g g. \tag{20.16}$$

This is an inhomogeneous differential equation of second order. Its general solution is given by *one* particular solution of the inhomogeneous differential equation plus the general solution of the homogeneous differential equation. The homogeneous differential equation reads

$$\ddot{z}_1 = \frac{g}{a}(1 + \mu_g) z_1$$

and, because of 20.8 and 20.9, the general solution is

$$z_1(t) = z_{\text{hom}}(t) = A e^{\sqrt{\frac{g}{a}(1+\mu_g)}\,t} + B e^{-\sqrt{\frac{g}{a}(1+\mu_g)}\,t}. \tag{20.17}$$

A particular solution of 20.16 is ($\ddot{z} = 0$)

$$z_2(t) = z_{\text{part}} = +\frac{\mu_g a}{1 + \mu_g} = \text{constant}. \tag{20.18}$$

One easily confirms that the sum $z(t) = z_1(t) + z_2(t)$ satisfies the inhomogeneous differential equation 20.16. The general total solution of 20.16 therefore reads

$$\begin{aligned} z_1(t) + z_2(t) &= z(t) \\ &= A e^{\sqrt{\frac{g}{a}(1+\mu_g)}\,t} + B e^{-\sqrt{\frac{g}{a}(1+\mu_g)}\,t} + \frac{\mu_g a}{1 + \mu_g}. \end{aligned} \tag{20.19}$$

As above, the initial conditions lead to the two equations

$$z(0) = A + B + \frac{\mu_g a}{1 + \mu_g} \stackrel{!}{=} b,$$

$$\dot{z}(0) = 0 = A\sqrt{\frac{g}{a}(1 + \mu_g)} - B\sqrt{\frac{g}{a}(1 + \mu_g)} = A - B,$$

with the solution

$$A = \frac{b}{2} - \frac{\mu_g a}{2(1+\mu_g)}, \qquad B = \frac{b}{2} - \frac{1}{2}\frac{\mu_g a}{(1+\mu_g)}.$$

Thus, the complete solution reads

$$z(t) = \left(\frac{b}{2} - \frac{\mu_g a}{2(1+\mu_g)}\right)\left[e^{\sqrt{\frac{g}{a}(1+\mu_g)}\,t} + e^{-\sqrt{\frac{g}{a}(1+\mu_g)}\,t}\right] + \frac{\mu_g a}{1+\mu_g}. \tag{20.20}$$

The time of sliding T is determined by the equation

$$z(T) = a = \left(\frac{b}{2} - \frac{\mu_g a}{2(1+\mu_g)}\right)\left[e^{\sqrt{\frac{g}{a}(1+\mu_g)}\,T} + e^{-\sqrt{\frac{g}{a}(1+\mu_g)}\,T}\right] + \frac{\mu_g a}{1+\mu_g}$$

$$\Rightarrow \quad a\left(\frac{1}{1+\mu_g}\right) = \frac{\left(b - \frac{\mu_g}{1+\mu_g}a\right)}{2}\left(x + \frac{1}{x}\right), \tag{20.21}$$

where

$$x = e^{\sqrt{\frac{g}{a}(1+\mu_g)}\,T}. \tag{20.22}$$

The quadratic equation 20.21 has the solution

$$x_{1,2} = \frac{a\left(\frac{1}{1+\mu_g}\right)}{b\left(1 - \frac{\mu_g}{1+\mu_g}\frac{a}{b}\right)} \pm \frac{1}{b\left(1 - \frac{\mu_g}{1+\mu_g}\frac{a}{b}\right)}\sqrt{a^2\left(\frac{1}{1+\mu_g}\right)^2 - b^2\left(1 - \frac{\mu_g}{1+\mu_g}\frac{a}{b}\right)^2}.$$

The solution x_2 drops out similarly as above. Therefore, the time of sliding T is evaluated as

$$T = \sqrt{\frac{a}{g(1+\mu_g)}} \times \ln\frac{a + \sqrt{a^2 - (b + \mu_g(b-a))^2}}{b + \mu_g(b-a)}. \tag{20.23}$$

We note that it may be seen from equation 20.16 that the chain begins to slide only then if $\ddot{z} > 0$, that is,

$$b > \frac{\mu_g}{1+\mu_g}a.$$

The time of sliding T increases under friction. This is not seen from 20.23 at first glance, and one needs to perform an expansion by μ_g, which will be saved here.

Problem 20.8: A disk on ice—the friction coefficient

A disk is sliding on ice. At a certain point of the straight path it has the velocity v_0. It comes to rest at the distance x_0 beyond this point. Determine the friction coefficient (e.g., for $v_0 = 40$ km/h; $x_0 = 30$ m).

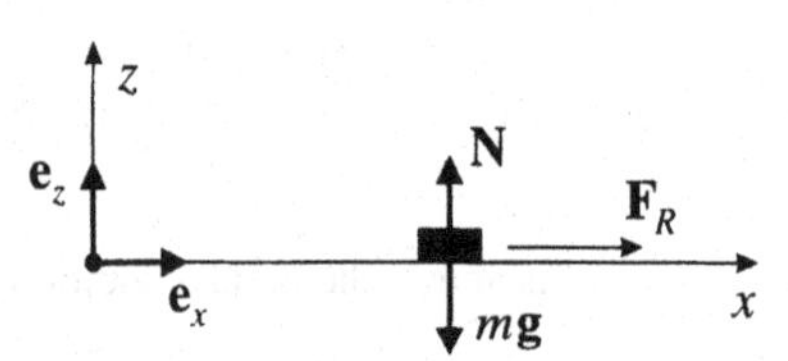

Forces acting on a sliding disk.

Solution The initial conditions are

$$t = 0, \quad x = 0, \quad v = v_0,$$
$$t = t_0, \quad x = x_0, \quad v = 0.$$

The individual forces are denoted as follows:

$$\begin{aligned}
\mathbf{W} &= -mg\mathbf{e}_z && \text{(weight)},\\
\mathbf{N} &= -\mathbf{W} && \text{(normal force)},\\
\mathbf{F}_R &= -\mu_g|\mathbf{N}|\,\mathbf{e}_x \Rightarrow && \\
\mathbf{F}_R &= -\mu_g mg\mathbf{e}_x && \text{(friction force)}.
\end{aligned}$$

For the equation of motion one then obtains

$$m\frac{d\mathbf{v}}{dt}\mathbf{e}_x = -\mu_g mg\mathbf{e}_x \qquad \Rightarrow \qquad \frac{dv}{dt} = -\mu_g g.$$

Separation of the variables and integration yield

$$\int_{v_0}^{v} dv' = -\mu_g g\int_0^t dt' \qquad \Rightarrow \qquad v = v_0 - \mu_g gt$$

and

$$\int_0^x dx' = \int_0^t (v_0 - \mu_g gt')\,dt' \qquad \Rightarrow \qquad x = v_0 t - \frac{1}{2}\mu_g gt^2.$$

The disk comes to rest if $v = 0$, namely at the time $t_0 = v_0/\mu_g g \Rightarrow$ inserted in x: $x_0 = (1/2)(v_0^2/\mu_g g)$, or solved for the friction coefficient: $\mu_g = (1/2)(v_0^2/x_0 g) \approx 0.21$.

Problem 20.9: A car accident

An accident happens on a straight-plane village street (allowed velocity 50 km/h). After activating the brakes, the car slides 39 m until it stops (friction coefficient: $\mu = 0.5$).

Find out whether the driver is guilty.

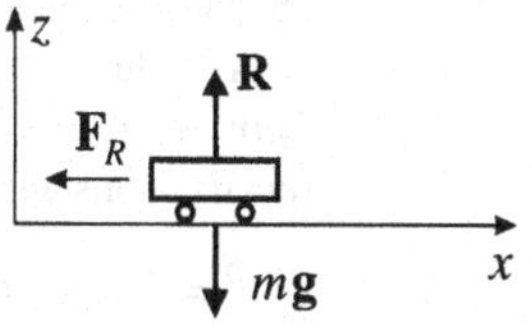

Solution Because the weight $m\mathbf{g}$ and the reactive force mutually cancel each other, only the friction force $\mathbf{F}_R$ acts on the car. The equation of motion then reads

Forces acting on a sliding car.

$$m\frac{d^2x}{dt^2} = -\mu mg \qquad \Rightarrow \qquad \frac{d^2x}{dt^2} = -\mu g.$$

We now have

$$\frac{d^2x}{dt^2} = \frac{d}{dt}\left(\frac{dx}{dt}\right) = \frac{d\dot{x}}{dx}\frac{dx}{dt} = v\frac{dv}{dx}$$

$$\Rightarrow \quad v\frac{dv}{dx} = -\mu g$$

$$\Rightarrow \quad \int_{v_0}^{0} v\,dv = -\mu g\int_{x_0}^{x_0+s} dx,$$

where x_0 is the position where full breaking begins, and s is the braking distance. Thus we get

$$\frac{1}{2}v_0^2 = \mu g s \quad \Rightarrow \quad v_0 = \sqrt{2\mu g s}.$$

With the numerical data it follows that

$$v_0^2 = 2 \cdot 0.5 \cdot 9.81 \cdot 39 \, \frac{\text{m}^2}{\text{s}^2}$$

$$\Rightarrow \quad v_0 = 19.56 \, \frac{\text{m}}{\text{s}} \quad \text{or} \quad v_0 = 70.42 \, \frac{\text{km}}{\text{h}}.$$

The driver drove too fast by about 20 km/h.

Problem 20.10: A particle on a sphere

Let a particle of mass m be positioned at the "north pole" of a frictionless smooth sphere of radius b. After a small displacement let it slide down at the sphere. At which time does it separate from the sphere, and what is its velocity in that moment?

Solution If the particle is at P, it is pressed to the sphere by the normal force

$$\mathbf{N} = -mg \sin\theta \, \mathbf{e}_r,$$

while the centrifugal force

$$\mathbf{Z} = (mv^2)/b \, \mathbf{e}_r$$

tries to pull it off the sphere. At the moment at which both forces are balancing each other,

$$\mathbf{N} + \mathbf{Z} = 0,$$

the particle separates from the sphere!

(a) Solution via the energy law: One has $\frac{1}{2}mv^2 + mgh = T + V = E = mgb$, where E remains constant in time. Then

$$v^2 = 2g(b - h) = 2gb(1 - \sin\theta)$$

and, therefore,

$$\mathbf{N} + \mathbf{Z} = -mg \sin\theta \, \mathbf{e}_r + \frac{mv^2}{b} \mathbf{e}_r$$

$$= [2mg(1 - \sin\theta) - mg \sin\theta] \, \mathbf{e}_r.$$

In order to have $\mathbf{N} + \mathbf{Z} = 0$, it must hold that

$$2mg(1 - \sin\theta) - mg \sin\theta = 0$$

or

$$3 \sin\theta = 2, \quad \text{i.e.,} \quad \sin\theta = \frac{2}{3},$$

$$\theta = 41.8°.$$

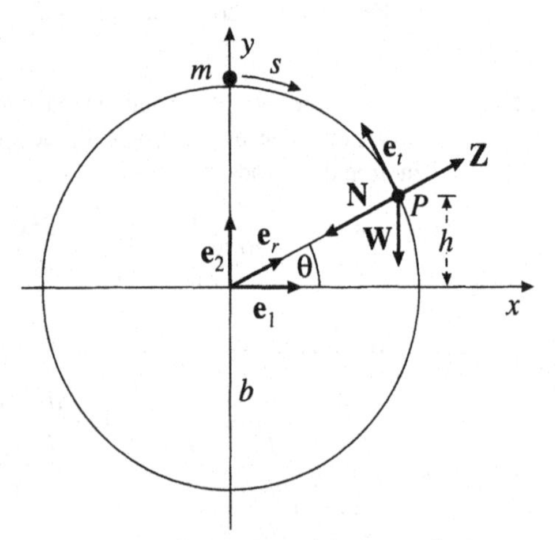

Forces acting on a particle on a sphere.

From there we immediately find that at the moment of separation

$$h = \frac{2}{3}b \qquad \text{and} \qquad v = \sqrt{\frac{2}{3}gb}\,.$$

(b) Solution via the equation of motion: Consider the sphere as (locally) an inclined plane. At P there acts the slope drift

$$\mathbf{H} = -mg\cos\theta\,\mathbf{e}_t\,.$$

We call s the distance of the particle from the north pole, as measured along the surface of the sphere. Because

$$s = b\left(\frac{\pi}{2} - \theta\right),$$

we have

$$\ddot{s} = -b\ddot{\theta}\,.$$

Hence, the equation of motion reads

$$-m\frac{d^2s}{dt^2}\mathbf{e}_t - \mathbf{H} = -m\frac{d^2s}{dt^2}\mathbf{e}_t + mg\cos\theta\mathbf{e}_t = 0$$

or

$$-\frac{d^2s}{dt^2} + g\cos\theta = b\ddot{\theta} + g\cos\theta = 0.$$

Multiplication by $\dot{\theta}$ yields

$$b\ddot{\theta}\,\dot{\theta} + g\cos\theta\,\dot{\theta} = 0.$$

Integration of this differential equation leads to

$$\frac{1}{2}b\dot{\theta}^2 + g\sin\theta = c.$$

For $t = 0$, we have $\dot{\theta} = 0$ and $\theta = 90°$; thus $c = g$

$$\Rightarrow \quad \dot{\theta}^2 = -2\frac{g}{b}(\sin\theta - 1).$$

For the moment of separation, we obtain

$$\mathbf{N} + \mathbf{Z} = m\left(\frac{v^2}{b} - g\sin\theta\right)\mathbf{e}_r = 0$$

or with $v = b\dot{\theta}$:

$$\dot{\theta}^2 b - g\sin\theta = -2g(\sin\theta - 1) - g\sin\theta = 0$$

$$\Rightarrow \quad 3\sin\theta = 2 \qquad \text{or} \qquad \sin\theta = \frac{2}{3}; \quad \theta \approx 41°\,49'.$$

For the velocity results,

$$v = \dot{\theta}b = \sqrt{2g(1-\sin\theta)b} = \sqrt{\frac{2}{3}gb}.$$

Problem 20.11: A ladder leans at a wall

A ladder of length l and mass m leans at a vertical wall, enclosing the angle θ with the wall. The gravity force operates on the center of the ladder (see figure). The friction coefficient between the ground and ladder is μ_h, while the friction between the wall and ladder is being neglected.

Determine the maximum angle θ at which the ladder may lean against the wall without sliding down.

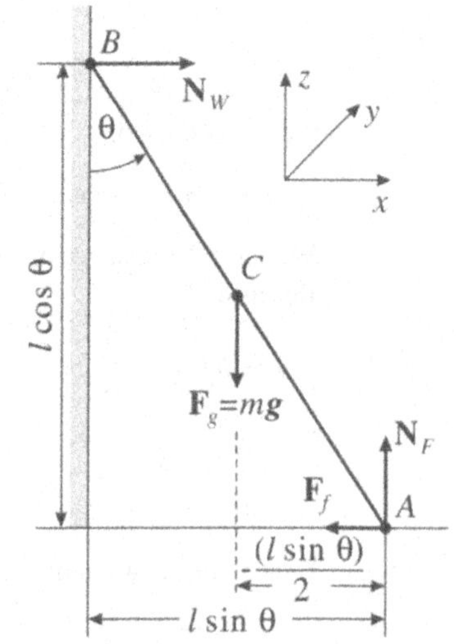

Forces acting on the ladder.

Solution The forces acting on the ladder are the reactive force $N_F\mathbf{e}_z$ at point A and the reactive force $N_W\mathbf{e}_x$ at point B.

In addition, at A the friction force $-F_f\mathbf{e}_x$ still acts in the negative x-direction. The gravitational force $-F_g\mathbf{e}_z = -mg\mathbf{e}_z$ acts on the center of the ladder.

The conditions that the system is in equilibrium are

(a) The sum over all forces must be zero.

(b) The sum over all torques with respect to a point must be zero.

From the figure we find the component representation:

$$\sum_i \mathbf{F}_i^x = 0: \qquad N_w - F_f = 0, \tag{20.24}$$

$$\sum_i \mathbf{F}_i^z = 0: \qquad -mg + N_F = 0. \tag{20.25}$$

The torques acting with respect to point A are caused by the forces $-F_g\mathbf{e}_z = -mg\mathbf{e}_z$ and $N_w\mathbf{e}_x$. We obtain

$$\mathbf{M}_A = \sum_i \mathbf{r}_i \times \mathbf{F}_i = \left(-mg\left(\frac{l}{2}\sin\theta\right) + N_w(l\cos\theta)\right)\mathbf{e}_y = 0, \tag{20.26}$$

or resolved for N_w:

$$N_w = \frac{mg}{2}\tan\theta. \tag{20.27}$$

According to equation 20.24, the friction force F_f is

$$F_f = \frac{mg}{2}\tan\theta. \tag{20.28}$$

Because the friction force cannot exceed the product of reactive force N_F and friction coefficient μ_h (see figure),

$$F_f(\max) = N_F\mu_h = mg\mu_h, \tag{20.29}$$

we obtain as equilibrium condition

$$N_w = \frac{mg}{2}\tan\theta = F_f < F_f(\max) = mg\mu_h, \tag{20.30}$$

or

$$\frac{mg}{2}\tan\theta < mg\mu_h\,,$$

and for the angle θ:

$$\tan\theta < 2\mu_h. \tag{20.31}$$

The maximum angle θ is independent of the mass and length of the ladder. It is only a function of the friction coefficient μ_h.

Problem 20.12: A mass slides under static and dynamic friction

Two masses of $m_1 = 6$ kg and $m_2 = 10$ kg are fixed to a nonstretchable rope that runs over a roller (see figure). The static friction coefficient for m_1 and the support has the value $\mu_h = 0.625$. The dynamic friction coefficient is $\mu_g = 0.33$.

(a) What is the minimum value of the mass m_3 such that m_1 does not move?

(b) What is the acceleration of the system if the mass m_3 is removed?

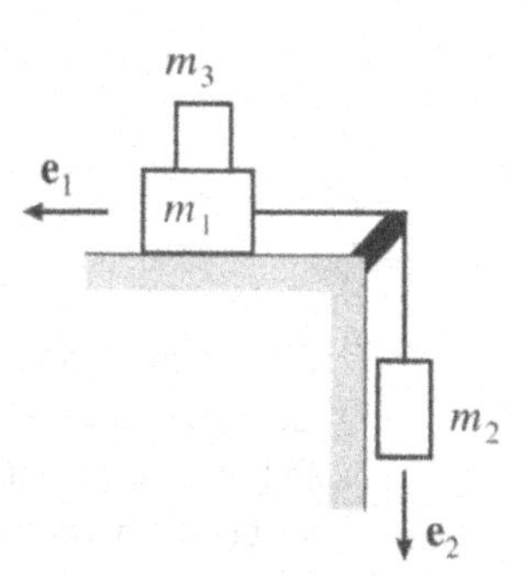

A sliding mass.

Solution (a) If the system is at rest, a static friction force $\mathbf{F}_H$ acts on the rope that is caused by the masses m_1 and m_3 and the support area:

$$\mathbf{F}_H = -\mu_H F_\perp \cdot \frac{\mathbf{v}_1}{v_1}.$$

$\mathbf{v}_1$ points, however, along $-\mathbf{e}_1$; hence $\mathbf{F}_H = \mu_H(m_1+m_3)g\mathbf{e}_1$. The forces acting along $-\mathbf{e}_1$ are

$$-(m_1+m_3)a\mathbf{e}_1 = -T\mathbf{e}_1 + \mu_H(m_1+m_3)g\mathbf{e}_1, \tag{20.32}$$

and the forces acting along $\mathbf{e}_2$ are

$$m_2 a\mathbf{e}_2 = -T\mathbf{e}_2 + m_2 g\mathbf{e}_2, \tag{20.33}$$

where T is the rope tension. Insertion of T from 20.33 in 20.32 yields

$$a = \frac{m_2 g - \mu_H(m_1+m_3)g}{m_1+m_2+m_3}. \tag{20.34}$$

We get as equilibrium condition

$$m_3 = \frac{m_2}{\mu_H} - m_1, \qquad m_3 = 10\ \text{kg}.$$

(b) If the system moves, μ_H must be replaced by μ_G ($m_3 = 0$):

$$a = g\frac{m_2 - \mu_G m_1}{m_1+m_2}; \qquad a = 0.5\,g.$$

---

21 The Harmonic Oscillator

The eminent meaning of the harmonic oscillator is due to the fact that it does not occur in mechanics only, but in an analogous manner governs extended sections of electrodynamics and atomic physics. Many complicated vibrational processes may be approximately described as harmonic oscillations and thus be treated simply in this way. The reason is the following: In the equilibrium (at $x = 0$) the forces acting on the mass point must vanish, that is, $\mathbf{F} = -\nabla V = 0$. If one expands the potential in a Taylor series

$$V(x) = V_0 + a_1 x + \frac{a_2}{2} x^2 + \ldots,$$

the equilibrium condition implies that $a_1 = 0$ must hold, just because $\mathbf{F}(0) = 0$. Therefore,

$$V(x) = V_0 + \frac{a_2}{2} x^2 + \ldots$$

must hold. For small displacements from the equilibrium the potential is therefore always harmonic. In mechanics we are dealing with a harmonic oscillator if a force acting on a body is proportional to, but oppositely directed to, its displacement from the rest position. This linear force law may be generated by a spring obeying *Hooke's law* (see also p. 162 in Section 18). To simplify the problem, we consider the harmonic oscillator only in the x-direction, i.e., the force law is

$$\mathbf{F} = -kx\mathbf{e}_1.$$

For the linear force law, obviously

$$\operatorname{curl} \mathbf{F} = -\mathbf{e}_2 \frac{\partial}{\partial z}(kx) + \mathbf{e}_3 \frac{\partial}{\partial y}(kx) = 0.$$

This implies: The force is conservative. Consequently, the energy law also holds:

$$\frac{1}{2} m v^2 + V(x) = E = \text{constant}.$$

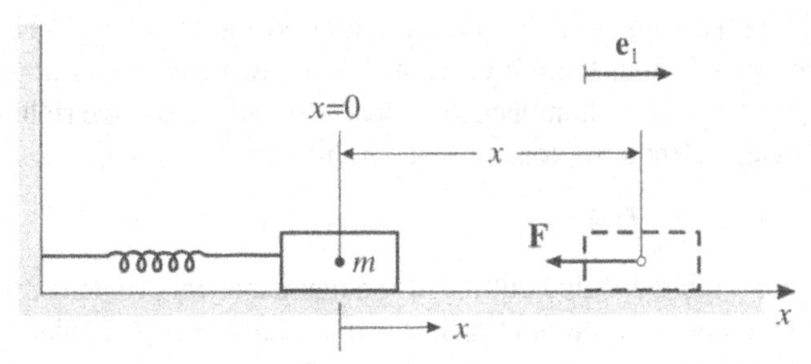

A mass coupled to a spring.

The potential is calculated as

$$V(x) = -\int_0^x \mathbf{F}\cdot d\mathbf{r} = -\int_0^x (-kx, 0, 0)\cdot(dx, dy, dz)$$

$$= \int_0^x kx\,dx = \frac{1}{2}kx^2.$$

Inserting $V(x)$ in the energy equation, we obtain

$$\frac{1}{2}mv^2 + \frac{1}{2}kx^2 = E.$$

We have already solved this equation as an example for the general potential motion (see Chapter 18). There we found

$$x(t) = a\cos(\omega t - \varphi), \tag{21.1}$$

with a being the maximum displacement (amplitude) and $\omega^2 = k/m$.

To get more experience in solving differential equations and to learn other solving methods, we shall use a second way of solving. For this purpose we start directly from the Newtonian basic equations:

$$m\frac{d^2x}{dt^2}\mathbf{e}_1 = \mathbf{F} = -kx\mathbf{e}_1.$$

We turn over to the scalar equation and divide by the mass m:

$$\frac{d^2x}{dt^2} = -\frac{k}{m}x = -\omega^2 x,$$

where we again have set $k/m = \omega^2$. We write this equation in the simpler form:

$$\ddot{x} + \omega^2 x = 0. \tag{21.2}$$

It is a *differential equation of second order.* That means that the highest derivative occuring in the differential equation is of the second order ($\ddot{x} = d^2x/dt^2$!). When solving this equation, two (integration) constants arise that are determined by the initial conditions.

The initial velocity $\dot{x}(0)$ and the initial position $x(0)$ must be arbitrarily selectable. The general solution, therefore, must involve two free constants. Moreover, the differential equation (21.2) is homogeneous since a zero arises on the right. In other words, there is no x-independent term, for example, of the form

$$\ddot{x} + \omega^2 x = f(t) .$$

For a more detailed outline of the mathematical problems, we refer to Chapter 25. The differential equation is also linear. If we have two particular solutions of the differential equation, for example, $x_1(t)$ and $x_2(t)$, then any linear combination

$$x(t) = A\,x_1(t) + B\,x_2(t) \tag{21.3}$$

also satisfies this differential equation. Here A and B are arbitrary, freely selectable constants. This is the characteristic feature of *linear* differential equations. This linear combination $x(t)$ involves two free constants A and B, that is, the linear combination $x(t)$ is already the general solution of equation (21.2). In order to check the correctness of our assumption, we imagine two particular solutions $x_1(t)$ and $x_2(t)$ of the differential equation (21.2), that is, there shall hold

$$\begin{aligned} &\ddot{x}_1 + \omega^2 x_1 = 0, \\ &\ddot{x}_2 + \omega^2 x_2 = 0. \end{aligned} \tag{21.4}$$

Inserting $x(t) = Ax_1(t) + Bx_2(t)$ in the differential equation (21.2), we obtain

$$\begin{aligned} \ddot{x} + \omega^2 x &= (A\ddot{x}_1 + B\ddot{x}_2) + \omega^2(Ax_1 + Bx_2) \\ &= (A\ddot{x}_1 + \omega^2 Ax_1) + (B\ddot{x}_2 + \omega^2 Bx_2) \\ &= A(\ddot{x}_1 + \omega^2 x_1) + B(\ddot{x}_2 + \omega^2 x_2) \\ &= 0. \end{aligned} \tag{21.5}$$

Hence, $x(t)$ solves the differential equation. This is the proof of validity of the *superposition principle* for the solutions of the harmonic oscillator: From two solutions one may generate other solutions by linear combination. In order to solve the differential equation (21.2), we need two solutions (x_1 and x_2). The solutions are, for example,

$$x_1(t) = \cos\omega t, \tag{21.6}$$

$$x_2(t) = \sin\omega t. \tag{21.7}$$

We form the second derivatives of the solutions (21.6 21.7):

$$\ddot{x}_1(t) = -\omega^2\cos\omega t, \tag{21.8}$$

$$\ddot{x}_2(t) = -\omega^2\sin\omega t, \tag{21.9}$$

and insert (21.6) and (21.8), or (21.7) and (21.9) in the differential equation (21.2), and so we obtain

$$\begin{aligned} &\ddot{x}_1 + \omega^2 x_1 = -\omega^2\cos\omega t + \omega^2\cos\omega t = 0, \\ &\ddot{x}_2 + \omega^2 x_2 = -\omega^2\sin\omega t + \omega^2\sin\omega t = 0. \end{aligned}$$

Both approaches fulfill our differential equation. Moreover, sine and cosine are linearly independent functions, that is, there is no constant C such that $C \sin \omega t = \cos \omega t$ holds for all times t.

The general solution of the differential equation of the harmonic oscillator therefore reads

$$x(t) = A \cos \omega t + B \sin \omega t. \tag{21.10}$$

The earlier form of the equation (21.1) has another form. We try to rewrite our solution (21.10) to this form and write

$$A \cos \omega t + B \sin \omega t = \sqrt{A^2 + B^2} \left(\frac{A}{\sqrt{A^2 + B^2}} \cos \omega t + \frac{B}{\sqrt{A^2 + B^2}} \sin \omega t \right).$$

By setting $A(A^2 + B^2)^{-1/2} = \cos \varphi$, then

$$\sin \varphi = \sqrt{1 - \cos^2 \varphi} = \sqrt{1 - \frac{A^2}{A^2 + B^2}} = \frac{B}{\sqrt{A^2 + B^2}}.$$

We thus obtain

$$x(t) = \sqrt{A^2 + B^2} (\cos \varphi \cos \omega t + \sin \varphi \sin \omega t).$$

We write this result as

$$x(t) = D \cos(\omega t - \varphi), \tag{21.11}$$

where $D = \sqrt{A^2 + B^2}$ and $\tan \varphi = B/A$. The symbols mean

$$\begin{aligned} \nu = \frac{\omega}{2\pi}&: \quad \text{frequency}, \\ T = \frac{1}{\nu} = \frac{2\pi}{\omega}&: \quad \text{vibration period}, \\ \omega&: \quad \text{angular frequency}, \\ D&: \quad \text{amplitude}, \\ \varphi&: \quad \text{phase angle}. \end{aligned}$$

The vibrational curve is obtained by superposing the sine and cosine curves of the vibration (superposition method), that is, the function values of both components are added for all times. The subsequent figure illustrates the approach; the components $A \cdot \cos \omega t$ and $B \cdot \sin \omega t$ are plotted in the upper part, and the sum of both in the lower part. The addition then yields (21.11).

In the vibration equation

$$x(t) = A \cos \omega t + B \sin \omega t$$

the free constants A and B do not yet have a physically evident meaning. But they are uniquely determined by the initial conditions. If we make the settings $x(0) = x_0$ and $v(0) = v_0$, A and B may be calculated:

$$x_0 = x(t=0) = A\cos\omega 0 + B\sin\omega 0 = A,$$
$$v_0 = v(t=0) = \dot{x}(t=0) = -A\omega\sin\omega 0 + B\omega\cos\omega 0 = B\omega;$$

hence:

$$x_0 = A \quad \text{and} \quad v_0 = B\omega.$$

Thus we may write our solution in the form

$$x(t) = x_0\cos\omega t + \frac{v_0}{\omega}\sin\omega t\,. \tag{21.12}$$

Transformation yields

$$x(t) = \sqrt{x_0^2 + \frac{v_0^2}{\omega^2}}\cos(\omega t - \varphi), \tag{21.13}$$

where $\tan\varphi = v_0/(\omega x_0)$. From this form we may immediately read off the vibration amplitude:

$$D = \sqrt{x_0^2 + \frac{v_0^2}{\omega^2}}.$$

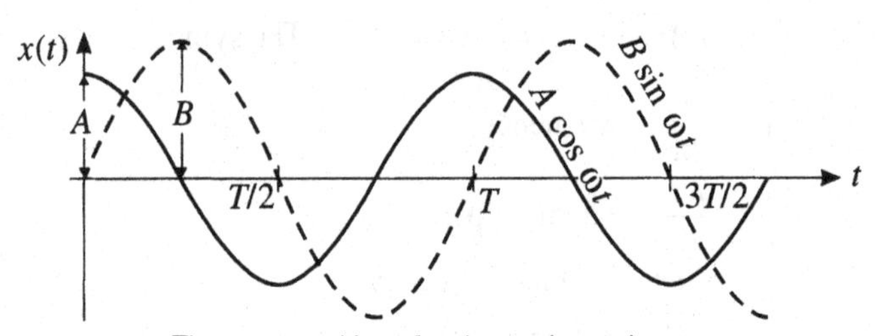

The superposition of a sine and a cosine ...

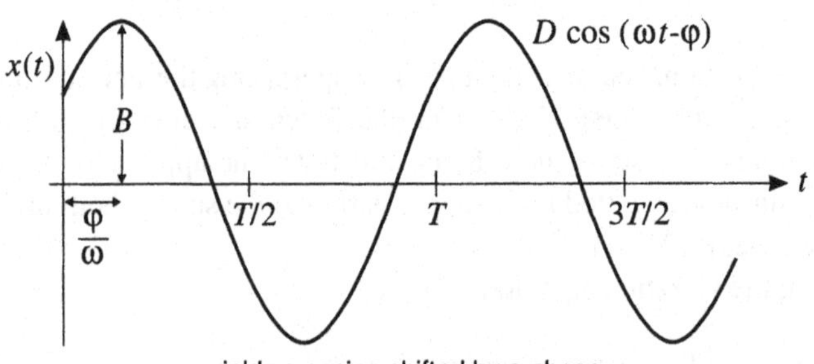

...yields a cosine shifted by a phase φ.

Finally, we shall investigate what the vibration equation looks like in several important special cases.

1. We displace the oscillator at the beginning by x_0, then release it and investigate its vibration. The initial conditions obviously are

$$x_0 = x(0), \quad v_0 = v(0) = 0.$$

By inserting them in the general equation (21.12), we find

$$x(t) = x_0 \cos \omega t.$$

The initial elongation is at the same time the amplitude of the vibration.

2. We apply an impulse to the body in its rest position, giving it instantaneously the velocity v_0. This case occurs (in higher order), for example, in the elastic collision (ballistic measuring instruments). The initial conditions then read

$$x_0 = x(0) = 0, \quad v(0) = v_0.$$

From (21.12) we obtain

$$x(t) = \frac{v_0}{\omega} \sin \omega t = \frac{v_0}{\omega} \cos\left(\omega t - \frac{\pi}{2}\right).$$

The amplitude of the vibration is $D = v_0/\omega$. This may also be derived from the energy law. One has

$$\frac{1}{2}mv^2 + \frac{1}{2}kx^2 = E = \frac{1}{2}mv_0^2.$$

When the body has reached the maximum displacement D, then $v = 0$. Hence

$$\frac{1}{2}kD^2 = \frac{1}{2}mv_0^2,$$

and therefore

$$D^2 = \frac{m}{k}v_0^2 = \omega^{-2}v_0^2 \qquad \text{or} \qquad D = \frac{v_0}{\omega}.$$

As was indicated already at the begin of this chapter, a large number of (vibrational) processes in physics obey the laws of the harmonic oscillator.

If, however, the corresponding potentials in the vicinity of an equilibrium configuration have a somewhat different form, they may frequently be described in the important ranges of small displacements by a harmonic approximation. Here we quote several examples of anharmonic potentials in mechanics and atomic physics together with the associated harmonic approximation.[1]

[1]The theory of rotation and vibration of atomic nuclei and nuclear molecules is described in detail in J.M. Eisenberg and W. Greiner, *Nuclear Theory, Vol 1: Nuclear Models*, 3rd ed., North Holland Publ. Company, Amsterdam and New York, 1987.

1. The pendulum

The potential of the mathematical pendulum has the form

$$V(x) = mgh = mgl(1 - \cos x) = c(1 - \cos x),$$

where $c = mgl$. It may be approximated by a harmonic potential centered at $x = 0$:

$$V(x) = \frac{c}{2}x^2 .$$

The zero point of the potential has been set to $x = 0$, that is, for the pendulum hanging vertically downward.

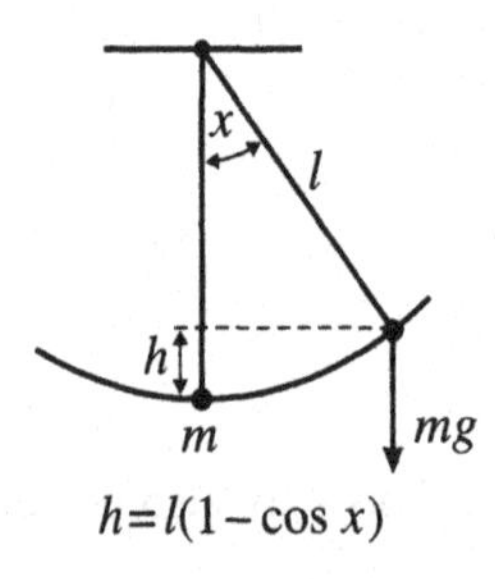

On the calculation of the potential of the pendulum.

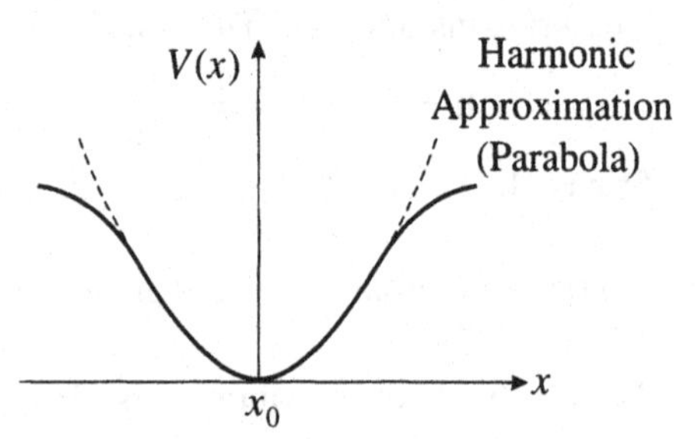

The potential of the pendulum.

2. Dumb-bell molecules

In a two-atomic molecule the individual atoms may vibrate along the longitudinal molecular axis. The mutual binding of the atoms is achieved by so-called molecular electrons, or electrons that are bound to both nuclei. "Atomic electrons," on the contrary, are hull electrons orbiting around the one or the other atomic nucleus (compare the schematic figures).

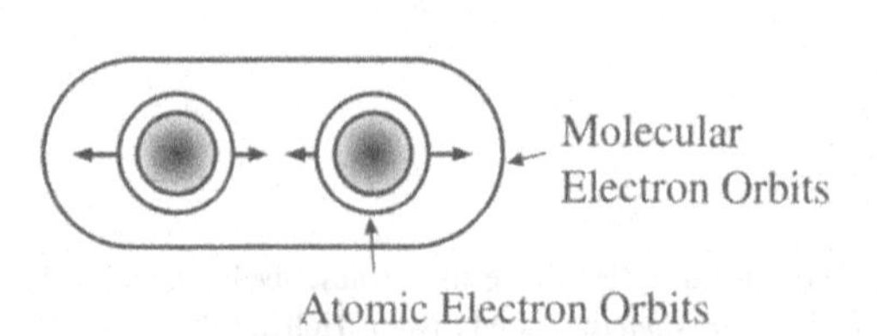

Schematic view of a dumb-bell molecule.

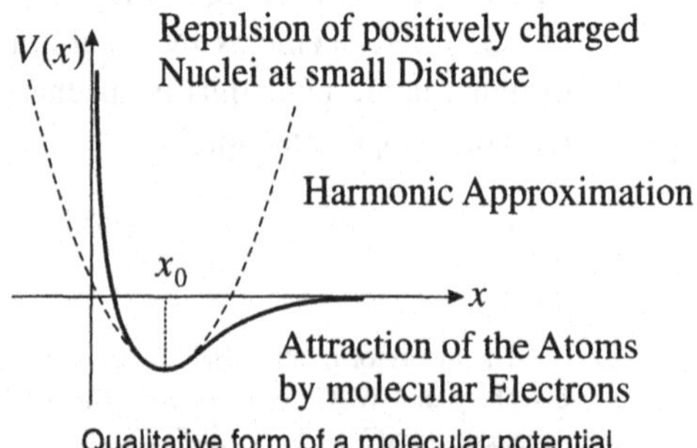

Qualitative form of a molecular potential.

3. Atomic nuclei

Some atomic nuclei (e.g., the rare earth elements Sm, Gd, Er, Yb) have the shape of a thick cigar. They may deform along their axis and in this way perform vibrations.

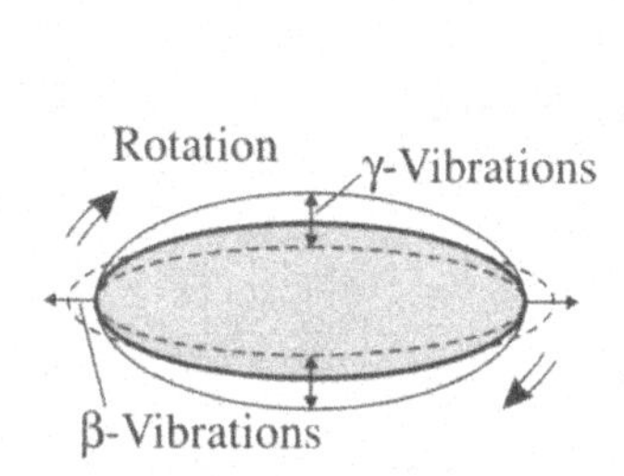

Illustration of the vibrations (β- and γ-vibrations) and rotations of a deformed nucleus.

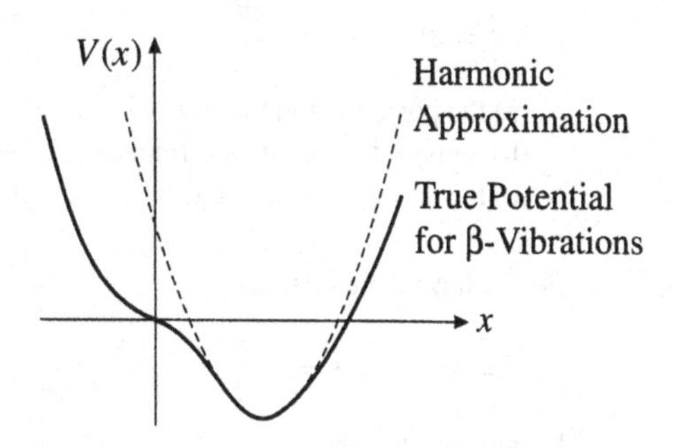

Qualitative form of the potential of β-vibrations of an atomic nucleus.

The contractions and extensions of the "cigar" are called β-vibrations. The contractions and thickenings of the "belly" are called γ-vibrations.

The cigar-shaped deformed nucleus also performs rotations. In doing so, γ quanta are emitted. The so-called rotational-vibrational spectra generated in this way are described by the so-called rotation-vibration model.[2]

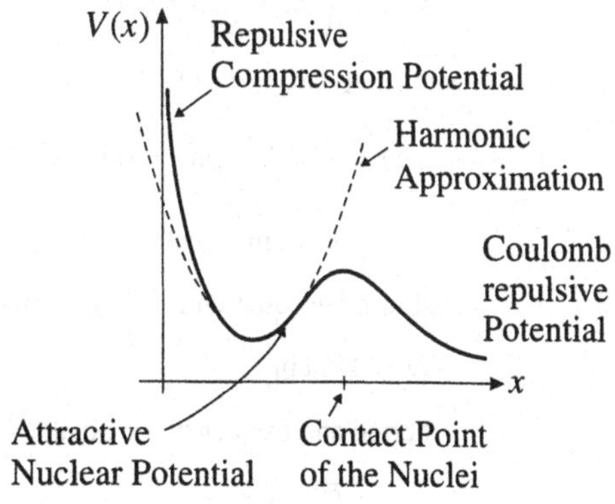

Short range nuclear forces yield a locally attractive potential, giving rise to nuclear molecules.

4. Nuclear molecules

If certain atomic nuclei (e.g., C^{12}, O^{16}) mutually penetrate each other, they may form short-lived but stable molecule-like states. The potential of the two nuclei plotted as a function of their distance follows the trend shown in the figure.

[2]For details, see again J.M. Eisenberg and W. Greiner, loc. cit.

Problem 21.1: Amplitude, frequency and period of a harmonic vibration

An object of mass $2 \cdot 10^4$ g performs harmonic vibrations along the x-axis. Find for the initial conditions

$$x(t=0) = 400 \text{ cm}, \qquad v(t=0) = -150 \,\frac{\text{cm}}{\text{s}},$$

$$a(t=0) = -1000 \,\frac{\text{cm}}{\text{s}^2}$$

(a) the position at the time t,
(b) amplitude, period, and frequency of the vibration,
(c) the acting force at the time $t = \pi/10$ s.

Solution (a) We have the equations:

$$F = ma, \qquad k = -\frac{F}{x}, \qquad \omega^2 = \frac{k}{m}.$$

From there, we get

$$\omega^2 = -\frac{a}{x} = -\frac{a(t=0)}{x(t=0)} = 2.5 \text{ s}^{-2},$$

or

$$\omega = \frac{1}{2}\sqrt{10}\text{ s}^{-1}.$$

By inserting

$$x(t) = \sqrt{x_0^2 + \frac{v_0^2}{\omega^2}} \cos(\omega t - \varphi),$$

we obtain the course of vibration

$$x(t) = 130\sqrt{10}\text{ cm} \cdot \cos\left(\frac{t}{2}\sqrt{10}\text{ s}^{-1} + 0.237\right)$$

$$= 411 \text{ cm} \cdot \cos(t \cdot 1.58 \text{ s}^{-1} + 0.237).$$

(b) From the equation for $x(t)$ we read off the amplitude

$$D = 411 \text{ cm}.$$

Period and frequency are obtained as follows:

$$T = \frac{2\pi}{\omega} = 3.97 \text{ s}; \qquad \nu = \frac{1}{T} = 0.252 \text{ Hz}.$$

(c) One has $F = m\ddot{x} = -2.06 \cdot 10^2 \text{ N} \cdot \cos(t \cdot 1.58 \text{ s}^{-1} + 0.237)$.
For our particular time value, we obtain

$$F\left(t = \frac{\pi}{10}\text{ s}\right) = m\ddot{x}\left(t = \frac{\pi}{10}\text{ s}\right) = -1.53 \cdot 10^2 \text{ N}.$$

Problem 21.2: Mass hanging on a spring

A mass of 20 g hangs on a massless spring and thereby stretches it by 6 cm.

(a) Determine its position at arbitrary time if is pulled down at time $t = 0$ by 2 cm and then is released.

(b) Find the amplitude, period, and frequency of vibration.

Solution (a) Again we have $k = -F/x$ and $\omega^2 = k/m$.
Because $F = -mg$, we find

$$\omega^2 = \frac{g}{x} = 981\ \text{cm} \cdot \text{s}^{-2}\,\frac{1}{6\ \text{cm}} = 163.5\ \text{s}^{-2}$$

and thus $\omega = 12.8\ \text{s}^{-1}$. With $v_0 = 0$, we obtain from

$$x(t) = x_0 \cos\omega t + \frac{v_0}{\omega}\sin\omega t$$

the vibration equation

$$x(t) = x_0 \cos\omega t = -2\ \text{cm} \cdot \cos(t \cdot 12.8\ \text{s}^{-1}).$$

(b) Amplitude, period, and frequency are obtained as in the last problem:

$$D = 2\ \text{cm}; \qquad T = \frac{2\pi}{\omega} = 0.491\ \text{s}; \qquad \nu = \frac{1}{T} = 2.035\ \text{Hz}.$$

Problem 21.3: Vibration of a mass at a displaced spring

Solve the last problem with the assumption that the weight at time $t = 0$ was pulled down by 3 cm and was thrown downward with a velocity of 2 cm/s.

Solution (a) We use equation (21.13): There are $x(t = 0) = -3$ cm and $v(t = 0) = -2$ cm/s and therefore,

$$x(t) = -3.004\ \text{cm} \cdot \cos(t \cdot 12.8\ \text{s}^{-1} - 0.052).$$

(b) Only the amplitude is changed. We now get

$$D = 3.004\ \text{cm}.$$

Problem 21.4: Vibration of a swimming cylinder

A cylinder swims with vertical axis in a liquid of density σ and has weight W and cross-sectional area A. What is the vibration period if the cylinder is slightly pressed down and then released?

Solution The body is pressed down by the distance $-z$. Two forces are then acting on the cylinder: The weight

$$\mathbf{W} = -mg\mathbf{e}_3,$$

and the buoyancy

$$\mathbf{B} = -\sigma A g(z_0 + z)\mathbf{e}_3,$$

where z_0 is the immersion depth in equilibrium. But in the equilibrium state it holds that

$$\mathbf{W} = -\mathbf{B}(z_0), \qquad \text{i.e.,} \quad mg = -\sigma A g z_0.$$

Hence, for an arbitrary position,

$$\mathbf{B} = -(-mg + \sigma A g z)\mathbf{e}_3.$$

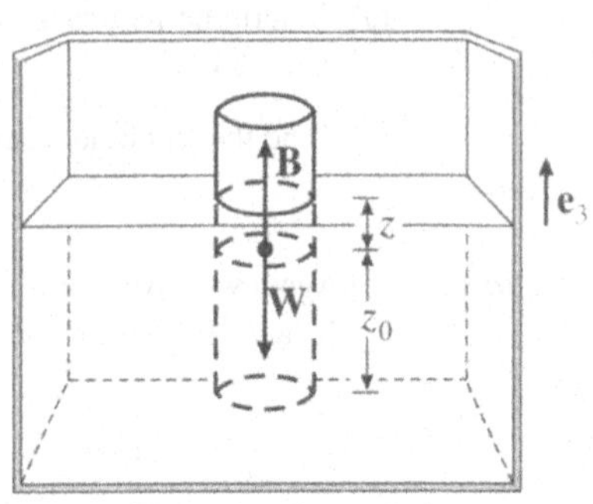

The swimming cylinder.

Therefore, the equation of motion reads

$$m\ddot{z} = W + B = -mg - (-mg + \sigma A g z) = -\sigma A g z$$

or

$$\ddot{z} + \frac{\sigma A g}{m} z = 0.$$

Thus, we find $\omega^2 = \sigma A g/m = (\sigma A/W)g^2$, and further $T = 2\pi/\omega = 2\pi/g\sqrt{W/(\sigma A)}$ as vibration period.

Problem 21.5: Vibrating mass hanging on two strings

Let a mass of 50 g be suspended by identical massless springs with elasticity constants of 0.5 N/m (see figure). In the rest position they form an angle of $\alpha_0 = 30°$ against the horizontal and have the length $l_0 = 2$m; outside the rest position the angle is $\alpha = \alpha_0 + \Delta\alpha$. Determine the period of the vibration that occurs when pulling the mass down by Δx and then releasing it.

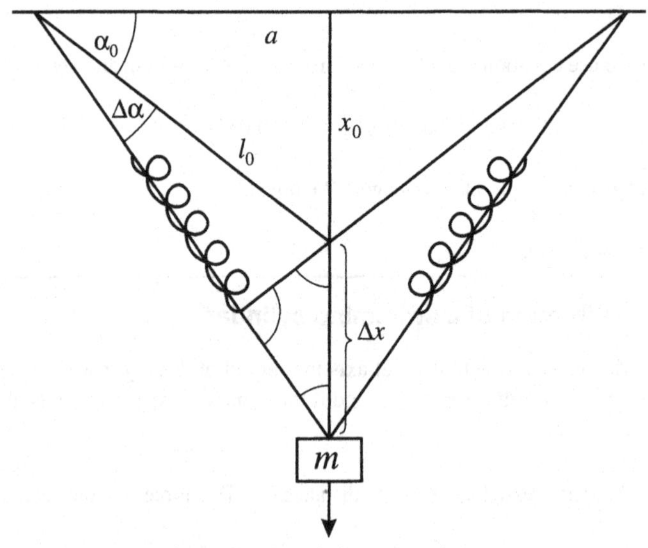

A mass suspended on two strings.

Solution The mass is under the action of the sum of weight force (along the x-direction) and the vertical projection of the backdriving force of the springs. Hence the equation of motion reads

$$m\ddot{x} = mg - 2k(l - \bar{l})\sin\alpha, \tag{21.14}$$

where $\bar{l}$ denotes the length of the springs at rest (absence of external forces). The equilibrium position (position at rest) is defined by the vanishing of the force $m\ddot{x}$, hence

$$mg = 2k(l_0 - \bar{l})\sin\alpha_0. \tag{21.15}$$

In order to solve the differential equation 21.14, both l as well as α must be expressed by the displacement x. It holds that

$$l = \sqrt{x^2 + a^2}, \tag{21.16}$$

$$\sin\alpha = \frac{x}{l} = \frac{x}{\sqrt{x^2 + a^2}}. \tag{21.17}$$

With

$$\bar{l} = l_0 - \frac{mg}{2k\sin\alpha_0}, \tag{21.18}$$

from 21.15 the equation of motion may be transformed to

$$\begin{aligned} m\ddot{x} &= mg - 2kx + 2kl_0\sin\alpha - mg\frac{\sin\alpha}{\sin\alpha_0} \\ &= mg - 2kx + 2kx\frac{l_0}{\sqrt{x^2 + a^2}} - mg\frac{x}{x_0}\frac{l_0}{\sqrt{x^2 + a^2}}. \end{aligned} \tag{21.19}$$

This is a very complicated nonlinear differential equation that has no simple analytic solution. But we are interested in vibrations of low amplitude

$$x = x_0 + \Delta x, \qquad \Delta x \ll x_0. \tag{21.20}$$

With this condition, 21.19 may be linearized by expanding the right side in a Taylor series about the point x_0. We employ the formula

$$\begin{aligned} \frac{l_0}{\sqrt{x^2 + a^2}} &= \frac{l_0}{\sqrt{(x_0 + \Delta x)^2 + a^2}} \approx \frac{l_0}{\sqrt{x_0^2 + 2x_0\Delta x + a^2}} \\ &= \frac{1}{\sqrt{1 + \dfrac{2x_0\Delta x}{l_0^2}}} \approx 1 - \frac{x_0\Delta x}{l_0^2}. \end{aligned} \tag{21.21}$$

Thus, up to the order $O((\Delta x)^2)$, 21.19 may be written out as follows:

$$\begin{aligned} m\Delta\ddot{x} &\approx mg - 2k(x_0 + \Delta x) + 2k(x_0 + \Delta x)\left(1 - \frac{x_0\Delta x}{l_0^2}\right) - mg\frac{x_0 + \Delta x}{x_0}\left(1 - \frac{x_0\Delta x}{l_0^2}\right) \\ &\approx \Delta x\left(-2k\frac{x_0^2}{l_0^2} - \frac{mg}{x_0} + mg\frac{x_0}{l_0^2}\right). \end{aligned} \tag{21.22}$$

Expressed in terms of $\sin\alpha_0 = x_0/l_0$, the linearized equation of motion finally reads

$$\Delta\ddot{x} + \left(\frac{2k}{m}\sin^2\alpha_0 + \frac{g}{l_0}\frac{\cos^2\alpha_0}{\sin\alpha_0}\right)\Delta x = 0. \tag{21.23}$$

The expression in brackets is the square of the angular frequency ω. Hence, the vibration period reads

$$T = \frac{2\pi}{\omega} = \frac{2\pi}{\sqrt{\dfrac{2k}{m}\sin^2\alpha_0 + \dfrac{g}{l_0}\dfrac{\cos^2\alpha_0}{\sin\alpha_0}}}. \tag{21.24}$$

For the given values of k, m, α_0, l_0 this leads to the value $T = 1.79$ s. In the limit $\alpha_0 \to 90^o$, the mass vibrates according to 21.24 just as if it were suspended on a spring with twice the spring constant:

$$T = 2\pi\sqrt{\frac{m}{2k}}. \tag{21.25}$$

The limit $\alpha_0 \to 0^o$, for fixed l_0, makes no sense, since according to 21.18 this would lead to a nonphysical negative value of $\bar{l}$.

Example 21.6: Composite springs

(a) Series connection

The figure illustrates the case of two springs with spring constants k_1 and k_2. The force **F** occurs in both of the springs and causes a variation of length $y_1 = F/k_1$ and $y_2 = F/k_2$. From $y_1 + y_2 = F/k$ there results for the "*effective spring constant*" k:

$$\frac{1}{k} = \frac{1}{k_1} + \frac{1}{k_2}, \qquad k = \frac{k_1 k_2}{k_1 + k_2},$$

such that $k < k_1$ and $k < k_2$. The generalization to n springs is trivial:

$$\frac{1}{k} = \frac{1}{k_1} + \frac{1}{k_2} + \cdots + \frac{1}{k_n}.$$

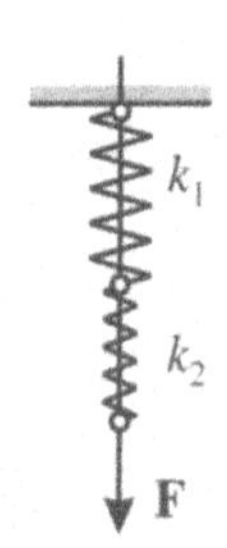

Series connection of springs.

(b) Parallel connection

Because now both springs undergo the same variation of length, namely, $y_1 = y_2 = y$, the resulting spring constant k is calculated from

$$F = k_1 y_1 + k_2 y_2 = ky$$

as

$$k = k_1 + k_2 .$$

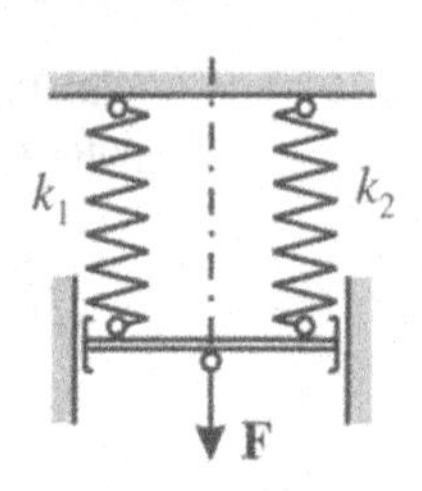

Parallel connection of springs.

The generalization to n springs in parallel connection is

$$k = k_1 + k_2 + \cdots + k_n .$$

The eigenfrequency is then

$$\omega = \sqrt{\frac{k}{m}}.$$

Problem 21.7: Vibration of a rod with pivot bearing

A weight mg is fixed to the upper end of a rod $\overline{AC}$ (assumed as massless), which is supported by a pivot bearing at point A, and is fixed at point B to a spring with the constant k (compare the figure).

(a) Determine the approximate eigenfrequency of the system for vibrations of small elongations φ.

(b) What is the maximum value of $G = mg$ in order to ensure harmonic motion for a small displacement?

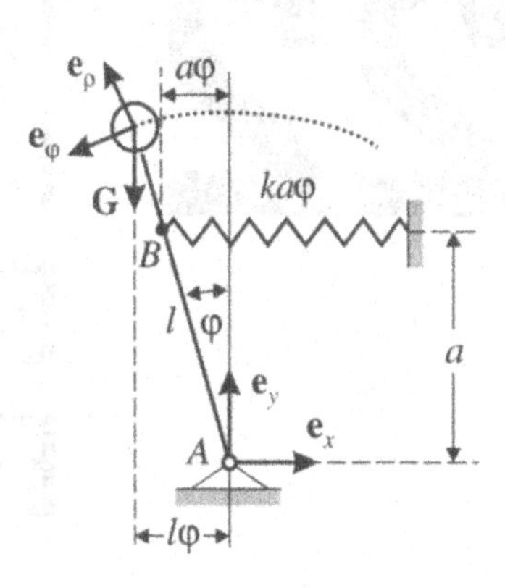

Vibration of a rod with pivot bearing.

Solution (a) The forces acting on the system in the limit $\sin\varphi \approx \varphi$, $\cos\varphi \approx 1$ are

$$\mathbf{G} = -mg\mathbf{e}_y, \qquad \text{weight force,}$$

and

$$\mathbf{F} = ka\varphi\mathbf{e}_x, \qquad \text{spring force,}$$

and the reactive force $\mathbf{F}_R$ along the connecting rod.

Hence,

$$m\ddot{\mathbf{r}} = -mg\mathbf{e}_y + ka\varphi\mathbf{e}_x\frac{a}{l} + \mathbf{F}_R, \qquad \mathbf{F}_R = R\mathbf{e}_\varrho\,,$$

or in polar coordinates

$$m(-\varrho\dot{\varphi}^2\mathbf{e}_\varrho + \varrho\ddot{\varphi}\mathbf{e}_\varphi) = -mg(\cos\varphi\mathbf{e}_\varrho - \sin\varphi\mathbf{e}_\varphi) + ka\varphi\left(\frac{a}{l}\right)(-\sin\varphi\,\mathbf{e}_\varrho - \cos\varphi\,\mathbf{e}_\varphi) + R\mathbf{e}_\varrho .$$

The components of the weight force and the spring force along the direction $\mathbf{e}_\varrho$ are neutralized by the reactive force $\mathbf{F}_R$, such that we obtain

$$m\varrho\ddot{\varphi} = mg\sin\varphi - k\frac{a^2}{l}\varphi\cos\varphi$$

and resolved for $\ddot{\varphi}$

$$\ddot{\varphi} = \frac{1}{\varrho}g\varphi - \frac{k}{m\varrho}\frac{a^2}{l}\varphi \qquad \text{or} \qquad \ddot{\varphi} + \frac{ka^2 - mgl}{ml^2}\cdot\varphi = 0.$$

This vibration equation may also be written as

$$\ddot{\varphi} + \omega_1^2\varphi = 0$$

with the eigenfrequency $\omega_1 = \sqrt{(ka^2 - mgl)/ml^2}$.

(b) The vibrational control remains harmonic while

$$\omega_1^2 > 0 \qquad \text{or} \qquad mg < \frac{ka^2}{l}.$$

22 Mathematical Interlude—Series Expansion, Euler's Formulas

In the following sections we need the series expansion of functions and the Euler relations, which shall be explained now: Many continuous, arbitrarily often differentiable functions $f(x)$ can be expanded in power series:

$$f(x) = a_0 + a_1 x + a_2 x^2 + \cdots = \sum_n a_n x^n. \tag{22.1}$$

The expansion coefficients a_n may be determined by inserting in equation (22.1) and its nth derivatives the corresponding values for $x = 0$; for example,

$$\begin{aligned} f(0) &= a_0, \\ f'(0) &= a_1, \\ f''(0) &= 1 \cdot 2a_2, \\ &\vdots \\ f^{(n)}(0) &= n!\,a_n, \end{aligned}$$

or generally $a_n = f^{(n)}(0)/n!$. f' denotes the first, $f^{(n)}$ the nth derivative of the function $f(x)$ with respect to x. Therefore, the series expansion (22.1) may also be written as follows:

$$f(x) = \sum_{n=0}^{\infty} \frac{f^{(n)}(0)}{n!} x^n. \tag{22.2}$$

This is the well-known *Taylor expansion*. We now give several examples:

1. Example $f(x) = e^x$.

$$f'(x) = f''(x) = \cdots = f^{(n)}(x) = e^x.$$

Thus, equation (22.2) just yields the series expansion of the exponential function, namely

$$e^x = \sum_{n=0}^{\infty} \frac{1}{n!} x^n = 1 + \frac{x}{1!} + \frac{x^2}{2!} + \frac{x^3}{3!} + \cdots . \tag{22.3}$$

By setting $x = i\varphi$ and taking into account $i^2 = -1$, $i^3 = -i$, $i^4 = 1$, etc., we immediately obtain

$$\begin{aligned} e^{i\varphi} &= \sum_{n=0}^{\infty} \frac{1}{n!} i^n \varphi^n \\ &= 1 - \frac{\varphi^2}{2!} + \frac{\varphi^4}{4!} - \frac{\varphi^6}{6!} + \frac{\varphi^8}{8!} - \cdots + i\left(\frac{\varphi}{1!} - \frac{\varphi^3}{3!} + \frac{\varphi^5}{5!} - \frac{\varphi^7}{7!} + - \cdots\right). \end{aligned} \tag{22.4}$$

2. Example $f(x) = \sin x$.

$$f(0) = 0; \; f'(0) = \cos 0 = 1, \; f''(0) = -\sin 0 = 0, \; f'''(0) = -\cos 0 = -1, \text{ etc.}$$

According to equation (22.2), this obviously yields

$$\sin x = x - \frac{x^3}{3!} + \frac{x^5}{5!} - \frac{x^7}{7!} \pm \cdots . \tag{22.5}$$

3. Example $f(x) = \cos x$.

$$f(0) = 1; \; f'(0) = -\sin 0 = 0, \; f''(0) = -\cos 0 = -1, \; f'''(0) = \sin 0 = 0, \text{ etc.}$$

According to equation (22.2), it therefore results that

$$\cos x = 1 - \frac{x^2}{2!} + \frac{x^4}{4!} - \frac{x^6}{6!} \pm \cdots . \tag{22.6}$$

Because $\sin(-x) = -\sin(x)$ and $\cos(-x) = \cos(x)$, (22.5) must involve only odd powers, and (22.6) only even powers x^n.

4. Example By comparing the results (22.4), (22.5), and (22.6), we arrive at the *Euler formulas:*[1]

$$e^{i\varphi} = \cos\varphi + i\sin\varphi; \qquad e^{-i\varphi} = \cos\varphi - i\sin\varphi;$$
$$\cos\varphi = \frac{e^{i\varphi} + e^{-i\varphi}}{2}; \qquad \sin\varphi = \frac{e^{i\varphi} - e^{-i\varphi}}{2i}. \tag{22.7}$$

Problem 22.1: Various Taylor series

Taylor series: In many cases, a function that is arbitrarily often differentiable in an interval I (with $0 \in I$) can be represented by expansion about the point 0 in a power series of the form

$$f(x) = \sum_{n=0}^{\infty} \frac{f^{(n)}(0)}{n!} x^n .$$

Let $f^{(n)}(0)$ be the nth derivative at the point $x = 0$, $f^{(0)}(0) = f(0)$, and $n!$ (n factorial) $= 1 \cdot 2 \cdot 3 \cdot \dots \cdot n$ $(0! = 1)$.

Expand the following functions according to this prescription:

$$\text{(a) } a^x, \qquad \text{(b) } \frac{1}{1-x}, \qquad \text{(c) } \ln(1+x).$$

Solution (a) Equation (22.3) states that

$$e^x = 1 + x + \frac{1}{2!}x^2 + \frac{1}{3!}x^3 + \cdots = \sum_n \frac{x^n}{n!},$$

[1] *Leonhard Euler*, b. April 15, 1707, Basel as son of a priest with extended mathematical interests—d. Sept. 18, 1783, St. Petersburg. Euler studied in Basel, since 1720 philosophy and since 1723 theology. Moreover, he attended private lectures by Johann Bernoulli. In 1727 Euler went to St. Petersburg, there in 1730 he became professor of physics, and in 1733 professor of mathematics at the Academy. In 1741 he was called to Berlin as professor of mathematics and director of the class of mathematics at the Academy. Later on in Berlin the relation between Euler and Friedrich II. went to the worse; he returned to St. Petersburg in 1766. Even his complete blindness in the same year could not stop his mathematical creative power, and already in his last years he was considered as a legendary phenomenon.

The total opus of Euler comprises 886 titles, among them many voluminous treatises. In many branches his kind of representation became final, and all eminent mathematicians of the following era took it over. This concerns the *Introductio in Analysin Infinitorum* (1748), in which, for example, the theory of series, trigonometry, analytic geometry, elimination theory, and the zeta function are outlined, and also the *Institutiones Calculi Differentialis* (1755) and the *Institutiones Calculi Integralis* (1768–1774), which not at all deal with elementary relations only. In 1736 his treatise of mechanics was published which contains the first analytic development of the Newtonian dynamics, and 1744 the first outline of *variational calculus*. Important personal achievements are the Euler polyhedron theorem, the Euler straight line, the Euler constant, the quadratic reciprocity law, and the solution of the Königsberg bridge problem, as well as the convention that the logarithm is infinitely ambiguous (1749). Euler made essential contributions also on astronomy, on the theory of moon and celestial mechanics, on construction of ships, cartography, optics, hydraulics, philosophy, and theory of music. His manner of approaching mathematical problems was characterized by intuitive realization of the essentials and by an eminent formal mastery. But Euler, like by the way all mathematicians before Gauss, often failed to give a fully correct reasoning for his conclusions.

and therefore,

$$a^x = e^{x \ln a} = 1 + x \ln a + \frac{1}{2!}x^2 \ln^2 a + \frac{1}{3!}x^3 \ln^3 a + \cdots = \sum_n \frac{(x \ln a)^n}{n!}.$$

(b)

$$\frac{1}{1-x} = 1 + x + x^2 + x^3 + x^4 + \cdots = \sum_n x^n,$$

because

$$f'(x) = \frac{1}{(1-x)^2}, \qquad f''(x) = \frac{2}{(1-x)^3}, \qquad f'''(x) = \frac{6}{(1-x)^4}, \qquad \ldots.$$

This is, of course, nothing else but the infinite geometric series.

(c)

$$\ln(1+x) = 0 + x - \frac{1}{2}x^2 + \frac{1}{3}x^3 - \frac{1}{4}x^4 + \cdots = \sum_n \frac{(-1)^{n+1}}{n}x^n,$$

because

$$f'(x) = \frac{1}{1+x}, \qquad f''(x) = \frac{-1}{(1+x)^2},$$
$$f'''(x) = \frac{2}{(1+x)^3}, \qquad f''''(x) = \frac{-6}{(1+x)^4}, \qquad \ldots.$$

23 The Damped Harmonic Oscillator

As an example of a damped harmonic oscillator, we again consider a mass m connected to a spring. Let the mass slide frictionless on the support, but the friction at the surrounding medium shall add a velocity-dependent friction force (e.g., air resistance). For the latter one we adopt the Stokes *ansatz*:

$$\mathbf{F}_R = -\beta\mathbf{v}.$$

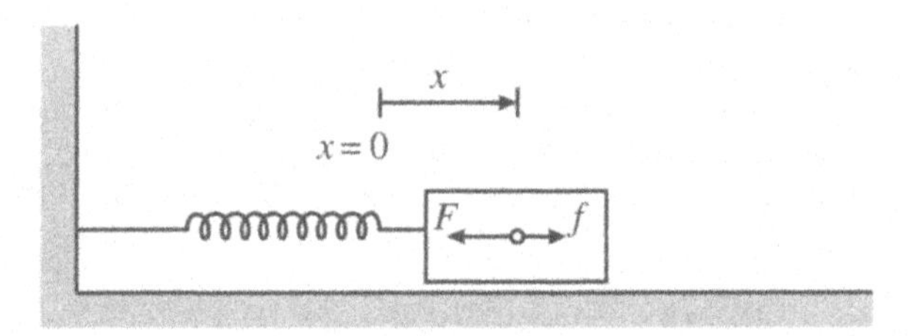

A mass fixed to a spring and sliding on a plane is an example of a damped harmonic oscillator.

Hence, we arrive at the equation of motion

$$m\frac{dv}{dt} = -kx - \beta v. \tag{23.1}$$

Putting all quantities to the left side and writing for the velocity $\dot{x}$ instead of v, the equation reads

$$m\ddot{x} + \beta\dot{x} + kx = 0. \tag{23.2}$$

When dividing by m and setting $2\gamma = \beta/m$, $\omega^2 = k/m$, the equation takes the form

$$\ddot{x} + 2\gamma\dot{x} + \omega^2 x = 0. \tag{23.3}$$

It is a linear differential equation that may easily be checked, similar to the case of the nondamped harmonic oscillator (see equations (21.3), (21.4) ff.). Moreover, the equation is homogeneous and of second order. To solve this differential equation, we first have to look for two linear independent solutions $x_1(t)$ and $x_2(t)$, and then obtain the most general solution of the differential equation by an arbitrary choice of the coefficients A and B. Because the equation, apart from constant coefficients, contains only derivatives of $x(t)$, and because the exponential function remains unchanged under differentiation—apart from constant coefficients—we try the *ansatz*

$$x(t) = e^{\lambda t}$$

and obtain

$$\lambda^2 e^{\lambda t} + 2\gamma\lambda e^{\lambda t} + \omega^2 e^{\lambda t} = 0. \tag{23.4}$$

We divide by $e^{\lambda t}$, because $e^{\lambda t} \neq 0$ always, and obtain the following conditional equation for λ:

$$\lambda^2 + 2\gamma\lambda + \omega^2 = 0.$$

This is called the *characteristic equation*. It is fulfilled by the two values

$$\lambda_{1,2} = -\gamma \pm \sqrt{\gamma^2 - \omega^2}. \tag{23.5}$$

Thus we have found two particular solutions:

$$\begin{aligned} x_1(t) &= e^{\lambda_1 t} = e^{-\gamma t} e^{\sqrt{\gamma^2-\omega^2}\,t}, \\ x_2(t) &= e^{\lambda_2 t} = e^{-\gamma t} e^{-\sqrt{\gamma^2-\omega^2}\,t}. \end{aligned} \tag{23.6}$$

The general solution of our equation is therefore

$$x(t) = A\,e^{\lambda_1 t} + B\,e^{\lambda_2 t}. \tag{23.7}$$

There are three cases of the vibrational equation, depending on the value of the expression $\sqrt{\gamma^2 - \omega^2}$:

(a) $\gamma^2 < \omega^2$: the root is imaginary.

(b) $\gamma^2 = \omega^2$: The root vanishes; the *ansatz* yields only one solution.

(c) $\gamma^2 > \omega^2$: The root is real.

(a) Weak damping

In this case, $(\gamma^2 < \omega^2)$, the general solution is

$$x(t) = e^{-\gamma t}\left(Ae^{i\sqrt{\omega^2-\gamma^2}\,t} + Be^{-i\sqrt{\omega^2-\gamma^2}\,t}\right). \tag{23.8}$$

It seems that this general solution is a complex one. But for an appropriate choice of A and B, this is not so. To get a real form we remind ourselves of the Euler formulas

$$e^{i\varphi} = \cos\varphi + i\sin\varphi, \qquad e^{-i\varphi} = \cos\varphi - i\sin\varphi. \tag{23.9}$$

By addition of these two equations we obtain

$$e^{i\varphi} + e^{-i\varphi} = 2\cos\varphi, \tag{23.10}$$

and by subtracting the second equation from the first one:

$$e^{i\varphi} - e^{-i\varphi} = 2i\sin\varphi. \tag{23.11}$$

Using these results we now rewrite the solutions of the differential equation as follows: First we set $\Omega^2 = \omega^2 - \gamma^2$; then we obtain from our two special solutions

$$x_1(t) = e^{-\gamma t}\cdot e^{i\Omega t}, \qquad x_2(t) = e^{-\gamma t}\cdot e^{-i\Omega t}, \tag{23.12}$$

two other solutions as a linear combination:

$$x_1'(t) = \frac{1}{2}e^{-\gamma t}(e^{i\Omega t} + e^{-i\Omega t}), \qquad x_2'(t) = -\frac{i}{2}e^{-\gamma t}(e^{i\Omega t} - e^{-i\Omega t}). \tag{23.13}$$

The solutions (23.12) are just as useful as the other solutions (23.13). By means of the formulas ((23.9)–(23.11)) obtained above, we may write these solutions also in the form

$$x_1'(t) = e^{-\gamma t}\cos\Omega t, \qquad x_2'(t) = e^{-\gamma t}\sin\Omega t.$$

From there it immediately follows the most general form of the vibration equation:

$$x(t) = e^{-\gamma t}\left(\overline{A}\cos\Omega t + \overline{B}\sin\Omega t\right),$$

where $\Omega^2 = \omega^2 - \gamma^2$. In this equation the coefficients $\overline{A}$ and $\overline{B}$ are real, contrary to the form we started from.

This equation—analogous to equation (21.11)—may also be written in the form

$$x(t) = De^{-\gamma t}\cos(\Omega t - \varphi),$$

where again $D^2 = \overline{A}^2 + \overline{B}^2$ and $\tan\varphi = \overline{B}/\overline{A}$ (see (21.10), (21.11)).

The graphical representation of the solution displays a damped harmonic vibration confined between two exponential curves:

Let x_n and x_{n+1} be two successive maximum elongations belonging to the times t_n and $t_n + T = t_n + (2\pi/\Omega)$, respectively. One obtains $x_n/x_{n+1} = e^{\gamma T} = e^{\gamma 2\pi/\Omega}$, and therefore

$$\ln\frac{x_n}{x_{n+1}} = \gamma T = \gamma\frac{2\pi}{\Omega}.$$

This is the *logarithmic decrement*, which may be used for experimental determination of the decay constant γ and the damping constant β by measuring x_n and x_{n+1}.

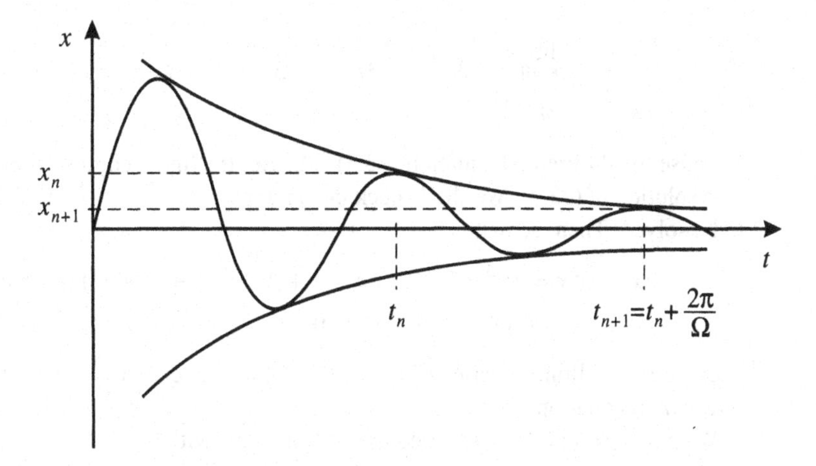

Graphical representation of the amplitudes of a weakly damped oscillator with the initial conditions $x(0) = 0$, $\dot{x}(0) > 0$.

(b) Critical damping

If in the case of damped vibration (see above) the friction continues to increase, already the second elongation may become relatively small. Finally, the mass no longer passes the rest position but so to speak comes to rest just at the moment when reaching the rest position. This particular case occurs for $\gamma^2 = \omega^2$.

However, we have to state that in this case the two solutions obtained above coincide. Hence, only one solution is at disposal, namely

$$x_1(t) = e^{-\gamma t}\,.$$

To get a *second solution*, we don't consider our limiting case but a somewhat stronger damped vibration:

$$\gamma^2 = \omega^2 + \varepsilon^2.$$

Then, according to (23.7) there exist two solutions that may be expanded into a Taylor series:

$$e^{\lambda_1 t} = e^{-\gamma t} \cdot e^{\varepsilon t} = e^{-\gamma t}\left(1 + \varepsilon\, t + \frac{\varepsilon^2}{2!}t^2 + \frac{\varepsilon^3}{3!}t^3 + \cdots\right);$$

$$e^{\lambda_2 t} = e^{-\gamma t} \cdot e^{-\varepsilon t} = e^{-\gamma t}\left(1 - \varepsilon\, t + \frac{\varepsilon^2}{2!}t^2 - \frac{\varepsilon^3}{3!}t^3 + \cdots\right).$$

We subtract the second solution from the first one and divide by ε. Then we let ε approach 0:

$$\lim_{\varepsilon\to 0} \frac{x_1 - x_2}{\varepsilon} = \lim_{\varepsilon\to 0} \frac{e^{-\gamma t}}{\varepsilon}\left(2\varepsilon\, t + 2\frac{\varepsilon^3}{3!}t^3 + 2\frac{\varepsilon^5}{5!}t^5 + \cdots\right)$$

$$= \lim_{\varepsilon \to 0} e^{-\gamma t}\left(2t + 2\frac{\varepsilon^2}{3!}t^3 + 2\frac{\varepsilon^4}{5!}t^5 + \cdots\right)$$
$$= 2te^{-\gamma t}. \qquad \textbf{(23.14)}$$

Because the differential equation (23.3) is linear, the linear combination (23.14) also must be a solution of (23.3). We shall check that and insert $x = te^{-\gamma t}$ in the differential equation to be solved. Then, actually,

$$\begin{aligned}\ddot{x} + 2\gamma\dot{x} + \omega^2 x &= (\gamma^2 te^{-\gamma t} - 2\gamma e^{-\gamma t}) + 2\gamma(e^{-\gamma t} - \gamma te^{-\gamma t}) + \omega^2 te^{-\gamma t}\\ &= (\omega^2 - \gamma^2)te^{-\gamma t} = 0,\end{aligned}$$

because in our limiting case $\gamma^2 = \omega^2$, that is, in this case $x = te^{-\gamma t}$ is a solution of the differential equation.

We now again have two particular solutions, and with

$$x_1(t) = e^{-\gamma t},$$
$$x_2(t) = te^{-\gamma t},$$

we may immediately write down the general solution:

$$x(t) = (A + Bt)e^{-\gamma t}. \qquad \textbf{(23.15)}$$

(c) Overdamped system

If the damping becomes even stronger than in the case just discussed, that is, if $\gamma^2 > \omega^2$, the mass returns much slower to the rest position.

The general solution is then

$$x(t) = e^{-\gamma t}(Ae^{\sqrt{\gamma^2-\omega^2}\,t} + Be^{-\sqrt{\gamma^2-\omega^2}\,t}).$$

In this case the mass after the first elongation creeps gradually back to the rest position, namely, the oscillator performs a creeping motion.

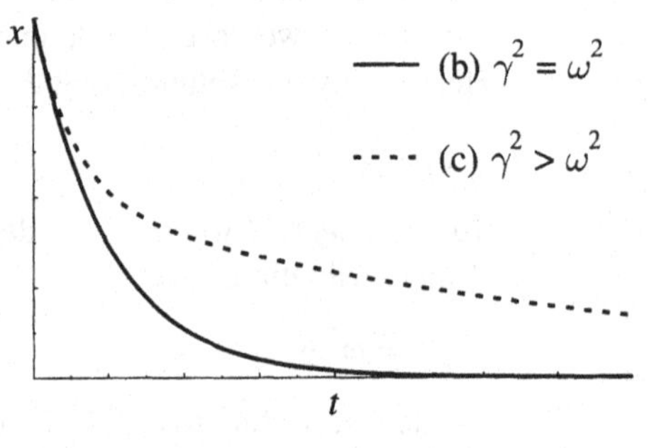

Illustration of the motion in the case of critical damping (b) and creeping motion (c).

We now consider the graphical representation of the last two cases, namely

(b) critical damping,

(c) creeping motion.

For critical damping the oscillator obviously returns most quickly to the rest position. Therefore, this case is very important for the damping of measuring instruments (e.g., mirror galvanometer): In the limit of critical damping the measured value is displayed most quickly, because the measuring instrument (the damped oscillator) performs a vibration but due to the damping "gets stuck" after the first quarter of the period.

Finally we still investigate the energy content of the vibrating system with damping. To this end we start directly from the differential equation:

$$\ddot{x} + \omega^2 x = -2\gamma \dot{x}.$$

We multiply the entire equation by $\dot{x}$:

$$\ddot{x}\dot{x} + \omega^2 \dot{x} x = -2\gamma \dot{x}^2.$$

The left side represents a complete differential, namely

$$\frac{d}{dt}\left(\frac{1}{2}\dot{x}^2 + \frac{\omega^2}{2}x^2\right) = -2\gamma \dot{x}^2.$$

If the equation is still multiplied by m, the left side just represents the time derivative of the total energy of the vibrating system:

$$\frac{d}{dt}\left(\frac{m}{2}\dot{x}^2 + \frac{k}{2}x^2\right) = \frac{d}{dt}(T+V) = \frac{d}{dt}E = -\beta \dot{x}^2 \leq 0. \tag{23.16}$$

Hence the time derivative of the total energy of the spring is negative, that is, the total energy of the system permanently decreases due to damping, as energy is permanently converted to heat by friction and is released to the environment.

Damped vibration with a periodic external force

Let a mass m be suspended via an elastic spring with the spring constant k and rigidly connected to a damping piston immersed into a liquid.

If the spring is displaced by a periodically acting external force $F = F_0 \cdot \cos\alpha t$, the system performs a variation of the position depending on the time which corresponds to the graph of a damped vibration. A downward motion of the mass is related with a spring force pointing upward and proportional to the displacement

$$F_f = -kx,$$

and moreover with a friction or damping force F_r that is proportional to $\mathbf{v}$:

$$F_r = -\beta \dot{x}.$$

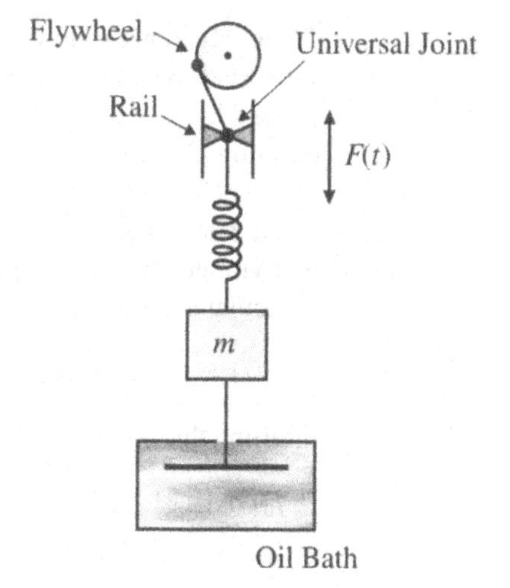

Visualization of a damped system with periodic external force.

Together with the periodic external force $F(t) = F_0 \cos\alpha t$, there results the following differential equation for this system:

$$m\frac{d^2x}{dt^2} = -kx - \beta\dot{x} + F_0 \cos\alpha t, \tag{23.17}$$

or rewritten

$$\ddot{x} + 2\gamma\dot{x} + \omega^2 x = f_0 \cos\alpha t \tag{23.18}$$

with the abbreviations:

$$2\gamma = \frac{\beta}{m}; \qquad \omega^2 = \frac{k}{m}; \qquad f_0 = \frac{F_0}{m}.$$

This differential equation is *inhomogeneous* (there occurs a term independent of x, namely $f_0 \cos\alpha t$, in the differential equation) and describes a damped *forced vibration.* The general solution of an inhomogeneous differential equation is composed of the general solution of the homogeneous differential equation $x_1(t), x_2(t)$ and a particular solution $x_0(t)$ of the inhomogeneous differential equation, such that the general solution has the form

$$x(t) = x_0(t) + Ax_1(t) + Bx_2(t)\,. \tag{23.19}$$

Thus the general solution again involves two free constants A and B that are needed to fulfill the initial conditions (initial position and initial velocity).

These three solving approaches obey the differential equations

$$\ddot{x}_0 + 2\gamma\dot{x}_0 + \omega^2 x_0 = f_0 \cos\alpha t, \tag{23.20}$$

$$\ddot{x}_{1,2} + 2\gamma\dot{x}_{1,2} + \omega^2 x_{1,2} = 0. \tag{23.21}$$

These equations follow directly from the meaning (definition) of the various solutions: $x_0(t)$ shall be a particular solution of the inhomogeneous differential equation, as is expressed by (23.20), while $x_1(t)$ and $x_2(t)$ shall be solutions of the homogeneous differential equation (23.21).

To get a particular solution $x_0(t)$, we make the following consideration:

After termination of the *initial transient process*, ("Einschwingvorgang") the mass m will vibrate with the frequency α of the acting force. We therefore try the *ansatz* for the particular solution

$$x_0(t) = C_1 \cos\alpha t + C_2 \sin\alpha t. \tag{23.22}$$

Inserting this *ansatz* in (23.20) yields

$$\begin{aligned} f_0 \cos\alpha t = {} & -\alpha^2(C_2 \sin\alpha t + C_1 \cos\alpha t) + 2\gamma(C_2\alpha \cos\alpha t - C_1\alpha \sin\alpha t) \\ & + \omega^2(C_2 \sin\alpha t + C_1 \cos\alpha t) \end{aligned}$$

By combining and rearranging, we obtain

$$\sin\alpha t\,(-\alpha^2 C_2 - 2\gamma\alpha C_1 + \omega^2 C_2) + \cos\alpha t\,(-C_1\alpha^2 + 2\gamma\alpha C_2 + C_1\omega^2) = f_0 \cos\alpha t.$$

As sine and cosine are linearly independent, a comparison of coefficients yields

$$\begin{aligned} C_1(2\gamma\alpha) + C_2(\alpha^2 - \omega^2) &= 0, \\ -C_1(\alpha^2 - \omega^2) + C_2(2\gamma\alpha) &= f_0. \end{aligned} \tag{23.23}$$

From there it follows for C_1 and C_2 that

$$C_1 = \frac{-(\alpha^2 - \omega^2) f_0}{4\gamma^2\alpha^2 + (\alpha^2 - \omega^2)^2},$$

$$C_2 = \frac{f_0 2\gamma\alpha}{4\gamma^2\alpha^2 + (\alpha^2 - \omega^2)^2}. \tag{23.24}$$

Inserting the values found for C_1 and C_2 in the *ansatz*, we then obtain the particular solution:

$$x_0(t) = f_0\left[\underbrace{-\frac{\alpha^2 - \omega^2}{(\alpha^2 - \omega^2)^2 + 4\gamma^2\alpha^2}}_{\overline{A}} \cos\alpha t + \underbrace{\frac{2\gamma\alpha}{(\alpha^2 - \omega^2)^2 + 4\gamma^2\alpha^2}}_{\overline{B}} \sin\alpha t\right], \tag{23.25}$$

or rewritten, we obtain with

$$\overline{A}\cos\alpha t + \overline{B}\sin\alpha t = \sqrt{\overline{A}^2 + \overline{B}^2}\cos(\alpha t - \varphi),$$

$$\tan\varphi = \frac{\overline{B}}{\overline{A}}:$$

$$x_0(t) = f_0\sqrt{\frac{4\gamma^2\alpha^2 + (\alpha^2 - \omega^2)^2}{((\alpha^2 - \omega^2)^2 + 4\gamma^2\alpha^2)^2}}\cos(\alpha t - \varphi), \tag{23.26}$$

$$x_0(t) = \frac{f_0}{\sqrt{(\alpha^2 - \omega^2)^2 + 4\gamma^2\alpha^2}}\cos(\alpha t - \varphi), \qquad \tan\varphi = \frac{-2\gamma\alpha}{\alpha^2 - \omega^2}.$$

Because the solutions of the homogeneous differential equation (23.21) for weak damping are $x_1(t) = e^{-\gamma t}\sin\Omega t$ and $x_2(t) = e^{-\gamma t}\cos\Omega t$, the complete solution of the differential equation is

$$x(t) = \frac{f_0}{\sqrt{(\alpha^2 - \omega^2)^2 + 4\gamma^2\alpha^2}}\cos(\alpha t - \varphi) + e^{-\gamma t}(A\sin\Omega t + B\cos\Omega t)$$

$$= \frac{f_0}{\sqrt{(\alpha^2 - \omega^2)^2 + 4\gamma^2\alpha^2}}\cos(\alpha t - \varphi) + De^{-\gamma t}\cos(\Omega t - \vartheta) \tag{23.27}$$

with $D^2 = A^2 + B^2$, $\Omega^2 = \omega^2 - \gamma^2$, and $\vartheta = \arctan(B/A)$.

Whatsoever the initial conditions are, for a nonvanishing damping ($\gamma > 0$) after sufficiently long time only the first term, the particular solution of the differential equation $x_0(t)$, survives. The second term in (23.27) that decays proportional to $e^{-\gamma t}$ depends on the constants A, B, which are fixed by the initial conditions. This second term obviously describes the *initial transient process*, which is "forgotten" after some time.

For the particular excitation frequency

$$\alpha = \sqrt{\omega^2 - 2\gamma^2}, \tag{23.28}$$

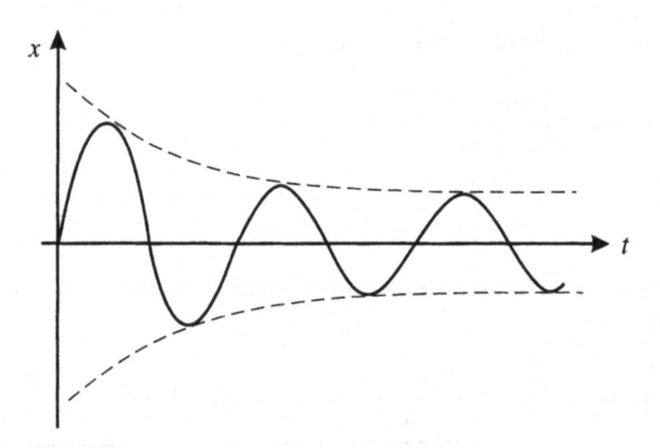

Graphical representation of the motion (4) of a weakly damped oscillator with periodic external force. The initial transient process ("Einschwingvorgang") depends on the initial conditions.

a maximum elongation is reached. The damping constant γ determines also the half-width of the resonance. It can, however, not become bigger than $\gamma = \omega/\sqrt{2}$, as can be seen from equation (23.28).

The amplitude of the forced vibration (23.27) is plotted in the folowing figure as a function of the forced frequency α for various damping values. Near the eigenfrequency ω of the oscillator at $\alpha = \sqrt{\omega^2 - 2\gamma^2}$, the system is resonating (a resonance occurs). In the case without damping ($\gamma = 0$), the amplitude at the resonance becomes infinitely large (the spring breaks—*resonance catastrophe*). In the case of very strong damping, the resonance is barely visible.

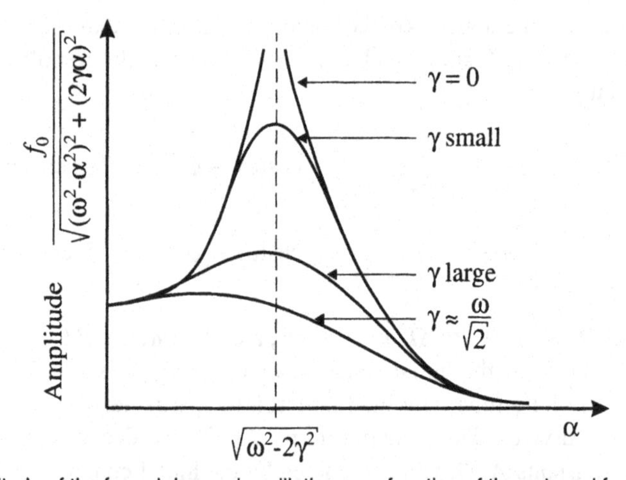

The amplitude of the forced damped oscillation as a function of the external frequency α.

The associated phase of vibration is plotted for various damping values in the second figure. At very low frequency α ($\alpha \ll \omega$) of the imposed force, the phase shift φ between

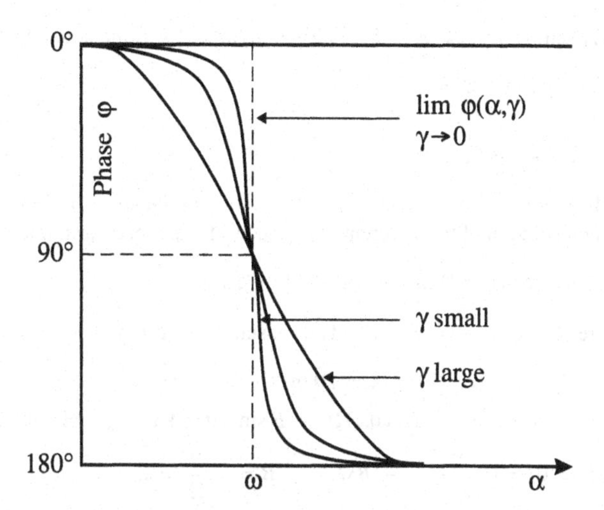

The phase shift of the damped oscillator against the extarnal force as a function of the external frequency α, thus a plot of $\varphi = -\arctan[2\gamma\alpha/(\omega^2 - \alpha^2)]$.

force and motion of the mass vanishes. At very high frequency ($\alpha \gg \omega$), the corresponding phase shift is 180°. Both results are plausible.

Problem 23.1: Damped vibration of a particle

A particle of mass 5 kg moves along the x-direction under the influence of two forces:

1. A force toward the origin with the value $40\,\frac{\mathrm{N}}{\mathrm{m}} \cdot x$, and
2. A velocity-proportional friction force of, e.g., 200 N for $v = 10\,\mathrm{m/s}$. Let $x(t = 0) = 20$ m, $\dot{x}(t = 0) = 0$.

Find

(a) the differential equation of the motion,

(b) $x(t)$ analytically and graphically,

(c) amplitude, period, and frequency of the vibration, and

(d) the ratio of two successive amplitudes (logarithmic decrement).

Solution (a) The equation of motion reads

$$m\ddot{x} = -kx - \beta\dot{x},$$

where $k = 40\,\mathrm{N/m}$. The friction coefficient β may be determined from the condition $F_{\mathrm{reib}} = -\beta v$. One finds

$$\beta = \frac{200\ \mathrm{N}}{10\ \mathrm{m/s}} = 20\,\frac{\mathrm{N\,s}}{\mathrm{m}}\,.$$

By setting $\omega^2 = k/m = 8\ \mathrm{s}^{-2}$, $2\gamma = \beta/m = 4\ \mathrm{s}^{-1}$, the equation of motion turns into

$$\ddot{x} + 2\gamma\dot{x} + \omega^2 x = 0$$

or

$$\ddot{x} + 4\dot{x} + 8x = 0.$$

(b) From $\omega^2 = 8\ \mathrm{s}^{-2}$ and $\gamma^2 = 4\ \mathrm{s}^{-2}$ it follows that $\omega^2 > \gamma^2$, that is, there is a weak damping. The general solution of the differential equation of a damped harmonic motion is given by [1]

$$x(t) = \exp(-\gamma t)\left[A\cos(\Omega t) + B\sin(\Omega t)\right],$$

where $\Omega = \sqrt{\omega^2 - \gamma^2} = 2\ \mathrm{s}^{-1}$. The constants A and B may be determined from the initial conditions:

$$x_0 = x(t = 0) = A = 20\ \mathrm{m},$$
$$\dot{x} = -\gamma e^{-\gamma t}\left(A\cos(\Omega t) + B\sin(\Omega t)\right) + e^{-\gamma t}\left(-A\Omega\sin(\Omega t) + B\Omega\cos(\Omega t)\right),$$

$$\dot{x}(t = 0) = 0 = -\gamma x_0 + B\Omega, \qquad B = x_0\frac{\gamma}{\Omega} = x_0 = 20\ \mathrm{m}.$$

Hence

$$x(t) = 20\left(\cos\Omega t + \sin\Omega t\right)e^{-\gamma t}\ \mathrm{m}. \tag{23.29}$$

Because

$$A\cos\Omega t + B\sin\Omega t = \sqrt{A^2 + B^2}\cos(\Omega t - \varphi),$$

with

$$\tan\varphi = \frac{B}{A},$$

for $x(t)$ it results that

$$x(t) = 20\sqrt{2}\cos\left(\Omega t - \frac{\pi}{4}\right)e^{-\gamma t}\ \mathrm{m}$$

or

$$x(t) = 20\sqrt{2}e^{-\gamma t}\cos\left(\Omega t - \frac{\pi}{4}\right)\ \mathrm{m}.$$

When setting $\dot{x}(t) = 0$, we obtain a necessary condition for extrema: $t = k\pi/2$ s, with k an integer number. The zeros follow from $\cos(\Omega t - \frac{\pi}{4}) = 0$. Thus, we get the following table:

t	0	$\frac{3\pi}{8} = 1.18$	$\frac{\pi}{2} = 1.57$	$\frac{7\pi}{8} = 2.75$	$\pi = 3.14$	$\frac{11\pi}{8} = 4.32$
$x(t)$	20	0	−0.86	0	0.04	0

Obviously this vibration is damping out rapidly. Actually, the parameters $\gamma = 2/\mathrm{s}$ and $\omega = \sqrt{8}/\mathrm{s}$ are close before the critical damping.

(c) I. The amplitudes are therefore

$$a(t) = 20\sqrt{2}e^{-\gamma t}\ \mathrm{m}.$$

[1] Here we use a frequently adopted notation for the exponential function: $\exp(x) \equiv e^x$.

II. The frequency is

$$\Omega = \sqrt{\omega^2 - \gamma^2} = 2\ \mathrm{s}^{-1}.$$

III. For the period, it results that

$$T = 2\pi \frac{1}{\Omega} = \pi\ \mathrm{s}.$$

(d) For two successive maximal elongations, we obtain

$$x_n = 20\sqrt{2} e^{-\gamma t}\ \mathrm{m},$$
$$x_{n+1} = 20\sqrt{2} e^{-\gamma (t + 2\pi/\Omega)}\ \mathrm{m},$$

from which it follows that

$$\frac{x_n}{x_{n+1}} = e^{\gamma T} \left(\text{where } T = \frac{2\pi}{\Omega}\right).$$

Therefore,

$$\ln\left(\frac{x_n}{x_{n+1}}\right) = \gamma T \tag{23.30}$$

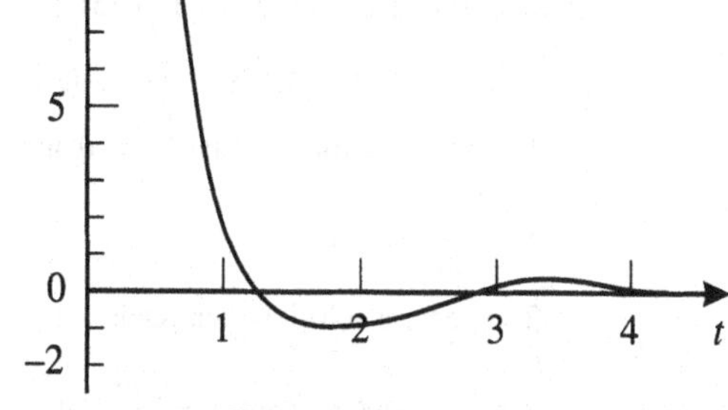

The solution function.

is the *logarithmic decrement*. The meaning of this quantity is due to the fact that according to equation 23.30 the damping constant γ may be determined directly by measuring the ratio of successive maximal elongations.

Problem 23.2: The externally excited harmonic oscillator

(a) An oscillator with the eigenfrequency ω be undamped and excited by a harmonic external force of the same frequency ω (e.g., by a balance wheel). The amplitude of the oscillator then increases as a function of time according to the equation

$$x = A\cos\omega t + B\sin\omega t + \frac{f_0 t}{2\omega}\sin\omega t.$$

Check that!

(b) Give a physical interpretation!

Solution (a) The force law reads

$$m\frac{d^2x}{dt^2} = -kx - \beta\frac{dx}{dt} + F_0\cos\alpha t$$

and there must be $\beta = 0$ because the oscillator shall be undamped. By rewriting it follows that

$$\ddot{x} + \omega^2 x = f_0\cos\alpha t, \qquad \text{where} \quad \alpha = \omega \quad \text{and} \quad \omega^2 = \frac{k}{m}. \tag{23.31}$$

To get the general solution of the equation, we add to the general homogeneous solution, that is, to the solution of

$$\ddot{x} + \omega^2 x = 0, \tag{23.32}$$

a particular solution of 23.31. The general solution of 23.32 now reads

$$x = A\cos\omega t + B\sin\omega t. \tag{23.33}$$

It is convenient to adopt the following *ansatz* for the particular solution:

$$x = t(C_1\cos\omega t + C_2\sin\omega t). \tag{23.34}$$

Here C_1 and C_2 are so far unknown coefficients. Differentiation yields

$$\dot{x} = t(-\omega C_1\sin\omega t + \omega C_2\cos\omega t) + (C_1\cos\omega t + C_2\sin\omega t) \tag{23.35}$$

and

$$\ddot{x} = t\left(-\omega^2 C_1\cos\omega t - \omega^2 C_2\sin\omega t\right) + 2(-\omega C_1\sin\omega t + \omega C_2\cos\omega t). \tag{23.36}$$

We insert equations (23.34), (23.35), and (23.36) in (23.31) and obtain after simplifying

$$-2\omega C_1\sin\omega t + 2\omega C_2\cos\omega t = f_0\cos\omega t.$$

From there it follows that $C_1 = 0$ and $C_2 = f_0/2\omega$. Thus the particular solution 23.34 reads

$$x = \frac{f_0}{2\omega}t\sin\omega t. \tag{23.37}$$

The general solution then reads

$$x = A\cos\omega t + B\sin\omega t + \frac{f_0}{2\omega}t\sin\omega t. \tag{23.38}$$

(b) The constants A and B are determined from the initial conditions. Because there is no damping, the terms proportional to A and B do not become small at large times. But for large times ($t \to \infty$), the term proportional to t increases beyond any limits such that the spring finally will break. A drawing of the latter term shows the increase of the vibration amplitudes with time: This is the typical case of "amplification" of a vibration as is well known from everyday life, for instance, on swinging, periodic pulling of a cut-in tree to cause its breaking, etc.

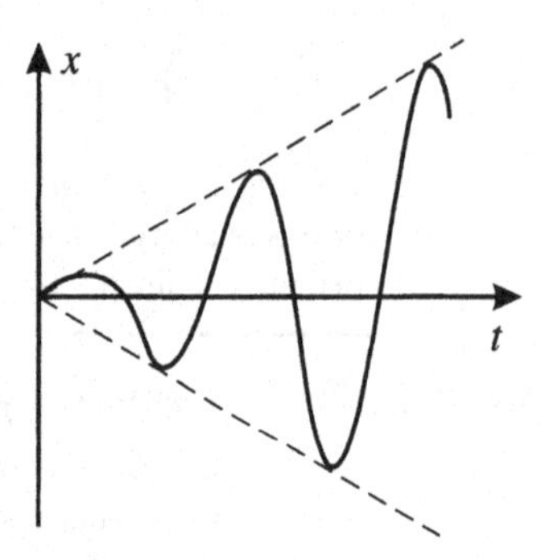

Amplification of an externally driven oscillator.

Problem 23.3: Mass point in the x, y-plane

A mass moves in the x, y-plane. In the x-direction the harmonic force $F_x = -m\omega^2 x$ and the additional force $K_x = \alpha m\omega^2 y$ ($\alpha > 0$) act, in the y-direction only the harmonic force $F_y = -m\omega^2 y$ acts.

(a) Solve the equations of motion with the initial conditions

$$x(0) = y(0) = 0, \qquad \dot{x}(0) = 0, \qquad \dot{y}(0) = A\omega.$$

(b) Draw a qualitative figure of the path of the mass point.

Solution (a) The equations of motion read

$$m\ddot{x} = -\omega^2 mx + \alpha m\omega^2 y,$$

$$\ddot{x} = -\omega^2(x - \alpha y), \tag{23.39}$$

and

$$\ddot{y} = -\omega^2 y. \tag{23.40}$$

Equation 23.40 is solved by the general *ansatz*

$$y(t) = a \sin \omega t + b \cos \omega t .$$

The initial conditions yield

$$y(0) = b = 0, \qquad \dot{y}(0) = a\omega = A\omega.$$

Hence the solution for $y(t)$ reads

$$y(t) = A \sin \omega t. \tag{23.41}$$

For 23.39 we then get with the help of 23.41

$$\ddot{x} = -\omega^2 (x - \alpha A \sin \omega t). \tag{23.42}$$

We guess a particular solution of the inhomogeneous equation

$$\begin{aligned} x_s(t) &= ct \cos \omega t, \\ \ddot{x}_s(t) &= -2c\omega \sin \omega t - c\omega^2 t \cos \omega t, \\ &\stackrel{!}{=} -\omega^2 ct \cos \omega t + \alpha A \omega^2 \sin \omega t \\ \Rightarrow \quad -2c &= \alpha A \omega, \qquad c = -\frac{\alpha A \omega}{2}. \end{aligned}$$

The general solution of 23.42 is then

$$x(t) = d \cos \omega t + e \sin \omega t - \frac{\alpha A \omega}{2} t \cos \omega t. \tag{23.43}$$

The initial conditions yield

$$x(0) = d = 0, \qquad \dot{x}(0) = e\omega - \frac{\alpha A \omega}{2} = 0$$

$$\Rightarrow \quad e = \frac{\alpha A}{2}.$$

Hence, the solution of the equations of motion reads

$$\begin{aligned} x(t) &= \frac{\alpha A}{2} [\sin \omega t - \omega t \cos \omega t], \\ y(t) &= A \sin \omega t. \end{aligned}$$

(b) In the y-direction one obviously observes a harmonic vibration with amplitude A:

$$y\left(t_n^A\right) = \pm A \qquad \text{for} \quad t_n^A = \frac{(2n+1)\pi}{2\omega}, \qquad n = 0, 1, 2, \dots .$$

The associated x-coordinate reads

$$x\left(t_n^A\right) = \pm \frac{\alpha A}{2}.$$

The zero passages of y are obtained from

$$y\left(t_n^0\right) = 0 \qquad \text{for} \quad t_n^0 = \frac{n\pi}{\omega}, \qquad n = 0, 1, 2, \dots \tag{23.44}$$

$$\Rightarrow \quad x\left(t_n^0\right) = -\frac{\alpha A}{2} n\pi (-1)^n. \tag{23.45}$$

The motion of the particle is shown in the following figure.

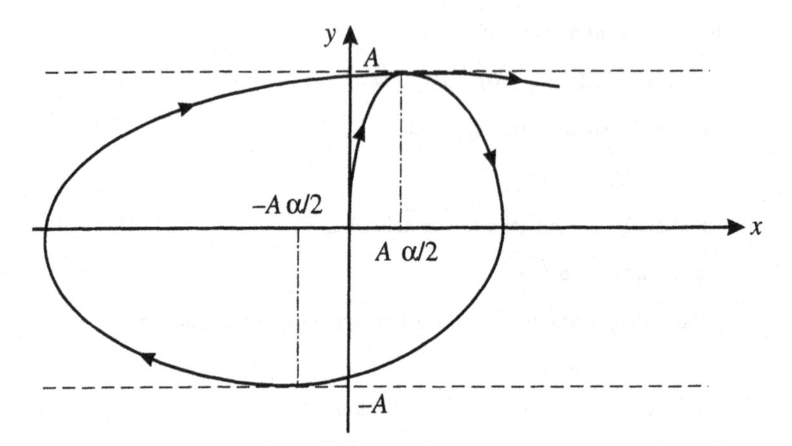

The motion of the particle. The path of the particle is stretching more and more in a cigar shape along the x-direction, while its width approaches the maximum value $2A$.

24 The Pendulum

A mass m vibrating in a plane, suspended on a string of length l (let the mass of the string be negligibly small), is called a *mathematical pendulum*. The vibration period of the pendulum shall be calculated.

(a) Without damping

The backdriving force F_R after displacing the mass by the angle φ is the component of the earth attraction along the direction of motion of the pendulum

$$F_R = -mg \sin\varphi.$$

Hence, the differential equation for the pendulum without damping is

$$m\ddot{s} = -mg \sin\varphi,$$

$$\ddot{s} + g \sin\varphi = 0,$$

$$s = l\varphi,$$

$$\ddot{s} = l\ddot{\varphi},$$

$$l\ddot{\varphi} + g \sin\varphi = 0,$$

$$\ddot{\varphi} + \frac{g}{l} \sin\varphi = 0,$$

$$\ddot{\varphi} + \omega^2 \sin\varphi = 0. \qquad \textbf{(24.1)}$$

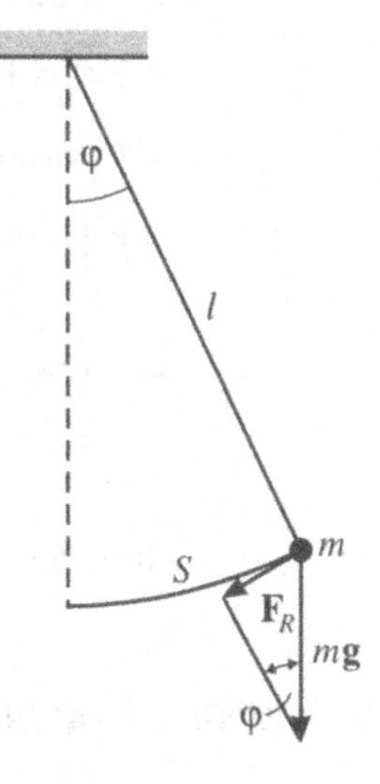

The pendulum: S gives the lenght of the arc, $\mathbf{F}_R$ the acting force.

This differential equation is nonlinear. Howver, for small angles φ the sine of the angle can be replaced by the angle itself, that is, for $\varphi \ll 1$ we can use $\sin\varphi = \varphi$. The differential equation for the pendulum vibration *for small* displacements thus reads

$$\ddot{\varphi} + \omega^2\varphi = 0.$$

This is a linear differential equation. Its general solution is

$$\varphi = A\cos\omega t + B\sin\omega t, \qquad \omega = \sqrt{\frac{g}{l}},$$

from which we get the vibration period

$$T = \frac{2\pi}{\omega} = 2\pi\sqrt{\frac{l}{g}}. \tag{24.2}$$

(b) Vibration of the pendulum with friction but for small elongations

The differential equation reads

$$m\ddot{s} = -mg\sin\varphi - \beta\dot{s}.$$

The last term $-\beta\dot{s}$ represents the friction force. After division by $m \cdot l$ and by $2\gamma = \beta/(ml)$, we have

$$\ddot{\varphi} + \omega^2\sin\varphi + 2\gamma\dot{\varphi} = 0. \tag{24.3}$$

For small vibration amplitudes this turns into

$$\ddot{\varphi} + \omega^2\varphi + 2\gamma\dot{\varphi} = 0.$$

The general solution is now (compare to Chapter 23)

$$\varphi = \left(A\cos\sqrt{\omega^2-\gamma^2}\,t + B\sin\sqrt{\omega^2-\gamma^2}\,t\right)e^{-\gamma t} \qquad \text{(weakly damped vibration)},$$

or

$$\varphi = \left(Ae^{-\sqrt{\gamma^2-\omega^2}t} + Be^{\sqrt{\gamma^2-\omega^2}\,t}\right)e^{-\gamma t} \qquad \text{(strong damping)},$$

or

$$\varphi = (At+B)e^{-\gamma t} \qquad \text{(critical damping)}.$$

In all of these cases the pendulum comes to rest at some time ($t \to \infty$).

(c) Solution of the pendulum equation without friction, but for large elongations

We begin with the nonlinear differential equation (24.1)

$$\frac{d^2\varphi}{dt^2} + \omega^2\sin\varphi = 0$$

and substitute the *angular velocity* $u = d\varphi/dt$:

$$\frac{du}{d\varphi}\cdot\frac{d\varphi}{dt} + \omega^2\sin\varphi = 0, \qquad \text{hence} \quad u\frac{du}{d\varphi} + \omega^2\sin\varphi = 0.$$

Separation of the variables and integration yield

$$\int u\,du = -\int \omega^2 \sin\varphi\,d\varphi \qquad \text{or} \qquad \frac{u^2}{2} = \omega^2\cos\varphi + C.$$

With the boundary condition that for $\varphi = \varphi_0$, $u = 0$, we obtain

$$0 = \omega^2\cos\varphi_0 + C, \qquad C = -\omega^2\cos\varphi_0,$$

or

$$\frac{u^2}{2} = \omega^2(\cos\varphi - \cos\varphi_0),$$

$$\frac{d\varphi}{dt} = u = \sqrt{2}\omega\sqrt{\cos\varphi - \cos\varphi_0}.$$

Another separation of variables and integration yield

$$\int\limits_{\varphi_1}^{\varphi} \frac{d\varphi}{\sqrt{\cos\varphi - \cos\varphi_0}} = \int \sqrt{2}\omega\,dt = \sqrt{2}\omega t\,.$$

φ_1 is an arbitrary initial angle. It is determined such that for $t = 0$ we get $\varphi = 0$. This means

$$t = +\sqrt{\frac{l}{2g}}\int\limits_0^{\varphi} \frac{d\varphi}{\sqrt{\cos\varphi - \cos\varphi_0}}, \tag{24.4}$$

and in particular

$$\frac{T}{4} = +\sqrt{\frac{l}{2g}}\int\limits_0^{\varphi_0} \frac{d\varphi}{\sqrt{\cos\varphi - \cos\varphi_0}}$$

or

$$T = 4\sqrt{\frac{l}{2g}}\int\limits_0^{\varphi_0} \frac{d\varphi}{\sqrt{\cos\varphi - \cos\varphi_0}}. \tag{24.5}$$

To evaluate the integrals (24.4, 24.5), we substitute $\cos\varphi = \cos(\varphi/2+\varphi/2) = \cos^2\varphi/2 - \sin^2\varphi/2 = 1 - 2\sin^2\varphi/2$, which yields

$$T = \frac{4}{2}\sqrt{\frac{l}{g}}\int\limits_0^{\varphi_0} \frac{d\varphi}{\sqrt{-\sin^2\varphi/2 + \sin^2\varphi_0/2}}.$$

The further substitution of

$$\sin\frac{\varphi}{2} = \sin\frac{\varphi_0}{2}\sin\phi$$

means a stretching of the variable φ that varies between $0 \le \varphi \le \varphi_0$ over the range $0 \le \phi \le \pi/2$. Then

$$\cos\phi = \sqrt{1 - \frac{1}{\sin^2\varphi_0/2}\sin^2\varphi/2}\,. \tag{24.6}$$

Furthermore, we have

$$\frac{1}{2}\cos\frac{\varphi}{2}d\varphi = \sin\frac{\varphi_0}{2}\cos\phi\,d\phi,$$

and therefore

$$d\varphi = \frac{2\sin\frac{\varphi_0}{2}\cos\phi\,d\phi}{\sqrt{1-\sin^2\frac{\varphi_0}{2}\sin^2\phi}}.$$

With the abbreviation $k^2 = \sin^2\varphi_0/2$, we get

$$T = 2\sqrt{\frac{l}{g}}\cdot\int_0^{\pi/2}\frac{2\sin\frac{\varphi_0}{2}\cos\phi\,d\phi}{\sqrt{1-k^2\sin^2\phi}\cdot\sqrt{\sin^2\frac{\varphi_0}{2}\left(1-\frac{1}{\sin^2\frac{\varphi_0}{2}}\sin^2\frac{\varphi}{2}\right)}},$$

or according to (24.6)

$$T = 4\sqrt{\frac{l}{g}}\int_0^{\pi/2}\frac{\cos\phi\,d\phi}{(\sqrt{1-k^2\sin^2\phi})\cos\phi} = 4\sqrt{\frac{l}{g}}\int_0^{\pi/2}\frac{d\phi}{\sqrt{1-k^2\sin^2\phi}}.$$

For

$$\varphi_0 \ll \frac{\pi}{2} \quad\Rightarrow\quad T = 4\sqrt{\frac{l}{g}}\int_0^{\pi/2}d\phi = 2\pi\sqrt{\frac{l}{g}},$$

that is, for small pendulum elongations the result known from equation (24.2) is reproduced. For larger elongations ϕ, the equation for the vibration period T with $x(\phi) = -k^2\sin^2\phi$ reads

$$T = 4\sqrt{\frac{l}{g}}\int_0^{\pi/2}\frac{d\phi}{\sqrt{1+x(\phi)}}. \tag{24.7}$$

This is an *elliptic integral.* Such types of integrals arise, for example, on calculating the arc length of an ellipse, which explains the name. It may be evaluated approximately by expansion. Using the *general binomial theorem*

$$(1+x)^p = 1 + \binom{p}{1}x + \binom{p}{2}x^2 + \binom{p}{3}x^3 + \cdots,$$

thus

$$(1+x)^p = 1 + px + \frac{p(p-1)x^2}{1\cdot 2} + \frac{p(p-1)(p-2)x^3}{1\cdot 2\cdot 3} + \cdots,$$

which may also be proved by means of a Taylor expansion (Chapter 22), follows for $1/\sqrt{1+x}$, which may also be written as $(1+x)^{-1/2}$:

$$(1+x)^{-1/2} = 1 + \left(-\frac{1}{2}x\right) + \frac{-1/2(-3/2)}{2}x^2 + \cdots,$$

$$(1+x)^{-1/2} = 1 - \frac{1}{2}x + \frac{3}{8}x^2 - \cdots,$$

$$(1-k^2\sin^2\phi)^{-1/2} = 1 + \frac{1}{2}k^2\sin^2\phi + \frac{3}{8}k^4\sin^4\phi + \cdots,$$

$$T = 4\sqrt{\frac{l}{g}}\int_0^{\pi/2}\left(1 + \frac{1}{2}k^2\sin^2\phi + \frac{3}{8}k^4\sin^4\phi + \cdots\right)d\phi.$$

By using the recursion formula

$$\int \sin^m x\,dx = -\frac{1}{m}\sin^{m-1}x\cdot\cos x + \frac{m-1}{m}\int\sin^{m-2}x\,dx \qquad \text{for } m \neq 0,$$

which is obtained by partial integration, we obtain

$$\int_0^{\pi/2}\sin^{2n}\varphi\,d\varphi = \frac{1\cdot 3\cdot 5\cdot\ldots(2n-1)}{2\cdot 4\cdot 6\ldots(2n)}\cdot\frac{\pi}{2}.$$

Then we get for the vibration period

$$T = 4\sqrt{\frac{l}{g}}\left[\frac{\pi}{2} + \frac{1}{2}k^2\frac{\pi}{4} + \frac{3}{8}k^4\frac{3}{8}\frac{\pi}{2} + \cdots\right]$$

or

$$T = 2\pi\sqrt{\frac{l}{g}}\left[1 + \frac{1}{4}k^2 + \frac{9}{64}k^4 + \cdots\right].$$

With $k^2 = \sin^2\varphi_0/2$, this expression finally turns into

$$T \approx 2\pi\sqrt{\frac{l}{g}}\left[1 + \frac{1}{4}\sin^2\frac{\varphi_0}{2} + \ldots\right]$$

$$= T_0\left(1 + \frac{1}{4}\sin^2\frac{\varphi_0}{2} + \ldots\right), \qquad \text{where } T_0 = 2\pi\sqrt{\frac{l}{g}}.$$

If $\varphi_0 \ll 1$, we obviously obtain the old formula. If φ_0 becomes larger, the vibration period increases over T_0. This is plausible as the backdriving forces are $\sim \sin\varphi$. Harmonic approximation means $\sin\varphi \approx \varphi$. For larger φ the backdriving forces become smaller than $\sim \varphi$, and therefore $T > T_0$.

Problem 24.1: The cycloid

A circle of radius a rolls on a straight line. A given point on this circle then performs a cycloid. Find the parameter representation of this cycloid.

Solution One has (see figure)

$$\overline{OA} = a \cdot t, \quad \overline{OA} = x + a\sin t,$$
$$a = y + a\cos t,$$

and therefore,

$$x = at - a\sin t, \qquad y = a - a\cos t,$$
$$x = a(t - \sin t), \qquad y = a(1 - \cos t).$$

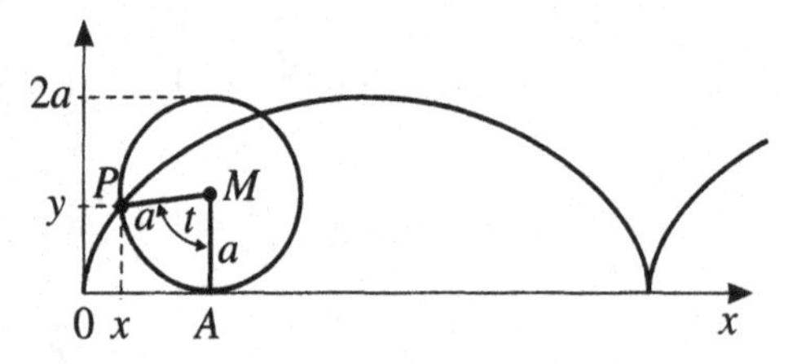

Rolling a circle yields a cycloid.

This is the wanted parameter representation of the cycloid. Elimination of t yields the trajectory in x-y-representation,

$$x(y) = -\sqrt{2ay - y^2} + a\arccos\left(\frac{a-y}{a}\right).$$

Problem 24.2: The cycloid pendulum

In a vibration of a pendulum of mass m the string shall osculate forth and back to the two branches OA and OC of a cycloid (cycloid pendulum). The length of the string shall be half of the length of the cycloid bow.

Show that the curve ABC again is a cycloid.

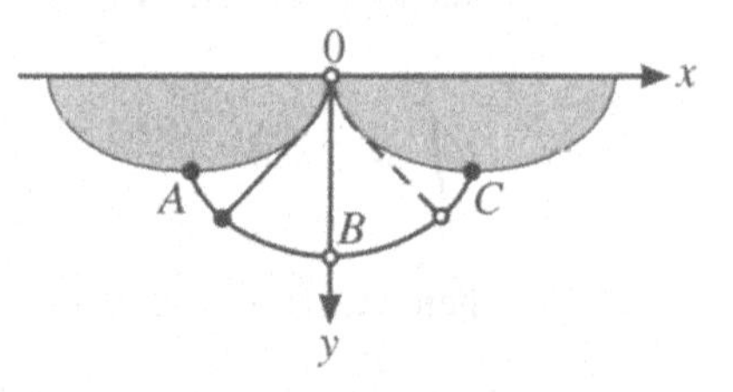

The string of the pendulum is warped along the two branches of the cycloid (gray). The mass m moves again on a cycloid.

Solution The equation of the cycloid branches reads

$$x = a(\phi - \sin\phi), \qquad y = a(1 - \cos\phi).$$

The equation of the curve generated by the pendulum is

$$x = x_1 + \Delta x, \qquad x_1 = a(\phi_1 - \sin\phi_1), \tag{24.8}$$
$$y = y_1 + \Delta y, \qquad y_1 = a(1 - \cos\phi_1).$$

(Equation of the cycloid. ϕ_1 is the curve parameter of the cycloid point where the string lifts off from the osculation curve.)

Moreover,

$$(\Delta y)^2 + (\Delta x)^2 = l_1^2 \qquad \text{and} \qquad l_1 = l - s_1. \tag{24.9}$$

The calculation of s_1 runs as follows:

$$\left(\frac{ds}{d\phi}\right)^2 = \left(\frac{dx}{d\phi}\right)^2 + \left(\frac{dy}{d\phi}\right)^2$$

$$= a^2\left[(1-\cos\phi)^2 + \sin^2\phi\right]$$
$$= a^2\left[1 - 2\cos\phi + \cos^2\phi + \sin^2\phi\right]$$
$$= 2a^2(1-\cos\phi),$$

$$\int_0^{s_1} ds = \int_0^{\phi_1} a\sqrt{2}\sqrt{(1-\cos\phi)}\,d\phi.$$

Moreover, we set $1-\cos\phi = 2\sin^2\frac{\phi}{2}$, $\frac{\phi}{2} = z$, hence $\frac{dz}{d\phi} = \frac{1}{2}$, and $d\phi = 2\,dz$. We then obtain

$$s_1 = -4a\cos\frac{\phi}{2}\bigg|_0^{\phi_1} = 4a\left(1-\cos\frac{\phi_1}{2}\right).$$

Hence, the total length of the cycloid bow is $8a$, and therefore the string length is $l = 4a$ and $l_1 = l - s_1$ (equation 24.9), that is,

$$l_1 = 4a\cos\frac{\phi_1}{2}. \qquad \textbf{(24.10)}$$

To get the equation of the trajectory of the vibrating mass, we now need the quantity Δx according to equation 24.9. It holds that (see figure)

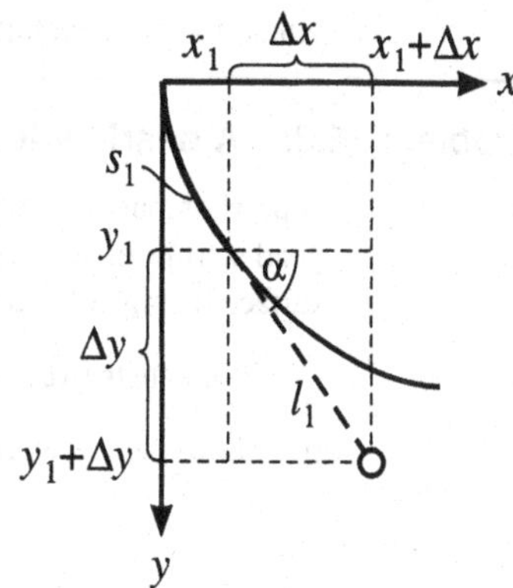

The determination of Δx.

$$\tan\alpha = \frac{dy_1}{dx_1}$$
$$== \frac{\sin\phi_1}{1-\cos\phi_1}$$
$$== \frac{\sin\phi_1}{2\sin^2\phi_1/2},$$

and therefore

$$\Delta x = l_1\cos\alpha = 4a\cos\frac{\phi_1}{2}\cdot\frac{1}{\sqrt{1+\tan^2\alpha}} = 4a\cos\frac{\phi_1}{2}\sin\frac{\phi_1}{2}.$$

The quantity Δy is calculated in a similar way, namely

$$\Delta y = l_1\sin\alpha = 4a\cos\frac{\phi_1}{2}\frac{\tan\alpha}{\sqrt{1+\tan^2\alpha}} = 4a\cos\frac{\phi_1}{2}\cos\frac{\phi_1}{2} = 4a\cos^2\frac{\phi_1}{2}.$$

From there the x- and y-coordinates of the path result according to equation 24.8 as

$$x = x_1 + \Delta x = a\left[(\phi_1 - \sin\phi_1) + 4\sin\frac{\phi_1}{2}\cos\frac{\phi_1}{2}\right]$$

and because

$$\frac{1}{2}\sin\phi_1 = \cos\frac{\phi_1}{2}\sin\frac{\phi_1}{2},$$
$$x = a[\phi_1 + \sin\phi_1],$$
$$x = a[\phi_1 - \sin(\phi_1 + \pi)],$$

$$y = y_1 + \Delta y = a\left[(1-\cos\phi_1) + 4\cos^2\frac{\phi_1}{2}\right]$$
$$= a\left[(1-\cos\phi_1) + 2(\cos\phi_1 + 1)\right],$$

$$y = a[3+\cos\phi_1] = a[1-\cos(\phi_1+\pi)+2],$$
$$y = a[1-\cos(\phi_1+\pi)] + 2a.$$

The trajectory of the vibrating mass again is a cycloid, namely

$$x = a[(\phi_1+\pi) - \sin(\phi_1+\pi)] - \pi a,$$
$$y = a[1-\cos(\phi_1+\pi)] + 2a.$$

It has the same form as the branches of the generating cycloid. The pendulum, however, is shifted with respect to the generating cycloid branches, namely by $2a$ in the y-direction, and by $-a\pi$ in the x-direction. Thus, one may ensure by this simple construction that a mass suspended by a string vibrates along a cycloid. Such a pendulum is called a cycloid pendulum.

Problem 24.3: A pearl slides on a cycloid

A pearl of mass m is forced to slide down on a frictionless wire with the contour of a cycloid. Let the pearl start from the rest position $x = y = 0$. The wire hangs in the gravitational field near the earth's surface (compare figure).

(a) Calculate the velocity of the pearl at the point $y = 2a$.

(b) Show that the vibration period of this motion equals that of a pendulum of length $4a$.

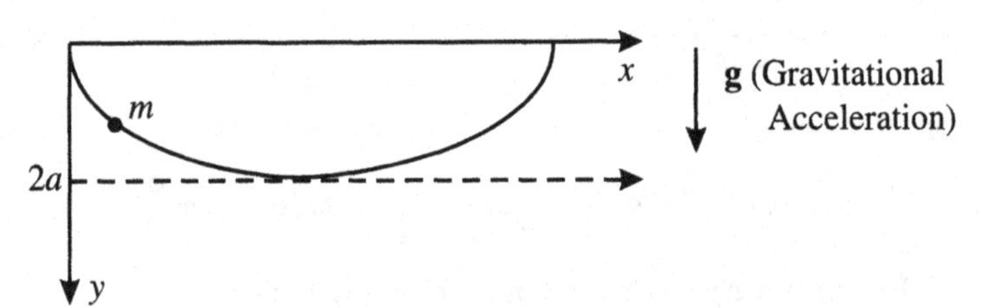

Motion of a pearl along a cycloid.

Solution (a) According to the energy law, the balance for an arbitrary point P on the wire reads as follows:

$$E_{\text{pot}}(P) + E_{\text{kin}}(P) = E_{\text{pot}}(0,0) + E_{\text{kin}}(0,0),$$

that is,

$$mg(2a-y) + \frac{1}{2}m\left(\frac{ds}{dt}\right)^2 = mg(2a) + 0$$

or

$$2mga - mgy + \frac{1}{2}mv^2 = 2mga$$

or

$$v^2 = 2gy$$

and finally,

$$v = \sqrt{2gy}.$$

We ask for the velocity v at the position $y = 2a$:

$$v(2a) = \sqrt{2g \cdot 2a} = \sqrt{4ga} = 2\sqrt{ga}.$$

This result is so far independent of the special curve of the wire.
(b) From the first part of the problem $\Rightarrow \quad (ds/dt)^2 = 2gy$.
The square of velocity along the cycloid reads

$$\left(\frac{ds}{dt}\right)^2 = \left(\frac{dx}{dt}\right)^2 + \left(\frac{dy}{dt}\right)^2$$
$$= a^2(1-\cos\beta)^2\dot{\beta}^2 + a^2\sin^2\beta \cdot \dot{\beta}^2 = 2a^2(1-\cos\beta)\dot{\beta}^2,$$

because the cycloid is given by $x = a(\beta - \sin\beta)$, $y = a(1-\cos\beta)$. Therefore,

$$2a^2(1-\cos\beta)\dot{\beta}^2 = 2ga(1-\cos\beta),$$

namely

$$\dot{\beta}^2 = \frac{g}{a} \qquad \Rightarrow \qquad \dot{\beta} = \frac{d\beta}{dt} = \sqrt{\frac{g}{a}} \qquad \Rightarrow \qquad \beta = t\sqrt{\frac{g}{a}} + C_1.$$

The last step is performed by integration after separating the variables. The initial conditions are $\beta = 0$ for $t = 0$, $\beta = 2\pi$ for $t = T/2$, T period of vibration. Therefore $T = 4\pi\sqrt{a/g} = 2\pi\sqrt{4a/g}$.

By comparison with the formula for the simple pendulum, we find

$$T_{\text{pendulum}} = 2\pi\sqrt{\frac{l}{g}} \quad \text{and} \quad T_{\text{cycloid}} = 2\pi\sqrt{\frac{4a}{g}} \qquad \Leftrightarrow \qquad l = 4a\,.$$

Problem 24.4: The search for the tautochrone

The problem of the tautochrone[1] is the search for that curve for which the vibration period is independent of the elongation: Which trajectory must be passed by a mass point m to ensure that the vibration period T of a frictionless vibrational motion becomes independent of the value of the initial elongation h?

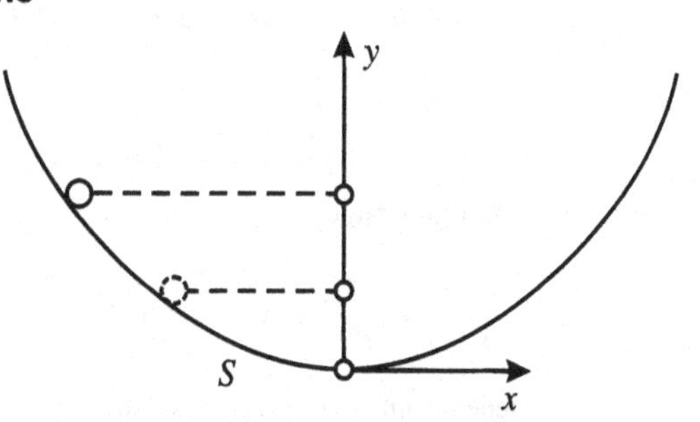

The quantities used to find the tautochrone.

Solution Let s be the arc length on the wanted tautochrone (see figure). With $v = ds/dt$, the energy law yields for the initial position and an arbitrary intermediate position

$$\frac{m}{2}\left(\frac{ds}{dt}\right)^2 + mgy = mgh,$$

[1] Greek: *tautos chronos* = equal time.

or

$$\frac{m}{2}\left(\frac{ds}{dt}\right)^2 = mg(h-y),$$

from which, using

$$ds = \frac{ds}{dy}dy = s'(y)\,dy,$$

after separation of the variables it follows that

$$dt = \frac{1}{\sqrt{2g}}\frac{s'(y)\,dy}{\sqrt{h-y}},$$

or after integration along a quarter of vibration

$$\frac{1}{4}T = \frac{1}{4}T(h) = \frac{1}{\sqrt{2g}}\int\limits_{y=0}^{h}\frac{s'(y)\,dy}{\sqrt{h-y}},$$

and with the transformation $y/h = u$

$$T(h) = \sqrt{\frac{8}{g}}\int\limits_0^1 \frac{s'(hu)\sqrt{h}\,du}{\sqrt{1-u}}. \tag{24.11}$$

Here T obviously is still a function of the parameter h occuring under the integral. In the sense of the formulated problem (T = constant) we now have to require

$$\frac{dT(h)}{dh} = 0 = \sqrt{\frac{8}{g}}\int\limits_0^1 \frac{d}{dh}\left[\frac{s'(hu)\sqrt{h}}{\sqrt{1-u}}\right]du = \sqrt{\frac{8}{g}}\int\limits_0^1 \frac{\sqrt{h}\cdot us''(hu) + \frac{1}{2\sqrt{h}}s'(hu)}{\sqrt{1-u}}\,du\,.$$

This is definitely fulfilled if the numerator of the integrand vanishes, that is, if the differential equation

$$2hus''(hu) + s'(hu) = 0 = 2ys''(y) + s'(y)$$

or

$$\frac{s''(y)}{s'(y)} = -\frac{1}{2y} \tag{24.12}$$

is fulfilled. Now

$$\frac{s''(y)}{s'(y)} = \frac{d}{dy}[\ln s'(y)] \qquad \text{and} \qquad -\frac{1}{2y} = \frac{d}{dy}\ln\sqrt{\frac{C_1}{y}}\,.$$

Hence from 24.12 it follows that

$$\frac{d}{dy}\left[\ln s'(y) - \ln\sqrt{\frac{1}{y}} - \ln\sqrt{C_1}\right] = 0$$

or after integration

$$\ln s'(y) = \ln\sqrt{\frac{C_1}{y}} \qquad \text{or} \qquad s'(y) = \sqrt{\frac{C_1}{y}}.$$

From 24.12 it follows with $ds = \sqrt{1+(dx/dy)^2}\,dy = (ds/dy)dy = s'(y)\,dy$, hence $\sqrt{C_1/y} = \sqrt{1+(dx/dy)^2}$, from which after separation of variables

$$dx = \sqrt{\frac{C_1}{y} - 1}\,dy = \sqrt{C_1 y - y^2}\,\frac{dy}{y},$$

and after integration with the new integration constant C_2 it finally results in

$$\begin{aligned} x &= \int \sqrt{C_1 y - y^2}\,\frac{dy}{y} \\ &= \sqrt{C_1 y - y^2} - \frac{C_1}{2}\arccos\frac{2y - C_1}{C_1} + C_2 . \end{aligned} \tag{24.13}$$

We check: With

$$f = \sqrt{C_1 y - y^2},$$

it follows that

$$f'(y) = \frac{C_1 - 2y}{2\sqrt{C_1 y - y^2}},$$

and with

$$g(y) = \arccos\left(\frac{2y - C_1}{C_1}\right),$$

it follows that

$$g'(y) = -\frac{2}{C_1\sqrt{1 - \left(\frac{2y - C_1}{C_1}\right)^2}} = -\frac{2}{\sqrt{-4y^2 + 4yC_1}} = -\frac{1}{\sqrt{C_1 y - y^2}},$$

such that in fact

$$\left(f(y) - \frac{C_1}{2}g(y)\right)' = \frac{C_1 - 2y}{2\sqrt{C_1 y - y^2}} + \frac{C_1}{2\sqrt{C_1 y - y^2}} = \frac{C_1 - y}{\sqrt{(C_1 - y)y}} = \frac{\sqrt{C_1 y - y^2}}{y}.$$

Hence, the curve 24.13 is a cycloid (compare to Problem 24.1). This result becomes even more obvious if we determine the integration constants C_1 and C_2 from the boundary conditions $y(x = 0) = 0$, $y(x = \pi a) = 2a$ as $C_1 = 2a$ and $C_2 = a\pi$, so that for 24.13 it finally follows that

$$\begin{aligned} x &= \sqrt{2ay - y^2} - a\arccos\frac{y - a}{a} + a\pi \\ &= \sqrt{2ay - y^2} + a\left(\pi - \arccos\frac{y - a}{a}\right). \end{aligned}$$

We still check the vibration period, using 24.11. With $s'(y) = \sqrt{C_1/y} = \sqrt{2a/y}$, we get according to 24.13:

$$T = \sqrt{\frac{8}{g}}\int_0^h \frac{s'(y)\,dy}{\sqrt{h - y}} = \sqrt{\frac{8}{g}}\int_0^h \frac{\sqrt{2a}\,dy}{\sqrt{y(h - y)}} = 2\sqrt{\frac{4a}{g}}\int_0^h \frac{dy}{\sqrt{y(h - y)}},$$

$$= -2\sqrt{\frac{4a}{g}} \cdot \arcsin\left(1 - \frac{2y}{h}\right)\Bigg|_0^h = 2\pi\sqrt{\frac{4a}{g}} = 2\pi\sqrt{\frac{l_r}{g}}, \qquad (l_r = 4a),$$

namely, actually a value that is independent of the initial elongation h. We shall prove the uniqueness of the solution in Vol. 2 of the lectures, after we have become familiar with the Fourier series.

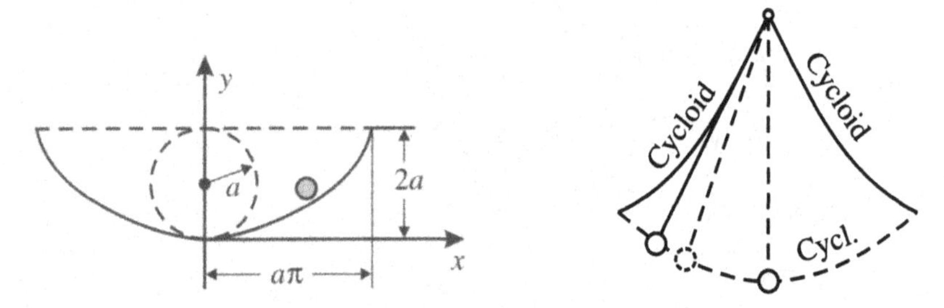

The cyloid as the tautochrone (left), and used as jaws in Huygens' construction of a pendulum whose period does not depend on the amplitude (right).

Historical remark: The treatment of this problem may be traced back to Huygens,[2] who aimed at the construction of a pendulum, with the vibration period being independent of the amplitude. Because the evolvent of a cycloid again represents a cycloid, a cycloid string pendulum may be constructed by forcing the pendulum motion of the mass m into a cycloid trajectory by an appropriate assembly of two cycloid jaws (compare the figure and Problem 24.2). Such a construction was realized in 1839 by the Austrian engineer Stampfer for the clock of the City Hall tower in Lemberg. This clock excelled by a very high accuracy of performance until its destruction by lightning.

[2] *Christiaan Huygens*, physicist and mathematician, b. April 14, 1629, Den Haag—d. there July 8, 1695. After first studying jurisprudence, he turned to mathematical research and published among others in 1657 a treatise on probability calculus. At the same time he invented the pendulum clock. In March 1655 he discovered the first Saturn moon; in 1656 the Orion nebula and the shape of the Saturn ring. Already then he was also familiar with the laws of collisions and those of central motion, but published them—without proofs—only in 1669. In 1663 Huygens became elected as member of the Royal Society. In 1665 he settled in Paris, as a member of the newly founded French Academy of Sciences, from where he returned to the Netherlands in 1681. After publishing his *Systema Saturnium, sive de causis mirandorum Saturni phaenomenon* already in 1657, his main work *Horologium oscillatorium* (The pendulum clock) emerged in 1673, which, besides the description of an improved clock construction, contains a theory of the physical pendulum. Also contained there are treatises on the cycloid as an isochrone and important theorems on central motion and the centrifugal force. From 1675 dates Huygens' invention of the spring watch with balance spring; from 1690 the *Tractatus de lumine*, the treatise on light where a first version of the wave theory of double refraction of Iceland spar is developed. The spherical propagation of the action about the light source is explained there by means of Huygens' principle. The French edition of the *Traité de la lumière* (Leiden, 1690) also includes a *Discours de la cause de la pesanteur* as a supplement [BR].

25 Mathematical Interlude: Differential Equations

On treating mechanical problems we became familiar with differential equations. An (ordinary) differential equation is a relation between an independent variable (t), a function $x(t)$, and one or several of its derivatives ($\dot{x}, \ddot{x}, \ldots$) from which the wanted function $x(t)$ shall be calculated. The differential equation is said to be of first order if only the first of its derivatives is involved. Such a differential equation may be written as

$$F(t, x, \dot{x}) = 0 \tag{25.1}$$

or, when solving for $\dot{x}$, as

$$\dot{x}(t) = f(t, x). \tag{25.2}$$

A differential equation is of second order if no higher derivative than the second one occurs. A differential equation of second order therefore has the form

$$F(t, x, \dot{x}, \ddot{x}) = 0$$

or, resolved for $\ddot{x}$,

$$\ddot{x} = f(t, x, \dot{x}). \tag{25.3}$$

The meaning of a differential equation of first order (25.2) is understood as follows: $\dot{x}$ determines the direction of the curve $x(t)$ in the t, x-plane. The differential equation (25.2) assigns a direction to any point t, x; it defines a *direction field.* We may visualize this field, for example, by plotting in a sufficiently dense lattice of points t, x the direction at each lattice point by a short dash (see the figure). The differential equation is solved by plotting curves into this direction field; the directions of these curves at any point correspond to the direction field. If $f(t, x)$ is a *reasonable* function, one may interpolate between the plotted directions in the direction field. In this way one obtains a set of curves. In other words: The differential equation (25.2) allows a set of solving functions $x(t)$. An individual curve of the set is specified by prescribing the value of x belonging to a fixed value of t (in the figure, the value x_0 for $t = 0$). Such a set of curves in which the individual curve is determined by a single number (a parameter) is called a *one-parametric set of curves*. We therefore may state:

A differential equation of first order (25.2)—with a reasonable function $f(t, x)$—determines a one-parametric set of curves. The general solution contains an arbitrary integration constant ($x(0) = x_0$).

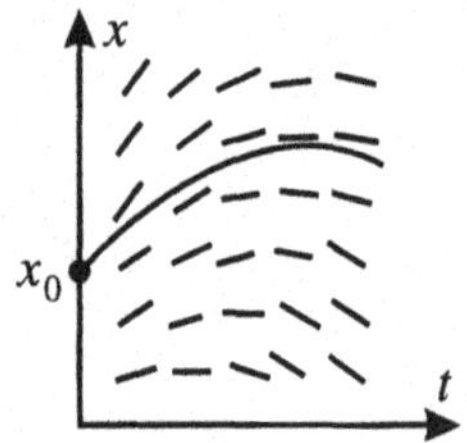

Direction field of a differential equation of first order.

This also holds vice versa: To any (reasonable) one-parametric set of nonintersecting curves in the t, x-plane there corresponds a differential equation of first order. The curves of the set may namely be described by the equation

$$\varphi(t, x) = c, \tag{25.4}$$

where c for each curve takes a distinct value. The function φ is not uniquely determined by the set of curves, as any possible function φ may be replaced by a function of φ, that is, by $F(\varphi) = F(c) = C$ and nevertheless describes the same set of curves. For the direction of the curves it follows that

$$\frac{\partial \varphi}{\partial t} dt + \frac{\partial \varphi}{\partial x} dx = 0, \tag{25.5}$$

or (assuming $\partial F/\partial \varphi \neq 0$)

$$\frac{\partial F}{\partial \varphi}\left(\frac{\partial \varphi}{dt} dt + \frac{\partial \varphi}{\partial x} dx\right) = 0 \quad \Rightarrow \quad \frac{\partial \varphi}{\partial t} dt + \frac{\partial \varphi}{\partial x} dx = 0, \tag{25.6}$$

that is, always the relation (25.5). From this relation it then follows that

$$\dot{x} = -\frac{\varphi_t(t, x)}{\varphi_x(t, x)} \equiv f(t, x), \tag{25.7}$$

where φ_t and φ_x denote the partial derivatives with respect to t and x, respectively. If φ is replaced by a function of φ, we obtain according to (25.6) the same equation (25.7). A one-parametric set of curves therefore essentially corresponds to a single differential equation of first order. We therefore may state: *A one-parametric set of curves* (25.4) *is equivalent to a differential equation.* Particularly simple differential equations of first order are of the type

$$\dot{x} = f(t) \tag{25.8}$$

and

$$\dot{x} = f(x). \tag{25.9}$$

In these cases the direction field depends only on one of the variables t or x, respectively. The solution of (25.8) may be obtained immediately:

$$x(t) = \int_0^t f(t')\,dt' + x_0. \tag{25.10}$$

Obviously all solutions originate from a single solution by adding an arbitrary constant to $x(t)$ (by shifting the solution curves along the x-direction). The solution of (25.9) is obtained via the transformation

$$dt = \frac{dx}{f(x)} \tag{25.11}$$

by the integral

$$t(x) = \int_0^x \frac{dx'}{f(x')} + t_0. \tag{25.12}$$

In this case all solutions are generated from a single (fixed) solution by adding an arbitrary constant to t (by shifting the solution curve in the t-direction). A differential equation of first order may be solved easily also then if it may be put into the form

$$g(x)\,dx = h(t)\,dt, \tag{25.13}$$

that is, if the variables may be separated. We then get

$$\int_{x_0}^x g(x')\,dx' = \int_0^t h(t')\,dt'. \tag{25.14}$$

We now turn to the discussion of a differential equation of second order. The function $f(t, x, \dot{x})$ in (25.3) ascribes to each point t, x and to each given direction ($\dot{x}$) through this point a defined change of direction. For a reasonable function $f(t, x, \dot{x})$, we may find graphical solutions as follows: We begin at an arbitrary point of the t, x-plane with an arbitrary direction of the curve, and then calculate the associated value of $\ddot{x}$ according to (25.3). The curve is then continued as a parabola in the assumed direction ($\dot{x}$) with the calculated value of $\ddot{x}$ (a parabola with a vertical axis has the same value of $\ddot{x}$ everywhere). After a certain piece of continuation we have a new point t, x and a new direction $\dot{x}$. There we again calculate $\ddot{x}$ according to (25.3) and continue the curve by the corresponding new parabola, etc. The solution curve obtained this way depends on the choice of the position and direction when starting the procedure. In total, we obtain an entire set of solutions. The individual solution curve is thus determined by specifying two numbers, for example, the values of x and $\dot{x}$ at a certain time point (t-value). A set of curves in which the individual curve is determined by giving two numbers is called a *two-parametric set of curves* (see figure). Thus we may state:

A differential equation of second order (25.3) with a reasonable function $f(t, x, \dot{x})$ determines a two-parametric set of curves.

The general solution contains two arbitrary integration constants.

Particularly simple differential equations of second order (which we met already) are

$$\ddot{x} = f(t), \tag{25.15}$$

$$\ddot{x} = f(\dot{x}), \tag{25.16}$$

$$\ddot{x} = f(x). \tag{25.17}$$

In the first case, (25.15), the acceleration is given as a function of time; in the second case, (25.16), the acceleration is given as a function of the velocity; and in the third case, (25.17), as a function of the position. (25.15) may be solved by a twofold integration. (25.16) is of first order in $\dot{x}$, thus it may be rewritten with $v = \dot{x}$ into $\dot{v} = f(v)$ and may then be solved as (25.9). (25.17) is transformed to

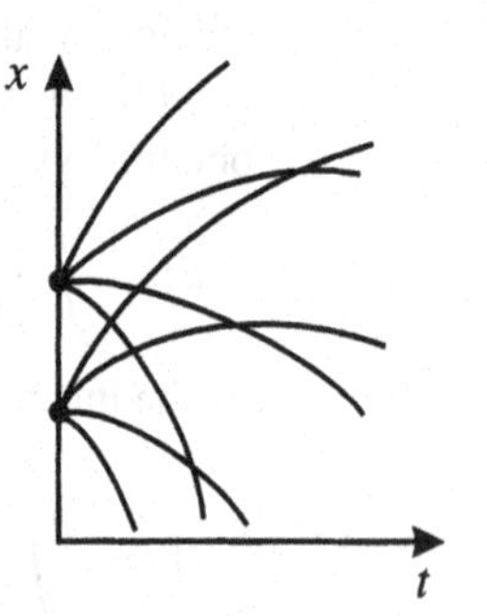

Set of solutions of a differential equation of second order.

$$\begin{aligned} \dot{x}\ddot{x} &= f(x)\dot{x}, \\ \dot{x}\,d\dot{x} &= f(x)\,dx, \\ \frac{1}{2}\dot{x}^2 &= \int_{x_0}^{x} f(x')\,dx' + c, \end{aligned} \tag{25.18}$$

and we thereby obtain a differential equation

$$\dot{x} = \varphi(x),$$

which may be solved as (25.9). In physics the *linear differential equations* are of particular importance because the phenomena described by these equations obey the *superposition principle* (compare equations (21.4) and (21.5) ff.). We shall outline this point of view for a differential equation of second order; the reader may extend that to other orders. The differential equation is linear if $x, \dot{x}, \ddot{x}$ occur linearly, that is, if the equation has the form

$$A\ddot{x} + B\dot{x} + Cx + D = 0, \tag{25.19}$$

where A, B, C, D may be functions of t. If the term D is missing, the equation is called *homogeneous.* If $x_1(t)$ solves a homogeneous linear differential equation, then cx_1 with c being a constant is also a solution. If $x_1(t)$ and $x_2(t)$ are solutions, then $c_1x_1 + c_2x_2$ with arbitrary constants c_1 and c_2 is also a solution (compare equations (21.4) and (21.5) ff.). Because the general solution of a differential equation of second order contains two and only two arbitrary constants, a homogeneous linear differential equation of second order has been solved generally if two distinct (linearly independent) solutions are known. If we know a solution $x_1(t)$ of an inhomogeneous linear differential equation (25.19), that is,

$$A\ddot{x}_1(t) + B\dot{x}_1(t) + Cx_1(t) + D = 0, \tag{25.20}$$

and if $x_0(t)$ is a solution of the homogeneous equation that arises by omitting the term D, that is,

$$A\ddot{x}_0(t) + B\dot{x}_0(t) + Cx_0(t) = 0, \tag{25.21}$$

then $(x_0 + x_1)$ is again a solution of equation (25.19). We have namely

$$A(\ddot{x}_0 + \ddot{x}_1) + B(\dot{x}_0 + \dot{x}_1) + C(x_0 + x_1) + D$$
$$= \underbrace{A\ddot{x}_0 + B\dot{x}_0 + Cx_0}_{=0} + \underbrace{A\ddot{x}_1 + B\dot{x}_1 + Cx_1 + D}_{=0} = 0.$$

Hence, an inhomogeneous linear equation is solved generally if one has solved generally the homogeneous equation and then adds a particular solution of the inhomogeneous equation. We have already used this statement in the context of the forced vibration (Chapter 23). One may convince oneself by means of (25.20) that two possibly distinct solutions of the inhomogeneous differential equation, $x_1(t)$ and $x_2(t)$, must be equal to each other, apart from a solution of the homogeneous equation (25.21). From $A\ddot{x}_1 + B\dot{x}_1 + Cx_1 = -D = A\ddot{x}_2 + B\dot{x}_2 + Cx_2$, it follows namely that

$$A\ddot{x}_1 + B\dot{x}_1 + Cx_1 = A\ddot{x}_2 + B\dot{x}_2 + Cx_2;$$

hence

$$A(x_1 - x_2)^{\cdot\cdot} + B(x_1 - x_2)^{\cdot} + C(x_1 - x_2) = 0,$$

that is, the difference $x_1 - x_2$ of the two particular solutions must be a solution of the homogeneous equation. Homogenous linear equations with constant coefficients (A, B, C) are solved by means of the *ansatz*

$$x(t) = e^{\lambda t}.$$

From the differential equation

$$A\ddot{x} + B\dot{x} + Cx = 0$$

results the algebraic equation (it is called *characteristic equation*)

$$A\lambda^2 + B\lambda + C = 0$$

for λ. Its two solutions yield, if they don't just coincide, two solutions of the differential equation and thus the general solution

$$x = c_1 e^{\lambda_1 t} + c_2 e^{\lambda_2 t}.$$

If the quadratic equation in λ has only one solution, then

$$x = c_1 e^{\lambda t} + c_2 t e^{\lambda t}$$

is the general solution of the differential equation, as may be checked easily (directly or by a limiting process).

26 Planetary Motions

In this chapter we shall investigate the motion in a central force field. As usual in physics, we begin with experimental observations—in our case the Kepler laws of planetary motion—and deduce the forces–in ur case the gravitational force between two masses. Later we shall reverse this process and start our reasoning with the forces and Newton's equations of motion, and we shall then derive Kepler's laws, As will become evident, theory then predicts new phenomena not contained in Kepler's laws, for example the orbits of comets, perhelion motion, and other facts.

Accordingsly, let us now consider in particular the planetary motion and start from the three Kepler laws, which were derived by Johannes Kepler[1] from the observations of planets made by Brahe.[2] These three laws are as follows:

[1] *Johannes Kepler*, b. Dec. 27, 1571, Weil der Stadt—d. Nov. 15, 1630, Regensburg. Kepler was son of a trader who often also served in military, first attended the school in Leonberg, and later the monastic school in Adelberg and Maulbronn. In 1589 Kepler began his studies in Tübingen, to become a theologian, but in 1599 he took the position of professor of mathematics in Graz offered to him. In 1600 Kepler had to leave Graz, because of the counterreformation, and he went to Prague. After the death of Tycho Brahe (Oct. 24, 1601) Kepler as his successor became imperial mathematician. After the death of his patron, emperor Rudolf II, Kepler left Prague and went in 1613 to Linz as a land surveyor. From 1628 Kepler lived as employee of the powerful Wallenstein mostly in Sagan. Kepler died fully unexpectedly during a visit to the meeting of electors in Regensburg.

Kepler's main fields were astronomy and optics. After extraordinarily lengthy calculations he found the *fundamental laws of planetary motion*: the Kepler's first and second laws were published 1609 in *Astronomia Nova*, the third one 1619 in *Harmonices Mundi*. In 1611 he invented the astronomical telescope. His *Rudolphian Tables* (1627) continued to be one of the most important tools of astronomy until the modern age. In the field of mathematics he developed heuristic infinitesimal considerations. His best-known mathematical writing is the *Stereometria Doliorum* (1615) where, e.g., Kepler's tub rule is given.

[2] *Tycho Brahe*, Danish astronomer, b. Dec. 14, 1546, Knudstrup on Schonen—d. Oct. 24, 1601, Prague. He first studied law, secretly dealt with astronomy until he inherited a considerable fortune, and then continued his study in Germany. In 1572 he became known by the discovery of a new star, the Nova Cassiopeiae, which was in fact a supernova. He lectured in Copenhagen and, by recommendation by Wilhelm IV, count of Hessen-Kassel, who dealt with astronomy, he got the support of the Danish king Friedrich II, who in 1576 transferred to him the island Ven in the Sound near Copenhagen. At the observatory "Uranienborg" built there Brahe dealt with research and education and tutoring his numerous scholars and assistants. The troubles he met after the death of Friedrich II (1588) forced him to leave the country in 1597. After a two-year stay with the count Rantzau in Wandsbek near Hamburg, he served as imperial astronomer with Rudolf II. In Prague he again gathered a couple of coworkers, among them Christian Ljöngberg (Longomontanus) and first of all Johannes Kepler.

1. All planets are moving on ellipses. The sun stands in one of their focal points.
2. The radius vector sun–planet covers equal areas in equal times (area theorem).
3. The squares of the revolution periods of two planets are related to each other as the cubes of the large semi-axes of their trajectories.

Let us denote the large semi-axis by a_ν and the revolution period of the νth planet by T_ν; then

$$\frac{T_1^2}{a_1^3} = \frac{T_2^2}{a_2^3}.$$

This means for a planet: $T^2 \sim a^3$.

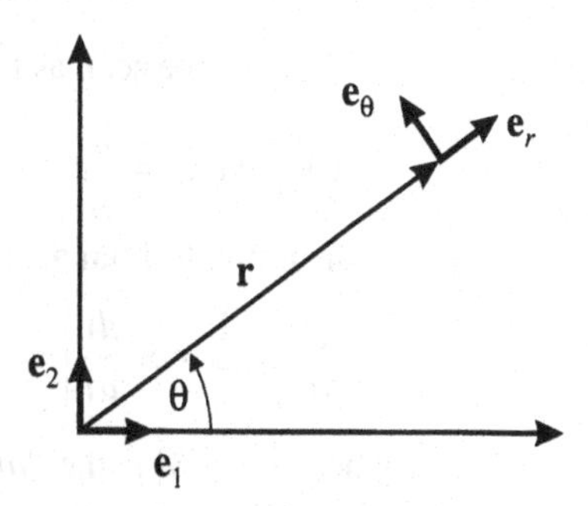

Unit vectors for Cartesian and polar coordinates.

We adopt two approaches: First we try to find out the properties of the force field from the Kepler laws. Later on we shall start from the force field that will be assumed to be given, and deduce the properties of the path. In order to formulate the motion and the force law, it is appropriate to formulate the equations of motion in polar coordinates. According to the first of Kepler's laws, the motion must be a *planar motion.*

We therefore introduce the local unit vectors $\mathbf{e}_r$ and $\mathbf{e}_\theta$ at each point. They are defined by the equations

$$\mathbf{e}_r = \cos\theta\,\mathbf{e}_1 + \sin\theta\,\mathbf{e}_2,$$
$$\mathbf{e}_\theta = -\sin\theta\,\mathbf{e}_1 + \cos\theta\,\mathbf{e}_2.$$

We know them already from Chapter 10 but shall briefly remind the essentials. The orientation of these unit vectors is time-dependent. Therefore,

$$\dot{\mathbf{e}}_r = (-\sin\theta\,\mathbf{e}_1 + \cos\theta\,\mathbf{e}_2)\dot{\theta} = \dot{\theta}\mathbf{e}_\theta,$$
$$\dot{\mathbf{e}}_\theta = (-\cos\theta\,\mathbf{e}_1 - \sin\theta\,\mathbf{e}_2)\dot{\theta} = -\dot{\theta}\mathbf{e}_r.$$

We now express the velocity and acceleration in terms of these coordinates. Twofold differentiation yields

$$\mathbf{r} = r\mathbf{e}_r\,,$$
$$\dot{\mathbf{r}} = \dot{r}\mathbf{e}_r + r\dot{\mathbf{e}}_r = \dot{r}\mathbf{e}_r + r\dot{\theta}\mathbf{e}_\theta \equiv \mathbf{v},$$

Brahe was the most significant observing astronomer before the invention of the telescope. He practically reached the possible accuracy of observations with the bare eye. The observations of Brahe and his coworkers formed the prerequisite for Kepler's works on the orbits of planets. Brahe tried to substitute the Copernican world system by his own system, according to which sun and moon are orbiting about the earth resting at the center of the universe, while the remaining planets orbit about the sun. The *Tychonian system* was favored in the 17th century since the assumption of the incredibly large distances of the fixed stars that had to be presupposed by Copernicus were not needed in Brahe's system. Brahe proved that the comets are not phenomena caused by the earth's atmosphere, as was assumed, for example, by Aristotle [BR].

$$\dot{\mathbf{v}} = \ddot{r}\mathbf{e}_r + \dot{r}\dot{\mathbf{e}}_r + \dot{r}\dot{\theta}\mathbf{e}_\theta + r\ddot{\theta}\mathbf{e}_\theta + r\dot{\theta}\dot{\mathbf{e}}_\theta$$
$$= \left(\ddot{r} - r\dot{\theta}^2\right)\mathbf{e}_r + \left(r\ddot{\theta} + 2\dot{r}\dot{\theta}\right)\mathbf{e}_\theta. \tag{26.1}$$

The area theorem now reads simply

$$r^2\dot{\theta} = h \qquad (h = \text{constant}). \tag{26.2}$$

This may be seen as follows: Let the force center be at the coordinate origin; then

$$dA = |d\mathbf{A}| = \frac{1}{2}|\mathbf{r} \times d\mathbf{r}|$$

is the infinintesimal area element, and furthermore

$$\frac{dA}{dt} = \frac{1}{2}\left|\mathbf{r} \times \frac{d\mathbf{r}}{dt}\right| = \frac{1}{2}|\mathbf{r} \times \mathbf{v}| = \frac{1}{2}h = \text{constant}, \tag{26.3}$$

where $\frac{1}{2}|\mathbf{r} \times \mathbf{v}|$ is the *"area velocity"* of the radius vector. Hence:

$$|\mathbf{r} \times \mathbf{v}| = r^2\dot{\theta} = h.$$

From the area theorem found empirically by Kepler, it now follows that

$$\frac{d(r^2\dot{\theta})}{dt} = r(2\dot{r}\dot{\theta} + r\ddot{\theta}) = 0.$$

A comparison with (26.1) yields for the wanted force field

$$\ddot{\mathbf{r}} \cdot \mathbf{e}_\theta = 0, \tag{26.4}$$

that is, no acceleration and hence no force is acting along the $\mathbf{e}_\theta$-direction. The area theorem thus implies that we are dealing with a central force field. This is already known from earlier (Chapter 17). And vice versa, a central force field requires the area theorem to be valid: For central forces the torque vanishes, $\mathbf{D} = \mathbf{r} \times \mathbf{F} = 0$. Hence, for central forces conservation of the angular momentum generally holds:

$$\dot{\mathbf{L}} = \mathbf{r} \times \mathbf{F} = 0, \qquad \mathbf{L} = \overrightarrow{\text{constant}};$$

hence,

$$\mathbf{L} = \mathbf{r} \times \mathbf{p} = (\mathbf{r} \times \mathbf{v})m = \overrightarrow{\text{constant}}.$$

From there one may immediately derive

$$|\mathbf{L}| = r^2\dot{\theta}m = hm. \tag{26.5}$$

Mathematical interlude: consideration of conic sections in polar coordinates—ellipse, parabola, hyperbola:

The equation in polar coordinates

$$r = \frac{k}{1 + \varepsilon\cos\theta} \tag{26.6}$$

describes

circles	(for $\varepsilon = 0$),
ellipses	(for $\varepsilon < 1$),
parabolas	(for $\varepsilon = 1$),
hyperbolas	(for $\varepsilon > 1$).

Equation (26.6) is therefore the *general equation for conic sections* in polar coordinates. We make that clear in detail now:

(a) Ellipse

It is the set of all points whose distances from two fixed *focal points* F and F' in a distance of $2c$ (see figure) have a constant sum $2a$, which is larger than $\overline{FF'}$. Thus (compare figure), $r + r' = 2a$, $c^2 + b^2 = a^2$, where a and b are the major and minor semi-axes of the ellipse, respectively. Further it holds that

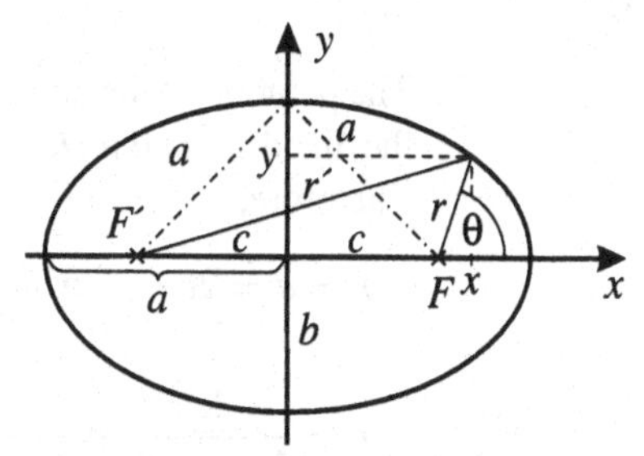

The geometry of the ellipse.

$$c = \sqrt{a^2 - b^2} = \varepsilon \cdot a, \qquad \varepsilon < 1.$$

ε is called *eccentricity*. For the circle $\varepsilon = 0$ (both focal points coincide, i.e., $c = 0$). Obviously (compare figure)

$$r + \sqrt{(2c)^2 + r^2 + 2(2c)r\cos\theta} = 2a$$

or

$$4\varepsilon^2 a^2 + r^2 + 4\varepsilon ar\cos\theta = (2a - r)^2,$$

$$r = \frac{a(1-\varepsilon^2)}{1+\varepsilon\cos\theta} \equiv \frac{k}{1+\varepsilon\cos\theta},$$

where

$$k = a(1-\varepsilon^2) = a\left(1 - \frac{a^2 - b^2}{a^2}\right) = \frac{b^2}{a^2}.$$

We still give the equation of an ellipse in Cartesian coordinates. From the figure one may immediately read off

$$r = \sqrt{(x-c)^2 + y^2}, \qquad r' = \sqrt{(x+c)^2 + y^2},$$

such that the defining equation for the ellipse reads

$$r + r' = \sqrt{(x-c)^2 + y^2} + \sqrt{(x+c)^2 + y^2} = 2a.$$

Forming twice the square, together with $b^2 = a^2 - c^2$, then leads to

$$\frac{x^2}{a^2} + \frac{y^2}{b^2} = 1.$$

(b) Circle

Circles casually fit in as special cases of ellipses ($\varepsilon = 0$).

(c) Parabola

The parabola is the geometric locus of all points P of a plane that have equal distance from the fixed *guideline* L and the fixed focal point F.
Therefore,

$$r = d = 2c - r\cos\theta$$

or

$$r = \frac{2c}{1+\cos\theta} \equiv \frac{k}{1+\varepsilon\cos\theta},$$

where $\varepsilon = 1$ and $k = 2c$. We shall also write the parabola in Cartesian coordinates. From the figure we read

$$r = \sqrt{(c+x)^2 + y^2}\,,$$

The geometry of the parabola.

such that from

$$r = d = c - x$$

after squaring follows

$$y^2 = -4cx\,.$$

(d) Hyperbola

The hyperbola is the geometric locus of all points of a plane whose distances from two fixed points on the plane (the focal points) F and F' (with distance $2c$) have a constant difference. Hence

$$r - r' = 2a < \overline{FF'}$$

or

$$r - \sqrt{r^2 + 4c^2 + 4rc\cos\theta} = 2a.$$

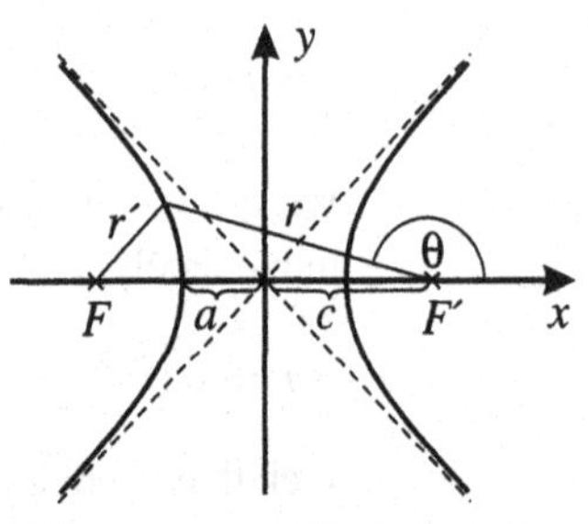

The geometry of the hyperbola.

With $c = \varepsilon a$ ($\varepsilon > 1$, see figure) it follows that

$$r = \frac{a(1-\varepsilon^2)}{1+\varepsilon\cos\theta} \equiv \frac{k}{1+\varepsilon\cos\theta}, \qquad k = a(1-\varepsilon^2).$$

In Cartesian coordinates the hyperbola equation follows from the defining equation

$$r - r' = 2a$$

or

$$\sqrt{(c-x)^2+y^2} - \sqrt{(c+x)^2+y^2} = 2a$$

after squaring twice and using $b^2 = c^2 - a^2$ as

$$\frac{x^2}{a^2} - \frac{y^2}{b^2} = 1.$$

Thus, the general form (26.6) of conical sections is founded.

We now continue our physical considerations and return to the further investigation of Kepler's laws. In order to derive the special form of the force law from the Kepler laws, we now take into account that the trajectory is an ellipse with the sun at one focal point. The equation of the ellipse reads in polar coordinates

$$r = \frac{k}{1+\varepsilon\cos\theta}, \tag{26.7}$$

with the parameter

$$k = a(1-\varepsilon^2) = a^2\frac{1-\varepsilon^2}{a} = \frac{a^2-c^2}{a} = \frac{b^2}{a},$$

and the *eccentricity*

$$\varepsilon = \frac{\sqrt{a^2-b^2}}{a} < 1\,.$$

We already know from (26.4) that the force, and hence the acceleration, must be central, namely, proprtional to $\mathbf{e}_r$. We thus can calculate the central acceleration—see (26.1)—and taking into account (26.2), we get

$$\dot{r} = \frac{dr}{d\theta}\dot{\theta} = \frac{\varepsilon}{k}\sin\theta\, r^2\dot{\theta} = \frac{\varepsilon}{k}h\sin\theta,$$

$$\ddot{r} = \frac{\varepsilon}{k}h\cos\theta\,\dot{\theta} = \frac{\varepsilon h^2}{kr^2}\cos\theta,$$

and, using (26.2) and (26.7) for the component of (26.1) along $\mathbf{e}_r$, finally

$$\ddot{r} - r\dot{\theta}^2 = \frac{h^2}{r^2}\left(\frac{\varepsilon}{k}\cos\theta - \frac{1}{r}\right) = -\frac{h^2}{kr^2}. \tag{26.8}$$

The central force of the wanted field for a planet of mass m is therefore given by (see also (26.1))

$$\mathbf{F}(r) = m(\ddot{r} - r\dot{\theta}^2)\mathbf{e}_r = \frac{-h^2}{kr^2} m \frac{\mathbf{r}}{r}.$$

At first the quantity h^2/k appears as a constant that is specific for each planet. But keeping in mind the third Kepler law, we find that h^2/k has the same value for all planets. This may be seen as follows: Because $h/2$ is the area velocity of the radius vector for a defined planet, the area of the ellipse equals πab and $b^2 = ak$, it follows for the revolution period T that

$$\frac{h}{2} T = \pi ab,$$

$$h \cdot T = 2\pi ab = 2\pi \sqrt{a^3 k},$$

and

$$\frac{T^2}{a^3} = \frac{4\pi^2 k}{h^2} \quad \Rightarrow \quad \frac{h^2}{k} = \frac{4\pi^2 a^3}{T^2};$$

Because

$$\frac{a^3}{T^2} = \text{constant} \quad \Rightarrow \quad \frac{h^2}{k} = \text{constant}. \tag{26.9}$$

Because according to the third Kepler law a^3/T^2 is equal for all planets, the same obviously holds also for h^2/k. The quantity h^2/k is the same for all planets. Therefore, all planets obey the force law

$$\mathbf{F}(r) = -\text{ constant } \frac{m}{r^2} \frac{\mathbf{r}}{r}.$$

If, according to the principle of actio and reactio, the mass of the central star is still factorized out from the constant (finally the force must vanish if the sun mass M vanishes), the gravitational law thus takes the form

$$\mathbf{F} = -\gamma \frac{Mm}{r^2} \cdot \frac{\mathbf{r}}{r}. \tag{26.10}$$

It is remarkable how this fundamental force law may be deduced from Kepler's laws. As we have seen, it is completely contained in these laws. Already Newton realized that the acceleration a planet feels due to the attraction by the sun is of the same nature as the acceleration on a freely falling body by the earth gravitation. The factor const $= \gamma M$ in the law (26.10) is of course only then the same if the attracting body is the same in both cases, for example, the earth. Newton therefore compared the acceleration of fall near the earth's surface, roughly 10 m/s^2, with the central acceleration of the moon on its orbit about the earth. The latter one is

$$\omega^2 a = \frac{4\pi^2 a}{T^2} = \frac{40 \cdot 6370 \cdot 10^5 \cdot 60}{27^2 \cdot 24^2 \cdot 60^2 \cdot 60^2} \frac{\text{cm}}{\text{s}^2},$$

where the distance of moon a is set equal to 60 times the earth's radius (6370 km), and the circulation period of the moon equal to 27 days. Because

$$\frac{40 \cdot 6370 \cdot 10^2}{27^2 \cdot 24^2 \cdot 60} \approx 1,$$

it follows that

$$\omega^2 a \approx \frac{10^3}{60^2} \frac{\text{cm}}{\text{s}^2}$$

and

$$\omega^2 a \,/\, g \approx 1 \,/\, 60^2;$$

that is, the acceleration of the moon on circulating about earth is actually related to the acceleration of free fall near the earth's surface inversely as the squares of the distances from the earth's center.

Example 26.1: The Cavendish experiment

In principle the gravitational constant γ may be determined by measuring the attractive force between two bodies of known mass. In practice, however, the gravitational force is so weak that it becomes highly difficult to demonstrate it in the laboratory. In the so-called Cavendish experiment (Cavendish,[3] 1798) the force between two masses is determined from the torsion of an elastic suspension string (see figure).

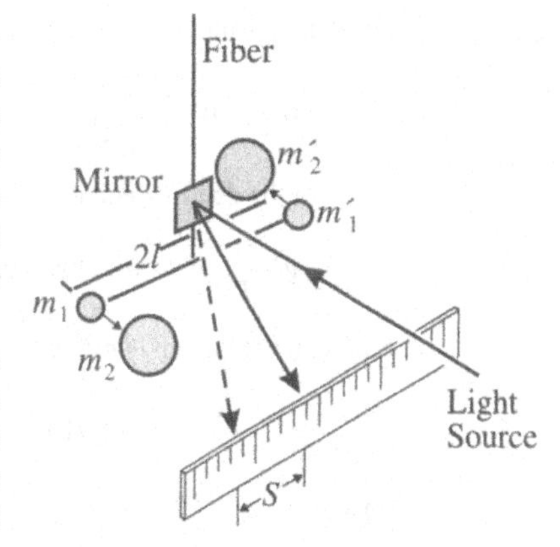

Torsional balance for the determination of the gravitational constant.

The masses m_1 and m'_1 are fixed to the ends of a light scale beam of length $2l$ suspended by a very thin quartz fiber. Already a very weak force may force the fiber to rotate about its axis (torsion), such that the torsion angle may provide a measure for very weak forces. To make the small torsion of the string visible, a mirror is attached to this string, which is hit by a light ray. Observation of the reflected ray allows us to measure any rotation of the string, and thus of the mirror.

In the measurement of the gravitational constant γ two large masses m_2 and m'_2 are positioned close to the masses m_1 and m'_1, as shown in the left figure below. Because of the gravitational attraction the small masses m_1, m'_1 move and thereby twist the string by the angle θ. After stabilization of this configuration, within several hours the masses m_2, m'_2 are brought into a new position, as represented in the right figure below.

[3] *Henry Cavendish*, chemist, b. Oct. 10, 1731, Nizza—d. Feb. 28, 1810, London. He investigated gases in detail, isolated carbon dioxide and hydrogen as distinct kinds of gases (1766); he realized the composition of air, discovered the explosionlike combination of hydrogen and oxygen (oxyhydrogen gas) and hence the composition of water. When working on nitrogen he discovered nitric acid. His determination of the gravitational constant by means of the torsional balance was of particular significance.

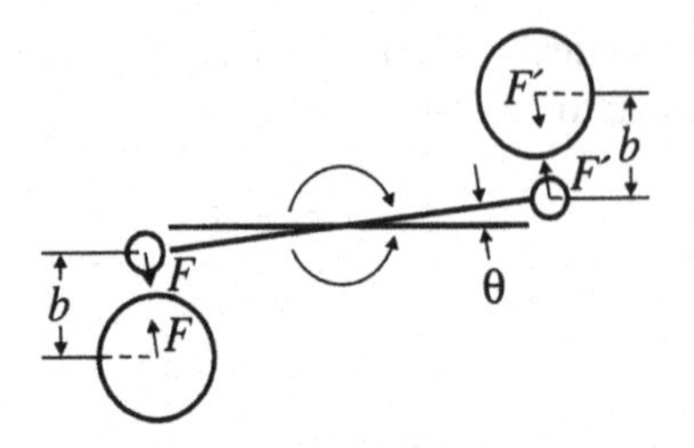

The first position of the masses m_2 and m_2' for the determination of the gravitational constant.

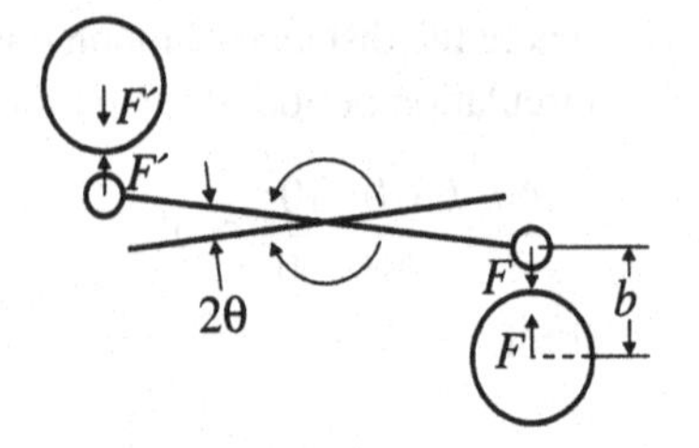

The second position of the masses m_2 and m_2'.

The string is now twisted anew by the gravitational force, namely in opposite direction by the angle 2θ. The system does not reach the final equilibrium state right now but rather oscillates with decreasing amplitude toward the final position (weakly damped oscillator—see figure on the next page). The period of oscillation amounts to about 8 min, and after about 30 min the system reaches the final equilibrium state. From these data the force between the spheres is determined, and with known mass and known distance between the centers of the spheres, we may calculate the gravitational constant γ from Newton's gravitational law:

$$\gamma = 6.67 \cdot 10^{-11} \frac{\mathrm{m}^3}{\mathrm{kg\,s}^2}.$$

From the defining equation for the gravitational acceleration on the earth's surface

$$g = \gamma \cdot \frac{m_E}{R_E^2} \qquad (m_E\text{: mass of earth, } R_E\text{: radius of earth})$$

and with the known constant γ we may now calculate the mass of earth. We obtain

$$m_E = \frac{g \cdot R_E^2}{\gamma} = 5.97 \cdot 10^{24}\,\mathrm{kg},$$

where $R_E = 6.37 \cdot 10^6$ m, $g = 9.81\mathrm{m/s}^2$ have been assumed. This implies a mean mass density of

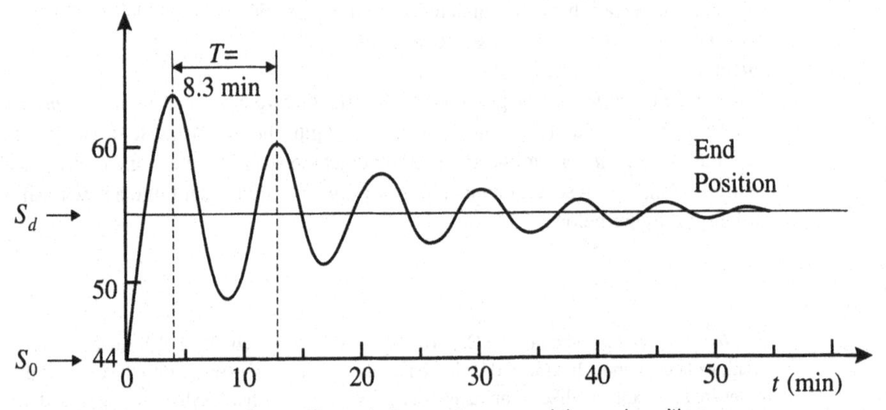

Oscillations with weak dampening around the end position.

earth of

$$\varrho = \frac{m}{V} = \frac{m_E}{\frac{4}{3}\pi R_E^3} \approx 5.5 \cdot 10^3 \,\frac{\mathrm{kg}}{\mathrm{m}^3}$$

or

$$\varrho \approx 5.5\ \mathrm{g/cm^3} \qquad \left(\varrho(\mathrm{iron}) = 7.86\ \mathrm{g/cm^3}\right).$$

Derivation of the Kepler laws from the force law

So far we have derived the gravitational law from the Kepler laws. Now we shall investigate central force fields in general. One may start from the assumption that the force field is known. The central force field has the following properties:

1. Central force fields $\mathbf{F} = f(r)\frac{\mathbf{r}}{r}$ with arbitrary radial dependence $f(r)$ are conservative, that is, the energy conservation law is valid because

$$\operatorname{curl}\mathbf{F} = \begin{vmatrix} \mathbf{e}_1 & \mathbf{e}_2 & \mathbf{e}_3 \\ \dfrac{\partial}{\partial x} & \dfrac{\partial}{\partial y} & \dfrac{\partial}{\partial z} \\ f(r)\dfrac{x}{r} & f(r)\dfrac{y}{r} & f(r)\dfrac{z}{r} \end{vmatrix} = 0.$$

With $r = \sqrt{x^2+y^2+z^2} = \sqrt{x_1^2+x_2^2+x_3^2}$ and $\partial r/\partial x_i = x_i/\sqrt{x_1^2+x_2^2+x_3^2} = x_i/r$ there holds, for example, for the $\mathbf{e}_1$-component

$$\begin{aligned} \frac{\partial}{\partial y}\left[f(r)\frac{z}{r}\right] - \frac{\partial}{\partial z}\left[f(r)\frac{y}{r}\right] &= z\frac{\partial}{\partial r}\left(\frac{f(r)}{r}\right)\frac{\partial r}{\partial y} - y\frac{\partial}{\partial r}\left(\frac{f(r)}{r}\right)\frac{\partial r}{\partial z} \\ &= \frac{\partial}{\partial r}\left(\frac{f(r)}{r}\right)\left(\frac{zy}{r} - \frac{yz}{r}\right) = 0. \end{aligned}$$

The vanishing of the the other components can be deduced in an analogous way. This is also vividly clear, as a central force field that points only toward the center or off center cannot have vortices.

2. If a body moves on an orbit in the central force field, then the orbital angular momentum is conserved. That means that the area theorem holds. For central force fields, we have

$$\mathbf{D} = \mathbf{r}\times\mathbf{F} = \mathbf{r}\times\frac{f(r)}{r}\mathbf{r} = 0 = \dot{\mathbf{L}};$$

thus

$$\mathbf{L} = \mathbf{r}\times\mathbf{p} = m(\mathbf{r}\times\mathbf{v}) = \overrightarrow{\text{constant}} = m\mathbf{h}$$

or

$$\frac{1}{2}|\mathbf{r}\times\mathbf{v}| = \frac{dA}{dt} = \frac{1}{2}h = \text{constant}.$$

3. A body in the central force field always moves in a plane, because from

$$\mathbf{r} \times \mathbf{v} = \mathbf{h} = \overrightarrow{\text{constant}}$$

it follows that

$$\mathbf{r} \cdot \mathbf{h} = \mathbf{r} \cdot (\mathbf{r} \times \mathbf{v}) = \frac{1}{m}\mathbf{r} \cdot \mathbf{L} = 0.$$

Hence, $\mathbf{r}$ points perpendicular to $\mathbf{L}$. Because $\mathbf{L}$ is constant, $\mathbf{r}$ always lies in a plane. In other words: The body moves only within a plane perpendicular to the angular momentum vector. Based on the conservation of energy E and of the angular momentum $\mathbf{L}$, we shall try to make statements on the orbital motion. The conservation laws concerning the angular momentum and the energy according to (26.5) read

$$mr^2\dot{\theta} = L, \tag{26.11}$$

$$\frac{1}{2}mv^2 + V(r) = E \tag{26.12}$$

with the gravitational potential

$$\begin{aligned} V(r) &= -\int \mathbf{F}(r) \cdot d\mathbf{r} \\ &= \gamma Mm \int_\infty^r \frac{\mathbf{r} \cdot d\mathbf{r}}{r^3} = \gamma Mm \int_\infty^r \frac{dr}{r^2} = \frac{-\gamma Mm}{r}. \end{aligned}$$

The gravitational potential has been chosen such as to vanish at infinity (i.e., for $r \to \infty$). This is always possible since we know that the potential is determined only up to an additive constant. Using

$$v^2 = \dot{r}^2 + r^2\dot{\theta}^2,$$

we rewrite the energy conservation law (26.12) into

$$\frac{m}{2}(\dot{r}^2 + r^2\dot{\theta}^2) + V(r) = E.$$

With the angular momentum (26.11), it follows that

$$\frac{m}{2}\dot{r}^2 + \frac{L^2}{2mr^2} + V(r) = E. \tag{26.13}$$

Hence, the total energy is composed of three components: a *radial kinetic energy* ($\frac{m}{2}\dot{r}^2$); a *rotational energy* ($L^2/2mr^2$); and a *potential energy* ($V(r)$). The rotational energy is usually written in the form $L^2/2J$, where $J = mr^2$ is the moment of inertia of the mass point (the planet) with mass m when revolving at a distance r from the axis of rotation. This will be treated in more detail in the second part of the mechanics course. From equation (26.13) we may now easily determine $r(t)$, because

$$\dot{r} = \sqrt{\frac{2}{m}(E - V(r) - L^2/2mr^2)}, \tag{26.14}$$

$$dt = \frac{dr}{\sqrt{\frac{2}{m}(E - V(r) - L^2/2mr^2)}} \qquad \text{(separation of variables)}, \tag{26.15}$$

$$t - t_o = \int_{r_0}^{r} \frac{dr}{\sqrt{\frac{2}{m}(E - V(r) - L^2/2mr^2)}}. \tag{26.16}$$

As already mentioned, the total energy (26.13) consists of three terms; here $\frac{1}{2}m\dot{r}^2$ is denoted as kinetic radial energy, and $L^2/2mr^2$ as rotational energy. This rotational energy may be incorporated into the potential, as L^2 is constant and $L^2/2mr^2$ therefore acts like a potential term in (26.13). The term $L^2/2mr^2$ is therefore also called rotational potential or *centrifugal potential*. Thus one is led to the *effective potential*

$$V_{\text{eff}} = V(r) + \frac{L^2}{2mr^2}$$

consisting of the gravitational potential $V(r)$ and the centrifugal potential $L^2/2mr^2$. From (26.11) we may calculate the orbit, using the expression (26.14) for $\dot{r}$. It then results that

$$\begin{aligned} d\theta &= \frac{L\,dt}{mr^2} = \frac{L\,dr}{r^2\sqrt{2m(E - V - L^2/2mr^2)}} \\ &= \frac{dr}{r^2\sqrt{2mE/L^2 - 2mV(r)/L^2 - 1/r^2}} \end{aligned}$$

or

$$\theta - \theta_0 = \int_{r_0}^{r} \frac{dr}{r^2\sqrt{(2m/L^2)(E - V) - 1/r^2}}. \tag{26.17}$$

The integrals (26.16) and (26.17) yield $t = t(r)$ and $\theta = \theta(r)$, respectively. The motion $r(t)$ and $r(\theta)$ may be determined by means of the inverse functions. There always enter four integration constants: E, L, r_0, and t_0 or θ_0. Energy and angular momentum may, of course, also be expressed by the initial velocities $\dot{r}_0$ and $\dot{\theta}_0$. In principle from (26.17), the function $\theta(r)$ or $r(\theta)$ may be determined. As will be seen later on, it is, however, easier to calculate $u(\theta) \equiv 1/r(\theta)$ directly from the dynamic basic law (force law). We now shall follow the second approach.

The equation for the orbit in the gravitational field

The path of a body in the Newtonian force field

$$\mathbf{F}(r) = -\gamma\frac{Mm}{r^2}\cdot\frac{\mathbf{r}}{r}$$

shall now be determined. We don't start from the integrals (26.16) and (26.17) but shall derive a differential equation for $r(\theta)$ and look for the possible solutions in the gravitational potential. The energy law is (see equation (26.13))

$$\frac{1}{2}m(\dot{r}^2 + r^2\dot{\theta}^2) + V(r) = E\,, \tag{26.18}$$

and the angular momentum conservation law reads

$$r^2\dot{\theta} = h.$$

Then

$$\dot{r} = \frac{dr}{d\theta}\dot{\theta} = \frac{dr}{d\theta}\frac{h}{r^2}\,,$$

and the energy law (26.18) thus can be written as

$$\frac{1}{2}m\frac{h^2}{r^4}\left(\left(\frac{dr}{d\theta}\right)^2 + r^2\right) + V(r) = E. \tag{26.19}$$

We expect the conic sections (26.6) as solutions. Therefore, it is obvious to consider the variable $u(\theta) = 1/r(\theta) = (1 + \varepsilon\cos\theta)/k$. For $u(\theta)$ one may expect a simple differential equation. By substituting $u = 1/r$, we have with $dr/du = -1/u^2$

$$\frac{dr}{d\theta} = \frac{dr}{du}\frac{du}{d\theta} = -\frac{1}{u^2}\cdot\frac{du}{d\theta},$$

$$\dot{r} = \frac{dr}{d\theta}\dot{\theta} = -\frac{1}{u^2}\frac{du}{d\theta}hu^2 = -h\frac{du}{d\theta}, \tag{26.20}$$

and we obtain for (26.19)

$$\frac{1}{2}mh^2u^4\left(\frac{1}{u^4}\left(\frac{du}{d\theta}\right)^2 + \frac{1}{u^2}\right) + V\left(\frac{1}{u}\right) = E,$$

or

$$\frac{1}{2}mh^2\left(\left(\frac{du}{d\theta}\right)^2 + u^2\right) = E - V\left(\frac{1}{u}\right). \tag{26.21}$$

These relations will be useful later on. The function wanted is $u = u(\theta)$. It could be calculated directly by integration, but another path is much easier. For this end we start from the Newtonian equation for the central force

$$F(r) = m(\ddot{r} - r\dot{\theta}^2)\,.$$

Replacing again r by u, then with the use of (26.20) it holds that

$$\begin{aligned}\ddot{r} &= -h\frac{d}{dt}\left(\frac{du}{d\theta}\right) = -h\frac{d^2u}{d\theta^2}\dot{\theta}\\ &= -h^2\frac{1}{r^2}\frac{d^2u}{d\theta^2} = -h^2u^2\frac{d^2u}{d\theta^2},\end{aligned}$$

and with $r^2\dot{\theta} = h$ it then results that

$$\frac{d^2u}{d\theta^2} + u = -\frac{1}{mu^2h^2}F\left(\frac{1}{u}\right). \tag{26.22}$$

$F(1/u)$ may now be determined from the gravitational law (26.10). We have

$$\mathbf{F} = F(r)\mathbf{e}_r = -\gamma\frac{Mm}{r^2}\mathbf{e}_r = -Hu^2\mathbf{e}_r\,, \tag{26.23}$$

where

$$H = \gamma Mm. \tag{26.24}$$

Hence, (26.22) turns into

$$\frac{d^2u}{d\theta^2} + u = \frac{H}{mh^2}. \tag{26.25}$$

This inhomogeneous differential equation is to be solved. The solution of the corresponding homogeneous differential equation

$$\frac{d^2u}{d\theta^2} + u = 0$$

is, however,

$$u(\theta) = A\cos\theta + B\sin\theta. \tag{26.26}$$

A particular solution of the inhomogeneous differential equation is easily found, namely

$$u = \text{constant} = \frac{H}{mh^2}. \tag{26.27}$$

The general solution of equation (26.25) therefore reads

$$u = \frac{H}{mh^2} + A\cos\theta + B\sin\theta, \tag{26.28}$$

or written in another form—see (21.10) and (21.11):

$$u = \frac{H}{mh^2} + C\cos(\theta - \phi), \tag{26.29}$$

where ϕ is a constant angle; its magnitude depends on the choice of the coordinate frame. As no assumptions on the coordinate frame were made yet, one may choose it now such that $\phi = 0$. One then obtains for $u(\theta)$:

$$u(\theta) = \frac{H}{mh^2} + C\cos\theta = \frac{1}{r(\theta)}. \tag{26.30}$$

Solving for $r(\theta)$ yields

$$r(\theta) = \frac{mh^2/H}{1 + (Cmh^2/H)\cos\theta}. \tag{26.31}$$

With a look at the equation of the conical sections (26.6) we introduce the constants $k = mh^2/H$ and $\varepsilon = Cmh^2/H$. We then obtain for the path equation

$$r(\theta) = \frac{k}{1+\varepsilon\cos\theta}, \qquad k = \frac{mh^2}{H}, \qquad \varepsilon = \frac{Cmh^2}{H}. \tag{26.32}$$

This is just the equation of a conic section. The particular shape of the path curve is determined by the eccentricity ε:

$\varepsilon = 0$: $r(\theta)$ describes a circle,

$0 < \varepsilon < 1$: an ellipse,

$\varepsilon = 1$: a parabola,

$\varepsilon > 1$: a hyperbola.

We shall now investigate on which physical quantities (e.g., energy, angular momentum) the eccentricity depends. For this purpose we first determine the constant C by means of the energy law:

$$u(\theta) = \frac{H}{mh^2} + C\cos\theta \tag{26.33}$$

is differentiated and inserted into the energy equation (26.21); hence

$$\frac{1}{2}mh^2\left(\left(\frac{du}{d\theta}\right)^2 + u^2\right) = E - V\left(\frac{1}{u}\right), \tag{26.34}$$

$$\frac{1}{2}mh^2\left(C^2\sin^2\theta + \left(\frac{H}{mh^2} + C\cos\theta\right)^2\right) = E - V\left(\frac{1}{u}\right), \tag{26.35}$$

where $V = V(r)$ is the potential. It reads

$$V(r) = -\int \mathbf{F}\cdot d\mathbf{r} = \int_\infty^r \frac{\gamma Mm}{r^2}dr = -\gamma\frac{Mm}{r} = -Hu = V\left(\frac{1}{u}\right).$$

We now insert $V(1/u) = -Hu$ in the energy equation, which leads to

$$\frac{1}{2}mh^2\left(C^2\sin^2\theta + \left(\frac{H}{mh^2} + C\cos\theta\right)^2\right) = E + H\left(\frac{H}{mh^2} + C\cos\theta\right).$$

With the intermediate calculation

$$\frac{1}{2}mh^2\left[C^2(\sin^2\theta + \cos^2\theta) + \left(\frac{H}{mh^2}\right)^2 + 2C\frac{H}{mh^2}\cos\theta\right]$$

$$= E + H\left(\frac{H}{mh^2} + C\cos\theta\right),$$

we may solve for C and obtain

$$C = \sqrt{\frac{H^2}{m^2h^4} + \frac{2E}{mh^2}}. \tag{26.36}$$

From there we calculate ε according to (26.32) as

$$\varepsilon = \sqrt{1 + \frac{2Emh^2}{H^2}}. \tag{26.37}$$

Hence, the shape of the path depends on the total energy E and the angular momentum $l = mh$ of the moving body, and it holds that

for a parabola:	$\varepsilon = 1,$	hence $E = 0,$
for an ellipse:	$0 < \varepsilon < 1,$	hence $E < 0, \quad -\gamma^2 \frac{M^2 m}{2h^2} < E < 0,$
for a circle:	$\varepsilon = 0,$	hence $E = -\frac{H^2}{2mh^2} = -\frac{\gamma^2 m M^2}{2h^2},$
for a hyperbola:	$\varepsilon > 1,$	hence $E > 0.$

The effective potential—overview on path types

If one writes the total energy in the form (26.13)

$$\frac{m}{2}\dot{r}^2 + V(r) + \frac{L^2}{2mr^2} = E$$

and introduces the effective potential

$$V_{\mathrm{eff}}(r) = V(r) + \frac{L^2}{2mr^2},$$

hence

$$\frac{m}{2}\dot{r}^2 + V_{\mathrm{eff}}(r) = E, \tag{26.38}$$

then this equation just corresponds to a one-dimensional motion under a force depending only on r; the potential energy of this one-dimensional motion is just the effective potential energy $V_{\mathrm{eff}}(r)$. We shall discuss its trend. Let the angular momentum L be given as fixed. Then V_{eff} consists of the attractive gravitational potential $\sim -1/r$, which dominates at large distances, and of the repulsive angular momentum barrier $\sim L^2/r^2$, which governs the motion at small distances. The superposition of both terms yields a potential, as shown in the sketch.

We now consider various energy values E. At the reversal points of the orbital motion one has $\dot{r} = 0$, that is, according to (26.38), $V_{\mathrm{eff}} = E$. These positions correspond to the points with the maximum and minimum distances from the central star. For the parabola and the circle there exists only one solution for a given V_{eff}, but there exist, on the contrary,

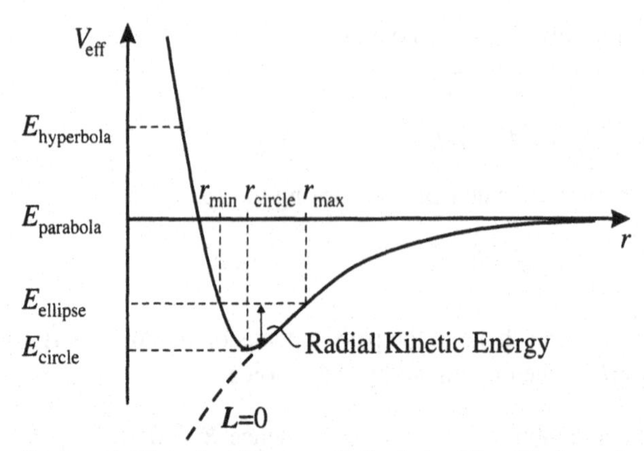

Classification of different orbit types with the help of the effective potential.

infinitely many choices for hyperbolic and elliptic orbits. For the parabola and hyperbola, there are no bound solutions (the kinetic energy is large). The bodies come in from the infinite, are then reflected by the effective potential, and disappear again to infinity. Such processes play an important role also in atomic and nuclear physics: For example, atomic nuclei scattered by other atomic nucei are moving on hyperbolic paths, the same holds for electrons scattered by atoms or nuclei. Electrons may also be bound (Bohr's model of atoms). These considerations, which were developed in the context of the gravitation law, may also be transferred to the Coulomb force law, as both types of forces have the same radial dependence and in both cases central forces are acting.

Path parameters, the third Kepler law, and the scattering problem

The semi-axes a and b of the elliptic orbit may be determined from the path equation (26.32). We have

$$\begin{aligned} a &= \frac{1}{2}\left[r(\theta=0)+r(\theta=\pi)\right] = \frac{1}{2}\left[\frac{k}{1+\varepsilon}+\frac{k}{1-\varepsilon}\right] = \frac{k}{1-\varepsilon^2} \\ &= \frac{mh^2/H}{-2Emh^2/H^2} = -\frac{H}{2E} = \frac{-\gamma Mm}{2E}, \end{aligned} \tag{26.39}$$

$$\begin{aligned} b &= \sqrt{a^2-c^2} = \sqrt{a^2-\varepsilon^2 a^2} = a\sqrt{1-\varepsilon^2} = \frac{k}{\sqrt{1-\varepsilon^2}} \\ &= \frac{mh^2/H}{\sqrt{-2Emh^2/H^2}} = \frac{mh}{\sqrt{-2Em}} = \sqrt{\frac{-m}{2E}}h. \end{aligned} \tag{26.40}$$

This allows us to calculate the period of revolution T, based on the constant area velocity $dA/dt = h/2$, equation (26.3), and equation (26.37):

$$T = \frac{\pi ab}{dA/dt} = \frac{\pi k^2}{(1-\varepsilon^2)^{3/2}h/2} = \frac{\pi(mh^2/H)^2}{(-2Emh^2/H^2)^{3/2}h/2},$$

such that

$$\begin{aligned}\frac{T^2}{a^3} &= \frac{\pi^2 b^2}{a\cdot(h/2)^2} = \frac{\pi^2 k^2/(1-\varepsilon^2)}{k/(1-\varepsilon^2)\cdot(h/2)^2}\\ &= \frac{4\pi^2 k}{h^2} = \frac{4\pi^2 mh^2}{h^2\cdot H}\\ &= \frac{4\pi^2 mh^2}{h^2\gamma Mm} = \frac{4\pi^2}{\gamma M}.\end{aligned}$$

Thus, T^2/a^3 depends only on the universal gravitational constant γ and the mass M of the central star. Therefore,

$$\frac{T^2}{a^3} = \text{constant} = \frac{4\pi^2}{\gamma M} \qquad \textbf{(26.41)}$$

for all planets. This is the third Kepler law. We note, however, that in the derivation of the Kepler laws recoil effects were neglected. Therefore, there result minor deviations of the order m/M. For the earth orbit, for example, such corrections are of the order $m/M \sim 1/3 \cdot 10^{-5}$. For the case of just two bodies interacting by gravitation, these recoil effects can be treated exactly with the help of the *reduced mass*; see Example 26.10.

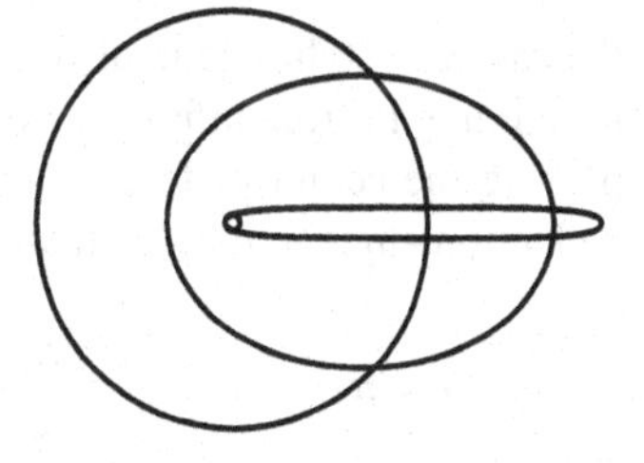

Orbits with the same energy have identical large half-axis.

According to (26.32) and (26.37) the constants of the elliptic orbits k, a, ε depend on the constants E (energy) and $h = L/m$ (angular momentum constant). In particular, according to (26.39) the major semi-axis of a planet of mass m depends only on the energy E, the quantity $k = mh^2/H = h^2/\gamma M$ only on the angular momentum constant.

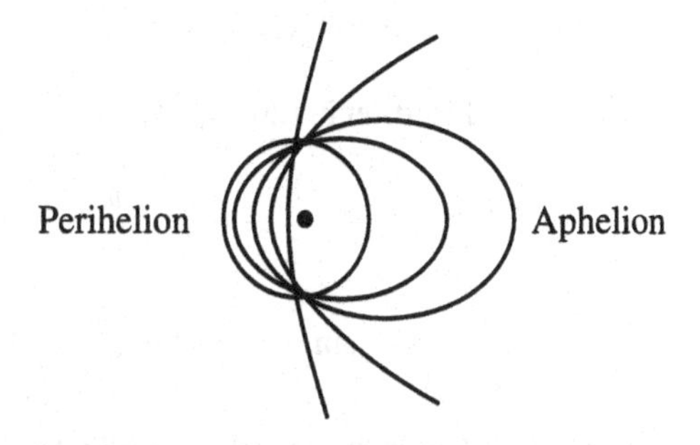

Orbits with the same area constant (constant angular momentum). All orbits with identical area constant h intersect at $r(\Theta = \pi/2) = k = h^2/(\gamma M)$.

The first statement is immediately evident from the discussion of planetary orbits in terms of the effective potential: For given V_{eff} (i.e., given angular momentum) both $r_{\max}$ and $r_{\min}$ depend only on E. If one initiates, for example, an elliptic motion by ejecting the mass m from a fixed position with a fixed initial velocity, then the direction of the initial velocity has no influence on the magnitude of the major semi-axis. The left-hand figure ahead shows elliptic trajectories of identical energy; the right-hand figure shows elliptic, parabolic, and hyperbolic trajectories of identical area constant (identical magnitude of angular momentum) h. Among the trajectories of equal energy the circular orbit has the maximum angular momentum constant; among the trajectories of equal area constant the circular orbit has the minimum energy.

Hyperbolic orbits—the scattering problem

We have already seen that for comets $E > 0$ may also occur. Because there are also other central force fields of the type

$$\mathbf{F} \sim \frac{1}{r^2}\frac{\mathbf{r}}{r}, \qquad \textbf{(26.42)}$$

for example, the electric forces between two charges q_1 and q_2

$$\mathbf{F}_{\text{el.}} = \frac{q_1 q_2}{r^2}\frac{\mathbf{r}}{r}, \qquad \textbf{(26.43)}$$

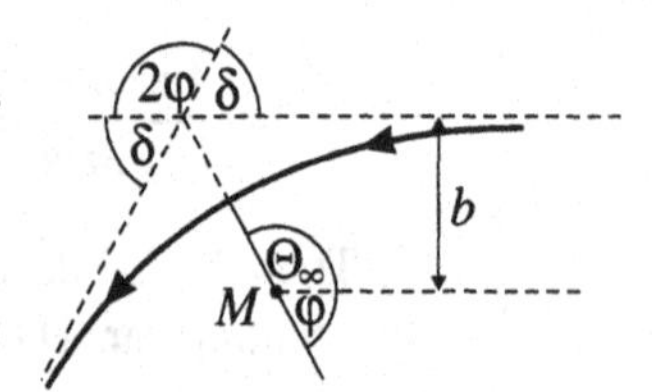

Typical orbit for the scattering of opposite electrical charges on each other or of two masses with gravitational interaction.

the case $E > 0$ has general significance. We ask for the deflection δ of a mass point (mass m) coming in from infinity with a velocity v_∞ and an *impact parameter b* ("distance") and passing the center of force (mass M) caused by the attractive force (see figure). For the deflection angle δ, it holds that

$$2\varphi + \delta = \pi \quad \Rightarrow \quad \frac{\delta}{2} = \frac{\pi}{2} - \varphi. \qquad \textbf{(26.44)}$$

The quantities φ and Θ_∞ are further related by

$$\varphi = \pi - \Theta_\infty \,. \qquad \textbf{(26.45)}$$

Insertion in (26.44) yields

$$\frac{\delta}{2} = \Theta_\infty - \frac{\pi}{2}$$

$$\Rightarrow \quad \sin\frac{\delta}{2} = \sin\left(\Theta_\infty - \frac{\pi}{2}\right) = -\cos\Theta_\infty. \qquad \textbf{(26.46)}$$

The radius in polar coordinates is given by

$$r(\Theta) = \frac{k}{1 + \varepsilon\cos\Theta}. \qquad \textbf{(26.47)}$$

For $r = \infty$, it follows from (26.47) that

$$1 + \varepsilon \cos \Theta_\infty = 0 \quad \Rightarrow \quad 1 - \varepsilon \sin \frac{\delta}{2} = 0 \quad \Rightarrow \quad \sin \frac{\delta}{2} = \frac{1}{\varepsilon}.$$

It then holds that

$$\tan\left(\frac{\delta}{2}\right) = \frac{\sin \delta/2}{\cos \delta/2} = \frac{\sin \delta/2}{\sqrt{1 - \sin^2 \delta/2}}$$
$$= \frac{1/\varepsilon}{\sqrt{1 - 1/\varepsilon^2}} = \frac{1}{\sqrt{\varepsilon^2 - 1}}. \tag{26.48}$$

ε is, however, given by

$$\varepsilon = \sqrt{1 + \frac{2Emh^2}{(\gamma M m)^2}}, \tag{26.49}$$

with the constants E (energy) and $h = L/m$ (constant of angular momentum):

$$E = \frac{1}{2} m v_\infty^2;$$
$$h = \frac{|\mathbf{L}|}{m} = \frac{m b v_\infty}{m} = b v_\infty. \tag{26.50}$$

Insertion of (26.49) and (26.50) into (26.48) then yields

$$\tan\left(\frac{\delta}{2}\right) = \frac{1}{\sqrt{(2Emh^2)/(\gamma^2 m^2 M^2)}} = \frac{\gamma M \sqrt{m}}{\sqrt{2\,\frac{1}{2} m v_\infty^2 b^2 v_\infty^2}} = \frac{\gamma M}{b\, v_\infty^2}. \tag{26.51}$$

For the deflection angle δ, we thus obtain

$$\delta = 2 \arctan\left(\frac{\gamma M}{b v_\infty^2}\right). \tag{26.52}$$

If v_∞ increases from 0 to ∞, $\delta/2$ decreases from $\pi/2$ to 0, or δ from π to 0, respectively. We still briefly consider the case of a repulsive force of the form (26.42). The calculation follows the same lines, only the coupling constant γ changes its sign, and the deflection angle is given by the same equation (26.51) but with $\gamma = -|\gamma|$.

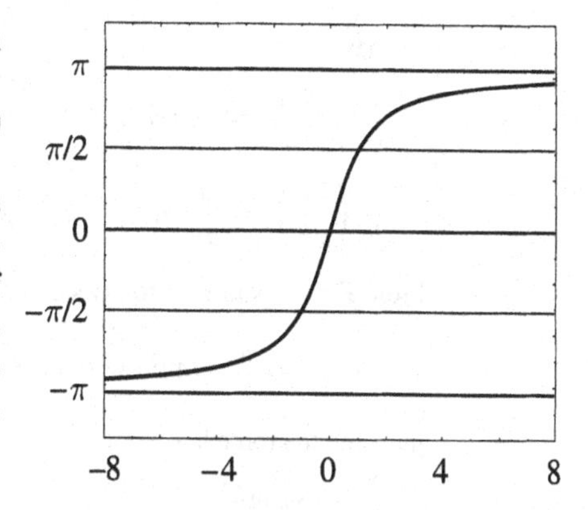

Scattering angle δ as a function of $x = \gamma M/(b\, v_\infty^2)$.

These scattering problems play an important role in particle physics. Also in the modern heavy-ion physics heavy nuclei may be interpreted as classical particles that are scattered by the central force field of another nucleus. This so-called Coulomb scattering gets important, both in Coulomb excitations of nuclei (the nuclei scatter by each other but don't get in touch—nevertheless the individual nuclei are excited by the electric (Coulomb) forces), as well as in peripheral nucleus–nucleus collisions (the nuclear forces hardly play a role in a grazing touch of nuclei—only very few nucleons may be excanged between nuclei). (see, e.g., J.M. Eisenberg and W. Greiner, *Nuclear Theory*, Vols. 1–3, North Holland, Amsterdam, 1985).

Problem 26.2: Force law of a circular path

A particle moves on a circular path through the origin under the action of a force pointing to the origin. Find the force law

$$\mathbf{F} = +f(r)\mathbf{e}_r. \tag{26.53}$$

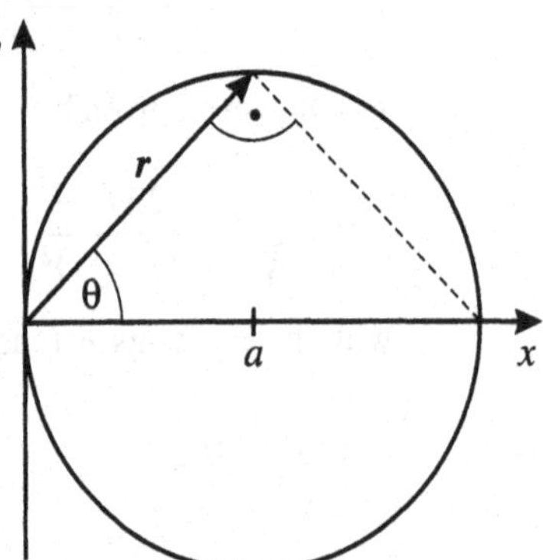

Illustration of the orbit and the coordinates used.

Solution 1. (using the energy law):
The path equation expressed in plane polar coordinates reads

$$r = 2a\cos\theta. \tag{26.54}$$

For central forces holds the energy equation (26.19):

$$E = \frac{mh^2}{2r^4}\left(\left(\frac{dr}{d\theta}\right)^2 + r^2\right) + V(r) = \text{constant},$$

$$h = \frac{L}{m}.$$

When differentiating the total energy with respect to r, we get

$$-2\frac{mh^2}{r^5}\left(\left(\frac{dr}{d\theta}\right)^2 + r^2\right) + \frac{mh^2}{2r^4}\left(\frac{d}{dr}\left(\frac{dr}{d\theta}\right)^2 + 2r\right) + \frac{dV}{dr} = 0. \tag{26.55}$$

We have

$$\frac{dr}{d\theta} = -2a\sin\theta,$$

$$\frac{d}{dr}\left(\frac{dr}{d\theta}\right)^2 = \frac{d}{dr}(4a^2 - r^2) = -2r. \tag{26.56}$$

From $\mathbf{F} = -\operatorname{grad} V$, it follows that $dV/dr = -f(r)$. Inserting (26.56) in (26.55) hence yields

$$f(r) = -\frac{2mh^2}{r^5}\left(4a^2\sin^2\theta + 4a^2\cos^2\theta\right) = -\frac{8a^2mh^2}{r^5}, \tag{26.57}$$

meaning the force law reads

$$\mathbf{F} = -\frac{8a^2mh^2}{r^5}\mathbf{e}_r. \tag{26.58}$$

Solution 2. (using equation (26.22)):
We may obtain the force $\mathbf{F}$ also by taking into account equation (26.22) because

$$f(r) = f\left(\frac{1}{u}\right) = -mu^2h^2\left(\frac{d^2u}{d\theta^2} + u\right), \qquad \text{where} \quad u = \frac{1}{2a\cos\theta}. \tag{26.59}$$

To get $f(r)$, we first have to form $d^2u/d\theta^2$:

$$\frac{du}{d\theta} = \frac{1}{2a} \cdot \frac{\sin\theta}{\cos^2\theta}, \tag{26.60}$$

$$\frac{d^2u}{d\theta^2} = \frac{1}{2a} \cdot \frac{\cos^3\theta + 2\sin^2\theta\cos\theta}{\cos^4\theta} = \frac{1}{2a}\left(\frac{1}{\cos\theta} + \frac{2\sin^2\theta}{\cos^3\theta}\right). \tag{26.61}$$

Inserting 26.61 in 26.59, we obtain

$$\begin{aligned} f(r) &= -mh^2 \frac{1}{4a^2\cos^2\theta}\left(\frac{2}{2a\cos\theta} + \frac{2\sin^2\theta}{2a\cos^3\theta}\right) \\ &= -mh^2 \cdot \frac{2}{8a^3\cos^3\theta}\left(1 + \frac{1-\cos^2\theta}{\cos^2\theta}\right) \\ &= -\frac{2mh^2 4a^2}{r^3} \cdot \frac{1}{4a^2\cos^2\theta} = \frac{-2mh^2 4a^2}{r^5} = \frac{-8mh^2a^2}{r^5}. \end{aligned}$$

From there it again follows that

$$\mathbf{F} = -\frac{8a^2mh^2}{r^5}\mathbf{e}_r.$$

Problem 26.3: Force law of a particle on a spiral orbit

A particle in a central force field with the center at the origin of the coordinate frame moves on a spiral path of the form $r = e^{-\theta}$. What is the force law?

Solution Central forces obey equation (26.22) with $\mathbf{F}(r) = f(r)\mathbf{e}_r$:

$$f\left(\frac{1}{u}\right) = -mh^2u^2\left(\frac{d^2u}{d\theta^2} + u\right),$$

with $u = 1/r$. Here $u = e^{\theta}$, $u = u''$. By insertion we find $f(1/u) = -2mh^2u^3$; hence

$$f(r) = \frac{-2mh^2}{r^3}.$$

Problem 26.4: The lemniscate orbit

Determine the force field that forces a particle to follow the lemniscate path $r^2 = 2\,a^2\cos(2\theta)$.

Solution For central forces again equation (26.22) holds:

$$f\left(\frac{1}{u}\right) = -mh^2u^2\left(\frac{d^2u}{d\theta^2} + u\right),$$

where $u = 1/r$. The path equation then implies

$$r = a\sqrt{2\cos 2\theta}, \qquad u = \frac{1}{a\sqrt{2\cos 2\theta}}, \qquad \frac{du}{d\theta} = \frac{\sin 2\theta}{a\sqrt{2}(\cos 2\theta)^{3/2}},$$

$$\frac{d^2u}{d\theta^2} = \frac{1}{a\sqrt{2}}\left(\frac{3\sin^2 2\theta}{(\cos 2\theta)^{5/2}} + \frac{2}{\sqrt{\cos 2\theta}}\right) = \frac{1}{a\sqrt{2}}\left(\frac{3}{(\cos 2\theta)^{5/2}} - \frac{1}{\sqrt{\cos 2\theta}}\right).$$

Insertion into the above equation yields

$$f\left(\frac{1}{u}\right) = -12mh^2a^4u^7$$

$$\Rightarrow \quad f(r) = -\frac{12mh^2a^4}{r^7}.$$

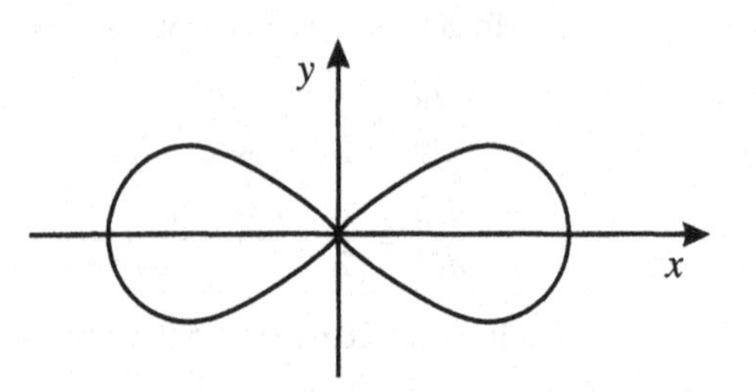

The lemniscate $r^2 = 2a^2 \cos 2\theta$. For $45° < \theta < 135°$, $\cos\theta$ is negative, and hence there is no "standing figure 8" part in the lemniscate.

This lemniscate path is illustrated in the figure.

For the sake of completeness we still remind ourselves of the definition of a lemniscate: The lemniscate is a particular *Cassini*[4] *curve* that is defined as the set of all points P of a plane for which the product of the distances $r_1 = |PF_1|$ and $r_2 = |PF_2|$ from two fixed points F_1 and F_2 have a constant value a^2 (see figure),

$$r_1 \cdot r_2 = a^2.$$

The distance of the two fixed points is $|F_1F_2| = 2e$. If $a = e$, the Cassini curve turns by definition into a lemniscate. Let F_1 and F_2 have the coordinates $(+e, 0)$ and $(-e, 0)$ in a Cartesian coordinate frame. Then $r_1^2 = (x-e)^2 + y^2$ and $r_2^2 = (x+e)^2 + y^2$ hold. From $r_1 \cdot r_2 = a^2$ we get after squaring the equation

$$(x^2+y^2) - 2e^2(x^2-y^2) = a^4 - e^4$$

of the Cassini curve. It is a fourth-order curve. In the special case of the lemniscate, $a = e$, hence

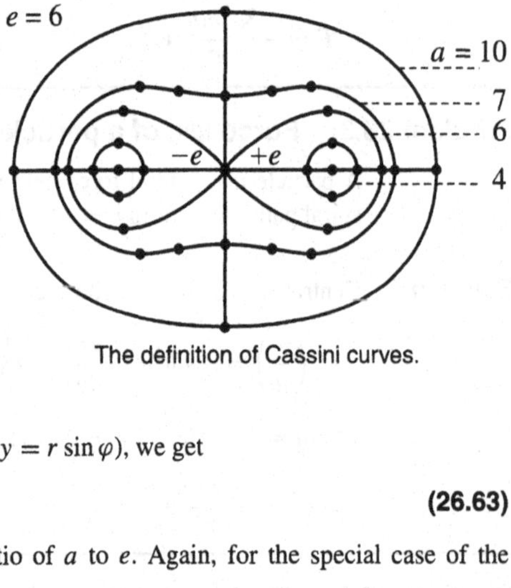

The definition of Cassini curves.

$$(x^2+y^2) - 2e^2(x^2-y^2) = 0. \qquad \textbf{(26.62)}$$

When changing to polar coordinates ($x = r\cos\varphi$, $y = r\sin\varphi$), we get

$$r^2 = e^2\cos 2\varphi \pm \sqrt{e^4\cos^2 2\varphi + (a^4 - e^4)}. \qquad \textbf{(26.63)}$$

The shape of a Cassini curve depends on the ratio of a to e. Again, for the special case of the lemniscate, $a = e$, we get

$$r^2 = 2\,e^2\cos 2\varphi\,. \qquad \textbf{(26.64)}$$

Problem 26.5: Escape velocity from earth

What must be the initial velocity of a projectile to leave the earth? The air friction is to be neglected!

[4] *Giovanni Domenico Cassini*, b. June 8, 1625, Parinaldo—d. Sept. 14(?), 1712, Paris. Cassini was professor of astronomy in Bologna, at the same time fortress architect and appointed to work on river regulation. From 1667 he was director of the observatory in Paris. He mostly published astronomic papers. The *Cassini curves* were supposed to replace the Kepler ellipses. However, they were published only in 1740 by his son, Jacques Cassini (1677–1756), in his book *Elements d'Astronomie*.

Solution The attractive force of earth is $F = -\gamma mM/r^2$. At the earth's surface $r = R$, there is $-F = mg = \gamma mM/R^2$, that is, $g = \gamma M/R^2$.

The equation of motion reads

$$m\ddot{r} = -\gamma \frac{mM}{r^2}.$$

With $\ddot{r} = dv/dt = (dv/dr)(dr/dt) = v(dv/dr)$, it follows that

$$\int_{v_0}^{v} v \cdot dv = -\int_{R}^{r} \frac{\gamma M}{r^2} dr;$$

hence

$$\frac{1}{2}(v^2 - v_0^2) = +\gamma M \left(\frac{1}{r} - \frac{1}{R} \right).$$

This is nothing else but the energy law, which we could have written down immediately:

$$\frac{1}{2}mv^2 - \gamma mM/r = \frac{1}{2}mv_0^2 - \frac{\gamma mM}{R}.$$

If the missile shall leave earth, that means $r \to \infty$. The minimum initial velocity results if the velocity of the missile arriving at $r = \infty$ just became equal to zero—$v(r \to \infty) = 0$—and therefore

$$v_0^2 = \frac{2\gamma M}{R} = 2gR, \qquad v_0 \approx 11\frac{\text{km}}{\text{s}}.$$

This is called the *escape velocity* that a body (independent of its mass) must have to leave the earth's gravitational field.

Problem 26.6: The rocket drive

A rocket of initial mass m_0 per unit time expels the quantity of gas $\alpha = \Delta m/\Delta t > 0$ with the constant velocity v_0. We look for the equation of motion. The gravitational force shall be assumed as constant. That means that the rocket problem shall be considered only near the earth's surface.

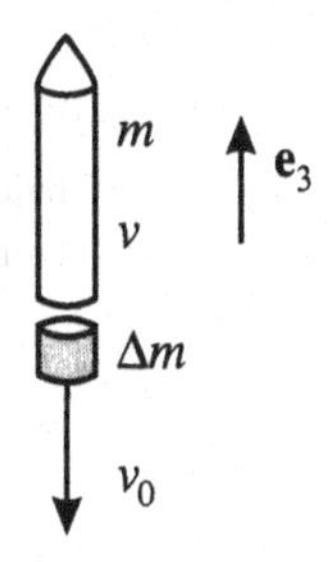

The rocket problem.

Solution The rocket of mass $m(t)$ moves upward with velocity $v(t)$. Thereby the mass Δm is expelled downward with the constant velocity v_0 (relative to the rocket).

To describe the motion of the rocket, we must adopt the Newtonian force law in its original form

$$\frac{d\mathbf{p}}{dt} = \mathbf{F}$$

because the mass of the rocket is variable. It therefore holds that

$$\frac{d\mathbf{p}}{dt} = m\frac{d\mathbf{v}}{dt} + \mathbf{v}\frac{dm}{dt},$$

where

$$\mathbf{v} = v\mathbf{e}_3$$

means the vertical velocity. Within the time interval Δt, the expelled gases carry off the momentum

$$\Delta \mathbf{p}' = \Delta m(v - v_0)\mathbf{e}_3 = \alpha(v - v_0)\Delta t\, \mathbf{e}_3\,.$$

This implies a force on the rocket (recoil force) of the magnitude

$$\mathbf{F}' = -\frac{\Delta \mathbf{p}'}{\Delta t} = -\alpha(v - v_0)\mathbf{e}_3.$$

In addition, there acts the gravity force $-mg\mathbf{e}_3$. Hence, the Newtonian force law reads

$$m\frac{dv}{dt} + v\frac{dm}{dt} = -\alpha(v - v_0) - mg = -mg - \frac{dm}{dt}(v_0 - v).$$

This balance holds in the inertial system that is tightly fixed to earth. With $m = m_0 - \alpha t$ and $dm/dt = -\alpha$, it follows that

$$m\frac{dv}{dt}\mathbf{e}_3 = +\alpha v_0\mathbf{e}_3 - mg\mathbf{e}_3.$$

The term αv_0 on the right side represents the recoil force. We further conclude that

$$\int_0^v dv = \int_0^t \left(\frac{\alpha v_0}{m_0 - \alpha t} - g\right) dt,$$

$$\begin{aligned} v(t) &= -gt + v_0 \cdot \int_0^t \frac{\alpha/m_0\, dt}{1 - (\alpha/m_0)\, t} \\ &= -gt - v_0 \left[\ln\left(1 - \frac{\alpha}{m_0}t\right)\right]_0^t \\ &= -gt - v_0 \ln\left(1 - \frac{\alpha}{m_0}t\right). \end{aligned}$$

Obviously, the rocket velocity depends linearly on the exit velocity v_0 of the recoil gases. A further integration yields the altitude $h(t)$ the rocket reached:

$$h = \int_0^h v\, dt = -\frac{1}{2}gt^2 - v_0 \int_0^t \ln\left(1 - \frac{\alpha}{m_0}t\right) dt.$$

With the substitution $u = 1 - (\alpha/m_0)\, t$, $du = -(\alpha/m_0)\, dt$, we get

$$\begin{aligned} \frac{v_0 m_0}{\alpha}\int_{t=0}^t \ln u\, du &= \frac{v_0 m_0}{\alpha}\Big[u \ln u - u\Big]_{t=0}^t \\ &= \frac{v_0}{\alpha}m_0\left[\left(1 - \frac{\alpha}{m_0}t\right)\ln\left(1 - \frac{\alpha}{m_0}t\right) - \left(1 - \frac{\alpha}{m_0}t\right)\right]_{t=0}^t \\ &= \frac{v_0 m_0}{\alpha}\left[\left(1 - \frac{\alpha}{m_0}t\right)\ln\left(1 - \frac{\alpha}{m_0}t\right) + \frac{\alpha}{m_0}t\right]. \end{aligned}$$

Hence, the altitude of the rocket after time t is

$$h = -\frac{1}{2}gt^2 + \frac{v_0 m_0}{\alpha}\left(1 - \frac{\alpha}{m_0}t\right)\ln\left(1 - \frac{\alpha}{m_0}t\right) + v_0 t.$$

To determine the moment T of the burnout, we introduce the mass of the casing m_1. We then have $m_0 = m_1 + \alpha T$, where αT is the mass of fuel.

$$T = \frac{m_0 - m_1}{\alpha}.$$

At the moment of burnout, the rocket has the velocity

$$v_1 = v(T) = -g\frac{m_0 - m_1}{\alpha} - v_0 \ln\frac{m_1}{m_0} = -g\frac{m_0 - m_1}{\alpha} + v_0 \ln\frac{m_0}{m_1}$$

and the altitude

$$h_1 = h(T) = -\frac{1}{2}g\left(\frac{m_0 - m_1}{\alpha}\right)^2 + v_0\left[\frac{m_0 - m_1}{\alpha} + \frac{m_1}{\alpha}\ln\frac{m_1}{m_0}\right].$$

The final velocity depends linearly on the exit velocity v_0 of the recoil gases and is proportional to the logarithm of the ratio of initial to final mass. The further motion of the rocket follows according to the energy law:

$$\frac{1}{2}m \cdot v_1^2 = m \cdot g \cdot h_2 .$$

The additional altitude h_2 the rocket reaches after burnout is calculated as

$$h_2 = \frac{v_1^2}{2g}.$$

The total ascension altitude of the rocket is therefore

$$h = h_1 + h_2 = h_1 + \frac{v_1^2}{2g},$$

$$\begin{aligned} h &= \frac{1}{2}g\left(\frac{m_0 - m_1}{\alpha}\right)^2 - \frac{1}{2}g\left(\frac{m_0 - m_1}{\alpha}\right)^2 + v_0\left(\frac{m_0 - m_1}{\alpha}\right)\ln\frac{m_1}{m_0} \\ &\quad + \frac{v_0^2}{2g}\ln^2\frac{m_1}{m_0} + v_0\frac{m_0 - m_1}{\alpha} + v_0\frac{m_1}{\alpha}\ln\frac{m_1}{m_0}, \\ &= \left(\ln\frac{m_1}{m_0} + 1\right)v_0\frac{m_0 - m_1}{\alpha} + v_0\ln\frac{m_1}{m_0}\left(\frac{v_0}{2g}\ln\frac{m_1}{m_0} + \frac{m_1}{\alpha}\right). \end{aligned}$$

Problem 26.7: A two-stage rocket

Establish the equation of motion of a two-stage rocket in the homogeneous gravitational field of earth.

Solution Let T be the burnout time of the first stage. For $t \le T$, the quantities $s(t)$ and $v(t)$ are obtained as in Problem 26.6. The mass at start is

$$m_0 = m_1 + \alpha T + \alpha T' + m_2;$$

m_1: casing of first stage, αT fuel of first stage,

m_2: casing of second stage, $\alpha T'$ fuel of second stage.

For $t = T$, the mass is $m'_0 = m_2 + \alpha T'$. For $t > T$ ($h'(t)$ and $v'(t)$ have the analogous meaning as in 26.6) we have

$$h(t) = h'(t - T) + h(T) + v(T)(t - T),$$
$$v(t) = v'(t - T) + v(T).$$

Problem 26.8: Condensation of a water droplet

A dust particle of negligible mass being under the influence of gravitation begins to fall at time $t = 0$ through saturated water vapor. The steam thereby condensates with constant rate λ [gram per centimeter] on the dust particle and forms a water droplet of steadily increasing mass.

(a) Calculate the acceleration of the droplet as a function of its velocity and the traversed path.

(b) Determine the equation of motion of the droplet by integrating the expression for the acceleration. Neglect friction, collisions, etc.

Solution (a) The only external force acting on the droplet is the gravitational force

$$F_g = mg.$$

According to Newton's law,

$$mg = \frac{dm}{dt}v + m\frac{dv}{dt} \qquad \text{and} \qquad \frac{dm}{dt} = \frac{dm}{dx}\frac{dx}{dt} = \lambda v, \tag{26.65}$$

because the increase of mass $dm/dx = \lambda$ shall be a constant. According to equation 26.65, the acceleration is

$$a = \frac{dv}{dt} = \frac{mg - \lambda v^2}{m}, \tag{26.66}$$

and, because the mass of the dust particle at time $t = 0$ and position $x = 0$ is assumed to be negligible, we have $m = \lambda x$, and for equation 26.66 we get the acceleration

$$a = g - \frac{v^2}{x}. \tag{26.67}$$

(b) The equation of motion for the dust particle shall now be determined by integration of equation 26.67. From 26.67 it follows for $x \neq 0$ that

$$x \cdot \frac{d^2x}{dt^2} + \left(\frac{dx}{dt}\right)^2 - gx = 0. \tag{26.68}$$

To solve this nonlinear differential equation, we try the *ansatz*

$$x = At^n$$

and substitute it in equation 26.68. We obtain

$$(At^n)n(n-1)At^{n-2} + (Ant^{n-1})^2 - gAt^n = 0,$$
$$A^2n(n-1)t^{2n-2} + A^2n^2t^{2n-2} - gAt^n = 0. \tag{26.69}$$

Equation 26.69 is fulfilled for $n = 2$, that is, if the powers of t are equal. By insertion equation 26.69 yields

$$A(2n^2 - n) = g \qquad \text{or} \qquad A = g/6$$

and therefore as solution for x

$$x = \frac{g}{6}t^2. \tag{26.70}$$

The proposed *ansatz* may be verified by inserting this solution into equation 26.68. Differentiation of equation 26.70 yields

$$v = (g/3)t \qquad \text{and} \qquad a = g/3,$$

that is, the acceleration of the droplet is constant and, independent of x, equal to $g/3$.

Problem 26.9: Motion of a truck with variable load

An empty truck of mass M_0 moves frictionless with velocity V_0 on a stretch of track. At the position $x = 0$ at time $t = 0$, the truck is loaded with sand with the load rate λ kg/s (see figure). Determine the position of the truck as a function of time.

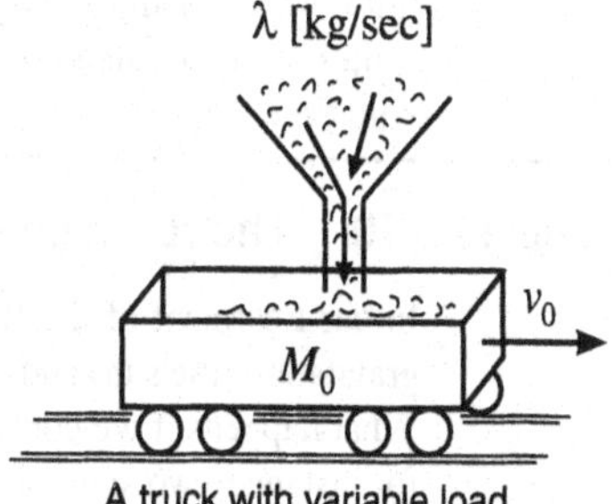

A truck with variable load.

Solution As no external forces are acting on the truck, the change of momentum is

$$\frac{d}{dt}(mv) = 0 \qquad \text{or} \qquad mv = \text{constant}, \tag{26.71}$$

although both m and v are functions of the time. With the initial conditions at the time $t = 0$ ($m = M_0$ and $v = V_0$), equation 26.71 becomes

$$mv = M_0 V_0. \tag{26.72}$$

Because the truck is being loaded with constant rate, the mass change $dm/dt = \lambda$ is a constant, and we have

$$m = M_0 + \lambda t.$$

Insertion into equation 26.72 yields for the velocity

$$v = \frac{M_0 V_0}{M_0 + \lambda t}. \tag{26.73}$$

With $v = dx/dt$, equation 26.73 after separation of the variables yields

$$dx = M_0 V_0 \frac{dt}{M_0 + \lambda t} \tag{26.74}$$

$$= \frac{M_0 V_0}{\lambda} \frac{d(M_0 + \lambda t)}{M_0 + \lambda t} = \frac{M_0 V_0}{\lambda}\left(\frac{dk}{k}\right), \tag{26.75}$$

where $k = M_0 + \lambda t$, and $dt = dm/\lambda = d(M_0 + \lambda t)/\lambda$. Integration of equation 26.74 leads to

$$x = \frac{M_0 V_0}{\lambda} \ln k + c \tag{26.76}$$

$$= \frac{M_0 V_0}{\lambda} \ln(M_0 + \lambda t) + c. \tag{26.77}$$

From the initial conditions $x = 0$ at time $t = 0$, the constant c is evaluated as

$$c = -\frac{M_0 V_0}{\lambda} \ln(M_0), \tag{26.78}$$

and thus equation 26.76 becomes

$$x = \frac{M_0 V_0}{\lambda} \ln\left(\frac{M_0 + \lambda t}{M_0}\right). \tag{26.79}$$

Equation 26.79 is plotted in the opposite figure as a function of the dimensionless quantities $t\lambda/M_0$ and $x\lambda/M_0 V_0$. The coordinate x thus increases steadily but logarithmically with time.

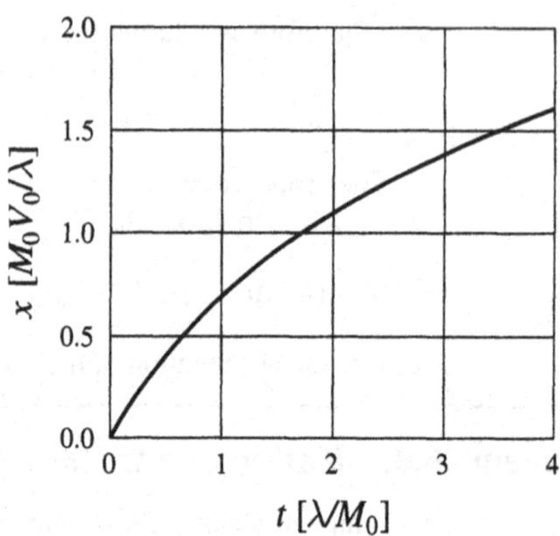

The position of the truck as a function of time.

Example 26.10: The reduced mass

In our treatment of the Kepler problem, we had considered up to now one fixed, massive center of gravitation (the sun) and small bodies (the planets) orbiting in the field of the immobile central mass. What happens if we consider two bodies of comparable mass that are bound together by gravitation, for instance, two stars in a double-star system?

Solution In a system that consists of two interacting masses m_1 and m_2 and that is not influenced by exterior forces, the force $\mathbf{F}$ between the masses can depend only on the distance vector $\mathbf{r} = \mathbf{r}_1 - \mathbf{r}_2$ and possibly on time derivatives of $\mathbf{r}$. This is due to the assumption of homogeneity of space. The equations of motions for the two masses thus read

$$m_1\ddot{\mathbf{r}}_1 = \mathbf{F}(\mathbf{r}, \dot{\mathbf{r}}, t), \tag{26.80}$$

$$m_2\ddot{\mathbf{r}}_2 = -\mathbf{F}(\mathbf{r}, \dot{\mathbf{r}}, t). \tag{26.81}$$

According to Newton's third law meaning "action equals minus reaction," the force on m_2 is opposite equal to the force on m_1. One defines the position vector of the center of mass of the two-body system by

$$\mathbf{R} = \frac{m_1\mathbf{r}_1 + m_2\mathbf{r}_2}{m_1 + m_2}. \tag{26.82}$$

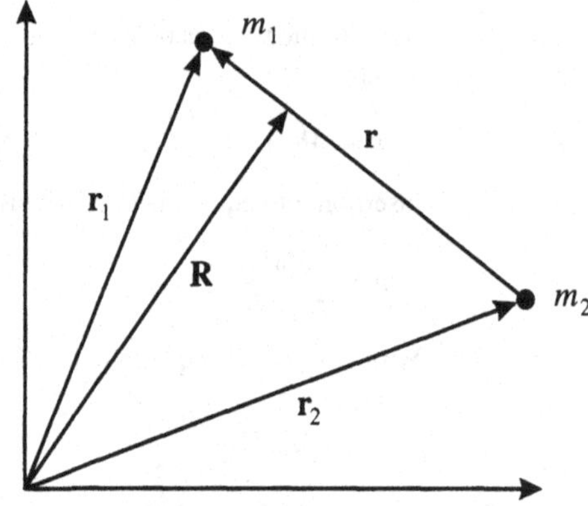

The definition of the center of mass.

With this definition, we can decompose the motion of the two-body system into the motions of the center of mass $\mathbf{R}$ and of the distance vector $\mathbf{r}$.

Adding equations 26.80 and 26.81 shows that the second time derivative of $\mathbf{R}$ vanishes, that is, the center of mass moves uniformly along a straight line through space:

$$m_1\ddot{\mathbf{r}}_1 + m_2\ddot{\mathbf{r}}_2 = 0, \tag{26.83}$$

$$\ddot{\mathbf{R}} = \mathbf{0}. \tag{26.84}$$

If one multiplies equation 26.80 by $m_2/(m_1+m_2)$ and subtracts equation 26.81, multiplied by $m_1/(m_1+m_2)$, one obtains the important relation

$$\frac{m_1 m_2}{m_1+m_2}\ddot{\mathbf{r}} = \mathbf{F}(\mathbf{r}, \dot{\mathbf{r}}, t)\,. \tag{26.85}$$

Thus a clear, physical picture emerges for the motion of the two-body system: The motion of the center of mass is, according to equation 26.82, uniform along a straight line and completely independent of the relative motion of the two masses, which is described by equation 26.85. According to 26.85, the relative motion of the two masses m_1 and m_2 (on which no extarnal forces are acting) corresponds to the motion of one single mass, called the *reduced mass* μ,

$$\mu = \frac{m_1 m_2}{m_1+m_2}\,, \tag{26.86}$$

in a force field described by $\mathbf{F}(\mathbf{r}, \dot{\mathbf{r}}, t)$. Thus, the two-body problem has been reduced to an effective one-body problem for the reduced mass μ. Upon computation of the relative motion $\mathbf{r}(t)$ by integrating equation 26.85, the position vectors of the masses m_1 and m_2 of the two-body system can be obtained as

$$\mathbf{r}_1(t) = \mathbf{R}(t) + \frac{m_1}{m_1+m_2}\mathbf{r}(t), \tag{26.87}$$

$$\mathbf{r}_2(t) = \mathbf{R}(t) - \frac{m_2}{m_1+m_2}\mathbf{r}(t). \tag{26.88}$$

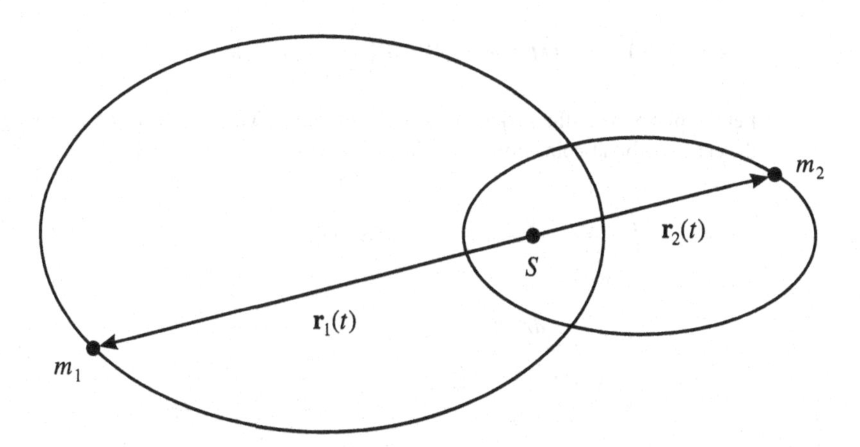

The motion of two masses m_1 and m_2 under the influence of their mutual gravitational interaction. Both masses move along elliptic orbits whose focal points lie in the center of mass of the two bodies.

Problem 26.11: Path of a comet

A comet moves on a parabolic path in the gravitational field of the sun being at rest. Its orbital plane coincides with the orbital plane (idealized as a circle) of earth. The perihelion distance is a third of the radius of the earth's orbit ($R_E = 1.49 \cdot 10^{11}$ m). What is the time of flight of the comet within the earth's orbit (a perturbation of the comet path by the planets shall be neglected)?

Solution The comet moves on a parabola, namely $E = 0$, $\varepsilon = 1$, with the path equation

$$r = \frac{k}{1+\cos\theta}, \qquad k = \frac{L^2}{\gamma M m \mu}.$$

We have

L: angular momentum of

reduced mass $\mu = \dfrac{Mm}{M+m}$,

M: sun mass,

m: comet mass,

γ: gravitational constant.

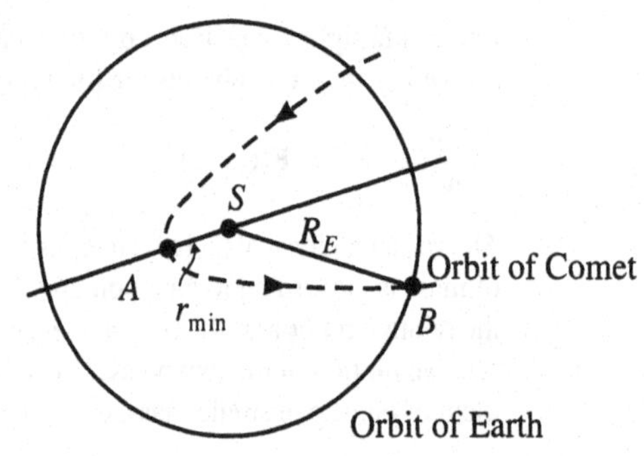

The orbit of a comet crossing the orbit of the earth at B.

According to (26.13), the energy law reads

$$\frac{\mu}{2}\dot{r}^2 + \frac{L^2}{2\mu r^2} + V(r) = E = 0.$$

We evaluate this relation at the point A where the term $(\mu/2)\dot{r}^2$ vanishes because there is no radial kinetic energy (the orbit is symmetric with respect to the straight line through A and S):

$$\frac{L^2}{2\mu(r_{\min})^2} = -V(r_{\min}) = \gamma\frac{Mm}{r_{\min}} \tag{26.89}$$

$$\Rightarrow \quad \left(\frac{L}{\mu}\right) = (2\gamma(M+m)r_{\min})^{1/2} = \left(\frac{2}{3}\gamma(M+m)R_E\right)^{1/2}. \tag{26.90}$$

Let the planet be at the perihelion A at the instant t_0. The flight time until leaving the radius R_E of the earth's orbit at point B is, according to (26.16),

$$t - t_0 = \int\limits_{r_{\min}}^{R_E} \frac{dr}{\sqrt{\frac{2}{\mu}\left(-V(r) - \frac{L^2}{2\mu r^2}\right)}} \qquad (E=0) \tag{26.91}$$

$$= \int\limits_{R_E/3}^{R_E} \frac{dr\cdot r}{\sqrt{\frac{2}{\mu}\left(\gamma M m r - \frac{L^2}{2\mu}\right)}}$$

$$= \sqrt{\frac{\mu}{2\gamma M\cdot m}} \int\limits_{R_E/3}^{R_E} \frac{dr\cdot r}{\sqrt{r - \frac{L^2}{2\mu\gamma M m}}}. \tag{26.92}$$

However, according to 26.89,

$$\frac{L^2}{2\mu\gamma Mm} = \left(\frac{2}{3}\gamma(M+m)R_E\right)\frac{1}{2}\mu\frac{1}{\gamma Mm}$$

$$= \frac{1}{3}R_E = r_{\min},$$

and therefore,

$$t - t_0 = \sqrt{\frac{\mu}{2\gamma M \cdot m}} \int\limits_{R_E/3}^{R_E} \frac{dr \cdot r}{\sqrt{r - R_E/3}}\,; \tag{26.93}$$

with the substitution $x := \sqrt{r - R_E/3}$, $dx = (1/2x)\,dr$, 26.93 may be solved:

$$\begin{aligned} t - t_0 &= \sqrt{\frac{\mu}{2\gamma M \cdot m}} \int\limits_0^{(\frac{2}{3}R_E)^{1/2}} \frac{2x(x^2 + R_E/3)\,dx}{x} \\ &= \sqrt{\frac{\mu}{2\gamma M \cdot m}} \left[\frac{2}{3}x^3 + \frac{2}{3}R_E x\right]_0^{(\frac{2}{3}R_E)^{1/2}} \\ &= \frac{10}{9}\sqrt{\frac{2}{3}}R_E^{3/2}\sqrt{\frac{\mu}{2\gamma M \cdot m}} \\ &= \frac{10}{9}R_E^{3/2}\,(3\gamma(M + m))^{-1/2}\,. \end{aligned}$$

For the total residence time of the comet within the earth's orbit ($T_{\mathrm{tot}} = 2(t - t_0)$), we then obtain

$$T_{\mathrm{tot}} = \frac{20}{9}R_E^{3/2}\,(3\gamma(M + m))^{-1/2}\,. \tag{26.94}$$

For an "ordinary" comet, the mass m may be neglected against the sun's mass M ($M \approx 330\,000$ earth masses). Insertion of the values $R_E = 1.49 \cdot 10^{11}$ m, $\gamma M = 1.32 \cdot 10^{20}\ \mathrm{m^3 s^{-2}}$ yields

$$T_{\mathrm{tot}} = 74.34 \text{ days}.$$

Problem 26.12: Motion in the central field

A mass m moves in the central force field with the potential

$$U(r) = \frac{-\alpha}{r} \qquad (\alpha > 0).$$

(a) Show that any orbit of finite motion (hence not at infinity) is closed. What happens if an additional term of the form β/r^3 is added to $U(r)$?

(b) Show that the vector (Lenz vector)

$$\mathbf{V} = \frac{1}{m\alpha}\,[\mathbf{L} \times \mathbf{p}] + \frac{\mathbf{r}}{r}$$

is a conserved quantity. How may it be interpreted?

Solution (a) Because we are dealing with a central force, we may use the expression (26.17) for the variation of the angle as a function of the radius:

$$\theta = \int \frac{L\,dr}{r^2\sqrt{2m(E - U(r)) - L^2/r^2}}\,.$$

Here L is the angular momentum. A revolution of the mass is characterized by the fact that the radius varies, for example, from r_{max} over r_{min} back to r_{max}. The corresponding variation of the angle is then

$$\Delta\theta = 2\int_{r_{\text{min}}}^{r_{\text{max}}} \frac{L\,dr}{r^2\sqrt{2m(E-U(r))-L^2/r^2}}. \tag{26.95}$$

We now insert $U(r) = -\alpha/r$ and obtain

$$\begin{aligned}
\Delta\theta &= 2\int_{r_{\text{min}}}^{r_{\text{max}}} \frac{L\,dr}{r^2\sqrt{2mE+2m\alpha/r-L^2/r^2}} \\
&= 2\int_{r_{\text{min}}}^{r_{\text{max}}} \frac{L\,dr}{r^2\sqrt{-(L/r-m\alpha/L)^2+m^2\alpha^2/L^2+2mE}} \\
&= 2\int_{r_{\text{min}}}^{r_{\text{max}}} \frac{L\,dr}{Cr^2\sqrt{1-(L/r-m\alpha/L)^2/C^2}},
\end{aligned}$$

where we set $(m^2\alpha^2)/L^2+2mE = C^2$. This may be integrated immediately $\left((\arccos x)' = -1/\sqrt{1-x^2}\right)$:

$$\Delta\theta = 2\arccos\frac{(L/r-m\alpha/L)}{C}\bigg|_{r_{\text{min}}}^{r_{\text{max}}}. \tag{26.96}$$

Because the motion shall be finite, the mass moves on a Kepler ellipse

$$r = \frac{k}{1+\varepsilon\cos\theta}, \qquad \text{with} \quad k = \frac{L^2}{m\alpha}.$$

r_{min} and r_{max} are then uniquely fixed (for given total energy):

$$r_{\text{min}} = \frac{k}{1+\varepsilon}, \qquad r_{\text{max}} = \frac{k}{1-\varepsilon}, \qquad 0 \le \varepsilon < 1, \quad \varepsilon = \sqrt{1+\frac{2EL^2}{m\alpha^2}}.$$

Insertion yields

$$\Delta\theta = 2\arccos\left(-\frac{m\alpha\varepsilon}{L\cdot C}\right) - 2\arccos\left(\frac{m\alpha\varepsilon}{L\cdot C}\right).$$

Because

$$C = \sqrt{2mE+\frac{m^2\alpha^2}{L^2}} = \frac{m\alpha}{L}\sqrt{1+\frac{2EL^2}{m\alpha^2}} = \frac{m\alpha}{L}\varepsilon,$$

we obtain

$$\Delta\theta = 2\Big(\Big[\pi+n\cdot 2\pi\Big]-n\cdot 2\pi\Big) = 2\pi.$$

But this just means that the path is closed.

If one adds to the potential $U(r)$ an additional term in the form of a small perturbation β/r^3, then $\Delta\theta \neq 2\pi$ or $\Delta\theta = 2\pi+\delta\theta$; one then observes rosette orbits (perihelion motion).

(b) We show that $\dot{\mathbf{V}} = 0$. We have

$$\mathbf{V} = \frac{1}{m\alpha}\left[\mathbf{L}\times\mathbf{p}\right]+\frac{\mathbf{r}}{r},$$

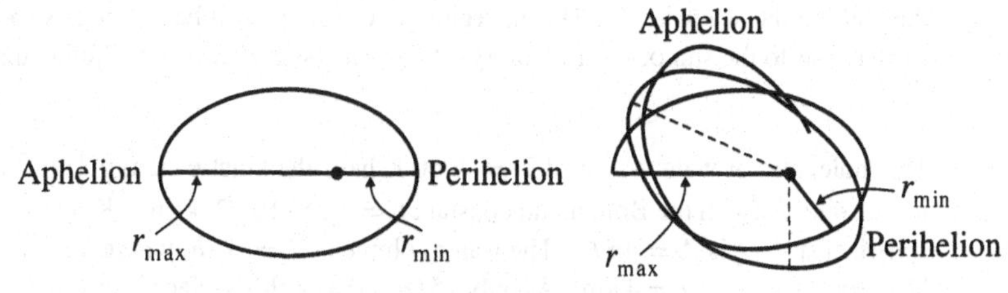

Kepler ellipse (left) and rosette orbit.

and therefore

$$\dot{\mathbf{V}} = \frac{1}{m\alpha} \underbrace{\dot{\mathbf{L}}}_{\substack{=0,\text{ since}\\ \mathbf{L}=\text{constant}}} \times \mathbf{p} + \frac{1}{m\alpha}\mathbf{L} \times \dot{\mathbf{p}} + \frac{1}{r}\dot{\mathbf{r}} - \frac{\dot{r}}{r^2}\mathbf{r}.$$

Using $\mathbf{L} = m(\mathbf{r} \times \dot{\mathbf{r}})$ and $\dot{\mathbf{p}} = m\dot{\mathbf{v}} = -\alpha\mathbf{r}/r^3$, we get

$$\begin{aligned}\dot{\mathbf{V}} &= \frac{1}{\alpha}(\mathbf{r} \times \dot{\mathbf{r}}) \times \left(-\frac{\alpha}{r^3}\mathbf{r}\right) + \frac{1}{r}\dot{\mathbf{r}} - \frac{\dot{r}}{r^2}\mathbf{r} \\ &= \frac{1}{r^3}\left[r^2\dot{\mathbf{r}} - (\dot{\mathbf{r}} \cdot \mathbf{r})\,\mathbf{r} - (\mathbf{r} \times \dot{\mathbf{r}}) \times \mathbf{r}\right] \\ &= \frac{1}{r^3}\left[r^2\dot{\mathbf{r}} - (\dot{\mathbf{r}} \cdot \mathbf{r})\mathbf{r} + (\mathbf{r} \cdot \dot{\mathbf{r}})\mathbf{r} - (\mathbf{r} \cdot \mathbf{r})\,\dot{\mathbf{r}}\right] = 0.\end{aligned}$$

Here we made use of the relation $(\mathbf{a} \times [\mathbf{b} \times \mathbf{c}]) = (\mathbf{a} \cdot \mathbf{c})\mathbf{b} - (\mathbf{a} \cdot \mathbf{b})\mathbf{c}$. Hence, it follows that

$$\mathbf{V} = \text{constant} \tag{26.97}$$

Both $[\mathbf{L} \times \mathbf{p}]$ as well as $\mathbf{r}$ are within the orbital plane; we now calculate the angle ϑ enclosed by the Lenz vector and the radius vector:

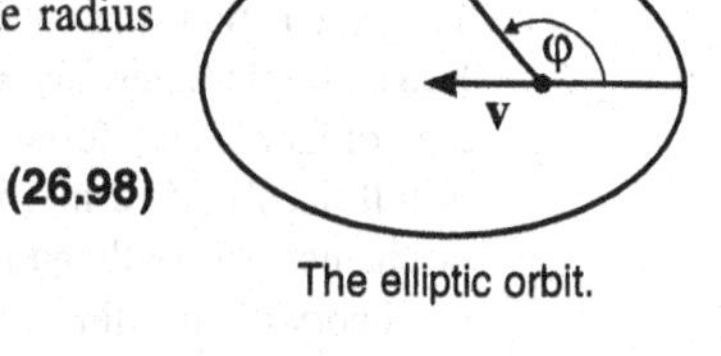

The elliptic orbit.

$$\mathbf{V} \cdot \mathbf{r} = V \cdot r\cos\vartheta = \frac{1}{m\alpha}[\mathbf{L} \times \mathbf{p}\,]\,\mathbf{r} + \frac{\mathbf{r} \cdot \mathbf{r}}{r} \tag{26.98}$$

$$= -\frac{1}{m\alpha}\mathbf{L} \cdot [\mathbf{r} \times \mathbf{p}\,] + r = r - \frac{1}{m\alpha}L^2$$

$$\Rightarrow \quad r = \frac{L^2/m\alpha}{1 - V\cos\vartheta}. \tag{26.99}$$

This is exactly the path curve of a conical section if $V = |\mathbf{V}|$ is identified with the eccentricity ε and ϑ with $(\varphi + \pi)$: For an ellipse the Lenz vector points from the focal point to the center, and its magnitude equals the eccentricity of the orbit. The figure illustrates this result.

Problem 26.13: Sea water as rocket drive

(For pre-Christmas entertainment)

In the near future the sun will cool down so much that no life will be possible any more on earth. A desperate physicist proposes to drill a hole until reach the hot earth's interior ($T = 4000$ K) and then letting the sea water in. The emerging jet of steam shall be utilized as rocket drive to move the earth closer to the sun or—if necessary—to another star. How do you judge that proposal?

Solution The molecules of water vapor of $T = 4000$ K have the kinetic energy $\frac{1}{2}mw^2 = \frac{3}{2}kT \Rightarrow$ velocity of $w \approx 2.4$ km/s (with the Boltzmann constant $k = 1.38 \cdot 10^{-23}$ kg m^2/K s^2, $m = 2.99 \cdot 10^{-26}$ kg). The spherical surface is $A = 4\pi R^2$. The water volume is $V = A \cdot h$, where h is the thickness of the water layer (ocean layer) ($h = 4$ km). As only 75 % of the earth's surface is covered by water, it follows that $V = 0.75 \cdot 4\pi R^2 \cdot h \approx 1.5 \cdot 10^9\ \text{km}^3 \mathrel{\widehat{=}} 1,5 \cdot 10^{21}$ kg $= m_\omega$; the mass of earth is $M = 5.6 \cdot 10^{24}$ kg, i.e., $m_w/M \approx 1/4000$. The velocity v of earth caused by expelling the water vapor is given by the momentum conservation as

$$vM = wm_w \quad \rightarrow \quad v = \frac{m_w}{M} \cdot w \approx \frac{1}{4000} w \approx 0.6\,\frac{\text{m}}{\text{s}}.$$

As compared to the orbital velocity of earth of 30 km/s, the recoil velocity of 0.6 m/s is negligible. Thus, the attempt would be useless—except for a depopulation of earth by an induced super-disaster (no life without water). But the attempt is bound to fail in any case, because the molecules of water vapor cannot leave earth at all: $w \approx 2.4$ km/s are significantly below the escape velocity $v_\infty \approx 11$ km/s. The vapor thus cannot be expelled at all by earth. The proposed rocket drive does not work, as earth and the water vapor form a closed system.

Example 26.14: Historical remark

One might wonder whether Kepler was denied discovering the force law (gravitational law—see equation (26.10)). After all, it seemingly follows "so easily" from his own laws. Of course, we cannot and will not accuse Kepler of any lack of brilliance and fantasy. He clearly was a master in empirical research and demonstrated fantasy in far-reaching speculations, sometimes even imaginations: for example, in his thoughts on the possible number of planets: Like the Pythagoreans, he, too, was convinced that God had created the world in number and size according to a definite law of numbers. The explanation is as follows: Kepler was a contemporary of Galileo, who survived him by 12 years. Hence, Kepler knew about Galileo's mechanics, in particular the central concept of acceleration, the laws of inertia and throw, by correspondence and hearsay, but probably did not realize their meaning in full. Kepler died in 1630, eight years before the appearance of Galileo's *Discorsi* in which his mechanics was outlined in 1638. Even more decisive is the fact that Kepler was not equipped with the theory of curvilinear motion. The elaboration of this theory was begun by Huygens for circular motion and was completed by Newton for general paths. Without the concept of acceleration for curvilinear motions, it is impossible to derive the form (26.8) of the radial acceleration from Kepler's laws by means of simple mathematical operations.

The Newtonian gravitational mechanics emerging from (26.8, 26.10) and the principle of action and reaction may be considered as a further development of the throw motion discovered by Galileo. Newton writes on this topic:

"That the planets are kept in their orbits by the central forces may be seen from the motion of thrown stones (stone-bullets). A (horizontally) thrown stone is deflected from the straight path since gravity is acting on it, and finally it falls to earth along a curved line. If it is thrown with higher velocity it flies further off, and so it might happen that it finally flies beyond the borders of earth and does no more fall back. Hence, the stones thrown off with increasing velocity from the top of a mountain would describe more and more wide parabola bows and finally—at a definite velocity—return to the top of mountain and by this way move about earth."

An explanation that convinces by intuition and logical conclusions! The "definite velocity" is today called orbital velocity. Its magnitude has been correctly given by Newton from $mv^2/R = mg$ for horizontal throw as $v = \sqrt{gR} = 7900 \text{ m sec}^{-1}$. For a vertical throw into the universe, the necessary velocity (escape velocity) results from the energy law (compare Problem 26.5) as $v = \sqrt{2gR} = 11\,200 \text{ m sec}^{-1}$. Both results hold without taking into account the friction losses by the air.

The English physicist Hooke (1635–1703), the founder of the law named after him in the theory of elasticity, also came close to the gravitational law. This is evident from the following of his statements: "I shall develop a world system which in every respect agrees with the known rules of mechanics. This system is founded on three assumptions: 1. All celestial bodies exhibit an attraction (gravity force) directed towards their center; 2. all bodies that are brought into a straight and uniform motion will move as long on a straight line, until they are deflected by some force and are forced into a curvilinear path; 3. the attractive forces are the stronger, the closer the body is on which they are acting. I could not yet establish by experiments what the various degrees of attraction are. But it is an idea that will enable the astronomers to determine all motions of the celestial bodies according to one law."

These remarks show that *Newton* did not at all create his monumental work *Principia* out of nothing: It took, on the contrary, his eminent mental power and bold ideas to summarize all that what *Galileo, Kepler, Huygens,* and *Hooke* had created in the fields of physics, astronomy, and mathematics in a unified manner, and in particular to realize that the force that lets the planets circulate on their orbits about the sun is identical with the force that causes the bodies on earth to fall to ground.

Mankind needed one-and-a-half millennia to realize this discovery if one considers that already *Plutarch* (46–120) in the *Moralia* ("De facie quae in orbe lunae apparet") stated that the moon is prevented from falling to earth because of the impetus of its circulation, just as does a body being "swung around" by a sling. It was the ingenious *Newton,* who realized what the "sling" of the planets is!

Some more remarks on the versatility and brillance of *Hooke:* In 1665 he wrote the prophetic lines: "I often thought that it should be possible to find an artificial glue-like substance being equal or superior to that excrement from which the silkworms produce their cocoon, and that may be spun to fibers by means of nozzles." It is the basic idea of the manmade fibers which—though two and a half centuries later—has revolutionized the textile industry! In the same year he anticipated the mechanical theory of heat (hence also the kinetic gas theory) in speculative thought: "That the particles of all bodies, whatsoever solid they may be, nevertheless are vibrating doesn't need to my opinion any other proof than the fact that all bodies include a certain amount of heat by themselves, and that an absolutely cold body never has been found yet."

27 Special Problems in Central Fields

The gravitational field of extended bodies

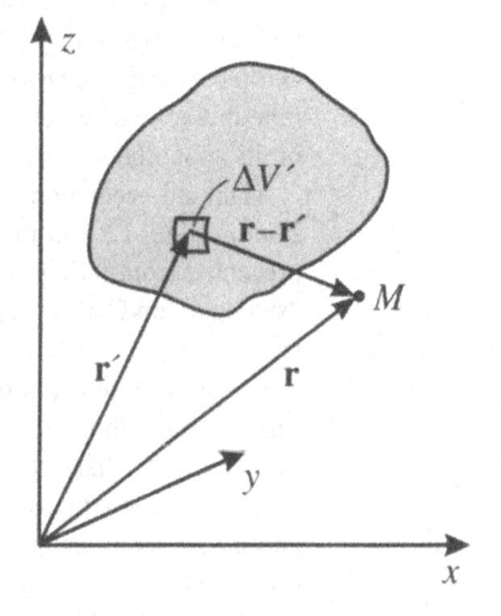

Calculation of the forces of an extended mass distribution on a point mass *M*.

So far only the interactions between pointlike masses have been considered. Now we shall investigate extended bodies concerning their gravitational action. Because of its linearity, the gravitational field of an extended body may be composed by superposition of the fields of individual (thought as pointlike in their action) partial bodies. When performing a limit transition with the volumes $\Delta V'$ of the individual partial bodies approaching zero, the problem is reduced to an integration. The force acting on a mass point M is

$$\mathbf{F} = \lim_{\Delta m_i \to 0} \sum_i \left(-\frac{\gamma M \Delta m_i}{|\mathbf{r}-\mathbf{r}'_i|^3} (\mathbf{r}-\mathbf{r}'_i) \right) = -\gamma M \int\limits_V \frac{\mathbf{r}-\mathbf{r}'}{|\mathbf{r}-\mathbf{r}'|^3}\, dm' .$$

Depending on the kind of mass distribution the differential dm is replaced by volume, area, or line densities (weight functions) multiplied by the corresponding space element dV, dF, or ds. In the three-dimensional case, the force is given by

$$\mathbf{F} = -\gamma M \int \varrho(\mathbf{r}') \frac{(\mathbf{r}-\mathbf{r}')}{|\mathbf{r}-\mathbf{r}'|^3}\, dV' ,$$

and correspondingly the potential energy is

$$V = \lim_{\Delta m_i \to 0} \sum_i \left(-\frac{\gamma M \Delta m_i}{|\mathbf{r}-\mathbf{r}'_i|} \right) = -\gamma M \int\limits_V \varrho(\mathbf{r}') \frac{1}{|\mathbf{r}-\mathbf{r}'|}\, dV' .$$

Here $\varrho(\mathbf{r}) = dm/dV$ is the mass density.

The attractive force of a spherical mass shell

A spherical shell of negligible thickness with the radius a is uniformly covered by mass (constant area density $\sigma = dm/df$). What is the force on a point of mass M at the distance R from the center of the shell?

Because the mass is distributed over an area, twofold integration suffices. We first decompose the spherical surface into circular rings (see following figure).

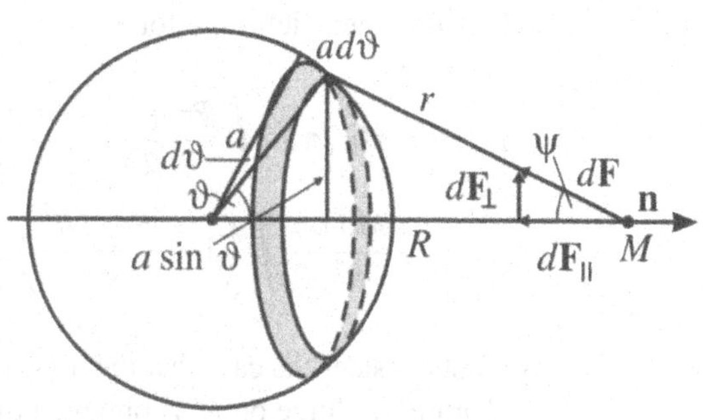

Calculation of the gravitational force of a spherical shell on a point mass M.

The radius of a ring is $a \sin\vartheta$, and the surface of the ring is $df = 2\pi a \sin\vartheta\, a\, d\vartheta$. The result of the first integration along the circumference may be given immediately by exploiting the axial symmetry of the mass distribution. To each section of the circular ring there is a second one, with the force component $d\mathbf{F}_\perp$ (perpendicular to $\mathbf{n}$) being equal but oppositely directed to the first one. Therefore, only the parallel components $-d\mathbf{F}_\parallel = dF \cos\psi\, \mathbf{n}$ are efficient, and the attractive force of the total mass ring is

$$d\mathbf{F} = -\frac{\gamma M \sigma\, df}{r^2} \cos\psi\, \mathbf{n}.$$

The total force of the spherical shell then follows by integration over all circular rings:

$$F = \int dF = -\gamma M \sigma 2\pi a^2 \int \frac{\cos\psi \sin\vartheta}{r^2} d\vartheta.$$

We replace the angles by the distance r via the following geometric relations:
(a)

$$r^2 = a^2 + R^2 - 2aR\cos\vartheta \qquad \text{(cosine law)},$$

and in differential form

$$2r\,dr = 2aR \sin\vartheta\, d\vartheta \qquad \text{or} \qquad \sin\vartheta\, d\vartheta = \frac{r\,dr}{aR},$$

and

$$\cos\vartheta = \frac{a^2 + R^2 - r^2}{2aR}.$$

(b)

$$a^2 = R^2 + r^2 - 2rR\cos\psi \qquad \text{(cosine law)}.$$

This yields

$$\cos\psi = \frac{R^2 + r^2 - a^2}{2Rr}.$$

Insertion then yields the force

$$F = -\gamma M\sigma 2\pi a^2 \int \frac{R^2 + r^2 - a^2}{2Rr^3}\frac{r\,dr}{aR}$$
$$= -\frac{\gamma M\sigma\pi a}{R^2}\int\left(1 + \frac{R^2 - a^2}{r^2}\right)dr.$$

We first consider the case that the mass point M is outside the sphere ($R \geq a$). The desired total attractive force on M is obtained by integration between the limits $R - a$ and $R + a$:

$$F = -\frac{\gamma M\sigma\pi a}{R^2}\int_{R-a}^{R+a}\left(1 + \frac{R^2 - a^2}{r^2}\right)dr$$
$$= -\frac{\gamma M\sigma\pi a}{R^2}\left(r\Big|_{R-a}^{R+a} - \frac{R^2 - a^2}{r}\Big|_{R-a}^{R+a}\right) = -\frac{\gamma M m}{R^2},$$

where $m = 4\pi a^2\sigma$ denotes the mass of the spherical shell. Hence: A hollow sphere (of low thickness) with a uniformly distributed mass is acting on the outer space with respect to its mass attraction so *as if its total mass were concentrated at the center.* This statement also holds for homogeneous full spheres (see Problem 27.3) and serves as the base for any calculations of celestial mechanics.

Now let the mass point M be inside the sphere ($R \leq a$). The integration is now performed between the limits $a - R$ and $a + R$:

$$F = -\frac{\gamma M\sigma\pi a}{R^2}\left(r\Big|_{a-R}^{a+R} - \frac{R^2 - a^2}{r}\Big|_{a-R}^{a+R}\right) = 0.$$

Inside a hollow sphere uniformly covered with mass there is no gravitational force. Because the electric force between two charges q_1 and q_2 is of a similar structure as the gravitational force, namely

$$\mathbf{F}_e = \frac{q_1 q_2}{|\mathbf{r}_1 - \mathbf{r}_2|^2}\frac{\mathbf{r}_1 - \mathbf{r}_2}{|\mathbf{r}_1 - \mathbf{r}_2|},$$

all results obtained here may immediately be transferred to the corresponding electrical charge distributions. In particular, one sees that a uniformly charged spherical shell does not admit fields (forces) in its interior.

The gravitational potential of a spherical shell covered with mass

Because the potential is a scalar, the potential of a circular ring is

$$dV = -\frac{\gamma M \sigma \, df}{r},$$

and the potential of the spherical shell is

$$V = \int dV = -\gamma M \sigma 2\pi a^2 \int \frac{\sin\vartheta \, d\vartheta}{r} = -\frac{\gamma M \sigma 2\pi a}{R} \int dr.$$

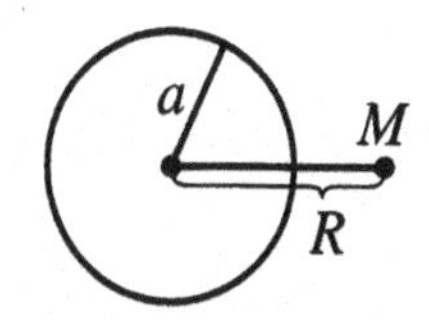

Case 1: Point outside the sphere

We again distinguish the two cases, namely:

1. Point outside the sphere ($R \geq a$)

The integration limits are then $(R-a)\ldots(a+R)$.

$$V = -2\pi \frac{\gamma \sigma a}{R} M((R+a)-(R-a)) = -\frac{\gamma M}{R} 4\pi a^2 \sigma = -\frac{\gamma M m}{R}.$$

2. Point inside the sphere ($R \leq a$)

The integration limits are then $(a-R)\ldots(a+R)$.

$$V = -2\pi \frac{\gamma \sigma a M}{R}((a+R)-(a-R)) = -4\pi \gamma \sigma M a = \frac{-\gamma M m}{a}.$$

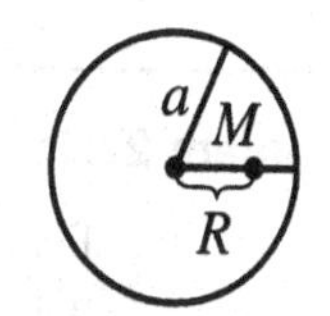

Case 2: Point inside the sphere

The same result may, of course, also be derived from $\mathbf{F}(r)$:

$R \geq a$:

$$V(R) = -\int_{\infty}^{R} \mathbf{F} \cdot d\mathbf{r} = \gamma M m \int_{\infty}^{R} \frac{1}{r^2} \frac{\mathbf{r}}{r} d\mathbf{r}$$

$$= \gamma M m \int_{\infty}^{R} \frac{1}{r^2} dr = -\frac{\gamma M m}{R}.$$

$R \leq a$: The contribution $\mathbf{F} \cdot d\mathbf{r}$ vanishes everywhere; therefore, the potential must be constant everywhere within the spherical shell. If one requires continuity for $R = a$ (otherwise the forces become infinite), then it follows that

$$V(R) = -\frac{\gamma M m}{a} \qquad \text{for} \qquad R \leq a.$$

The forces in a central field point along the radial direction. They are conservative and therefore in polar coordinates of the form

$$\mathbf{F}(r) = -\nabla V(r) = -\frac{\partial}{\partial r} V(r) \, \mathbf{e}_r.$$

The potential in the interior of a spherical shell is constant. As electrostatics is also governed by a $1/r^2$-force law, one there observes the same phenomenon: In the interior of a charged hollow body, no potential differences (voltages) and hence no forces may occur (Faraday cup).

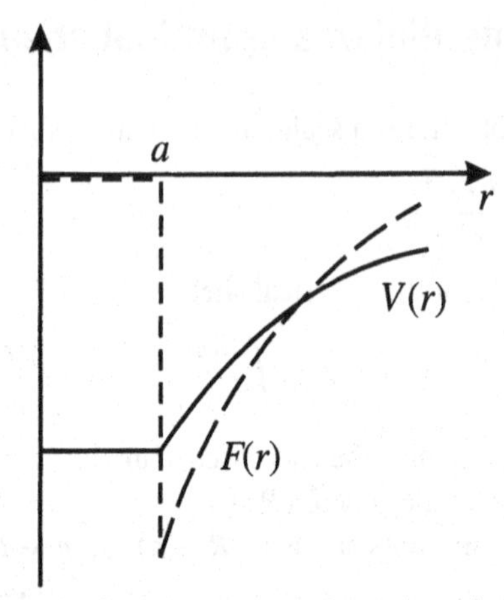

The potential and the force between a point mass and a hollow sphere.

Problem 27.1: Gravitational force of a homogeneous rod

Find the gravitational force of a homogeneous rod of length $2a$ and mass M on a particle of mass m that is positioned at distance b from the rod in a plane perpendicular to the rod through the rod center.

Solution We have

$$dF = -\frac{\gamma m\, dM}{r^2}$$

and

$$\cos\vartheta = \frac{b}{\sqrt{x^2+b^2}}\,.$$

Calculation of the interaction between a mass point and a rod.

dF may be decomposed into force components parallel and perpendicular to the rod. The components parallel to the rod mutually compensate each other. Only the force components perpendicular to the rod, $dF_\perp = dF\cos\vartheta$, are efficient.

$$dF_\perp = \frac{-\gamma m\, dM\cos\vartheta}{r^2} = \frac{-\gamma m\sigma\, dx\cos\vartheta}{x^2+b^2} = \frac{-\gamma m\sigma\, dx\, b}{(x^2+b^2)^{3/2}}\,,$$

$$F = \int\limits_{x=-a}^{a} dF_\perp = -2b\gamma m\sigma \int\limits_0^a \frac{dx}{(x^2+b^2)^{3/2}} = -\frac{2\gamma m\sigma a}{b\sqrt{a^2+b^2}}\,,$$

$$\mathbf{F} = -\frac{\gamma M m}{b\sqrt{a^2+b^2}}\mathbf{e}_2 \qquad \text{because} \quad M = 2a\sigma.$$

For $b \gg a$, a series expansion of the square root yields $F \sim 1/b^2$, as expected.

Problem 27.2: Gravitational force of a homogeneous disk

Let a particle of mass m be on the axis of a disk of radius a at the distance b from the center of the disk. Find the attractive force between the bodies. The disk is assumed to be homogeneously covered with mass.

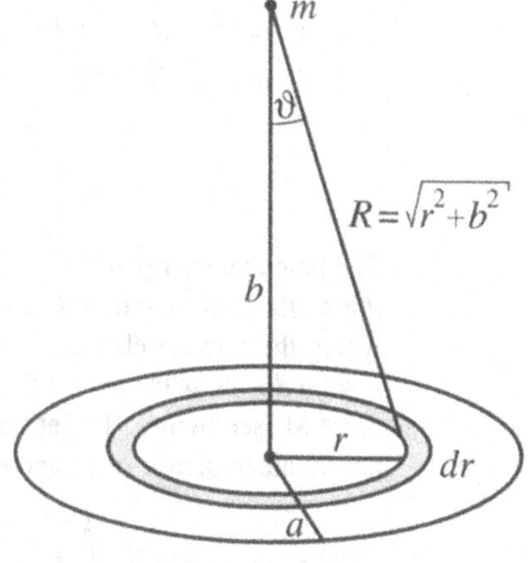

Calculation of the force between a mass point m and a homogeneous disk.

Solution The circular disk is decomposed into concentric rings. Only the force components perpendicular to the disk are efficient; the parallel components compensate each other.

$$dF_\perp = dF\cos\vartheta = -\frac{\gamma m\,dM\cos\vartheta}{R^2},$$

$$\cos\vartheta = \frac{b}{\sqrt{r^2+b^2}}, \qquad R = \sqrt{r^2+b^2},$$

$$dM = \sigma(2\pi r dr).$$

Hence the force between the circular ring and the mass is

$$dF_\perp = -\frac{\gamma m\sigma 2\pi r\,dr\,b}{(r^2+b^2)^{3/2}},$$

and the total attractive force is

$$F = \int dF_\perp = -2\pi\gamma m\sigma b\int_0^a \frac{r\,dr}{(r^2+b^2)^{3/2}}.$$

The integral is solved by substituting $u^2 = r^2+b^2$, $r\,dr = u\,du$,

$$F = -2\pi\gamma m\sigma b\int_b^{\sqrt{a^2+b^2}} \frac{u\,du}{u^3} = -2\pi\gamma\sigma m\left(1-\frac{b}{\sqrt{a^2+b^2}}\right).$$

For $b \gg a$, it follows by expansion of the square root that $F = -\gamma mM/b^2$ with $M = \pi a^2\sigma$, as it must be.

Problem 27.3: Gravitational potential of a hollow sphere

Show that the gravitational potential of a homogeneous hollow sphere with the outer radius b and the inner radius a has the form

$$V(R) = -2\pi\gamma M\varrho\cdot\begin{cases}\frac{2}{3}(b^3-a^3)R^{-1} \\ b^2-a^2 \\ b^2-\frac{2}{3}\frac{a^3}{R}-\frac{1}{3}R^2\end{cases} \quad\text{for}\quad \begin{cases}R\ge b, \\ a\ge R, \\ b\ge R\ge a.\end{cases}$$

Solution We call ϑ: polar angle, φ: azimuth (for rotation about the straight OM). According to the cosine law,

$$r^2 = r'^2 + R^2 - 2r'R\cos\vartheta,$$
$$2r\,dr = 2r'R\sin\vartheta\,d\vartheta,$$

or

$$\sin\vartheta\,d\vartheta = \frac{r\,dr}{r'R}. \qquad \textbf{(27.1)}$$

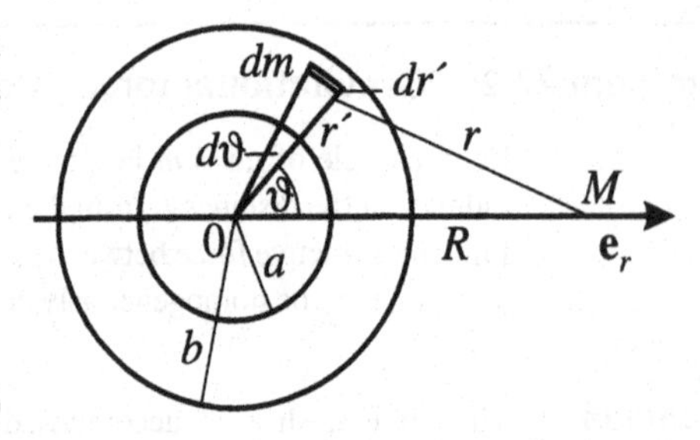

Calculation of the potential between a mass point m and a hollow sphere of homogenous mass density ρ.

The potential energy dV due to the mass element dm of the hollow sphere is $dV = -\gamma M\,dv\,\varrho/r$, where the volume element is $dv = dr' \cdot r'd\vartheta \cdot r'\sin\vartheta\,d\varphi$. M is the mass of the probe particle at point M (see figure). To get the total energy, one has to integrate over φ, ϑ, and r'.

$$V(R) = -\gamma M\varrho \int_a^b \int_0^\pi \int_0^{2\pi} \frac{r'^2 \sin\vartheta\,d\varphi\,d\vartheta\,dr'}{r}$$

$$= -\gamma M\varrho 2\pi \int_a^b \int_0^\pi \frac{r'^2 \sin\vartheta\,d\vartheta\,dr'}{r} \qquad \text{with 27.1}$$

$$= \frac{A}{R}\int_a^b \int_{r_{\min}}^{r_{\max}} r'\,dr\,dr' \qquad \text{with } A = -2\pi\gamma M\varrho.$$

We now distinguish three cases:
1. $R \geq b$: Then $r_{\min} = R - r'$, $r_{\max} = R + r'$, and we get

$$V(R) = \frac{A}{R}2\int_a^b r'^2 dr' = A\frac{2}{3}\frac{b^3 - a^3}{R}\left(= -\frac{\gamma M m}{R}\right).$$

2. $R \leq a$: $r_{\min} = r' - R$, $r_{\max} = r' + R$. Thus we obtain

$$V(R) = \frac{A}{R}2R\int_a^b r'dr' = A(b^2 - a^2).$$

3. $a \leq R \leq b$: The point M lies at the outer border of a spherical shell with the radii a and R, and at the same time at the inner border of a spherical shell between R and b. The energy may then be composed of the contributions according to cases 1 and 2:

$$V(R) = A\left(\frac{2}{3}\frac{R^3 - a^3}{R} + b^2 - R^2\right) = A\left(b^2 - \frac{2}{3}\frac{a^3}{R} - \frac{1}{3}R^2\right).$$

The forces are calculated by $\mathbf{F} = -\frac{\partial V}{\partial r}\mathbf{e}_r$.
1. $R \geq b$:

$$\mathbf{F} = -\frac{4}{3}\pi\gamma\varrho M\frac{b^3 - a^3}{R^2}\mathbf{e}_r = -\gamma\frac{mM}{R^2}\mathbf{e}_r.$$

In this case the force on M is such as if the total mass m of the hollow sphere were united at the center.

2. $R \leq a$: $\mathbf{F} = 0$. In the interior (empty) space of the hollow sphere, there is no gravitational force at all.

3. $a \leq R \leq b$:

$$\mathbf{F} = \gamma M \frac{4}{3}\pi\varrho\left(\frac{a^3}{R^2} - R\right)\mathbf{e}_r = \frac{-\gamma M\varrho\frac{4}{3}\pi(R^3 - a^3)}{R^2}\mathbf{e}_r = \frac{-\gamma Mm(R)}{R^2}\mathbf{e}_r\,,$$

where $m(R)$ is the mass of the spherical shell with inner radius a and outer radius R (position of the particle with mass M). The mass shell beyond R does not contribute to the force on M.

Problem 27.4: A tunnel through the earth

A tunnel for mail transportation is drilled from Frankfurt to Sydney (Australia). Determine the time needed for the freely falling air tube casing to cover this distance, assuming the earth is at rest and has a homogeneous mass distribution. Let the air friction be negligible.

Solution The gravitational force within a homogeneous sphere points to the center M and has the magnitude kr (compare Problem 27.3). At the surface of the sphere, one has $mg = kR$; hence $k = mg/R$. From the figure we read off $r = x/\sin\vartheta$. The component of the gravitational force along the tunnel is therefore $-kr\sin\vartheta = -(mg/R)\,x$.

Hence, the equation of motion is $\ddot{x} + (g/R)\,x = 0$, that is, the air tube casing performs a harmonic vibration between F and S with the period $T = 2\pi\sqrt{R/g}$.

The time needed between F and S is $\tau = T/2 = \pi\sqrt{R/g}$. With $R = 6370$ km and $g = 9.81$ m/s^2, it follows that $\tau = 42$ min. Note that this short time does not depend on the distance between F and S.

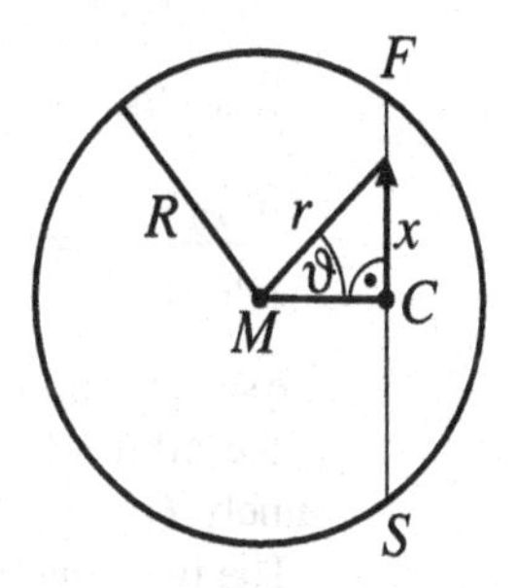

Illustration of the earth with Frankfurt (F) and Sydney (S) as endpoints of a tunnel.

Stability of circular orbits

In any attractive central force field, the attractive force and the centrifugal force may be brought into equilibrium, such that circular orbits are always possible. In practice (e.g., telecommunication satellites on geostationary orbits, particles in accelerators), however, it is moreover important that the circular motion is not destroyed by small elongations. We therefore investigate how a central force field must be structured to allow for stable circular orbits. We first consider the field with the particular force law

$$F(r) = -\frac{K}{r^n}$$

and look for the powers n for which stable orbits exist. By adding the centrifugal force we obtain the effective force

$$F_{\text{eff}}(r) = -\frac{K}{r^n} + m\dot{\theta}^2 r \qquad \text{with} \quad \dot{\theta} = \frac{L}{mr^2}$$
$$= -\frac{K}{r^n} + \frac{L^2}{mr^3},$$

and hence the effective potential

$$V_{\text{eff}}(r) = -\int_{\infty}^{r} F_{\text{eff}}\, dr = -\frac{K}{(n-1)r^{n-1}} + \frac{L^2}{2mr^2}; \qquad n \neq 1.$$

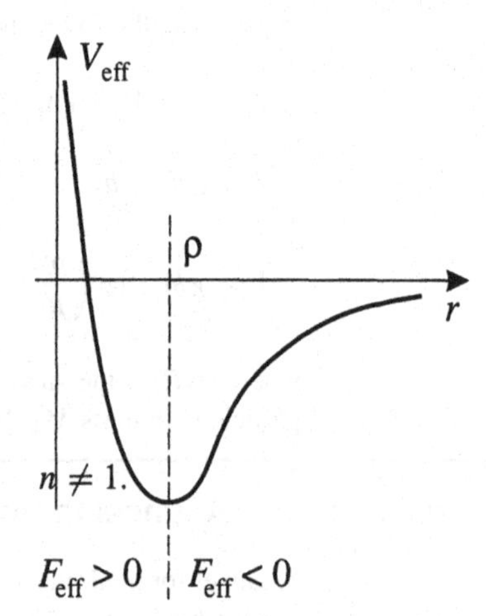

The effective potential around a stable orbit.

To get a stable circular orbit with the radius $r = \varrho$, the effective potential $V_{\text{eff}}(r)$ must have a minimum at this position:

$$\Rightarrow \quad F_{\text{centrifug}} = F_{\text{attr}}.$$

Thus, the following conditions shall be fulfilled:

$$\left.\frac{\partial V_{\text{eff}}}{\partial r}\right|_{r=\varrho} = 0 \qquad \text{and} \qquad \left.\frac{\partial^2 V_{\text{eff}}}{\partial r^2}\right|_{r=\varrho} > 0.$$

The second condition is essential for the path stability. It ensures that for small displacements of the orbit a backdriving force occurs, pushing the particle toward the stable radius ϱ, namely $F_{\text{eff}} > 0$ for $r < \varrho$ and $F_{\text{eff}} < 0$ for $r > \varrho$.

The two conditions lead us to

(a)

$$\left.\frac{\partial V_{\text{eff}}}{\partial r}\right|_{r=\varrho} = \frac{K}{\varrho^n} - \frac{L^2}{m\varrho^3} = 0, \qquad \varrho^{n-3} = \frac{mK}{L^2},$$

(b)

$$\left.\frac{\partial^2 V_{\text{eff}}}{\partial r^2}\right|_{r=\varrho} = -\frac{nK}{\varrho^{n+1}} + \frac{3L^2}{m\varrho^4} > 0,$$

which is equivalent to

$$-\frac{nK}{\varrho^{n-3}} + \frac{3L^2}{m} > 0.$$

Elimination of ϱ yields

$$(-n+3)\frac{L^2}{m} > 0.$$

The condition for stable circular orbits in a central force field of the form $F = -K/r^n$ is $n < 3$.

We now omit the restriction to the power law and investigate arbitrary central force fields. For all central motions it holds, using the angular momentum $L = mr^2\dot{\theta}$, that

$$F(r) = m(\ddot{r} - r\dot{\theta}^2) = m\ddot{r} - \frac{L^2}{mr^3} = m\left(\ddot{r} - \frac{L^2}{m^2r^3}\right).$$

We abbreviate:

$$g(r) = -\frac{F(r)}{m}, \qquad \text{then} \quad -g(r) = \ddot{r} - \frac{L^2}{m^2r^3}.$$

The particle circulates on the *reference orbit* with the radius ϱ. A *small perturbation* shall not displace it significantly from its path. After a *small elongation* x, the new orbit is

$$r = \varrho + x,$$

where

$$x \ll \varrho, \qquad \frac{x}{\varrho} \ll 1.$$

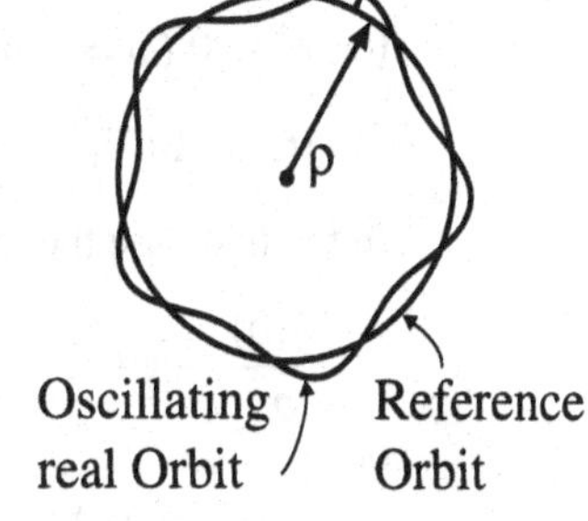

Reference orbit and (oscillating) real orbit.

Because $\varrho = \text{constant}$ (circular orbit), $\ddot{r} = \ddot{x}$.

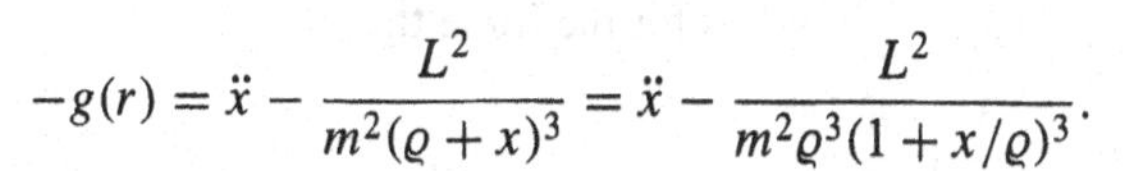

$$-g(r) = \ddot{x} - \frac{L^2}{m^2(\varrho + x)^3} = \ddot{x} - \frac{L^2}{m^2\varrho^3(1 + x/\varrho)^3}.$$

Because $x/\varrho \ll 1$, the last term may be expanded into a Taylor series. We now assume that $g(r)$ may also be represented by a Taylor expansion about $r = \varrho$:

$$g(\varrho + x) = g(\varrho) + xg'(\varrho) + \cdots .$$

Neglecting all terms with higher powers than x, we obtain

$$-(g(\varrho) + xg'(\varrho)) = \ddot{x} - \frac{L^2}{m^2\varrho^3}\left(1 - 3\frac{x}{\varrho}\right).$$

A consideration of the reference orbit $r = \varrho$, $x = 0$, $\ddot{x} = 0$ yields

$$g(\varrho) = \frac{L^2}{m^2\varrho^3}$$

which allows us to eliminate the angular momentum. This yields

$$\ddot{x} + \left(\frac{3g(\varrho)}{\varrho} + g'(\varrho)\right)x = 0.$$

This is just the equation $\ddot{x} + \omega^2 x = 0$ of the undamped harmonic oscillator with the angular frequency

$$\omega = \sqrt{\frac{3g(\varrho)}{\varrho} + g'(\varrho)}.$$

For $\omega^2 > 0$, the solution $x = Ae^{i\omega t} + Be^{-i\omega t}$ yields harmonic vibrations. For $\omega^2 < 0$ x tends to infinity as $Be^{|\omega|t}$. In the first case the particle "oscillates" on its actual orbit about the reference orbit. In the second case the particle in general runs away from the reference orbit.

The condition for stable circular orbits thus reads $\omega^2 > 0$. For $\omega^2 > 0$:

$$x(t) = Ae^{i\omega t} + Be^{-i\omega t} \quad \Rightarrow \quad x(t) = D\sin(\omega t + \varphi) \qquad \text{(stable)};$$

for $\omega^2 < 0$ $(\omega = i|\omega|)$:

$$x(t) = Ae^{-|\omega|t} + Be^{+|\omega|t} \qquad \text{(unstable)}.$$

In the first case the particle vibrates with the small amplitude x about its reference orbit if

$$\frac{3g(\varrho)}{\varrho} + g'(\varrho) > 0,$$

or

$$\frac{3}{\varrho} + \frac{g'(\varrho)}{g(\varrho)} > 0.$$

Because $mg(\varrho) = -F(\varrho)$, this implies for the force the

stability condition: $$\frac{3}{\varrho} + \frac{F'(\varrho)}{F(\varrho)} > 0.$$

When applied to the particular central force field $F(r) = -K/r^n$, the stability condition implies

$$F'(\varrho) = n\frac{K}{\varrho^{n+1}}; \qquad \text{hence} \qquad \frac{3}{\varrho} - n\frac{\varrho^n}{\varrho^{n+1}} > 0.$$

We obtain the condition $n < 3$, which agrees with our former calculation. This of course must be so, since by insertion of V_{eff} one easily realizes that the new stability condition is equivalent to $\partial^2 V_{\text{eff}}/\partial r^2 > 0$.

The investigation of path stability may refer, among others, in the atomic range to the electric field of the nuclei. The simple Coulomb potential $V(r) = -K/r$ allows for stable circular orbits, as was shown already. When taking into account the influence of an oppositely charged electron shell, this potential is weakened: An electron in the outer region "sees" only a small fraction of the nuclear charge. This phenomenon may be taken into account by multiplication by a correction factor < 1. An approximation for the "screened Coulomb potential" is

$$V(r) = -\frac{K}{r}e^{-r/a}.$$

a characterizes the exponential decay of the $1/r$-term. Stable circular orbits are also possible for the screened potential. One should note that this potential allows for closed orbits even for positive energies. The figures illustrate the trend of the potential and the effective potential. The closed orbits of positive energy are fully stable in classical mechanics we are treating here. In quantum mechanics, however, we shall see that these orbits decay, because the particles on such orbits of positive energy may "tunnel" through the potential barrier (tunnel effect).

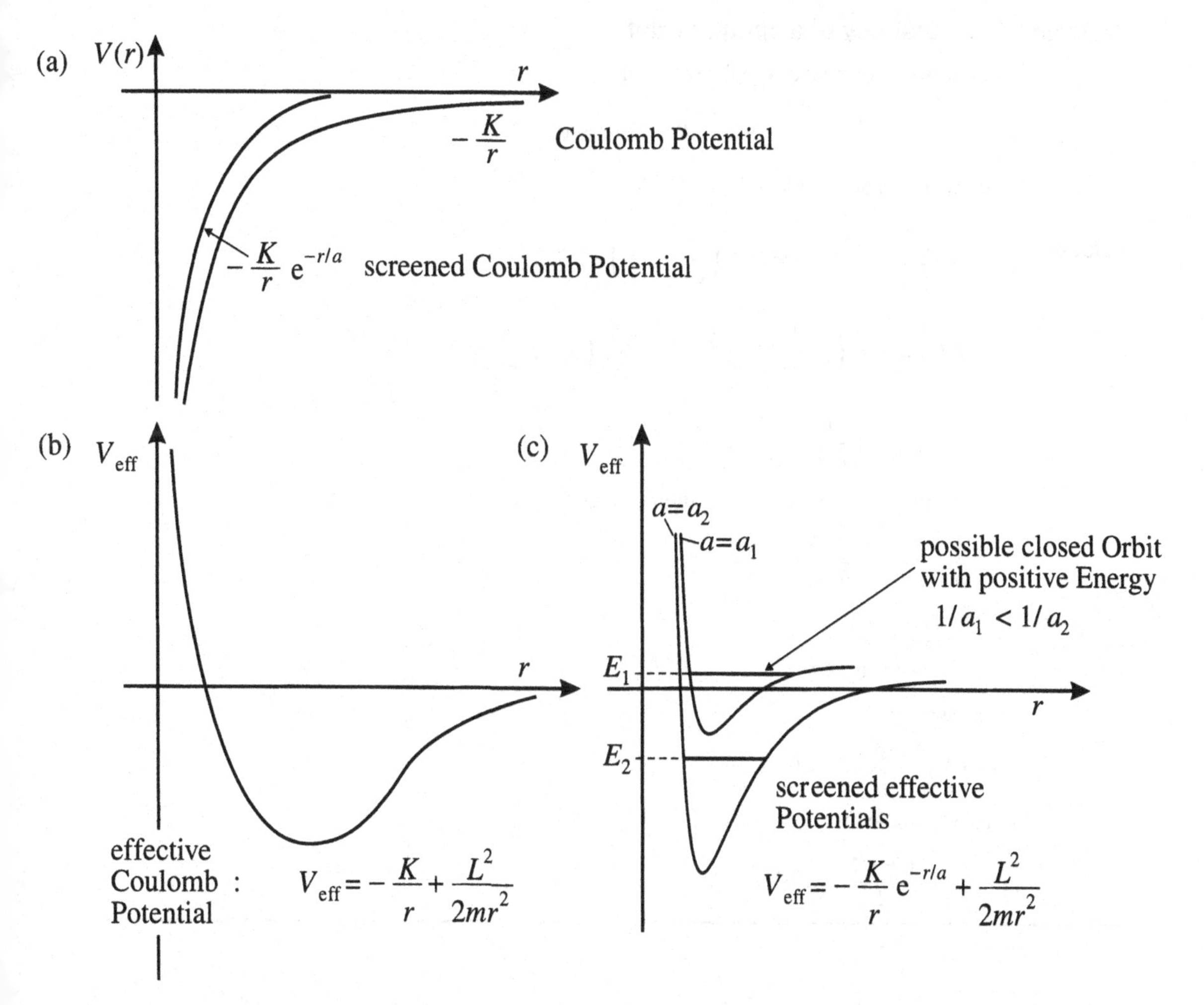

Shape of the potential: The screened Coulomb potential (a) decreases rapidly for distances $r > a$ and approaches zero faster than the unsceened potential. Effective Coulomb potential (b) and effective screened Coulomb potential (c).

Problem 27.5: Stability of a circular orbit

Show that for $\varrho^2 K > K'$ the circular orbit with $r = \varrho$ in the force field $F(r) = -K/r^2 - K'/r^4$ is stable (for $K, K' > 0$).

Solution The stability condition reads

$$\frac{3}{\varrho} + \frac{F'(\varrho)}{F(\varrho)} > 0;$$

hence

$$\frac{3}{\varrho} + \frac{2K/\varrho^3 + 4K'/\varrho^5}{-K/\varrho^2 - K'/\varrho^4} > 0, \qquad 3\varrho^2 K + 3K' > 2\varrho^2 K + 4K', \qquad \varrho^2 K > K'.$$

Problem 27.6: Stability of a circular orbit

Show that a force field with the potential

$$U(r) = -\frac{K}{r} e^{-r/a} \qquad \text{with} \quad K > 0,\ a > 0,$$

allows for stable circular orbits.

Solution

$$F(r) = -\frac{\partial}{\partial r} U(r) = -K \left(\frac{1}{ar} + \frac{1}{r^2} \right) e^{-r/a},$$

$$F'(r) = -K \left(-\frac{1}{ar^2} - \frac{2}{r^3} \right) e^{-r/a} + \frac{K}{a} \left(\frac{1}{ar} + \frac{1}{r^2} \right) e^{-r/a}$$

$$= K \left(\frac{1}{a^2 r} + \frac{2}{ar^2} + \frac{2}{r^3} \right) e^{-r/a}.$$

Insertion into the stability condition yields for $r = \varrho$

$$\frac{3}{\varrho} - \frac{1/a^2 + 2/\varrho a + 2/\varrho^2}{1/a + 1/\varrho} > 0.$$

This means

$$a^2 + a\varrho - \varrho^2 > 0$$

or rewritten

$$\left(\frac{\varrho}{a} \right)^2 - \frac{\varrho}{a} - 1 < 0.$$

This is fulfilled for

$$\frac{\varrho}{a} < \frac{1 + \sqrt{5}}{2} \approx 1.62.$$

28 The Earth and our Solar System

General notions of astronomy

Stars: Stars are celestial objects (suns) mostly of high mass concentration that emit light produced by nuclear reactions. In the core zone of our sun, for example, hydrogen (H) is burning to helium (^{4}He). In other, older stars, higher burning processes are going on, such as $3\,^4\text{He} \rightarrow {}^{12}\text{C}$, $^{12}\text{C} + {}^4\text{He} \rightarrow {}^{16}\text{O}$, etc. They are rather subtle in the details. A clear representation of these processes may be found in J. M. Eisenberg and W. Greiner, *Nuclear Theory 1: Nuclear Models*, 3rd ed., North Holland, Amsterdam (1987).

Planets: Planets are bodies circulating in the central force field of a star. They may reflect light (the ratio of reflected to incoming luminous flux is called albedo), but hardly emit any light by themselves (up to some thermal radiation corresponding to their temperatures). The point of maximum distance between a planet and its central body is called aphelion, th point of minimum distance is called perihelion.

Meteors: Collective noun for the light phenomena that are caused by penetration of solid particles (meteorites) into the earth's atmosphere. The meteorites that may have masses between 10^{-3} g and 10^6 kg enter the atmosphere with velocities between 10 and 200 km/s and usually burn out completely.

Comets: Comets are celestial bodies of low mass concentration (most likely all of them) moving in the central force field of a star. A comet has a core out of dust and ice grains. Under sufficient irradiation by the sun it develops a gas shell (coma) and a tail. The total length may reach up to 300 millions of km.

Satellites: Satellites are bodies circulating about planets. One may distinguish between natural satellites, the moons, and artificial ones (the first one was Sputnik I (10/14/1957)). In the case of earth satellites, the longest and shortest distance from earth is denoted as the apogee and perigee, respectively.

Asteroids and planetoids: These are pieces of rock. The size is small as compared to the usual planets. They are orbiting about the sun in the range between Mars and Jupiter and mostly have similar orbital data. Therefore, they were presumed to be the residues of a decayed planet (the orbits of the planetoids are crossing each other). There are also commensurability gaps within the belt of planetoids, presumably caused by Jupiter.

Period: The period denotes the time of a full course of any periodic motion. In astronomy one mostly means the sidereal period, namely, the time a mass needs for a complete revolution about its central body.

Solar system: The sun together with its associated planets and their moons, as well as the planetoids, comets, and swarms of meteors, in total constitute the solar system.

Ecliptic: The plane in which the center of mass of the system earth–moon orbits around the sun is called ecliptic.

Determination of astronomic quantities

We shall now briefly indicate how astronomic quantities are determined in practice.

The distance between planets and earth

(a) The distances may be determined by triangulation. From a measurement of the observation angles of the planet as seen from two distinct points and of the distance between these points, the distance of the planet may be calculated.

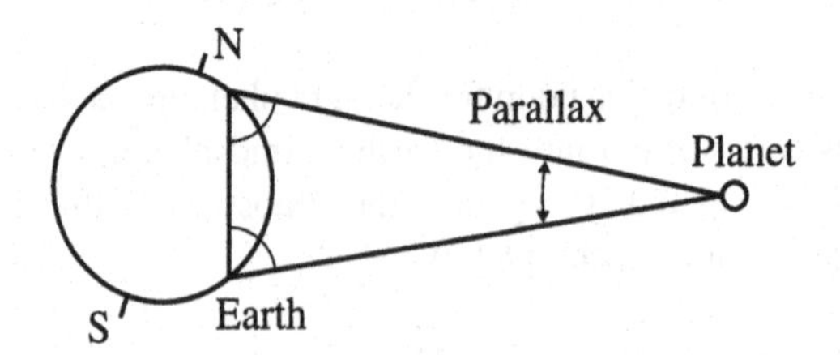

Principal scheme for measuring distances by triangulation.

(b) Distances may be measured by radar. Because the propagation speed of electromagnetic waves is known, one may conclude from the transit time of radar signals on the distance. This method works only for the immediate neighbors of the earth.

(c) In the sense of (a), the earth's orbit may also be used as a base for triangulation to measure the distance of the near fixed stars.

(d) The sun (and the planets) are moving uniformly by about 610 millions of km/year (or 4.09 astronomic units per year) toward the sun's apex in the constellation of Hercules (see later: "A model of the sun's environment" and "The spatial motion of the sun" on page 318). This may also be used for parallax measurements, and thus for measuring the distances of fixed stars up to more than 100 lightyears.

Determination of the distance of far away astronomical objects

The universe is expanding. The farther away the astronomical objects are, the larger their velocity is. This extraordinary discovery is due to Edwin Hubble,[1] who looked at the large-scale behavior of matter in the universe. The *Hubble law*

$$v = H_0 d$$

allows the determination of the distance d of extragalactic objects from their recessional velocity v if the numerical value of the constant H_0 is known. Within the theoretical framework of the Big Bang, the Hubble law is quite plausible. Matter that has been created with high initial velocity travels the longest distance within time T: $d = vT$; thus $v = 1/T \cdot d$. For nonrelativistic speeds, the recessional velocity v equals the product of the speed of light and the redshift z, which can be measured in the spectrum of the observed object,

$$z = \frac{\lambda - \lambda_0}{\lambda_0} .$$

Here, λ is the observed wavelength of a reference line in the line spectrum of the object and λ_0 is the wavelength of this line when the relative velocity between source and observer vanishes. If the period of the emitted light is T, we have $\lambda_0 = cT$ and $\lambda = (c + v)T$, or

$$\frac{\lambda_0}{c} = \frac{\lambda}{c + v} ,$$

from which we obtain $v = zc$.

In order to obtain the Hubble constant H_0, the distances of a suitable sample of galaxies have to be measured. Astronomical distances are usually measured step by step, progressing gradually from the solar system over nearby stars to ever more distant objects, finally reaching faraway galaxies.[2]

The first step is the determination of the size of the solar system and of the distances of the planets. This can be done today very accurately with the help of radar delay measurements.

[1] *Edwin Hubble* (1889–1953). American astronomer who determined the extragalactic distance scale by locating Cepheid variables in the galaxy M31 (the Andromeda galaxy) from the Mount Wilson Observatory in 1924 and NGC 6822 in 1925. Extending distance determinations by using the brightest star in galaxies, he proposed the Hubble law in 1929.

[2] For more details about the measurement of astronomical and cosmological distances, see, e.g., Rowan-Robinson, M.: *The Cosmological Distance Ladder*, W.H. Freeman and Company, 1985.

The only possible way for the direct determination of further distances is the method of triangulation. This method is suited for the determination of distances of stars in our Milky Way neighborhood. Here, the change of the direction to a star when observing from two different points at a distance d is measured. The line between the two observation points is called the baseline; the angle difference is called the parallax of the star. The parallax is the same angle under which the baseline would be seen when observed from the star. The distance to the star can then be calculated by simple trigonometry. In the ideal (and most simple) case, the observed star lies in the plane perpendicular to the baseline and cuts the baseline in half. Then, the distance is, to a very good approximation, given by

$$d = \frac{b}{\alpha},$$

where b is the length of the baseline and α is the parallax angle. The longest available baseline is the line between two opposite points of the earth's orbit around the sun. This is also the origin of the distance unit of Parsec (pc). One pc is the distance from which the orbit of the earth is seen under an angle of one arc-second, or, equivalently, the distance yielding a parallax of one arc-second. 1 Parsec corresponds to 3.26 lightyears.

The range of applicability of the parallax method is given by the error in the determination of the angle $\delta\alpha$ and by the restriction in the length of the available baseline. Gaussian error propagation yields a relative error for the distance from the parallax method of

$$\left|\frac{\delta d}{d}\right| = \left|\frac{b}{\alpha^2}\right| \delta\alpha \, \frac{1}{d} = \delta\alpha \frac{d}{b}\,.$$

This means that for a given error in the measurement of the angle α and with a given baseline b, not just the absolute error for the distance d, but also the relative error will increase.

The first parallax of a star was measured by the German astronomer Friedrich Wilhelm Bessel. In 1838, Bessel published his value of 0.314 arc-seconds for the parallax of the star 61 Cygni, corresponding to a distance of about 10 lightyears. The correct value of the parallax of 61 Cygni is 0.292 arc-seconds, or 11.2 lightyears.

During the 1990s, the Hipparcos satellite mission measured the parallax of 118,000 stars accurately down to 1 milliarc-seconds (mas), yielding a very exact picture of the distances in our Milky Way neighborhood.[3] But even before the advent of the Hipparcos data, there had been possibilities to measure distances beyond the range of the triangulation method. With the help of the *star drift parallax* (also called *convergent-point method*) one can measure the distance to nearby open star clusters and thus determine the absolute luminosity of main sequence stars.

The method of the star drift parallax is based on the determination of the two components of a star's motion that can be observed from the earth: The radial velocity (the velocity along the line of sight) can be measured from the Doppler shift in the spectrum of the star, while the proper motion of the star (the motion on the celestial sphere) can be converted to the transversal velocity if the distance of the star is known. Both velocity components taken together yield the compete, three-dimensional velocity vector of the star. If, on the

[3]See, e.g., Perryman, M.: "The Hipparcos astrometry mission," *Physics Today* (June 1998), http://astro.estec.esa.nl/Hipparcos/

other hand, the direction of the velocity vector and the radial velocity of a star are known, the distance of the star can be calculated from the proper motion. This is used for the determination of distances by the star drift parallax method.

If one knows the apparent convergence point of a cluster of stars in collective, parallel motion, one can deduce the transversal motion of the stars in the cluster from the measurement of their radial velocities. By comparison with the proper motion, one can determine the distance of the stars. The accuracy of this method relies on the large number of measured stars. The most prominent example for the use of the star drift parallax is the determination of the distances of the *Hyades*.[4] The distance thus obtained for the Hyades is 45 pc. The determination of the distance to the Hyades serves as a gauge point for methods reaching still farther out such as the *Cepheid method*.

When plotting in a diagram the absolute luminosity of stars versus their surface temperature, which can be inferred from their spectra, one finds a large class of stars showing a strong monotonic relation between these two observables. Such stars are called main sequence stars. The diagram is called the *Hertzsprung–Russell diagram* after its inventors. The fitting of large clusters of stars at the main sequence uses this relation between surface temperature and absolute luminosity in order to estimate the absolute luminosity of the stars and, by comparison with the measured apparent luminosities, the distance of the cluster. Thus, the ratio of apparent luminosities of stars in different clusters allows conclusions about the ratio of the distances of the clusters to the solar system.

The methods of star drift parallax and fitting to the main sequence thus allow the determination of the distance of faraway star clusters. When observing Cepheid stars in such clusters, one can gauge the period–luminosity relation of this class of variable stars and measure distances up to 4 Mpc, reaching beyond the Milky Way in extragalactic regions.

The Cepheid are a class of variable stars that show a definite relation between their absolute luminosities (i.e., the total amount of energy released as visible light) and the period of variation of their luminosity. They are named after the first known object of this type, the variable star δ Cephei. When observing a distant Cepheid, measuring the period of variation thus allows the calculation of the absolute luminosity. Comparing with the apparent luminosity (the light received in a telescope), one can determine the distance of the star. In order to obtain reliable results, one must take into account the attenuation of the light by interstellar matter.

The astrophysical mechanism responsible for the pulsation of Cepheid stars and the relation to absolute luminosity are quite well known. Modern astrophysics differentiates between classical Cepheids and W-Virginis stars, which show different light curves and spectra. Furthermore, one knows the relatively dim RR-Lyrae stars, which can be identified from their short periods. RR-Lyrae stars have a constant absolute luminosity, which can be used to determine their distance. However, since they are not as bright as Cepheids, they can be used only over shorter distances.

With the help of extragalactic Cepheids, one can determine the absolute diameter of the H-II regions of galaxies. Assuming that the diameters of large H-II regions of different

[4]The hyades, an association of several hundred stars, are an open star cluster in the constellation of Taurus (the bull). With a distance of about 145 lightyears, it is the second-closest star cluster to Earth. On the celestial globe, the hyades are centered around Aldebaran, the brightest star in Taurus, which, however, is not part of the hyades cluster.

galaxies are approximately equal, one can then determine the distance to other galaxies. Using this method, one reaches distances up to 25 Mpc.

H-II regions bear their name from the simply ionized hydrogen they consist of. The assumption that these regions—which are supposed to play an important role in the formation of stars—are all approximately equal in size relies on the hypothesis that the UV radiation from the core of their galaxies which makes them glow always has the same range. This would imply that the observable radius is constant.

The next step uses the distances of the H-II regions in order to determine the absolute luminosities of so-called Sc-I galaxies. In the Hubble classification of galaxies, Sc-I galaxies are a class of old spiral galaxies with wide, open spiral arms and a small core. All galaxies of this class have approximately the same absolute luminosity.

Measuring the apparent luminosity of far away Sc-I galaxies and using their known absolute luminosity, one can infer their distance. This finally allows us to determine the relation between distance and redshift and yields a value for the Hubble constant H_0.

For every step of the measurement of distances, there also exist alternative methods. Besides the well-established Cepheid calibration, one can look for novae, which have a definite relation between their maximal absolute luminosity and the time scale of the decrease of luminosity, or for bright main sequence stars, which can be identified from their spectra and whose absolute luminosity is well known. Supernovae can still be observed in the huge distance of 400 Mpc. While all supernovae approximately reach the same absolute luminosity, this value is not easy to calibrate. Still another method uses the third brightest galaxy in a small galaxy cluster, making the assumption that all such galaxies have nearly the same absolute luminosity. Experience has shown that the third-brightest galaxy is better suited for this purpose than the brightest or second-brightest galaxy. Another possibility for the determination of distances is the use of the brightest globular clusters of far galaxies. Finally, a radio-astronomical method uses the observed close relation between the half-width of the 21-cm line of hydrogen and the absolute luminosity of a galaxy in the blue spectral band.

The combination of all these different methods yields today a quite coherent picture of the distances in the universe. As for the Hubble constant, the accepted value from different measurements by the Hubble space telescope[5] is $H_0 = 72 \pm 8\,\text{km}\,\text{s}^{-1}\text{Mpc}^{-1}$.

The orbital velocity of the planets

(a) For circular orbits the velocity may be determined from the measurable quantities orbital radius and revolution time (period).

(b) For elliptic orbits the velocity may be determined from the measurable quantities semi-axes and period.

[5]Freedman, Wendy L. et al.: "Final Results from the Hubble Space Telescope Key Project to Measure the Hubble Constant," *The Astrophysical Journal* **553** (2001) 47-72.

The mass of the planets

(a) From the gravitational law and the equation for the centripetal force the relation $\gamma M = 4\pi^2 a^3 T^{-2}$ follows; see equation (26.41). This is the third Kepler law. M here means the mass of the central body, which is large as compared to the mass of the orbiting body. From this equation one may calculate the mass of the sun and the mass of every planet having moons.

(b) If planets don't have moons, their mass is determined from the orbital perturbations of the neighboring planets.

The rotational velocity of a planet or star

The rotational velocity of a planet may be determined by observation of marked points on its surface. For stars that are visible only as a pointlike light source, this method fails. For these objects the rotational velocity may be derived from their spectra and from the distortion of a spectral line due to the Doppler effect (distinct shift—red, blue—at opposite sides of the rotating star). The east border of the sun shows, for example, a red shift, and the west border a blue shift from which follows a rotational velocity of the surface of the sun of 2 km/s.

Detection of gases in the universe

Elements occuring in stars may be determined from the spectrum of the star light. In the case of planets one has to take into account that they only reflect or absorb light. The gases of the atmosphere may be identified by the absorption spectrum (Fraunhofer lines).

The tides

Two masses are moving in the gravitational field of a third mass M (see figure).

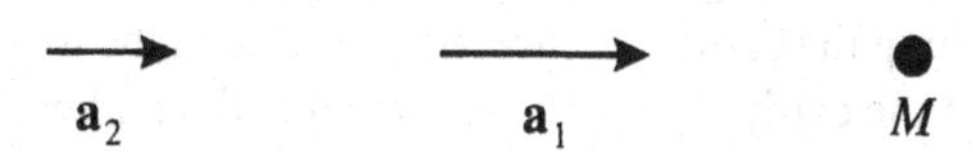

Two masses in the gravitational field of a mass M are subject to different accelerations $\mathbf{a}_1$ and $\mathbf{a}_2$ due to the inhomogeneities of the gravitational field.

The first mass is subject to an acceleration $a_1 = \gamma M/r_1^2$, the second mass is accelerated by $a_2 = \gamma M/r_2^2$. An observer on one of the masses therefore establishes that the other mass moves away from it with the acceleration $a_1 - a_2 = \gamma M(1/r_1^2 - 1/r_2^2)$. Hence, the distinct magnitude of the gravitational force implies a force between the two masses, which thereby are pulled apart from each other. Such a force always arises if the gravitational field is inhomogeneous; it is called a *tidal force* because the tides on earth are caused by the same effect.

Low tide and high tide are generated by the motion of the earth in the gravitational field of the moon (mass M_M). At point A or B (see figure) a body gets an acceleration $a = \gamma M_M/(r \pm R)^2$ due to the attractive force of the moon, where r is the distance between the centers of earth and moon, respectively, and R is the earth's radius.

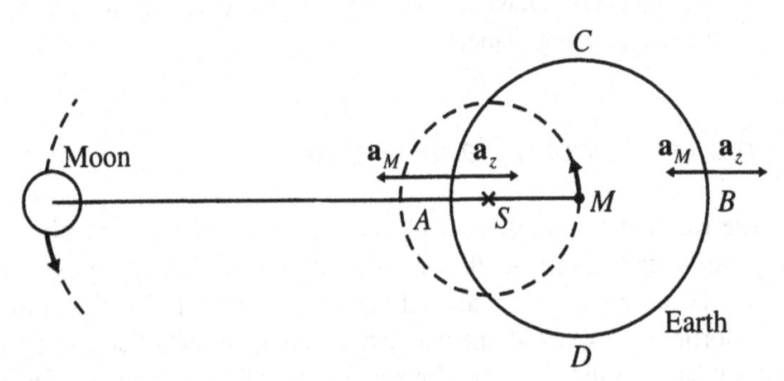

Explanation of the tides: Earth and moon orbit around the common center of mass S.

The Taylor expansion yields $a \approx (\gamma M_M/r^2)(1 \mp 2R/r)$. The acceleration at the earth's center is $a_z = \gamma M_M/r^2$, such that the difference is $a - a_z = a - \gamma M_M/r^2 = \mp 2\gamma M_M R/r^3$. This difference always points off the earth's surface and has the magnitude $8 \cdot 10^{-5}$ cm/s^2. At points A and B the earth's acceleration is thus reduced by this amount.

The common center of mass S of earth and moon is apart from the earth's center by about $\frac{3}{4}\,R$. Because the center of mass is conserved, both earth and moon are moving with the same angular velocity about this point S. The center of earth thus moves on a circle of radius $\frac{3}{4}\,R$ about S. This circular motion is the same for all points on earth and leads to a centrifugal acceleration a_z that points along the axis earth–moon off the center of the circle. At the earth's center the centrifugal acceleration and the gravitational acceleration $\gamma M_M/r^2$ just compensate each other.

The reduction of the earth acceleration at points A and B leads to formation of tide waves. Because the problem is symmetric about the axis moon–earth, one observes low-tide valleys in the ring through C and D perpendicular to this axis. The points A and B are floating along the earth's surface, in accord with the moon's circulation about earth and the rotation of the earth about its axis, such that the highest tide occurs twice within $24\frac{3}{4}$ h at a given position.

If the earth were completely covered by oceans, the height of the tide wave would amount to about 90 cm. By the various shapes of the coastlines, the times of the highest tide may shift, and tide waves with heights of several meters may evolve.

The gravitational field of the sun also causes tidal forces on earth that amount to about half of the lunar tidal forces. If sun, moon, and earth lie on a straight line (i.e., at full moon and at new moon, i.e., each $13\frac{1}{2}$ days), the tidal forces add up and a particularly high tide arises (spring tide); at half-moon one observes a neap tide (see figure).

The friction between the waters and earth implies a deceleration of the earth rotation, such that the day became longer by 0.0165 s during the last 1000 years. Because the total angular momentum of the earth–moon system is conserved, the decrease of the earth's angular momentum must be joined with an increase of the moon's angular momentum. The moon's angular momentum with repect to the earth's center is

$$L_{\mathrm{moon}} = M_{\mathrm{M}} v r.$$

Explanation of spring tide and neap tide.

The gravitational force just balances the centrifugal force:

$$\frac{\gamma M_{\mathrm{E}} M_{\mathrm{M}}}{r^2} = \frac{M_{\mathrm{M}} v^2}{r} \quad \Rightarrow \quad v = \sqrt{\frac{\gamma M_{\mathrm{E}}}{r}}.$$

Hence: $L_{\mathrm{moon}} = M_{\mathrm{M}} \sqrt{\gamma M_{\mathrm{E}} r}$. If L_{moon} increases, the distance earth–moon also increases. This increase amounts to about 3 cm per year.

The transfer of angular momentum from earth to moon is explained in the following somewhat simplified model. The friction between the waters of the oceans and the earth crust causes the two high-tide waves to flow with some delay behind the earth–moon axis (see figure). The differences in the gravitational forces **N** and **F** result in a torque that decreases the earth's angular momentum. The sum of the reactive forces acting on the moon has a component along the moon motion. Hence, there exists a torque that increases the angular momentum of the moon.

The tidal forces of the earth onto the moon over the ages have resulted in the moon always showing the same face toward earth: The eigenrotation of the moon is already decelerated so much that its period coincides with the revolution time of the moon about earth.[6]

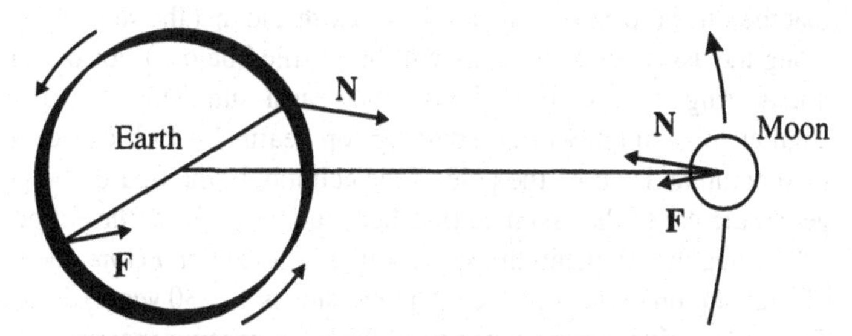

The high-tide waves are partly convected by the earth rotation.

[6]We recommend for further reading: Peter Brosche: The deceleration of the earth rotation, *Contemporary physics (Physik in unserer Zeit)* 20 no. 3 (1989) 70.

Precession and nutation of the earth

In the following considerations we always take into account that the celestial bodies (e.g., earth) have a finite spatial extension.

As the earth has no exact spherical shape but is a flattened rotational ellipsoid, and because the rotational axis of the earth is inclined against the ecliptic, the sun performs a torque $\mathbf{D}$ onto the earth that generates a change of angular momentum $d\mathbf{L}$ of the earth: $\dot{\mathbf{L}} = \mathbf{D}$ or $d\mathbf{L} = \mathbf{D}\,dt$. The torque $\mathbf{D}$ and hence also $d\mathbf{L}$ are perpendicular to the angular momentum $\mathbf{L}$. Because this relation holds at any time, the vector $\mathbf{L}$ moves along the surface of a cone, whose axis is the polar axis of the ecliptic. This cone is called the *precession cone*. The problem of motion of spinning bodies will be treated in more detail in connection with the theory of the top in *Classical Mechanics: Systems of Particles and Hamiltonian Dynamics*.

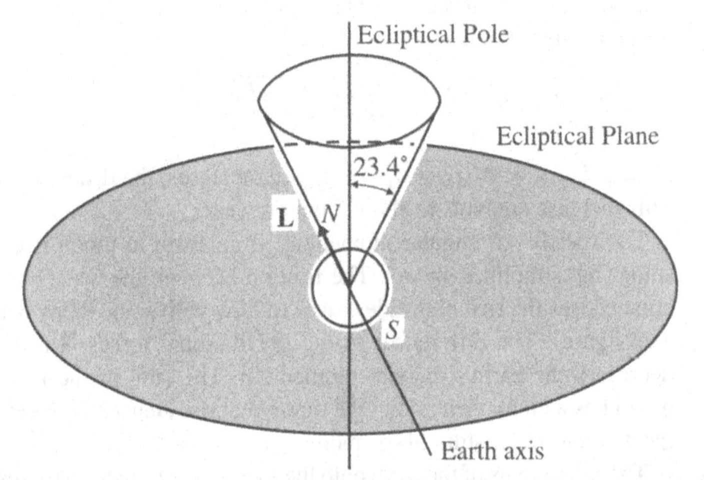

The geometry of the earth's precession. The bulge of the geoid is exaggerated.

We consider the problem from the earth's point of view and base it on the assumptions that the sun were circulating about the earth and that the sun's mass is uniformly distributed along the assumed path. (This will be justified below.) For our consideration there exists a mass ring about earth at the distance earth–sun. This mass ring generates a change of angular momentum for the spinning top "earth," which causes a rotation of the angular momentum axis about the pole of the ecliptic. In the figure, the pole of the ecliptic stands perpendicular to the assumed (hatched) sun orbit plane, the ecliptic.

The angular momentum axis describes a *precession cone* about the pole of the ecliptic. The revolution period of the earth precession is 25,730 years (the so-called "platonic year"). This now justifies our assumption of the homogeneous mass ring "sun" since the sun would have performed 25,730 turns about the earth during one period of the precession motion.

Besides the attraction by the sun, still other attractive forces by the moon and other planets are acting on earth, which also produce a change of the angular momentum.

The largest perturbations are caused by the moon; they result in precession motions with a period of 9.3 years.

Because of the flattening of the earth, the earth's axis and the angular momentum axis do not coincide exactly, such that the earth's axis moves about the angular momentum axis. These fluctuations of the earth's axis are called *nutations*. The measured period of the nutation motion of earth is 433 days.

A detailed quantitative discussion of these phenomena is given in the chapter on the theory of the spinning top (gyroscope) in the volume *Classical Mechanics: Systems of Particles and Hamiltonian Dynamics* of these lectures.

Small bodies in the solar system

The more thoroughly the astronomers investigate the solar system, the more difficult it becomes to maintain the classical subdivision into the various categories for the smaller celestial bodies. Several of the moons orbiting about the planets meanwhile have been uniquely identified as captured small planets (asteroids). Most of the asteroids, which presumably consist of the material of a "prevented" planet, are orbiting about the sun between the orbits of Mars and Jupiter. Several of them, however, on their flight also closely approach the earth's orbit.

It became possible by refined observation techniques to detect even small planets with a diameter of few meters in our neighborhood. Thus, in size they are comparable to meteorites.

At certain time intervals of the year, shooting stars are piling up in the sky, namely always when the earth is crossing the path of a comet. From that phenomenon the astronomers conclude that many meteorites are fragments of comets. Other meteorites display a composition that suggests an origin from small planets. It is also known that comets may split up and decay into debris. Initially intact comets later returned as twin comets.

Such a decay may obviously happen also among small planets. An English–Australian observation program has led to the discovery of an asteroid in 1991 denoted as 1991 RC and later dubbed "5786 Talos".[7] This object practically follows the same path as the small planet Icarus did, which had approached the earth in 1968 to only 6 million km.

In October 1990, the astronomers discovered a small planet with a diameter of only 60 to 120 m. One month earlier a telescope on the Kit Peak in Arizona had been set into operation for a systematic search for small planets in the close vicinity to earth, which raises serious problems for the classification of small cosmic objects. By means of this device, a "small planet" (1991 BA) of only 5- to 10-m diameter had been detected, which 12 hours later passed earth at a distance of 170, 000 km. At tihs time, it was the closest object to earth ever detected, and so small that it might also be a meteor.

The systematic search on the Kit Peak for cosmic rocks is performed for the first time by means of electronic detectors (CCD). Therefore, one has to expect such findings more frequently in the future. In just October and November 1991, four more objects were found with diameters less than 30 m each. Whether these were small planets or meteors could not be cleared in either case. For a meteor observed in 1972 in the west of the United States,

[7] D. Steel, Nature **354** 265-267 (1991).

it has been estimated that the glowing body had a diameter of 4 m—that is, not much less than the object 1991 BA.

By means of the telescope on the Kitt Peak, during only 10 months the astronomers found evidence for 15 formerly unknown "small planets" on their way toward earth, moreover 2000 further asteroids per month. The frequency of collision of such objects with earth will soon have to be calculated anew with additional data. At a conference held in 1991 in St. Petersburg ("The Asteroid Hazard") the participants still estimated the impact rate of a rock of 50-m diameter to be one event per century. This seems to be a major danger. Actually, meteors so far only rarely have caused noticable damages, because only a very small fraction of the earth's surface is inhabited.

If the object 1991 BA collided with earth, the impact energy—assuming a mass density of typical meteorite material—would be equivalent to about 40 kilotons of TNT. This is three times the energy of the Hiroshima bomb. For some time the American space agency NASA made plans and, indeed, has arranged for systematically localizing small objects moving toward earth and, if necessary, to destroy them before a collision. Whether such a project is meaningful and feasible with present-day means and wether it finds continous funding remains to be seen.

In this respect it is worthwhile to note that in December 2001, NASA took the decision to stop a routine search program for small nearby asteroids with the help of the 300-m radio telescope in Arecibo/Puerto Rico. The American Congress had instructed NASA to track down until 2008 all astronomical bodies larger than 1 km that may represent any danger to earth. However, Congress did not provide enough funding to accomplish this task, NASA says. The observations with the telescope in Aceribo are extremely important for the determination of the actual position, velocity, and orientation of the orbit of possibly dangerous small objects. Moreover, the telescope allows to take radar maps of some of these bodies. The only remaining radio telescope for the search for the "NEOs" (near-earth objects) is now the antenna of NASA's Deep Space Network at Goldstone/California. All other telescopes involved in the search for NEOs are optical telescopes.

Recent research on the solar system—Jupiter's large family of moons.

The exploration of our solar system is, obviously, far from complete. This view may be corroborated by the discovery of 11 hitherto unknown moons of Jupiter in December 2001, and of further 18 moons during the year 2002, bringing the total number of moons of the largest planet of our solar system to 58 (as of April 2003).

Jupiter has clearly captured several asteroids and minor planets within its gravitational field. The recently discovered moons had been found during a well-directed search program by a group of astronomers from Britain and Hawaii.[8] One expects that the overall number of satellites of Jupiter with a diameter of at least 1 km is well into the hundreds. Further discoveries will surely be made.

[8]See, e.g., the web page of the group leader, David Jewitt, from the University of Hawaii, at http://www.ifa.hawaii.edu/~sheppard/satellites/jup.html.

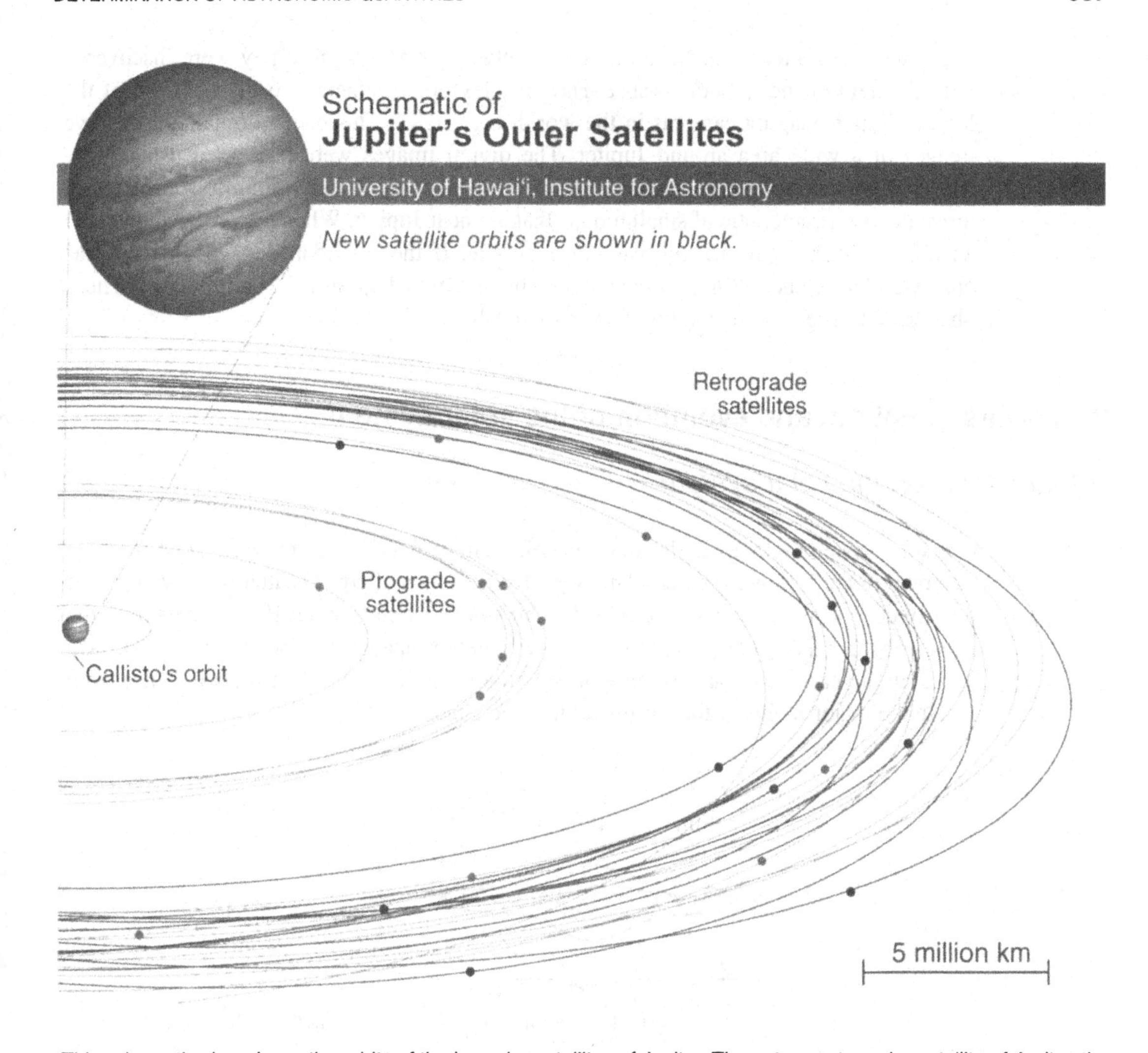

This schematic view shows the orbits of the irregular satellites of Jupiter. The outermost regular satellite of Jupiter, the Galilean moon Callisto, is shown for reference. The orbits of the new satellites are shown in black. (©University of Hawaii, reproduced with kind permission)

The newly discovered moons all are so-called irregular satellites of Jupiter, characterized by wide, elliptic orbits that do not lie inside the ecliptic. Many of those irregular satellites (including all the new ones) move along retrograde orbits, namely in a direction opposite to the direction of Jupiter's rotation.

The largest one of the irregular satellites, Himalia, was detected already in 1904. The retrograde orbit of these bodies is a clear hint that they are not primordial satellites of Jupiter, but captured objects. How Jupiter could capture and bind these small planetoids is not yet known. Astronomers cannot explain these events by celestial mechanics alone. It may be possible that Jupiter in the early stages of its history had a far-reaching atmosphere that could slow down the small planetoids.

Jupiter's new moons all have diameters between 2 and 4 km. They were discovered with the help of the Canada–France–Hawaii telescope (diameter 3.6 m) with one of the largest digital imaging cameras in the world, the "12K". This camera obtained sensitive images of a wide area around Jupiter. The digital images were processed using high-speed computers and then searched with an efficient computer algorithm for objects with movements characteristic of small moons that are near Jupiter. When the program detected an object, visual confirmation was made by eye. If the candidate looked good, it was observed during succeeding months at the University of Hawaii's 2.2-m telescope. These observations allowed the computation of their orbits.

Properties, position, and evolution of the solar system

General facts on the solar system

Our solar system belongs to the spiral nebula "Milky Way." A lateral view of Milky Way is shown in the following figure. The lines denote zones of equal matter density, with the density decreasing from inside to outside. Our solar system is about 10 kpc apart from the center of the galaxy. (The length unit *parsec* has the magnitude 1 pc $= 3.086 \cdot 10^{13}$ km $= 3.26$ lightyears. This value stems from the following definition: 1 pc is the distance from where the major radius of the earth's orbit is seen under $1''$.)

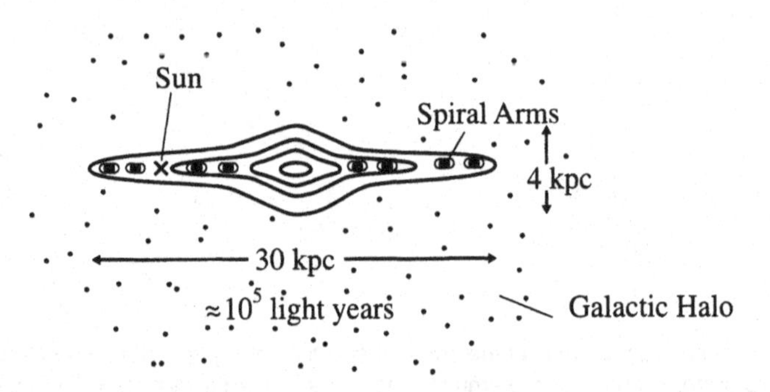

Schematic profile of the Milky Way galaxy. The galactic halo is scarcely occupied with old stars, but perhaps filled with so-called dark matter.

The two figures show how our Milky Way galaxy would look if we could see it from the top or from the side. This "synthetic photography" has been established by a computer from data measured within our galaxy.

Data on the solar system are compiled in the following figures. When considering the solar system, keep in mind that all planets have the same direction of revolution and almost the same orbital plane. Only Pluto displays larger deviations in its data, which led to the assumption that

Top view of the Milky Way galaxy.

Pluto was captured by the sun only after the evolution of the planetary system. In the context of the formation of the solar system the following, so far not yet explained empirical law for the major semi-axes of the planets deserves interest (the planetoids are well fitting in there). It is the called the *Titius–Bode relation* for the major semi-axes a_n of the planets: $a_n = a_0 k^n$. Thereby $a_0 = 1$ AU and $k \approx 1.85$. The abbreviation "AU" means "astronomical unit" = major semi-axis of the earth's orbit. The integer numbers n are associated to the planets (see the figure on p. 309).

Side view of the Milky Way galaxy.

Mercury	n = −2
Venus	−1
Earth	0
Mars	1
Planetoids	2
Jupiter	3
Saturn	4
Uranus	5
Neptune	6
Pluto	7

$\ln \frac{a_n}{a_0}$

Mercury Venus Earth Mars Ceres Jupiter Saturn Uranus Neptune Pluto

Illustration of the Titius–Bode relation.

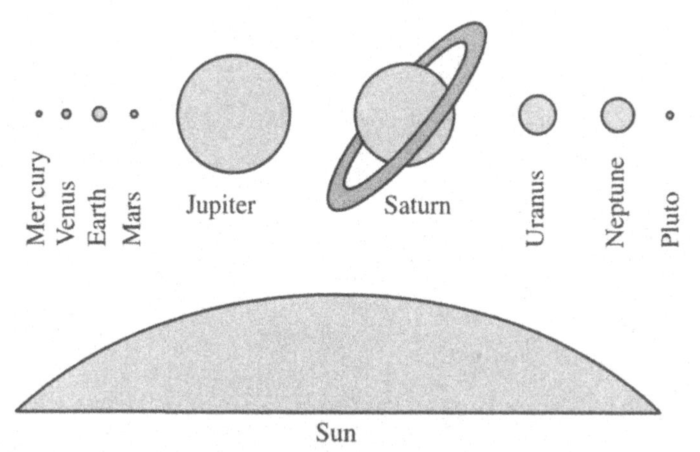

Illustration of the relative sizes of the planets and the sun.

Our solar system in figures.

Name	Sun	Mercury	Venus	Earth	Mars	Ceres (planetoid)
Symbol	☉	☿	♀	⊕	♂	—
Year of discovery	—	—	—	—	—	1801
Discoverer	—	—	—	—	—	Piazzi Gauss
Sidereal period (in earth years)	—	0.205	0.615	1	1.88	4.6
Mean distance Sun–planet in AE	—	0.387	0.723	1	1.524	2.767
Mean distance Sun–planet in 10^6 km	—	57.9	108.2	149.6	227.9	—
Eccentricity of orbit	—	0.206	0.007	0.017	0.093	0.076
Inclination of orbit	—	7°	3°	0°	1°51′	10°37′
Inclination of equator	—	~ 2°	~ 3°	23°27′	23°59′	—
Radius in earth radii	109	0.382	0.949	1	0.533	0.055
Mass in earth masses	$3.3 \cdot 10^5$	0.054	0.814	1	0.107	~ 0.0001
Surface gravity in *g*	—	0.4	0.0.9	1	0.4	—
Density (g/cm^3)	1.4	5.46	5.06	5.52	3.93	3.3
Sidereal rotation period	~ 25 d	58^d17^h	-243^d	23^h56^m	24^h37^m	$9^h0.5^m$
Moons	—	0	0	1	2	0
Mean surface temperature (in K)	5785	100–625	740	288	216	160
Spectroscopically found gases in atmosphere	H, He	He, H	CO_2, N_2, H_2O	N_2, O_2	CO_2, N_2, O_2, H_2O	—
Supposed chemical composition (main components)	H, He	Fe, Si	Fe, Si, O	Fe, Si, O	Fe, Si	—

$m_{Earth} = 5.976 \cdot 10^{24}$ kg, 1 AE = $1.496 \cdot 10^{6}$ km.

Name	Jupiter	Saturn	Uranus	Neptune	Pluto
Symbol	♃	♄	♅	♆	♇
Year of discovery	—	—	1781	1846	1930
Discoverer	—	—	Herschel	Leverrier Galle	Lowell Tombaugh
Sidereal period (in earth years)	11.8	29.45	84.015	164.78	247.7
Mean distance Sun–planet in AE	5.203	9.539	19.128	30.057	39.50
Mean distance Sun–planet in 10^6 km	779	1432	2888	4509	5966
Eccentricity of orbit	0.048	0.056	0.047	0.009	0.247
Inclination of orbit	1°18′	2°29′	0°46′	1°46′	17°10′
Inclination of equator	3°04′	26°44′	98o	29°	> 50°
Radius in earth radii	10.97	9.03	3.72	3.43	0.24 (?)
Mass in earth masses	317.45	95.21	14.9	17.2	0.002 (?)
Surface gravity in *g*	2.4	0.9	0.9	1.7	0.1
Density (g/cm^3)	1.33	0.71	1.55	2.41	0.8 (?)
Sidereal rotation period	9^h55^m	10^h40^m	-23^h50^m	17^h50^m	6^h23^m
Moons	39	21	5	8	1
Mean surface temperature (in K)	134	97	60	57	43 (?)
Spectroscopically found gases in atmosphere	H_2, He, CH_4, NH_3, H_2O	H_2, He, CH_4, NH_3	H_2, CH_4	H_2, He, CH_4, NH_3	(?)
Supposed chemical composition (main components)	H, He	H, He	H_2O, CH_4, NH_3	H, He	(?)

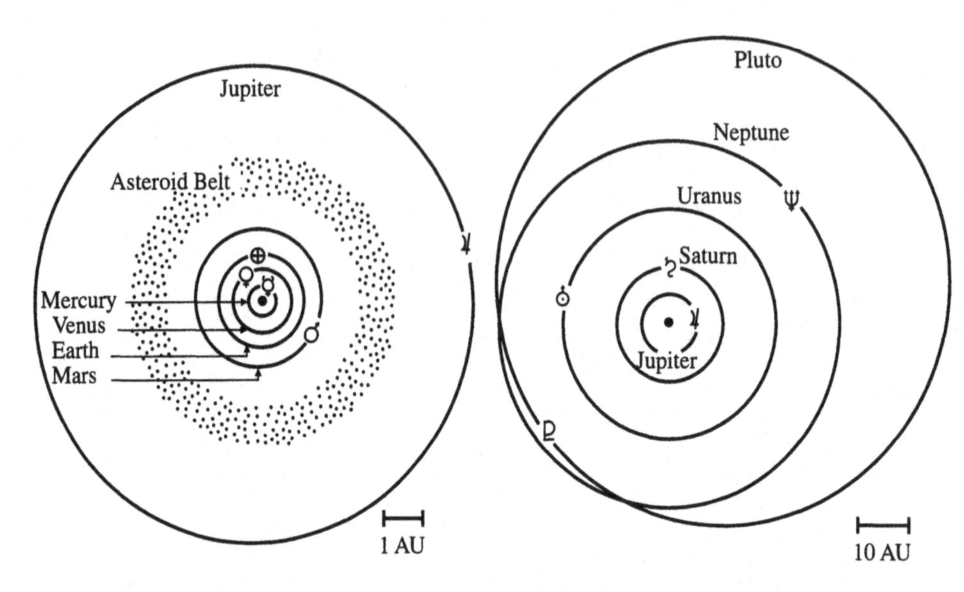

Maps of the solar system in two different scales. 1 AU (astronomical unit) is the radius of the earth's orbit. The symbol of each planet is given at the perihelion of its orbit.

Closed orbits and perihelion motion

As we have seen, there exist spatially fixed closed orbits in the $1/r$-force field. But if the gravitational potential differs somewhat from r^{-1}, hence $V(r) \neq r^{-1}$, for example,

$$V(r) = Ar^{-1} + Br^{-2} + Cr^{-3} + \cdots ,$$

a rosette motion may arise. The effective potential has a minimum as before, such that a minimum and a maximum radius exists. But in general the paths are no longer closed curves as in the case of the $1/r$-potential. They then must be rosette orbits. (We refer to Problem 26.12.)

Deviations from $V(r) \sim r^{-1}$, such that the potential differs from cr^{-1}, are caused by the influence of other planets on the path of a given planet, or by deformation (flattening) of the central star. These perturbations generate both a perihelion motion of the planets

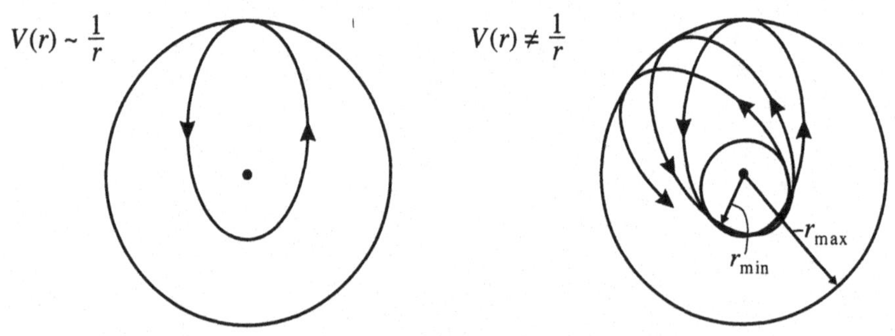

Closed orbits and rosette orbits in the force field of a central mass.

as well as the typical rosette path. The planetary orbits agree with the values calculated according to Newton, except for the case where the planet is very close to the sun. The normal mutual perturbations of the planets may be calculated by means of the tools of celestial mechanics. For Mercury, however, the observed value for the forward motion of the Mercury perihelion is too large to be traced back in full to perturbations by other planets and to the flattening of the sun. The calculated value is by 43″ per century smaller than the measured one. Einstein's theory of general relativity explains this effect.

For the mathematical treatment of the perihelion motion, we refer to Problems 26.12 and 28.4.

Evolution of the solar system

A sun is formed if a dense cloud of interstellar gas and dust contracts under the action of the gravitational force. Our sun is surrounded, however, by many other bodies forming the planetary system. The evolution of this planetary system is at present not yet fully understood. There are competing theories that always explain only some of the properties of the planetary system.

The multitude of theories may be grouped into three main classes that differ in the mechanism of formation of the planets.

1. Theories stating that the formation of planets is independent of the formation of the sun: The planets only emerged when the sun was already a normal star. This class includes, for instance, the tidal theories.

2. Theories stating that after the formation of the sun, the planets were generated from interstellar matter. These are the so-called accretion theories, which assume an increase of mass within a plane (the ecliptic).

3. Theories according to which the planets are formed out of the same nebula and by a similar process as the sun is formed (nebular hypotheses).

In the following paragraphs, several of the basic mechanisms of these theories will be further described.

1. Tidal theories (Bickerton, 1878; Chamberlain, 1901; Moulton, 1905; Jeans, 1916; Jeffreys, 1918)

Two suns pass each other but without mutually capturing each other. Due to the tidal forces, matter is pulled out of the suns that shall condensate to planets. Aside from the low probability of such an encounter, this theory has several further deficiencies. It could in no way explain the chemical composition of the planets, and the planet orbits should be strongly elliptic according to this theory. Moreover, some later calculations (Spitzer, 1939) showed that matter ejected by a star cannot condense to a planet, because of its high temperature. Therefore, the tidal theories meanwhile have been dropped.

2. Accretion theories (Hoyle and Littleton, 1939)

If the sun moves through a cloud of interstellar matter, it can bind particles by the gravitational force. Due to the attractive force between the particles and by collisions, larger masses may be formed that shall grow up to the size of the present planets. One also has to take into account the consequences of electromagnetic effects (Alfven, 1942). As shown in Example 28.2, the magnetic field of the sun prevents a particle with charge q and mass m to come closer to the sun than to a critical radius r_c, which is proportional to $(q/m)^{2/3}$. Therefore, the heavier particles pile up near the sun. By appropriate assumptions on the magnetic field of the sun, the chemical composition of the planets may be roughly explained.

3. Nebular theories (Descartes, 1644; Kant,[9] 1755; Laplace, 1796)

The gas nebula from which the sun originated was flattened by its rotation. Because of turbulences, parts of the nebula split off, which then begin to contract. They thereby rotate faster and faster because the angular momentum is conserved. The central part of the nebula forms the sun, while the peripheral region leads to many proto-planets. In the interior of these proto-planets a core evolves from the solid fractions of the nebula. The number of proto-planets may decrease by collisions.

In more recent time the following mechanism has been investigated: The solid fractions of the nebula are enriched in the middle plane of the disk-shaped gas nebula by the gravitational force (see figure). With increasing concentration, this dust disk becomes unstable and decays into regions of several kilometers of diameter. These regions are the cores for further mass accumulation. Larger and larger objects develop by attraction of further solid particles and by collisions, which grow to the size of planets.

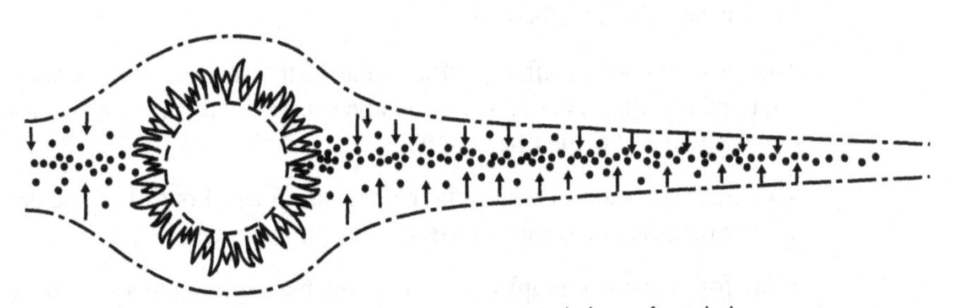

Motion of dust particles in the central plane of a nebula.

[9] *Immanuel Kant*, philosopher, b. April 22, 1724, Königsberg—d. there Feb. 12, 1804.

Kant originated from a workman's family. He attended the pietistic Friedrich gymnasium in his hometown and until 1746 studied there natural sciences, mathematics, and philosophy; from 1747 to 1754 he was a private tutor. In 1755 he did his *Habilitation* in Königsberg as magister of philosophy; he also served as subordinate librarian of the library of the castle. In 1763 he refused an offer for a professorship for poetry; in 1770 he became professor for logic and metaphysics. In 1786 and 1788 he administrated the principalship. In 1796 he stopped lecturing for health reasons. His life passed without striking external events: He never left East Prussia and rarely left Königsberg [BR].

If a certain size is exceeded, then the gaseous residues of the nebula (H_2, He) may be bound by gravitation; hence this theory may also explain the formation of Jupiter and Saturn.

There occurs a temperature gradient within the gas nebula such that the nonevaporating substances (dust particles) are condensing in the hot zone in the interior, while the gases (e.g., H_2O, NH_3, and CH_4) may condense only in the colder zones at larger distance from the young sun. This mechanism, in principle, can possibly explain the chemical composition of the planets.

The angular momentum of our solar system resides to a large extent in the planets. Our sun contains 99.87 % of the mass but only 0.54 % of the total angular momentum available in the solar system. If the total angular momentum were concentrated to the sun, the resulting value would be typical for young stars. Thus one may conclude that the sun must have transferred angular momentum to the planets. A mechanism for this process is provided by magneto-hydrodynamics (Hoyle, 1960; Edgeworth, 1962): In the plasma (ionized matter) of the gas nebula, very large perturbations may occur, and stabilized magnetic fields, "frozen" in the plasma, may be convected. The transfer of angular momentum from the center to the peripheral region may be explained in this way, similar to the principle of the eddy-current brake.

Only in the most recent time have detailed computer simulations of the evolution of a gaseous nebula been performed. One must take into account further physical effects (e.g., pressure, friction, solar wind, tidal forces, etc.). In due time one may judge whether these theories actually explain the presently observed properties of the planetary system.

World views

Geocentric—the Ptolemaic world view (about 140 AD)

The Ptolemaic[10] world view was the base of astronomy until the 17th century. It considers the earth as the world center being at rest. The moon, sun and the planets orbit about earth. The fact that this world view could survive undisputedly over such a long period is explained best by a sketch, showing that predictions on the position of the planets could actually be made, based on this view. It thus had "predictive power."

If one considers the actual situation (sun in the center of the planetary system), one gets the upper two figures for which hold $\mathbf{r}_p = \mathbf{R} + \mathbf{r}_E$ or $\mathbf{R} = \mathbf{r}_p - \mathbf{r}_E$, respectively.

[10] *Claudius Ptolemy*, b. after 83 AD, Ptolemais (middle Egypt)—d. after 161 AD. It is only known that he worked in Alexandria. He is considered as the most important astronomer of the Ancient World. He is the main representative of the geocentric world view. His *Great Astronomic System*—in the Arabic translation *Kitab al-magisti* known as *Almagest*—constituted the fundamental work on astronomy until Copernicus. In his representation, Ptolemy used the theory of epicycles of the Apollonios, a trigonometry of secants, and the stereographic projection. Ptolemy still published an *Optics*, the very influential astrologic work *Tetrabiblos*, and the most valuable *Introduction to Geography*, which was extraordinarily influential on science of the Middle Ages, just as astrology.

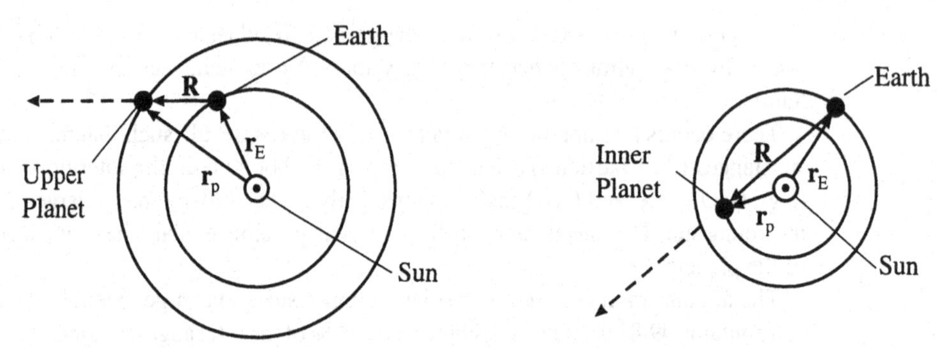

Inner and outer planets in a heliocentric world model.

Correspondingly, $\mathbf{r}_p$ circulates about the sun once in one planetary year, and $\mathbf{r}_E$ does the same in one earth year. For the geocentric world view, we obtain a different figure:

The equation $\mathbf{R} = \mathbf{r}_p - \mathbf{r}_E$ also holds in the geocentric world view, but here the *Ptolemaic deferent* has been introduced. It is an immaterial circle performed by $\mathbf{r}_p$ with the siderean revolution of the planet about the earth. Because one could not yet determine the distance of a planet, only the direction of $\mathbf{R}$ mattered but not its magnitude. This explains why the theory of epicycles describes the planetary motion correctly.

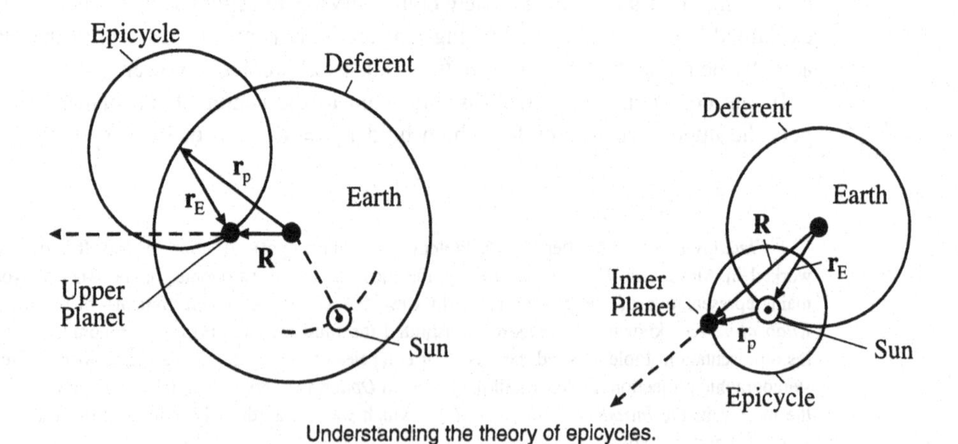

Understanding the theory of epicycles.

2. The heliocentric system—the Copernican world view

In the Copernican world view,[11] the sun is understood as the center (central body) of our planetary system. It culminated in Kepler's laws that allowed one to calculate all processes in the planetary system easily and exactly.

[11]*Nicolaus Copernicus*, German *Koppernigk*, Polish *Kopernik*, astronomer and founder of the heliocentric world view, named after him Copernican, b. Feb. 19, 1473, Thorn—d. May 24, 1543, Frauenburg (East Prussia). In 1491 he began humanistic, mathematical, and astronomical studies at Cracow University. From 1496–1500 he studied civil and clerical law in Bologna. At the instigations of his uncle, Bishop Lucas Watzelrode, he was admitted to the chapter of Ermland at Frauenburg in 1497. From autumn 1501 he studied in Padua and Ferrara, graduated there on May 31, 1503, as doctor of canonical law, and then studied medicine. After returning home in 1506, he lived in Heilsberg as secretary of his uncle from 1506 until his uncle's death in 1512 and was involved in administrating the diocese Ermland. As chancellor of the chapter Copernicus lived from 1512 mostly in Frauenburg. He resided as governor of the chapter from 1516–1521 in Mehlsack and Allenstein, and in 1523 he served as administrator of the diocese of Ermland. From 1522–1529 he represented the order chapter as deputy at the Prussian state parliaments and there in particular also supported a monetary reform.

His paternal family originates from the diocesian country Neiss in Silesia; hence his German origin may be considered as established, since in writing he utilized only the German and Latin languages. Copernicus was also considered as a famous physician, as is indicated by the lily of the valley in one of his woodcuts. As astronomer, Copernicus completed what Regiomontan had imagined: A revision of the doctrine of planetary motion, taking into account a series of critically evaluated observations. Only on such a basis could one then speculate on a calendar reform. The urgency of this reform was generally recognized at the beginning of the 16th century. Copernicus was presumably influenced by these considerations. In the course of his work he then decided to accept a heliocentric world system, inspired by vague antique writings. A brief, preliminary report on this topic is the "Commentariolus," presumably written before 1514. Already here the decisive assumptions are expressed: The sun is in the center of circular planetary orbits, and the earth also circulates about the sun; the earth rotates daily about its axis and in turn is orbited by the moon. The wider public got the first information on the Copernican doctrine by the *Narratio Prima* of G. J. Rheticus.

The main work of Copernicus, the *Six Books on the Orbits of Celestial Bodies (De Revolutionibus Orbium Coelestium Libri VI*, 1543, German 1879, new edition 1939), emerged only in the year Cpernicus died. It was dedicated to Pope Paul II, but the original foreword of Copernicus was replaced by a foreword of the Protestant theologist A. Osiander that inverted the meaning of the whole subject. The doctrines of Copernicus remained uncontested by the Church until the edict of the index congregation of 1616. The remaining imperfections of the Copernican theory of planets were removed by J. Kepler. But just as Copernicus, Kepler could also not prove in modern sense the correctness of the heliocentric system. Still at the time of I. Newton, the astronomic data were not precise enough to establish the very small "Copernicus effects." This was achieved only in 1728 by J. Bradley with the discovery of the aberration of the fixed stars, and in 1839 by F.W. Bessel by the first measurement of a fixed star's parallax. The objections of the opponents of the Copernican view are intelligible, since for most of the fixed stars the parallaxes are not detectable even by modern methods of measurement, due to the large distances from the sun. His opponents urged, for example, the famous observer T. Brahe to establish his own model of the planetary system, which represents a compromise between the geocentric and the heliocentric systems [BR].

A model of the sun's environment[12]

Already the nearest stars are so far away from earth that it is difficult to get an idea on the dimension. The following model shall assist on that point: The planetary system and the environment of the sun are reduced by the scale 1:100 billions. Then 1 cm in the model corresponds to 1 million km in nature. The solar system then could be accomodated on a schoolyard or on a large crossroads: The sun itself would have a diameter of 1.4 cm. At 1.5 m apart the earth of a size of 0.1 mm would be localized. At nearly 8-m distance from the sun follows Jupiter with a size of 1.4 mm, and at a distance of 59 m follows the outer planet Pluto with 0.05-mm size. Proxima Centauri would be apart from there by 410 km, Sirius by 820 km, etc. This scale model is shown in the following figure:

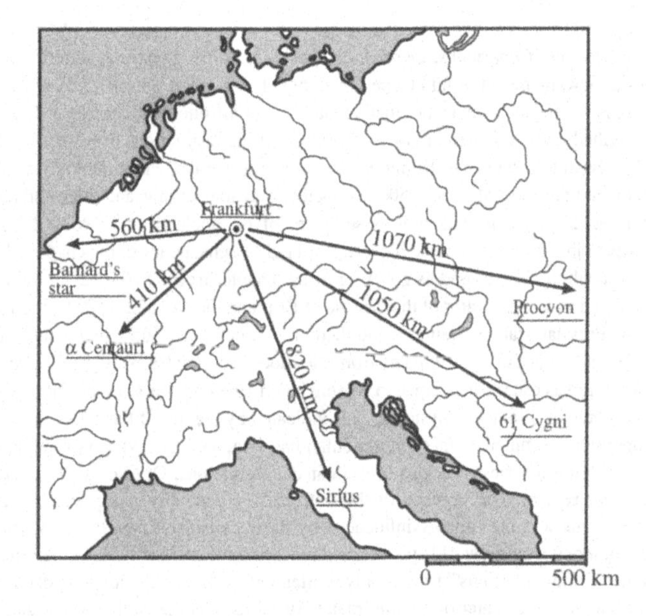

The closer environment of the sun in a model: Our sun is located in Frankfurt; its diameter on this scale ($1 : 10^{11}$) is only 1.5 cm. The next star is α Centauri at a distance of 410 km from Frankfurt, i.e., approximately in Paris.

The nearest stars to the sun are collected in the following table:

[12]We follow here the excellent booklet of J. Hermann: *dtv-Atlas zur Astronomie (Tafeln und Texte mit Sternatlas)*, Deutscher Taschenbuch Verlag München.

Star	*Constellation*	*Distance in lightyears*
α Centauri/Proxima Centauri	Centaurus	4.3
Barnard's arrow star	Ophiuchus	5.9
Wolf 359	Leo	7.7
Luyten 726-8	Cetus	7.9
Lalande 21 185	Ursa Maior	8.2
Sirius	Canis Maior	8.7
Ross 154	Sagittarius	9.3
Ross 248	Andromeda	10.3
ε Eridani	Eridanus	10.8
Ross 128	Virgo	10.9
61 Cygni	Cygnus	11.1
Luyten 789-6	Aquarius	11.2

Other planetary systems?

Due to the large distances between the stars and the fact that planets for themselves are very dim, there have only recently been successful attempts to obtain strong evidence for the existence of other planetary systems besides the solar system.[13] The star 51 Pegasi (in a distance of about 45 lightyears from the solar system in the constellation of Pegasus) showed periodic variations of its radial velocity (see figure).

This observation can be explained by the motion of the star and an assumed planet around their common center of mass. The radial motion of the star was deduced from observations of the Doppler shift of approximately 5000 absorption lines in the spectrum of the star. This methods allows results as precise as 15 m/s. In order to give a better impression of this velocity, we mention that the velocity of the sun, which is caused by the common motion of the sun and Jupiter is about 13 m/s. The indirectly observed planet, 51 Peg B, is supposed to have roughly the mass of Jupiter ($0.5M_\mathrm{J} \leq M \leq 2M_\mathrm{J}$) and a nearly circular orbit around its solar-type star ($\epsilon \approx 1$) with a radius of only 0.05 AU and the short period of $T \approx 2$d. This means that when compared to our solar system, the planet would move around clearly within the orbit of Mercury. Other possible explanations for the varying radial velocity of 51 Pegasi are very improbable.

Current models for the formation of planets do not foresee the emergence of such giant planets so close to their central stars, leaving the origin of 51 Peg B (and of similar planets that have been found since with similar methods) somewhat obscure. Beside the possible migration of a giant, Jupiterlike planet to such a close orbit, another explanation may be the possible capture of a so-called brown dwarf. Brown dwarfs are dim stars whose mass is not sufficient in order to ignite the thermonuclear burning of hydrogen in the center of

[13]M. Major, D. Queloz, "A Jupiter-mass companion to a solar-type star," *Nature* **378** (1995) 355; see also, e.g., *Europhysics News* **26** (1995) 123.

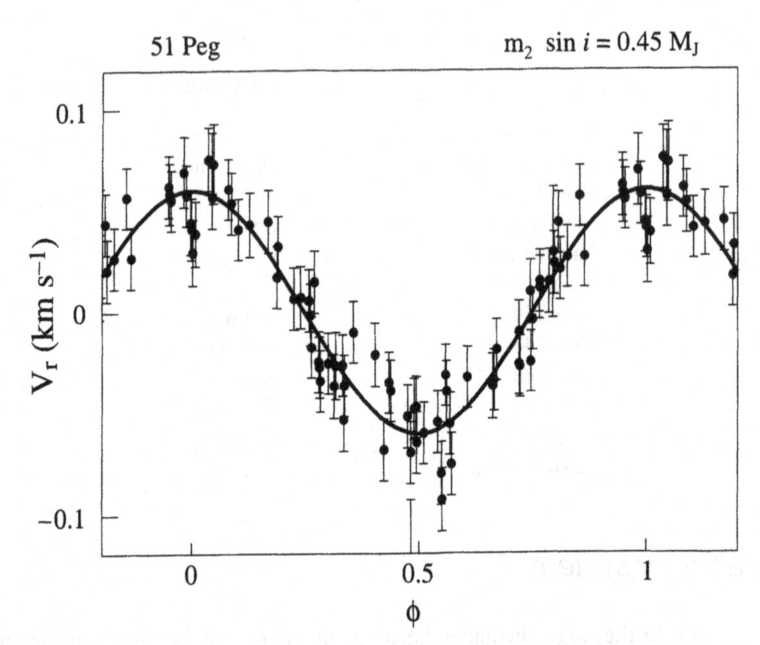

The orbital motion of the star 51 Pegasi, corrected for long-term variations of the velocity of the center of mass. The points, plotted as a function of the phase of the orbital rotation, correspond to experimental estimates of the radial velocity as determined from the spectroscopic data. The solid line is the theoretical curve fitted for a circular orbit with a period of 4.2293±0.0011 days. It shows that the data are remarkably stable and sinusoidal. (From M. Mayor, D. Queloz, Nature **378** 355 (1995) ©Nature Publishing Group, reproduced with kind permission)

the star.[14] The periodic variations of the radial velocity of 51 Pegasi show a superimposed periodic perturbation with a longer period, hinting at a farther planet that is less massive and orbits a greater distance from the star. This implies that one can talk about a real planetary system.

Before the discovery of the planet 51 Peg B orbiting a solar-type star, one had already found two planets with masses comparable to the earth's mass and periods of several months, orbiting, however, a pulsar.

A further discovery has been the periodic variation of the luminosity of some stars, due to the partial eclipse of the central star by the transit of an orbiting planet[15]. The following figure illustrates this method: Data taken with the Hubble space telescope show the light curve (the luminosity as a function of time) of the solar-type star HD 209458 in the constellation of Pegeasus. This star has a jupiter-like companion in a very close orbit with

[14] A good overview and an explanation of the different burning cycles in the interior of a star that is contracting by the gravitational forces can be found in Chapter 18 of J.M. Eisenberg and W. Greiner: *Nuclear Theory. Vol 1: Nuclear Models*, 3rd ed., North Holland, Amsterdam, 1987.

[15] L. R. Doyle, H.-J. Deeg, T. M. Brown: *Searching for shadows of other Earths*, Scientific American, Sep. 2000, p. 38

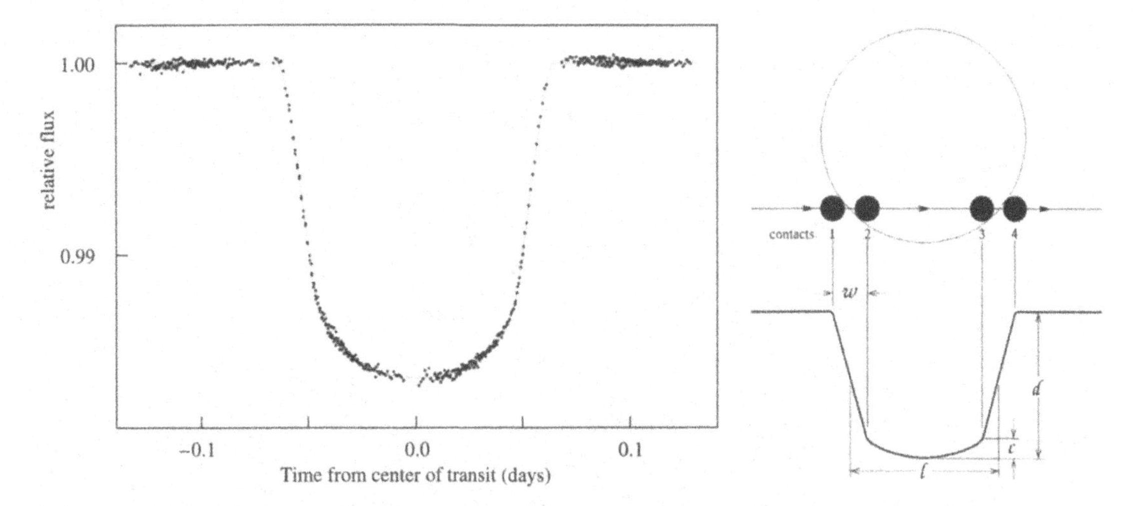

Left: The light curve of the star HD 209458. The star is partially eclipsed by the transit of its planet. Right: Schematic view of the partial eclipse of HD 209458 by its planet and the resulting light curve. (from T. M. Brown et al., Astrophysical Journal **552** 699-709 (2001), ©American Astronomical Society, reproduced by permission of the AAS.)

a period of about 3.5 days. Every transit of the planet in front of the star yields a dimming of the star, which is clearly visible in the light curve.

In the meantime (2001), more than 40 planets are believed to exist around stars in the vicinity of our solar system.

The spatial motion of the sun

From the spatial motion of the stars one may conclude that our sun also moves through the universe. The method of how to determine this motion is illustrated by the following example.

A driver moves by car along a straight road through the woods. If there were no chance to learn about the direction and speed of motion from other observations, one might derive it from the motion of the trees. When looking forward along the direction of motion, the trees seem to diverge. When looking perpendicular to the motion, the trees seem to pass the car in backward direction. If one looks in backward direction, the trees seem to converge (compare the following figure).

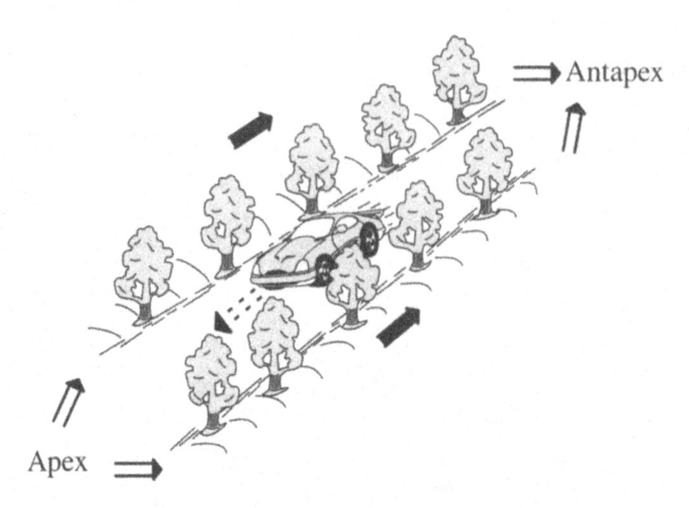

Motional effects that allow the determination of the speed and the direction when driving with constant speed on a road bordered by trees.

The same holds also for the motion of the sun through the universe: One has to observe the systematic effects of motion of the stars. A complication as compared to the case of the moving car, however, is due to the fact that the stars don't stand fixed as the trees, but are moving by themselves. But one may expect that in a statistical observation of very many stars the individual motions of the other stars will no longer show up too much, such that the effect described above manifests itself clearly.

Radial velocities have the largest negative value in the direction of the apex, the largest positive value in the direction of the antapex (dashed double arrows). The tangetial velocities of stars are largest in the direction vertical to the sun's motion (bold arrows), whereas the radial velocities there are smallest.

This will of course work only then if the observed stars display no systematic motions, i.e., their individual spatial motions are actually distributed in a purely random manner (statistically). If certain directions of motion show up with some preferences, there may arise faults when deriving the spatial motion of the sun.

This may easily be visualized by assuming, for example, that the trees in the above example move all in one direction, say from left ahead to right behind the car, looking along the direction of motion.

Actually the premise of arbitrary directions of motion of the stars is not strictly fulfilled, which makes an exact determination of the sun's motion rather difficult. Rough assignments, however, could be made already by W. Herschel,[16] who at that time investigated only 13

[16] *Sir (since 1816) Friedrich Wilhelm (Willam) Herschel*, b. Nov. 15, 1738, Hannover—d. Aug. 25, 1822, Slough near Windsor. At first musician, he went in 1765 as organist to Great Britain. The theory of music led him to mathematics and optics, and in 1766 he began to cut mirrors with such a success that no less than 400 mirrors

stars. Later on the investigations were extended to a much higher number of stars. The sun apex, the target point of the spatial motion of the sun, has the coordinates $\alpha = 18\,h\,04\,m$, $\delta = +30°$, that is, is localized in the constellation of Hercules.

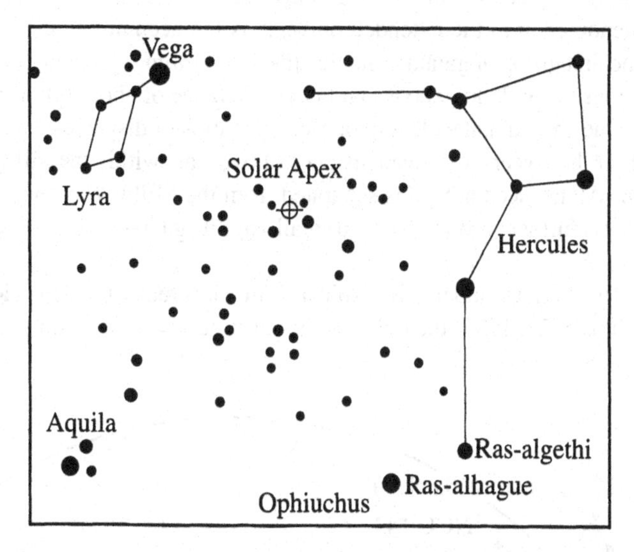

The position of the sun apex in the constellation of Hercules. Our solar system as a whole—located in the Orion branch of the Milky Way—moves toward the apex with a speed of 19.4 km/s $\approx$ 610 million km/year.

The velocity of the sun's motion may be derived from a systematic distribution of the radial velocities of the stars. The stars located in the direction toward the sun apex on the average show a negative radial velocity. As a result, one obtains a velocity of 19.4 km/s (610 million km/year) for the spatial motion of the sun. This motion relative to the neighboring stars is also denoted as peculiar motion (in contrast to the rotational velocity about the center of the Milky Way system).

In 1967 the peculiar motion of the sun could be determined for the first time also by radio-astronomic means, namely by the Doppler shift of the 21-cm radiation of interstellar neutral hydrogen. Taking into account the possible errors of measurement, this result agrees with the optical observations.

left his workshop. The largest one had a diameter of 1.22 m and 12-m focal length. The observations with his mirrors made him an astronomer. In 1781 he discovered the planet Uranus; in 1783 he established the motion of the solar system toward the constellation Hercules; in 1787 he found the two outer moons of Uranus; and in 1789 the two inner Saturn moons. His observations of double stars, nebula spots, and stellar clusters opened new fields of astronomy, and his star gauges founded the explorations of the structure of the Milky Way system.

Neighbourhood of our Milky Way

Our spiral nebula, the *Milky Way*, is embedded within the so-called local group, a cluster of about 9–10 galaxies. The Milky Way and the *Andromeda nebula*, just as the M33-galaxy, are spiral nebulas; all other galaxies are of a type of spherical clusters. A widespread phenomenon in the extended universe is the accumulation of galaxies to galaxy clusters. The first group of galaxies outside the *local group* is located toward the constellation Virgo; it consists of 2500 galaxies and is at a distance of about 60 million lightyears away.

One should make clear to oneself the ratios of distances: Our Milky Way has a diameter of 10^5 lightyears; the mean distance of two stars within the Milky Way is about 5 lightyears. The Andromeda nebula is separated from the Milky Way by $2 \cdot 10^6$ lightyears. The Milky Way is further "orbited" by two small satellite galaxies: The *Small* and the *Large Magellanic Cloud*.

The famous supernova explosion in the great Magellan cloud was seen on earth on February 24, 1987, the only one in our time whose light curve and neutrino showers have

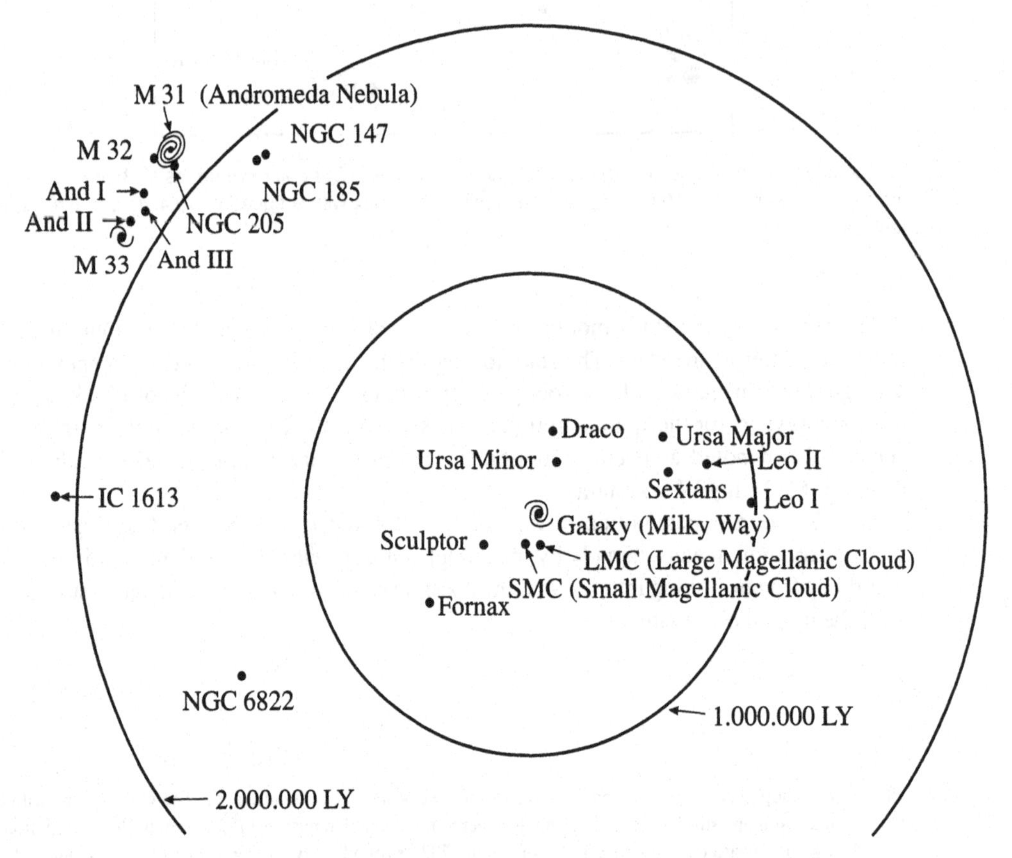

The local group. Only larger galaxies are shown. The circles indicate the distance from the Milky Way in millions of lightyears.

been recorded experimentally. Satellite galaxies are frequently observed. Andromeda also has two "small" satellite galaxies. The following figure illustrates our neighboring galaxies.

On the evolution of the universe

Our knowledge about the beginning of the universe is rather obscure, because it was born out of a state which cannot be described by any physical law we know of. We simply call these indescribable moments of birth of our universe the *Big Bang*.

Indeed, spectroscopic measurements have shown the existence of a relation between the redshift observed in all star spectra and the distance of the stars from earth. If we assume the Doppler effect as responsible for this shift, then the universe must expand in all directions. If all motions are now considered in a backward direction, then all bodies meet simultaneously in a certain space region. Here the cosmic "Big Bang" must have happened about $14 \cdot 10^9$ years ago. One imagines that matter (energy) was created by the transition from one state of the vacuum (true vacuum with zero energy) to an energetically deeper state of vacuum.

After the Big Bang, the universe rapidly expanded from an incredibly small region with dimensions of 10^{-33} cm and an unthinkably high energy density of 10^{94} g/cm^3—this initial phase of the Universe in known as the *Planck aera*. The Grand Unified Theories[17] of today suggest that physics was probably much simpler under the extreme conditions of the Planck aera, because all the forces we nowadays know of—gravitation, electromagnetism, weak interactions, and strong interactions—were one and the same, or indistinguishable. The world was then governed by an universal state of symmetry. However, while the universe was expanding rapidly, this symmetry was quickly broken into present–day forces with vastly different strength and range.

At the extremly short time of 10^{-23} s after the initial event, the entire now-existing matter at that time existed in the form of free elementary particles (photons, quarks, gluons, leptons—that is, electrons and neutrinos—, perhaps other, yet unknown elementary particles like supersymmetric particles) of enormous concentration ($\rho = 10^{55}$ g/cm^3) and temperature ($T = 10^{22}$ K). The expansion and the thereby implied cooling enabled the assembling of nucleons to nuclei, and finally the formation of complete atoms. Under the influence of gravitation, the cosmic primordial cloud then condensed to galaxies, and finally to individual stars.

In the following sections we shall give a short discussion of the modern ideas about the early universe.[18]

[17] see e. g. W. Greiner and B. Müller, *Gauge Theories of Weak Interactions*, Springer Verlag New York, 2000

[18] In this and the subsequent section about dark matter, we follow closely the excellent article by Klaus Pretzl, *In Search of the Dark Matter in the Universe*, Spatium **7**, May 2000, available from Association Pro ISSI at *http://www.issi.unibe.ch/spatium.html.*

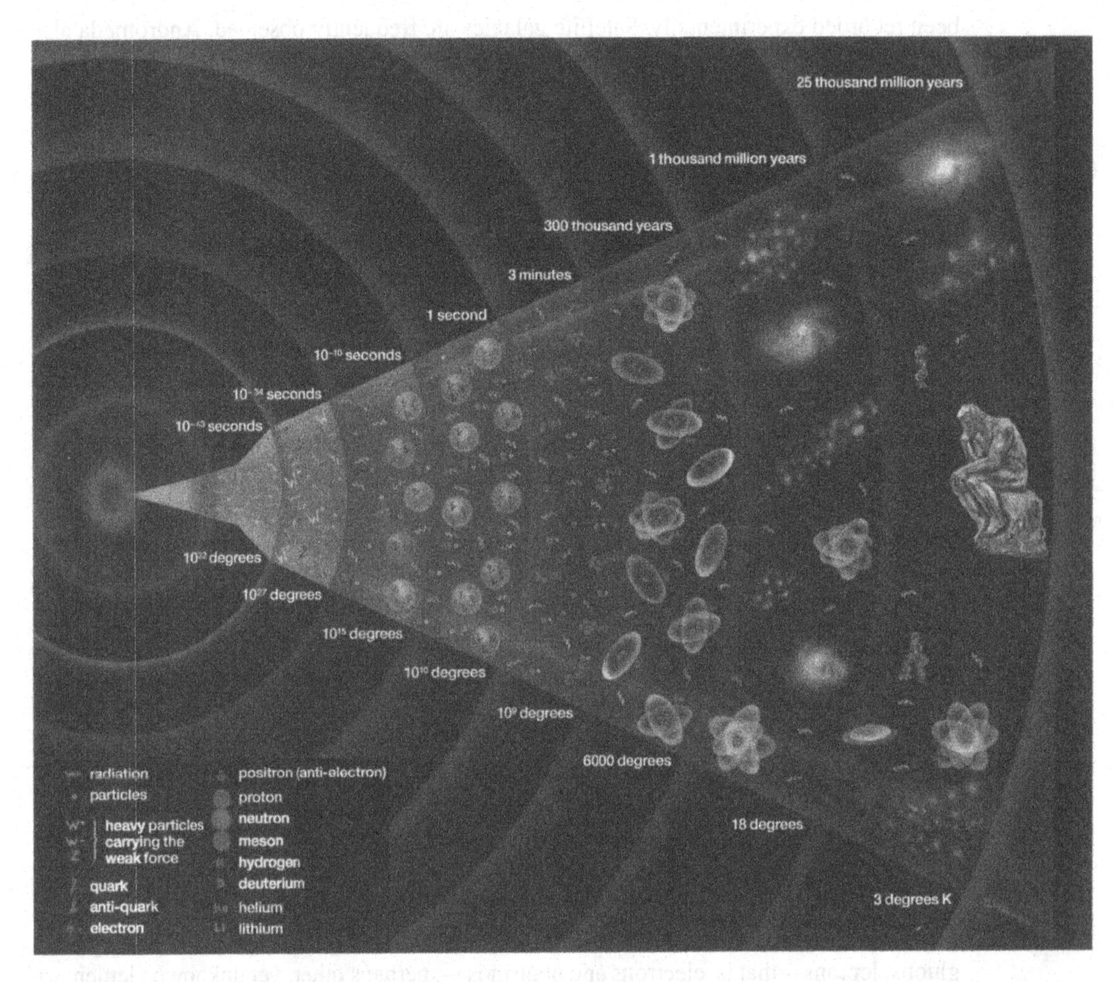

The evolution of the universe: Modern physics and experimental observations document the history of the universe from an incremental fraction of time after the Big Bang some 14 billions of years ago up to its present state. Dark matter is seen today as having played a key role in the formation of stars and galaxies. (©CERN Publications, July 1991, reproduced with kind permission)

Inflation and the very early universe.

One observes nowadays enormous homogeneity of the distribution of matter in the universe (averaged over large scales), and also within the *Cosmic Microwave Background* (CMB) radiation. This is very puzzling because there are regions in the expanding universe which have never been in causal contact, that is, light never had sufficient time to travel from one of these regions to another one.

To overcome this difficulty, some cosmologists (A. Guth, A. Linde, and others) suggested an exponentially rapid expansion of the universe, blowing up the universe by about a factor $3 \cdot 10^{43}$ between 10^{-36} s to 10^{-34} s after the Big Bang. This expansion is called *inflation* of the universe. All this reasoning seems utopic, but is helps to understand the present-day observations.

The inflationary phase ended abruptly due to the creation of photons and all the elementary building blocks of matter—quarks and leptons. Equal numbers of matter and antimatter were present, but most of the energy resided in radiation. The latter lost its energy faster due to the expansion, so that after 10^4 years the energy balance of the universe shifted on favour of matter.

The hadronic phase transition in the early universe and the CP problem

The quark gluon phase of matter ended about 10^{-6} seconds after the Big Bang, when the universe cooled to a temperature of $2 \cdot 10^{12}$ Kelvin. At that temperature a phase transition from a quark gluon plasma to a nucleonic phase of matter took place, where the protons and neutrons were formed.

In this process three quarks of different flavors (so-called up-quarks and down-quarks) combine together to form a proton (two up-quarks and one down-quark) or a neutron (one up-quark and two down-quarks) and similarly antiprotons (two antiup-quarks and one antidown-quark) or an antineutron (one antiup-quark and two antidown-quarks).

The gluons were given their name because they provide the glue for holding the quarks together in the nucleus. They mediate the strong force by exchange between the force centers (color charges) similarly as photons mediate the electromagnetic force between electric charges.

It is worthwhile to mention that in heavy nucleus–nucleus collisions at very high energies, one creates such a quark gluon plasma nowadays in the laboratory, for example, at CERN or at RHIC (Brookhaven). The idea is that in such collisions strong compression (up to 5–10 times nuclear density) and high temperatures ($\sim 10^{12}$ Kelvin) occur in a kind of nuclear shockwaves. Under such conditions, the protons and neutrons in the nuclei melt and set free quarks and gluons.

After the hadronic phase transition, one would expect to end up with the same number of nucleons and antinucleons, which annihilate each other after creation, leaving us not a chance to exist. Fortunately, this was not the case. The reason that we live in a world of matter with no antimatter is believed to be due to a very subtle effect, which treats matter and antimatter in a different way during the phase of creation. This effect, known as CP-violation (charge conjugation and parity violation), was first discovered in an accelerator experiment by V. Fitch[19] and J. Cronin[20] in 1964, for which they got the Nobel Prize in 1980, and was used by A. Sakharov[21] to explain the matter–antimatter asymmetry in the universe.

[19] *Val Logsdon Fitch*, American nuclear physicist, b. 1923, Merriman, Nebraska. He received his Ph.D. from Columbia University in 1954. After working on muons and muonic atoms, he started investigating the properties of kaons, where he found in 1964, together with Cronin, the CP violation in the decay of the neutral mesons. Fitch and Cronin shared the 1980 Nobel Prize in physics for this discovery.

[20] *James Watson Cronin*, American nuclear physicist, b. 1931, Chicago, Illinois. He received his Ph.D. from University of Chicago, 1955. His interest in strange particles was stimulated by Gell-Mann, and he worked mostly on kaons, first in Brookhaven, later in Princeton, joining the group of Fitch.

[21] *Andrei Sakharov*, Soviet physicist, b. 1921, Moscow—d. 1989. He was fascinated by fundamental physics and cosmology, but first he spent two decades designing nuclear weapons. He came to be regarded as the father of the Soviet hydrogen bomb. Gradually Sakharov became one of the regime's most courageous critics, a defender

Primordial nucleosynthesis

After further expansion the universe cooled to a temperature of 10^9 Kelvin, when protons and neutrons started to hang on to each other to form the light elements like helium, deuterium, lithium, and beryllium. This phase of nucleosynthesis began a few seconds after the Big Bang. The heavier elements were only formed many millions years later, mainly during star formation and supernova explosions. After their formation, the light nuclei had hundreds of thousands of years of time in order to catch electrons and build atoms.

The cosmic background radiation

About 300 thousand years after the Big Bang, radiation had not enough energy left to interact with matter because the excited states of atoms were appreciably higher than the photon energies contained in the cosmic radiation. Therefore, the universe became transparent for electromagnetic radiation. This radiation from the early universe was first discovered by R. Wilson[22] and A. Penzias[23] in 1965. They received the Nobel Prize for this finding in 1978. Their discovery was made by chance, since they were on a mission from Bell Laboratories to test new microwave receivers to relay telephone calls to earth-orbiting satellites. No matter in what direction they pointed their antenna, they always measured the same noise. At first, this was rather disappointing to them. But they happened to learn of the work of the physicist G. Gamow[24] and the astronomers R. Dicke[25] and P. Peebles,[26] thus

of human rights and democracy. His commitment as a "spokesman for the conscience of mankind" was honored by the Nobel Peace Prize in 1975.

[22] *Robert Woodrow Wilson*, b. 1936.

[23] *Arno Allan Penzias*, b. 1933, Munich, Germany, from where his family could escape to the United States in 1939.

[24] *George Gamow*, b. 1904, Odessa, Russia—d. 1968. Russian-American physicist who worked out the theory of alpha decay in terms of tunneling through the nucleus's potential barrier. Gamow showed that, as a star burns hydrogen, the star heats up. He supported the "Big Bang" theory of Lemaitre. He was also a popularizer of science, publishing many works including *Mr. Tompkins in Wonderland* (1937) and *Thirty Years that Shook Physics* (1966).

[25] *Robert Dicke*, b. 1916, St. Louis—d. 1997. Dicke received his his Ph.D. in 1941 from the University of Rochester. Dicke is widely known for his leadership in developing experimental tests of gravity physics and of the standard gravitational model for the large-scale evolution of our universe. Working at Princeton, he was responsible for the famous 1965 paper that proposed that radiation detected near one centimeter wavelength is left over from the hot Big Bang start of expansion of the universe. Dicke was building a radio antenna to test his theory when Penzias and Wilson discovered the echo by accident. Some physicists thought that Dicke had been unfairly excluded from sharing the 1978 Nobel Prize with them.

[26] *Philip James Edwin Peebles*, b. 1935, Winnipeg. He graduated from the University of Manitoba in 1958. He then went to Princeton University as a graduate student in physics, and he has been there ever since, currently as Albert Einstein Professor of Science Emeritus. With Robert Dicke and others he predicted the existence of the cosmic background radiation and planned to seek it just before it was found by Penzias and Wilson. He has investigated characteristics of this radiation and how it may be used to constrain models of the universe. He has led statistical studies of clustering and superclustering of galaxies. He has calculated the universal abundances of helium and other light elements, demonstrating agreement between Big Bang theory and observation. He has

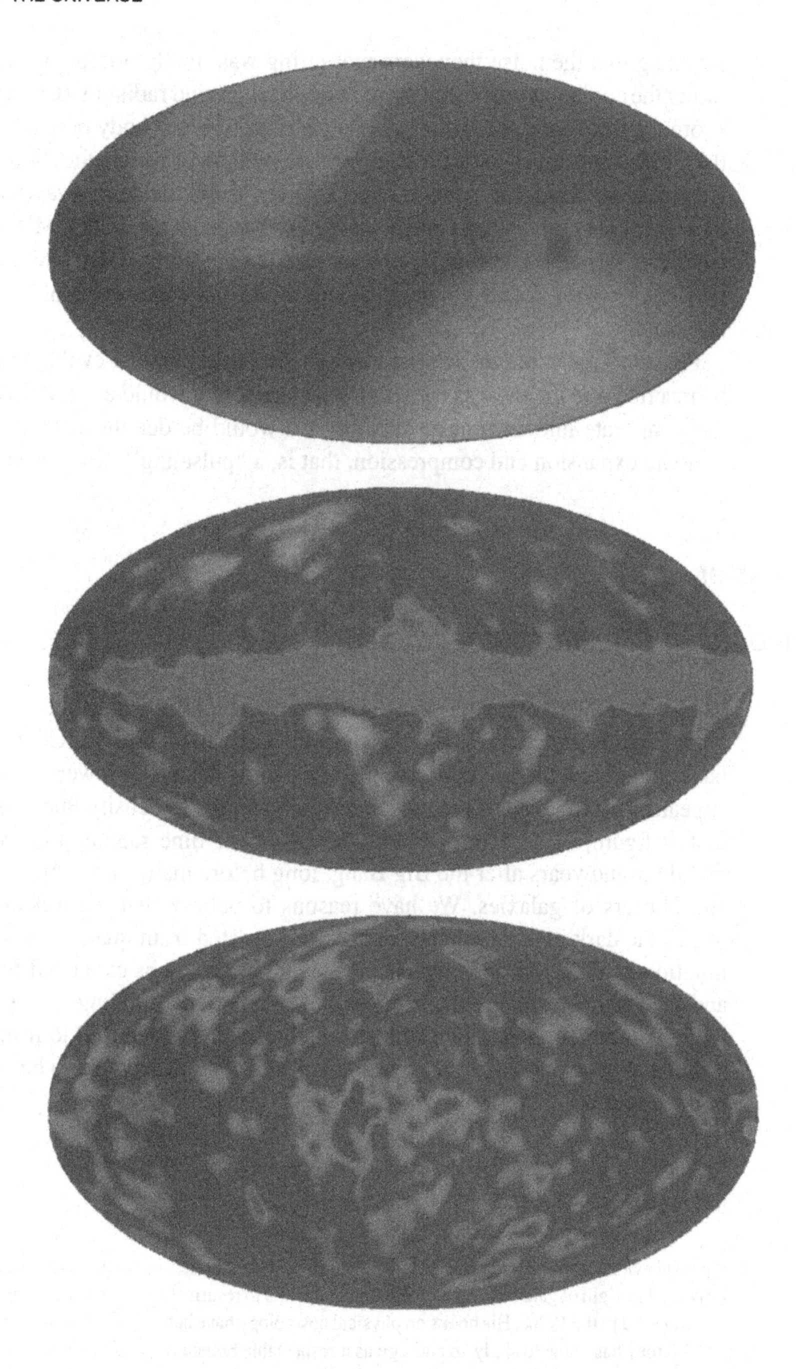

Full sky map of the cosmic background radiation as seen by the COBE mission. After subtraction of the dipole anisotropy, which is due to the motion of our solar system within the background radiation (still visible at top) and our own galaxy's emission (center), temperature variation of 0.01% unveils matter density fluctuations in the very early universe (bottom). (©NASA Goddard Space Flight Center and the COBE Science Working Group. Reproduced with kind permission.)

realizing that the noise they were measuring was finally not the noise of the receiver, but rather the cooled-down cosmic microwave background radiation (CMB) from the Big Bang. From the frequency spectrum and Planck's law of black body radiation, the temperature of the CMB was derived to be 2.7 Kelvin.[27] Regardless of which direction the cosmic radiation was received from, the temperature came out to be the same everywhere, demonstrating the enormous homogeneity of the universe. The most accurate CMB measurements come from the Cosmic Background Explorer satellite mission (COBE), which was sent into orbit in 1989.[28]. They found temperature variations only at a level of one part in a hundred thousand.

Presently it cannot yet be predicted whether the course of evolution observed so far will be inverted and the universe will collapse again. If it would expand more and more expand and evaporate into an infinite vacuum, we would be dealing with an "open universe." A periodic expansion and compression, that is, a "pulsating" universe, is also conceivable.

Dark Matter

Where does dark matter come from?

What about the dark matter? Why dark matter at all? When and how is it created? What is it made of? A partial answer to this question is given to us by the COBE cosmic microwave background radiation measurements. They show islands of lower and higher temperatures appearing on the map of the universe which are due to density fluctuations (see lower part of last figure). They were already present at the time radiation decoupled from matter, 300 thousand years after the Big Bang, long before matter was clumping to from galaxies and clusters of galaxies. We have reasons to believe that these density fluctuations are due to the dark matter, which was probably created from quantum fluctuations during the inflationary phase of the universe. These tiny fluctuations expanded first through inflation and then retarded their expansion due to gravitational binding forces. They then formed the gravitational potential wells into which ordinary matter fell to form galaxies and stars billions of years later. All galaxies and clusters of galaxies seem to be embedded into halos of dark matter.

provided evidence of the existence of large quantities of dark matter in the halos of galaxies, and he continues to work on the origin of galaxies. Peebles was one of the first to resurrect Einstein's cosmological constant, suggesting it was needed in the 1980s. His books on physical cosmology have had a significant impact in convincing physicists that the time has come to study cosmology as a respectable branch of physics.

[27] For more details, see W. Greiner, L. Neise, H. Stöcker, *Thermodynamics and Statistical Mechanics*, Springer, Berlin, New York, Tokyo, 1994.

[28] See, e.g., G. Smoot, *Wrinkles in Time*, New York, 1993, a popular account by the COBE leading scientist, the COBE homepage `http://space.gsfc.nasa.gov/astro/cobe/`, and Ch. L. Bennett, M. S. Turner, and M. White, "The cosmic Rosetta Stone," *Physics Today*, Nov. 1997, for a summary of the scientific results of COBE.

How much matter is in the universe?

At first this question seems to be highly academic. It is not. The fate of our universe depends on its mass and its expansion velocity.

In the 1920s the famous astronomer Edwin Hubble demonstrated that all galaxies are moving away from us and from each other—we have already mentioned this in the section on "Evolution of the universe." His discovery was the foundation stone of modern cosmology, which claims that the universe originated about 15 billion years ago in an unthinkably small volume with an unthinkably high-energy density, the so-called Big Bang, and is expanding ever since.

However, this expansion is counteracted by the gravitational pull of the matter in the universe. Depending on how much matter there is, the expansion will continue forever or come to a halt, which subsequently could lead to a collapse of the universe ending in a *Big Crunch*, the opposite of the Big Bang. The matter density needed to bring the expansion of the universe to a halt is called the critical mass density, which today would be roughly the equivalent of 10 hydrogen atoms per cubic meter. This seems incredibly small, like a vacuum, when compared to the density of our earth and planets, but seen on a cosmic scale it represents a lot of matter.

How can we find out how much matter there is? When estimating the visible matter in the universe, astronomers look in a very wide and very deep region in space and count the number of galaxies. Typical galaxies containing hundreds of billions of luminous stars have a brightness proportional to their mass.

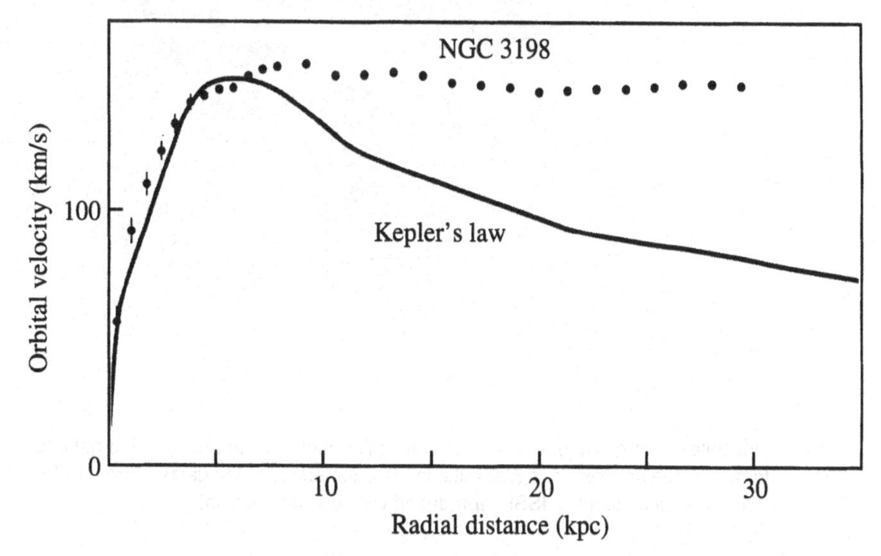

The observation of constant orbital velocities of stars around the galactic center (here the spiral galaxy NGC 3198) as a function of the radial distance provides convincing evidence for the presence of an extended halo of dark matter surrounding the galaxy. The expected curve from Kepler's law if there were no dark matter is also shown. (From K. Pretzl, Spatium 7, May 2000 ©Association Pro ISSI, reproduced with kind permission)

Thus, by simply counting galaxies over a large volume in space and by assuming that galaxies are evenly distributed over the entire universe, one can estimate the total mass they contribute in form of visible mass to the universe.

However, it turns out to be only 1% of the critical mass of the universe. Therefore, if the visible matter in the form of stars and galaxies were the only matter in the universe, the universe would expand forever. We neglected here the amount of matter in form of planets, because they contribute not more than a few percent of the mass of a star. However, it came as a surprise when Vera Rubin[29] and her team found out in the 1970s that the visible stars are not the only objects making up the mass of the galaxies. They measured the orbital speeds of stars around the center of spiral galaxies and found that they move with a constant velocity independent of their radial distance from the center (see figure). This is in apparent disagreement with Kepler's law, which says that the velocity should decrease as the distance

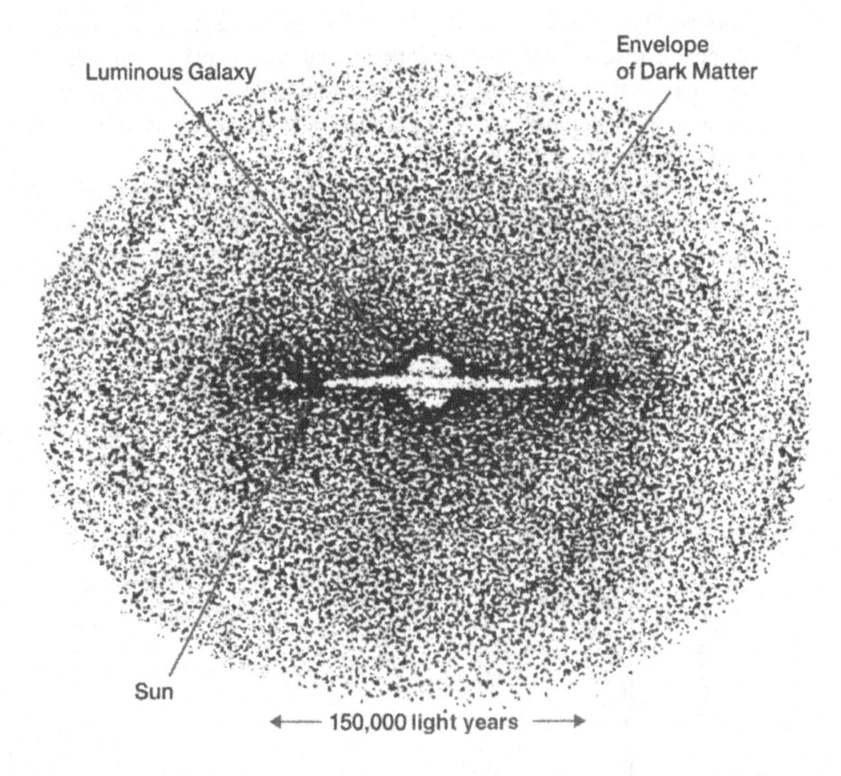

Our galaxy (Milky Way) as seen schematically from a distant point in the galactic plane. Dark matter forms a large halo extending far outside the outer edges of the galaxy. (From K. Pretzl, Spatium **7**, May 2000 ©Association Pro ISSI, reproduced with kind permission)

[29] *Vera Cooper Rubin*, b. 1928, Ph.D. from Georgetown University in 1954, working as an astronomer at the Carnegie Institution of Washington. Most of her work centers about the distribution and motion of galaxies.

of the star from the galactic center increases, provided that all mass is concentrated at the center of the galaxy, which seems to be the case if only the luminous matter is considered.

Indeed, starting from Kepler's third law,

$$\frac{T^2}{a^3} = \frac{4\pi^2}{\gamma M},$$

and assuming circular orbits for simplicity, we get

$$v = \frac{2\pi a}{T} = \sqrt{\frac{\gamma M}{a}} \sim \frac{1}{\sqrt{a}}.$$

This distance-dependence is indicated in the figure at the curve labeled "Kepler's law". Now, if Kepler's law, which describes the orbital motion of the planets in our solar system very correctly, is valid everywhere in the universe, then the rotational velocities of the stars can only be explained if the mass of the galaxy is increasing with the radial distance from its center. This is seen from our formula above, which tells us that v = constant implies $M \sim a$, that is, the mass inside the orbit has to grow proportionally to the radius a of the orbit.

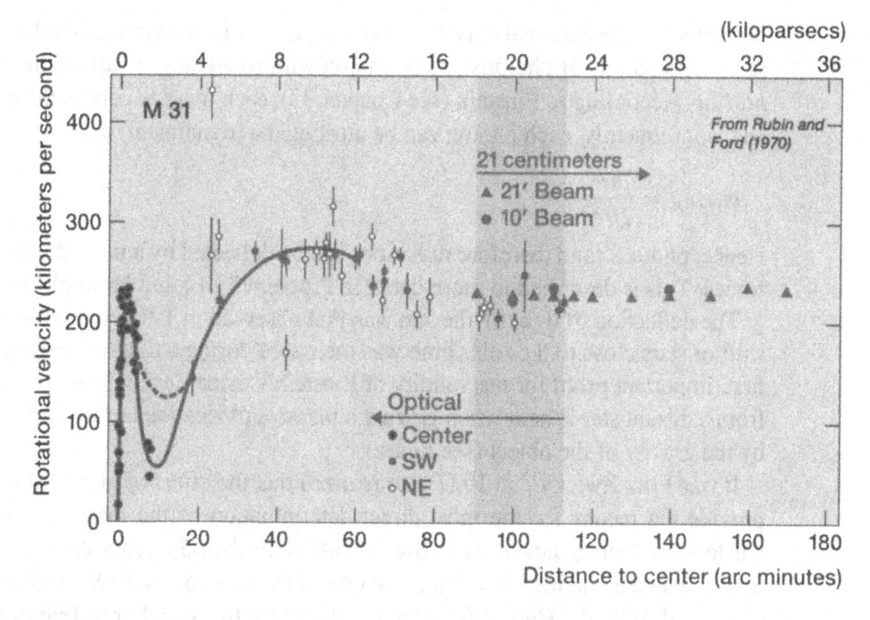

Experimental analysis of the rotational velocities in the Andromeda Galaxy M31 from optical observations (V. Rubin, W. Ford, Astrophysical Journal **159** 379-404 (1970)) and radio observations at 21-cm wavelength (M. Roberts, R. Whitehurst, Astrophysical Journal **201** 327-346 (1975)). (From K. Pretzl, Spatium **7**, May 2000 ©Association Pro ISSI, reproduced with kind permission)

Numerical calculations show that there must be at least an order of magnitude more matter in the galaxies than is visible. From measurements that were repeated on hundreds of different galaxies, it is conjectured that each galaxy must be embedded in an enormous halo of dark matter, which reaches out even beyond the visible diameter of the galaxy (see figure). Spiral galaxies are surrounded also by clouds of neutral hydrogen, which themselves do not contribute considerably to the mass of the galaxy, but which serve as tracers of the orbital motion beyond the optical limits of the galaxies. The hydrogen atoms in the clouds are emitting a characteristic radiation with a wavelength of 21 cm, which is due to a hyperfine interaction between the electron and the proton in the hydrogen atom and which can be detected. The Doppler shift of this characteristic radiation tells the velocity with which the hydrogen atoms (and thus the matter out there) are moving.

These measurements show that the dark matter halo extends far beyond the optical limits of the galaxies (see figure). But, how far does it really reach out? Very recently gravitational lensing observations seem to indicate that the dark matter halo of galaxies may have dimensions larger than 10 times the optical diameter. It is quite possible that the dark halos have dimensions that are already typical for distances between neighboring galaxies within galactic clusters.

Determining the mass in the universe

The effect of gravitational lensing is a consequence of Einstein's general relativity. Because radiation consists of photons, every photon with frequency ω carries the energy $E_{\text{photon}} = h\omega/2\pi$. According to Einstein (see Chapter 33), each mass m carries the energy $E = mc^2$, and, consequently, each photon can be attributed a (dynamical) mass

$$m_{\text{photon}} = \frac{h\omega}{2\pi c^2} .$$

Hence, photons (and therefore radiation) can be deflected by a mass M due to gravitational forces. This is described in more detail in Problem 33.14 and Example 34.4.

The deflection of light by the sun was first observed in 1919, when the apparent angular shift of stars close to the solar limb was measured during a total solar eclipse. This was the first, important proof for the validity of Einstein's theory, according to which light coming from a distant star is bent when grazing a massive object due to the space curvature caused by the gravity of the object (see figure).

It was Fritz Zwicky[30] in 1937 who realized that the effect of gravitational lensing would provide the means for the most direct determination of the mass of very large galactic clusters, including dark matter. But it took more than 50 years until his suggestion was finally realized and his early determination of the mass of the COMA cluster, in 1933, was confirmed. With the Hubble telescope in space and the Very Large Telescopes (VLT) at the

[30] *Fritz Zwicky*, (1898–1974). Swiss-American astronomer who was professor of astronomy at Caltech. He studied extragalactic supernovae and the distribution of galaxies in Coma Berenices. From his observations of the Coma galaxy cluster, he suggested already in 1933 that a large amount of matter in this cluster must be invisible in order to explain the dynmics of the galaxies in the cluster. In 1937 he was the first to consider gravitational lensing by extragalactic objects.

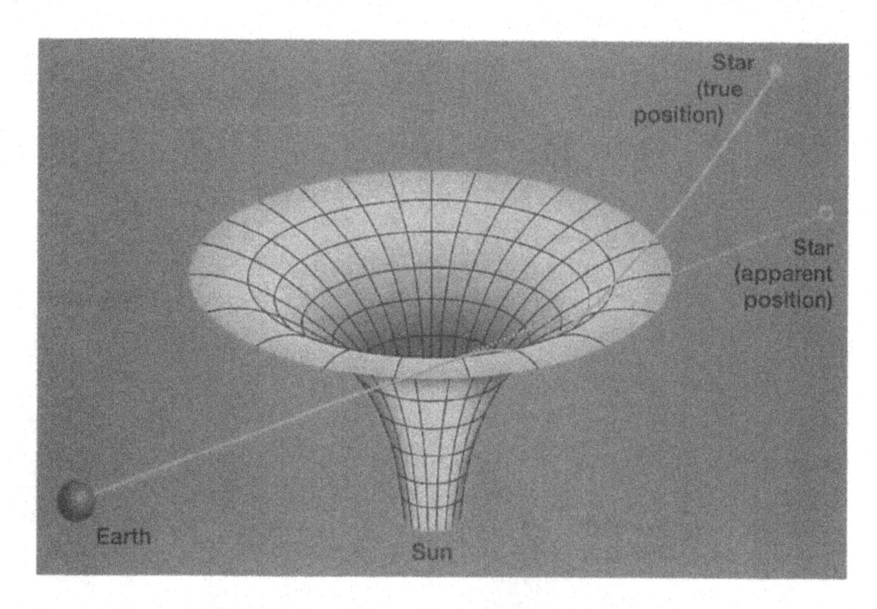

Space is curved by gravity. The light rays from a distant star are bent by the gravity field of the sun.The distant star therefore appears at a different position. (From K. Pretzl, Spatium **7**, May 2000 ©Association Pro ISSI, reproduced with kind permission)

Southern Observatory in Chile, astronomers now have very powerful tools, which allow them to explore not only the visible, but also the dark side of the universe with gravitational lensing.

An observer sees a distorted multiple image of a light source in the far background, when the deflecting massive object in the foreground is close to the line of sight. The light source appears to be a ring, the so-called Einstein ring, when the object is exactly in the line of sight (see figure). If one knows the distance of the light source and the object to the observer, one is able to infer the mass of the object from the lensing image. With this method it was possible to determine the mass of galactic clusters, which turned out to be much larger than the luminous matter. It seems that the gravitational pull of huge amounts of dark matter is preventing individual galaxies from moving away from each other and is keeping them bound together in large clusters, like, for example, the famous Coma cluster. By adding the total matter (dark and luminous matter) in galaxies and clusters of galaxies, one ends up with a total mass that corresponds to about 30% of the critical mass of the universe. With only 1% luminous mass, this would mean that there is 30 times more dark mass in the universe. In addition, the universe would be growing forever, since its total mass is subcritical to bring the expansion to a halt. But as we will see, this seems not to be the full story.

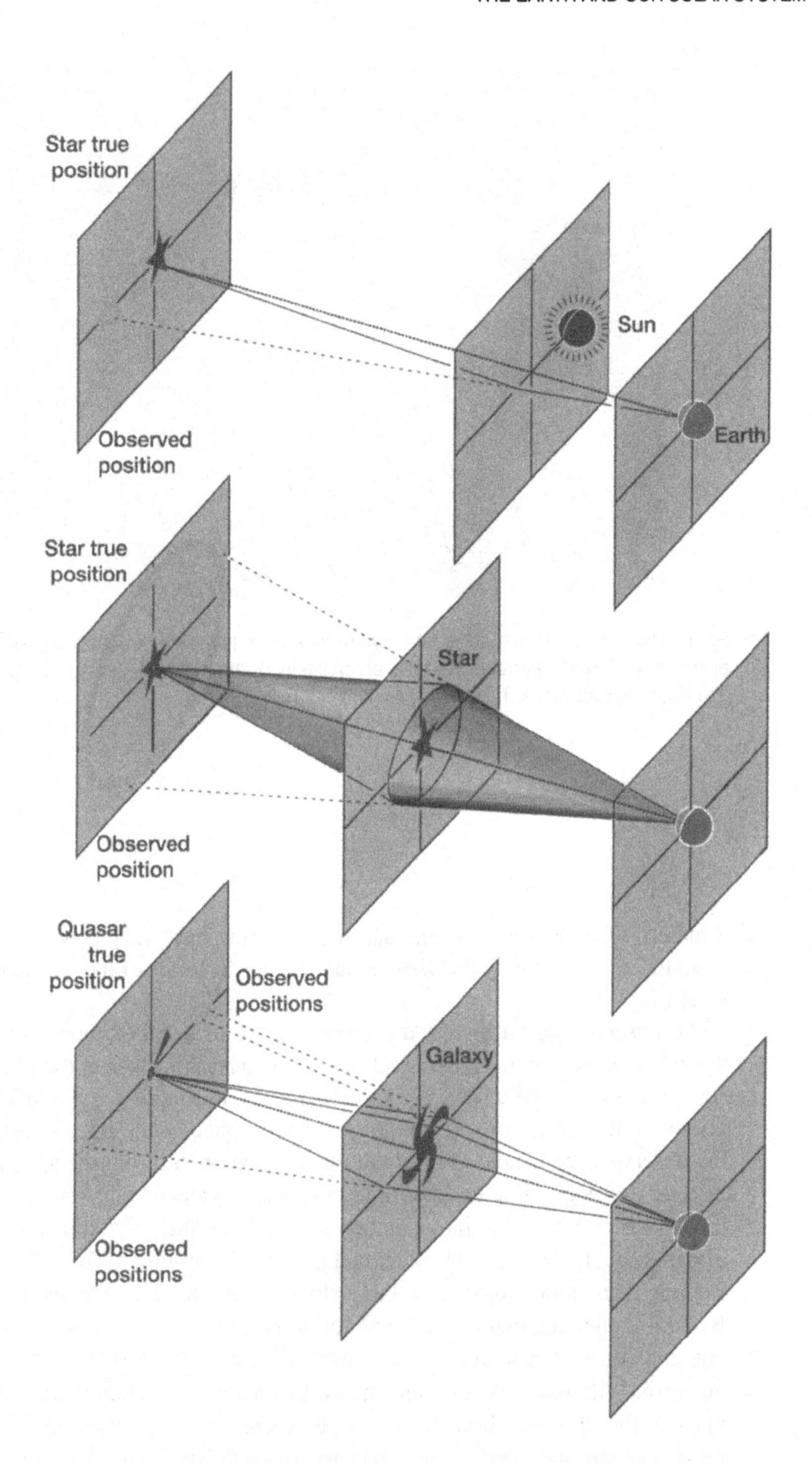

Gravitational lensing occurs when the gravity field of a massive celestial object bends the path of light emitted by a distant source. Einstein predicted the deflection of starlight by the sun (top) and the ring that would appear if the star and the celestial body were aligned perfectly (center). Lens systems found to date result from the alignment of extragalactic quasars and galaxies (bottom). (From Edwin L. Turner: "Gravitational Lenses", Scientific American, July 1988 ©Scientific American, Inc. reproduced with kind permission.)

The discovery of dark energy

The big surprise came in 1998 from a supernovae type 1a survey performed by the Super Cosmology Project (SCP) and the High z-Supernova Search (HZS) groups.[31] Supernovae of type 1a are 100 thousand times brighter than ordinary stars. They are still visible at very great distances, for which their light needed several million years to travel until it reached us. In principle, we experience now supernovae explosions that happened several million years ago. Since in every supernova type1a explosion there is always the same total amount of energy released, they all have the same brightness and therefore they qualify as *standard candles* in the cosmos. Their distance from us can then be inferred from the measurement of their apparent brightness. By probing space and its expansion with supernovae distance measurements, astrophysicists learned that the universe has not been decelerating, as assumed so far, but has rather been expanding with acceleration (see figure). More measurements are still needed to corroborate these astonishing findings of the supernovae survey. But it already presents a surprising new feature of our universe, which revolutionizes our previous views and leaves us with a new puzzle. In order to speed up the expansion of the universe, a negative pressure is needed, which may be provided by some unidentified form of dark energy.

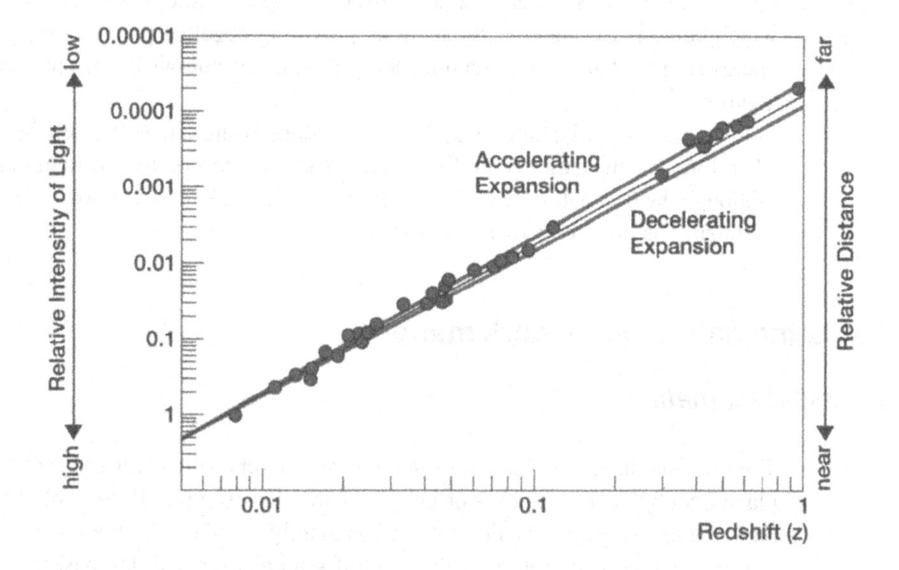

Recent supernovae distance measurements show that the expansion of the universe is accelerating rather than decelerating as assumed before. This observation suggests the presence of dark energy. (From Craig J. Hogan et al.: "Surveying Space-time with Supernovae", Scientific American, January 1999 ©Scientific American, Inc. reproduced with kind permission.)

[31]For a recent account, see, e.g., C. J. Hogan, R. P. Kirshner and N. B. Suntzef, "Surveying Space-time with Supernovae;" L. M. Krauss, "Cosmological Antigravity," both in *Scientific American*, Jan. 1999.

This ubiquitous dark energy amounts to 70% of the critical mass of the universe and has the strange feature that its gravitational force does not attract—on the contrary, it repels. This is hard to imagine since our everyday experience and Newton's law of gravity tell us that matter is gravitationally attractive. In Einstein's law of gravity, however, the strength of gravity depends not only on mass and other forms of energy, but also on pressure. From the Einstein equation, which describes the state of the universe, it follows that gravitation is repulsive if the pressure is sufficiently negative and it is attractive if the pressure is positive. In order to provide enough negative pressure to counterbalance the attractive force of gravity, Einstein originally introduced the *cosmological constant* to keep the universe in a steady state. At that early time all observations seemed to favor a steady-state universe with no evolution and no knowledge about its beginning and its end. When Einstein learned about the Hubble expansion of the universe in 1920, he discarded the cosmological constant by admitting that it was his biggest blunder.

For a long time cosmologists assumed the cosmological constant to be negligibly small and set its value to zero, as it did not seem to be of any importance in describing the evolution of the universe. This has changed very recently, because we know about the accelerated expansion of the universe. However, there remain burning questions like why is the cosmological constant so constant over the lifetime of the universe and has not changed similar to the matter density, and what fixes its value. Besides the cosmological constant, other forms of dark energy are also discussed by cosmologists, such as vacuum energy, which consists of quantum fluctuations providing negative pressure, or *quintessence*, an energy source, that, unlike vacuum energy and the cosmological constant, can vary in space and time.

In contrast to dark matter, which is gravitationally attractive, dark energy cannot clump. Therefore, it is the dark matter that is responsible for the structure formation in the universe. Although the true nature of the dark energy and the dark matter is not known, the latter can eventually be directly detected, while the former cannot.

What is the nature of the dark matter?

Baryonic dark matter

The obvious thing is to look for nonluminous or very faint ordinary matter in the form of planetary objects like jupiters or brown dwarfs, for example. If these objects represent the dark matter, our galactic halo must be abundantly populated by them.

Because they may not be visible even if searched for with the best telescopes, B. Paczynski[32] suggested to look for them by observing millions of individual stars in the Large and the Small Magellanic Cloud to see whether their brightness changes with time due to gravitational lensing when a massive dark object is moving through their line of sight (see figure).

[32]*Bohdan Paczynski*, Polish astronomer, b. 1940, Wilno, Poland, now professor of astrophysics, Department of Astrophysical Sciences, Princeton University. His main current interest and effort are in the work related to the Optical Gravitational Lensing Experiment.

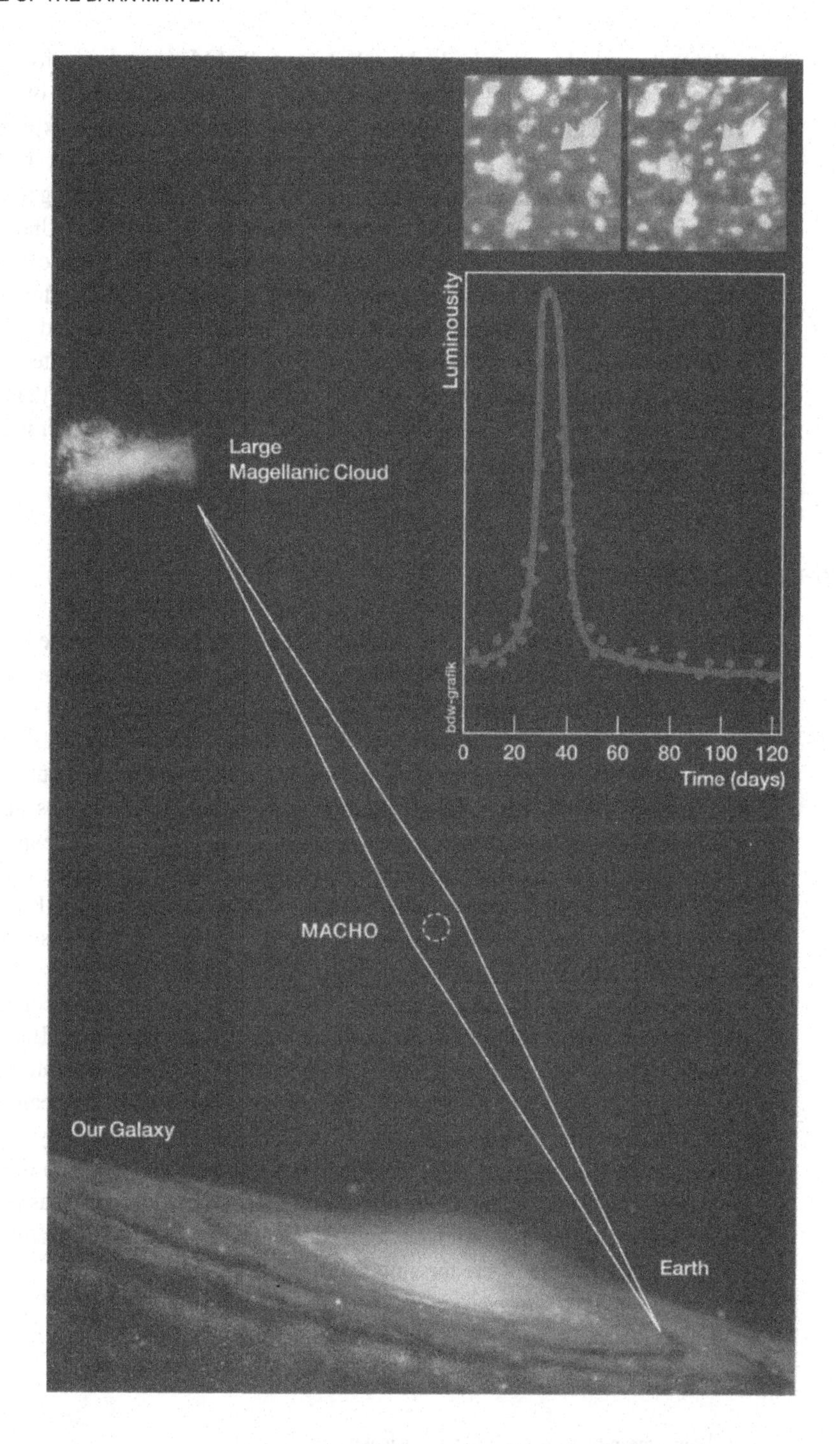

Massive dark objects (Massive Astrophysical Compact Halo Objects, MACHOs) moving through the line of sight between the observer and a distant star in the Large Magellanic Cloud cause the apparent luminosity to change. (From Bild der Wissenschaft 2/1997, ©bild der wissenschaft)

Several research groups looked for these so-called Massive Astrophysical Compact Halo Objects (MACHOs) using gravitational lensing. They found some of these dark objects with masses smaller than the solar mass, but by far not enough to explain the dark matter in the halo of our galaxy. Other objects like black holes or neutron stars could also have been detected by this method, but there are not many of them in the galactic halo.

Do we know how much ordinary matter exists in the universe? Under ordinary matter or so-called baryonic[33] matter, we understand matter in the form of chemical elements consisting of protons, neutrons, and electrons. About 3 minutes after the Big Bang, the light elements, like hydrogen, deuterium, and helium, were produced via nucleosynthesis. From the measurement of their present abundances, one can estimate the total amount of the baryonic matter density in the universe. This amounts to not more than 6% of the critical mass density of the universe. It shows that most of the baryonic matter is invisible and most of the dark matter must be of nonbaryonic nature.

Nonbaryonic dark matter

The most obvious candidates for nonbaryonic matter would be the neutrinos, if they had a mass. Neutrinos come in three flavors. If the heaviest neutrino had a mass of approximately 10^{-9} times the mass of a hydrogen atom, namely $m_{\text{neutrino}}c^2 = 10^{-9} \cdot 1\text{GeV} = 1\text{eV}$, it would qualify to explain the dark matter. This looks like an incredibly small mass, but the neutrinos belong to the most abundant particles in the universe and outnumber the baryons by a factor of 10^{10}. For a long time it was assumed that neutrinos have no mass. The standard model of particle physics[34] includes this assumption. All experimental attempts to determine the mass of the neutrinos ended in providing only upper limits.

However, in 1998 an underground detector with the name SUPER-Kamiokande in Japan observed anomalies in the atmospheric neutrino flux which is highly suggestive of neutrino oscillations, which can only occur if neutrinos have indeed a mass. These observations will have to be reproduced and further substantiated by planned accelerator experiments, like K2K in Japan, MINOS in the United States, and OPERA in Europe. The OPERA experiment will be constructed in the underground Gran Sasso laboratory, which is located about 100 km northeast of Rome. For this experiment, a neutrino beam will be sent from CERN to the Gran Sasso laboratory. If neutrinos have a mass, they would change their flavor during their journey over the 735-km distance from CERN to the Gran Sasso. The neutrinos would start as muon-neutrinos at CERN and would arrive as tau-neutrinos at the Gran Sasso. This change of flavor can be detected. Massive neutrinos may also provide the solution to the puzzle of the missing neutrinos from our sun.

[33]Barys meaning strong or heavy in ancient Greek.

[34]For more details about neutrinos, see, e.g., W. Greiner, B. Müller, *Gauge Theory of Weak Interactions*, Springer Verlag New York, 2000.

A cocktail of nonbaryonic dark matter, neutralinos, and WIMPs

Computer models allow us to study the development of small- and large-scale structures under the hypothesis of various nonbaryonic dark matter candidates. Two main categories are distinguished, namely the so-called hot and cold dark matter. Neutrinos would qualify under the category hot dark matter, since their velocities were very large when they decoupled from matter, a few milliseconds after the Big Bang. Because of their speed they were not able to clump on small, typical galactic scales, but their gravitational force would still allow for clustering on very large, typical supercluster scales. Thus in a hot dark matter-dominated universe, only the formation of large-scale superclusters would be favored. In such a model superclusters would fragment into smaller clusters at a later time. Hence galaxy formation would be a relatively recent phenomenon, which, however, is in contrast to observation. Cold dark matter candidates, on the other hand, would have small velocities at early phases and therefore would be able to aggregate into bound systems at all scales. A cold dark matter-dominated universe would therefore allow for an early formation of galaxies in good agreement with observations, but it would overpopulate the universe with small-scale structures, which does not fit our observations. Questions like how much hot and how much cold, or only cold dark matter, are still not answered. Some computer models yield results that come closest to observations when using a cocktail of 30% hot and 70% cold dark matter.

Exotic particles like neutralinos are among the most favored cold dark matter candidates. Neutralinos are stable elementary particles predicted to exist by Super Symmetry (SUSY), a theory that is an extension of the standard model of elementary particles. Thus if they exist, they would solve two problems at the same time, namely the dark matter as well as SUSY, which is a prerequisite for the unification of all forces in nature, the so-called grand unification theory (GUT). Experiments at the Large Hadron Collider (LHC) at CERN, which is under construction and will be operational in 2005, will also search for these particles.

If the dark matter consisted of neutralinos, which would have been produced together with other particles in the early universe and which would have escaped recognition because they only weakly interact with ordinary matter, special devices would have to be built for their detection. These detectors would have to be able to measure very tiny energies, which these particles transfer in elastic scattering processes with the detector material. Because of the very weak coupling to ordinary matter, these particles are also called WIMPs, for *Weakly Interacting Massive Particles*. They would abundantly populate the halo of our galaxy and would have a local density in our solar system equivalent to one hydrogen atom in 3 cm^3. Because they would be bound to our galaxy, allowing for an average velocity of 270 km/s, their flux (density times velocity) would be very large. However, because they only weakly interact with matter, the predicted rates are typically less than one event per day per kilogram detector material. WIMPs can be detected by measuring the nuclear recoil energy in the rare events when one of these particles interacts with a nucleus of the detector material. It is like measuring the speed of a billiard ball sitting on a pool table after it has been hit by another ball. Because of the background coming from the cosmic rays and the radioactivity of the material surrounding the detector, which yield similar signals in the detector as the WIMPs, the experiment must be carried out deep underground, where cosmic rays cannot penetrate, and must be shielded locally against the rest radioactivity of materials and the radioactivity in the rock.

There is presently a race for WIMPs, with several groups in the United States, in Europe and in Japan searching for WIMPs employing different techniques.

Problem 28.1: Mass accretion of the sun

Find the approximate accretion rate dM/dt of the sun if it moves with velocity v_s through a homogeneous gaseous cloud of density ϱ.

Solution A particle is captured by the sun if its velocity w in a coordinate frame convected with the sun is smaller than the escape velocity to leave the sun. According to Problem 26.5, the escape velocity reads

$$v_0^2 = \frac{2\gamma M}{R}. \tag{28.1}$$

For a given constant w all particles will be captured that are localized within a sphere about the sun with the critical radius

$$R_0 = \frac{2\gamma M}{w^2}. \tag{28.2}$$

This formula holds, of course, only if R_0 exceeds the sun's radius. To determine the accretion rate, one has to specify how many particles flow into the sphere of radius R_0 per unit time.

Let the mean thermal velocity of the gas molecules be v_G. We distinguish between two limits:

(a) $v_s \ll v_G$: In this case the motion of the sun may be neglected, and the mean velocity of the gas molecules in the coordinate frame fixed to the sun may be set equal to v_G.

The critical radius according to equation 28.2 is therefore

$$R_0 = \frac{2\gamma M}{v_G^2}. \tag{28.3}$$

If the sun were not existent, the numbers of particles flowing into and out of the sphere would be the same, provided that the velocity vectors $\mathbf{v}_G$ are distributed isotropically. Not only the molecules flowing into the sphere will be captured by the sun, but also the particles flying outward will be prevented from escaping and thus will also be captured. Therefore, the mean flow (= particles per unit area per unit time) of captured particles is approximately equal to ϱv_G. The accretion rate equals the flow multiplied by the surface of the sphere:

$$\frac{dM}{dt} = 4\pi R_0^2 \varrho v_G = \frac{16\pi\gamma^2 M^2 \varrho}{v_G^3}. \tag{28.4}$$

(b) $v_s \gg v_G$: In this case, the thermal motion of the gas molecules may be neglegted. In the coordinate frame of the sun, all particles then move with the velocity $\mathbf{v}_s$. The critical radius is therefore

$$R_0 = \frac{2\gamma M}{v_s^2}. \tag{28.5}$$

Because all gas molecules are moving in from the same direction, they "see" only the cross-sectional area of this sphere. The accretion rate therefore equals the flow ϱv_s multiplied by the area of a circle of radius R_0:

$$\frac{dM}{dt} = \pi R_0^2 \varrho v_s = \frac{4\pi\gamma^2 M^2 \varrho}{v_s^3} . \tag{28.6}$$

Numerical example

We set $v_s = 0$ and thus obtain an upper limit for the accretion rate. For v_G we assume a value of $10^3\ \mathrm{ms}^{-1}$ (this corresponds to a temperature of about 100 K for hydrogen molecules). A typical value for the density of an interstellar cloud is $\varrho = 10^{-18}\ \mathrm{kg\,m}^{-3}$. The sun mass is $M = 1.99 \cdot 10^{30}$ kg, and the gravitational constant is $\gamma = 6.67 \cdot 10^{-11}\ \mathrm{m}^3\,\mathrm{kg}^{-1}\,s^{-2}$.

According to equation 28.4, it then results that

$$\begin{aligned} \frac{dM}{dt} &= 8.86 \cdot 10^{14}\ \mathrm{kg}\,s^{-1} \\ &= 2.79 \cdot 10^{22}\ \mathrm{kg/year} \\ &= 4.67 \cdot 10^{-3}\ M_E/\mathrm{year} \end{aligned}$$

with the earth mass $M_E = 5.975 \cdot 10^{24}$ kg.

Example 28.2: Motion of a charged particle in the magnetic field of the sun

If the sun moves through a cloud of interstellar matter, one has to take into account also electromagnetic effects in the calculation of the mass accretion. These shall be estimated below in a simplified model.

The cloud shall contain both gases in ionized form as well as charged solid particles. We consider the motion of a charged particle of mass m and charge q that moves from far away toward the sun in the gravitational field and magnetic field of the sun.

For sake of simplicity we assume the magnetic field of the sun as being generated by a dipole with the magnetic dipole moment $\boldsymbol{\mu}$ (for a definition of the dipole moment, see Chapter III in Vol. 3: *Electrodynamics*). Moreover, we shall restrict ourselves to particles moving in the plane passing through the center of the sun and being perpendicular to $\boldsymbol{\mu}$.

The Lorentz force acting on the particle in the magnetic field $\mathbf{B}$ in this plane is (see volume 3, Electrodynamics):

$$\mathbf{F}_{\mathrm{magn}} = \frac{q}{c}\dot{\mathbf{r}} \times \mathbf{B} = \frac{q}{c}\frac{\dot{\mathbf{r}} \times \boldsymbol{\mu}}{r^3} , \tag{28.7}$$

where c is the speed of light.

According to (26.10), the gravitational force reads

$$\mathbf{F}_{\mathrm{grav}} = -\gamma M m \frac{\mathbf{r}}{r^3} \tag{28.8}$$

with the sun mass M. Hence, the equation of motion of the particle is

$$m\ddot{\mathbf{r}} = -\gamma M m \frac{\mathbf{r}}{r^3} + \frac{q}{c}\frac{1}{r^3}\dot{\mathbf{r}} \times \boldsymbol{\mu} .$$

In plane polar coordinates (r, φ) this equation, taking into account (10.11) and (10.12), reads

$$m\left(\left(\ddot{r} - r\dot{\varphi}^2\right)\mathbf{e}_r + \left(r\ddot{\varphi} + 2\dot{r}\dot{\varphi}\right)\mathbf{e}_\varphi\right) = -\gamma M m \frac{\mathbf{e}_r}{r^2} + \frac{q}{c}\frac{1}{r^3}\left(\dot{r}\mathbf{e}_r + r\dot{\varphi}\mathbf{e}_\varphi\right) \times \boldsymbol{\mu} . \tag{28.9}$$

Because $\boldsymbol{\mu}$ is perpendicular to $\mathbf{e}_r$ and $\mathbf{e}_\varphi$, this equation may be split with respect to the two components:

$$m(r\ddot{\varphi} + 2\dot{r}\dot{\varphi}) = -\frac{q}{c}\frac{\mu\dot{r}}{r^3} \tag{28.10}$$

$$m\ddot{r} = -\frac{\gamma Mm}{r^2} + \frac{q}{c}\frac{\mu\dot{\varphi}}{r^2} + mr\dot{\varphi}^2. \tag{28.11}$$

We begin with the first equation. The left side may be transformed such that the following holds:

$$\frac{m}{r}\frac{d}{dt}(r^2\dot{\varphi}) = -\frac{q}{c}\frac{\mu\dot{r}}{r^3}. \tag{28.12}$$

Integration of this equation yields

$$mr^2\dot{\varphi} = -\frac{q\mu}{c}\int\frac{\dot{r}}{r^2}dt = -\frac{q\mu}{c}\int\frac{dr}{r^2} = \frac{q\mu}{cr} + \text{constant} \tag{28.13}$$

The integration constant may be set to zero if we require the boundary condition that at large distances from the sun the particle shall have no angular momentum with respect to the sun (the left side of this equation just represents the angular momentum).

By inserting the result 28.13 in equation 28.11, we obtain

$$m\ddot{r} = -\frac{\gamma Mm}{r^2} + \frac{2q^2\mu^2}{mc^2r^5}. \tag{28.14}$$

Because

$$\ddot{r} = \frac{d\dot{r}}{dt} = \frac{d\dot{r}}{dr}\dot{r} \tag{28.15}$$

we get

$$\dot{r}\frac{d\dot{r}}{dr} = -\frac{\gamma M}{r^2} + \frac{2q^2\mu^2}{m^2c^2r^5}. \tag{28.16}$$

Integration of this equation yields

$$\dot{r}^2 = \frac{2\gamma M}{r} - \frac{q^2\mu^2}{m^2c^2r^4} + \text{constant} \tag{28.17}$$

With the boundary condition $\dot{r} = 0$ for $r \to \infty$, we may set the integration constant to zero. There is still another point r_c at which the radial velocity vanishes. Solving the equation

$$\frac{2\gamma M}{r_c} - \frac{q^2\mu^2}{m^2c^2r_c^4} = 0 \tag{28.18}$$

yields

$$r_c = \left(\frac{q^2\mu^2}{2\gamma Mm^2c^2}\right)^{1/3}. \tag{28.19}$$

Hence, a particle coming from outside can never approach the sun closer than to the radius r_c.

The only particle parameter entering the formula for r_c is the ratio q/m. The interstellar matter typically contains two kinds of particles: atoms (mainly hydrogen) and solid particles. Solid particles have a significantly smaller value for q/m than an ionized hydrogen atom and thus may approach the sun much closer than the hydrogen atoms may do.

An estimate of the magnetic field of the sun yields a value of r_c of about 10^{10} km for hydrogen. The actual value of r_c should be somewhat smaller because hydrogen atoms are ionized only at velocities of about $5 \cdot 10^4$ ms^{-1}, such that the boundary condition for equation 28.17 must be a distinct one. In any case the minimum distance for hydrogen atoms lies in the external regions of the planetary system where the large gas planets are actually localized.

For the solid particles one may assume that only their surface is ionized. One may then estimate their q/m-ratio to be proportional to the ratio of surface to volume, that is, inversely proportional to their radius. The radius of, for example, an interstellar dust particle is typically about 500 times larger than that of a proton, such that for $r_c \sim (q/m)^{2/3}$ there should result a value being by about a factor of 100 smaller. This is just the radius of the inner planetary orbits.

Example 28.3: Excursion to the external planets

Many new insights about our solar system have been collected by unmanned space probes such as Voyager I and II. The passage of Saturn by Voyager I (on Nov. 12, 1980) and of Voyager II (on Aug. 25, 1981) provided much new knowledge on this planet.[35]

The Cassini gap of the Saturn rings, caused by the largest moon Titan, is not empty but is also interspersed by a number of narrow rings. The Saturn rings consist of countless individual rings, the widths being about 2 km. Besides the classical 10 Saturn moons, 7 further ones with diameters of less than 100 km have been detected.

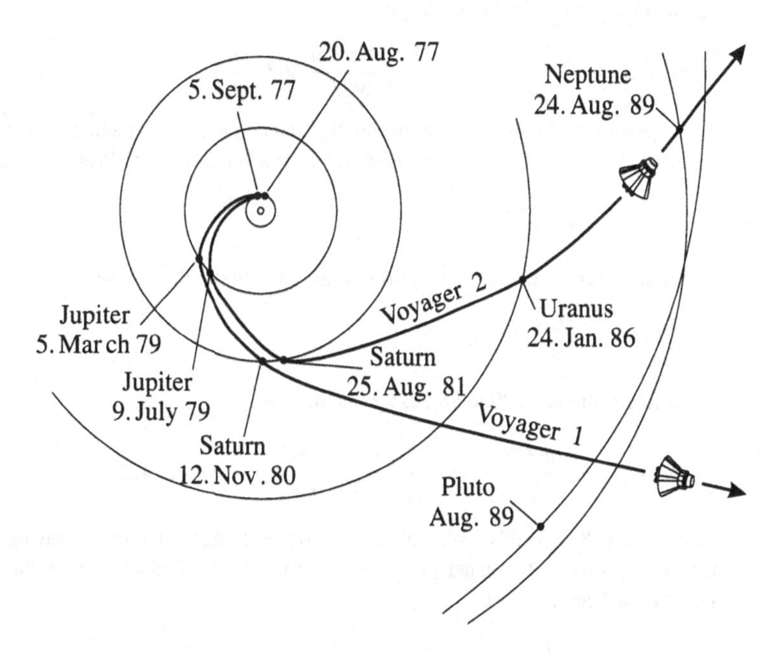

Typical pearl string configuration of the exterior planets with the orbits of the Voyager space probes shown. Note the "swing-by" maneuver, i.e., the optimized passages of the planets by the space probes in a kind of "planet swing."

[35]The discussion of this problem goes back to suggestions of students of the mechanics course in Frankfurt, using material from various sources.

The newly discovered outer ring, called the F-ring, is slightly eccentric, contrary to the other rings. Moreover, one could detect "spokes" in the Saturn rings. Their origin has not yet been explained, but presumably they may be traced back to a sun flare of the sunlight by tiny ice crystals. After about $4 \cdot 10^4$ years Voyager I will reach the vicinity of the star AC + 793 888 in the constellation Ursa Minor. Voyager II, after an encounter with Neptune on August 24, 1989, will travel a long way through innterstellar space, and after $3.58 \cdot 10^5$ years will pass Sirius, the main star of Canis Maior and the brightest fixed star of the firmament, at a distance of 0.8 lightyears. A difficulty in exploring the outer planets are the long flight times. These may, however, be shortened significantly if the probe exploits the gravitational field of a planet on its route for a calculated change of flight direction (swing-by). A rare constellation of the four largest planets Jupiter, Saturn, Neptune and Uranus that is particularly suited for this purpose arose in the 1980's: The planets didn't stand in a straight line but nevertheless along a flat curve. Such a "pearl string configuration" occurs only once in 175 years and allows Voyager II to pass our four largest planets. Due to the increase of kinetic energy by the various passages of planets ("swing-bys"), as is seen from the preceding figure, the program of Voyager II may be finished already after 12 years, while a direct flight with equivalent energy expense would last about 30 years.

The essential aspects of the calculated trajectories with gravitational support for such a mission may already be elaborated from the equations on planetary motion in Chapter 26.

For exploring the outer planets, the start should generally be performed in the direction of the earth's circulation about the sun. The velocity of earth v_E moving on an almost circular orbit of radius r_E and period τ_E about the sun is given by

$$v_E = \omega_E r_E = \frac{2\pi}{\tau_E} r_E = \frac{2\pi \cdot 1.5 \cdot 10^8 \text{ km}}{365 \cdot 24 \cdot 3600\text{s}} = 30 \frac{\text{km}}{\text{s}}. \qquad \textbf{(28.20)}$$

A spaceship of mass m with an initial distance r_E from the sun ($\odot$) needs a minimum escape velocity $v_{\text{Fl}}^{\odot}$ to leave the gravitational field of the sun (compare to Problem 26.5):

$$E = 0 = \frac{1}{2} m (v_{\text{Fl}}^{\odot})^2 - \frac{\gamma m M^{\odot}}{r_E}. \qquad \textbf{(28.21)}$$

On the other hand, the circular orbit of the earth about the sun obeys

$$\frac{M_E v_E^2}{r_E} = \frac{\gamma M_E M^{\odot}}{r_E^2}. \qquad \textbf{(28.22)}$$

From equations 28.20 to 28.22, we obtain

$$\frac{1}{2} m (v_{\text{Fl}}^{\odot})^2 = \frac{\gamma m M^{\odot}}{r_E} \quad \Leftrightarrow \quad v_{\text{Fl}}^{\odot} = \sqrt{\frac{2\gamma M^{\odot}}{r_E}} = \sqrt{2}\, v_E \cong 42 \text{ km/s}. \qquad \textbf{(28.23)}$$

Equation 28.23 yields a general relation for the escape velocity for leaving the solar system from a planetary orbit. The planet moves on a circular path of radius r with the velocity v_u about the gravitational center (sun).

$$v_{\text{Fl}}^{\odot}(r) = \sqrt{2}\, v_u(r). \qquad \textbf{(28.24)}$$

In a start from the earth moving with v_E, the escape velocity out of the gravitational field of the sun reduces according to 28.20 to

$$\tilde{v}_{\text{Fl}}^{\odot} = \left(\sqrt{2} - 1\right) v_E = 12 \text{ km/s}. \qquad \textbf{(28.25)}$$

The spaceship needs an additional initial velocity ($\sim$ 11 km/s) to leave the attraction by the earth.

For a direct flight to Uranus with minimum driving energy, the start should therefore be performed along the earth's orbit about the sun, such that the spaceship switches into a Kepler ellipse about the sun, with the earth standing in the perihelion and Uranus in the aphelion. Note that the shape of this ellipse (shown below) is uniquely fixed by two conditions:

(a) The distance between earth and Uranus fixes the major semi-axis, and therefore according to (26.39)

$$a = \frac{-\gamma Mm}{2E} = \frac{k}{1-\varepsilon^2},$$

also the energy.

(b) The condition that the probe shall enter the Kepler ellipse at the perihelion position and parallel to the earth's orbit uniquely fixes the angular momentum constant

$$k = \frac{L^2}{m^2\gamma M} = \frac{L^2}{mH}$$

and therefore also the eccentricity. Moreover, the launching time from the earth must be chosen such that the arrival of Uranus and that of the satellite in its aphelion position coincide in time. To calculate the trajectory in the following figure, we still need the orbit radii of earth and Uranus about the sun. These are

$$r_E = 1.5 \times 10^8 \text{ km} \equiv 1 \text{ AU (astronomic unit)},$$
$$r_U = 19.2 \text{ AU}.$$

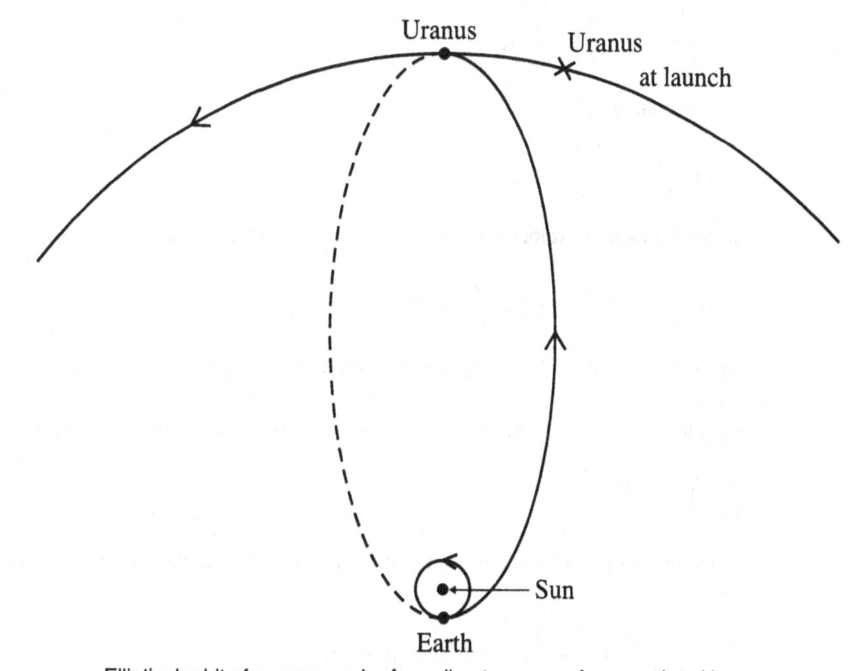

Elliptical orbit of a space probe for a direct passage from earth to Uranus.

According to (26.32), we obtain the following expression for the perihelion and aphelion positions of the ellipse:

$$r(\theta = 0) = r_E = \frac{k}{1+\varepsilon}, \tag{28.26}$$

$$r(\theta = \pi) = r_U = \frac{k}{1-\varepsilon},$$

$$\Leftrightarrow \quad \varepsilon = \frac{r_U - r_E}{r_E + r_U} = 0.9 \qquad \text{(ellipse)},$$

$$k = r_E(1+\varepsilon) = 1.9\ \text{AU}.$$

The resulting trajectory to Uranus reads

$$r(\theta) = \frac{1.9\ \text{AU}}{1 + 0.9\cos(\theta)}; \tag{28.27}$$

$$a = \frac{1}{2}(r_E + r_U) = 10.1\ \text{AU} \tag{28.28}$$

is the major semi-axis of the ellipse.

To get an expression for the velocity at an arbitrary point of the trajectory, we start from equation (26.39):

$$E = \frac{-\gamma m M^{\odot}}{2a} = \frac{1}{2}mv^2 - \frac{\gamma m M^{\odot}}{r}. \tag{28.29}$$

From there it follows that

$$v = \sqrt{2\gamma M^{\odot}\left(\frac{1}{r} - \frac{1}{2a}\right)},$$

and with equation 28.23,

$$v = v_{\text{Fl}}^{\odot}\sqrt{\frac{r_E}{r} - \frac{r_E}{2a}}, \tag{28.30}$$

such that the incident velocity at the perihelion of the ellipse to Uranus is given by

$$v_p = v_{\text{Fl}}^{\odot}\sqrt{1 - \frac{r_E}{2a}} = v_{\text{Fl}}^{\odot}\sqrt{\frac{192}{202}} \cong 41\ \text{km/s}. \tag{28.31}$$

By subtracting the orbital velocity of earth about the sun, we obtain the incident velocity $\widetilde{v}_p = 11$ km/s.

To calculate the flight time for approaching Uranus, we apply the third Kepler law

$$\left(\frac{\tau_1}{\tau_2}\right)^2 = \left(\frac{a_1}{a_2}\right)^3. \tag{28.32}$$

When denoting the circulation time of the earth by τ_E and the major semi-axis by $a_E \cong r_E$, we obtain

$$\frac{\tau}{2} = \frac{\tau_E}{2}\left(\frac{a}{r_E}\right)^{3/2} = \frac{1}{2}(10.1)^{3/2}a \cong 16\ \text{years}.$$

This flight time, with an equivalent energy expense, may be shortened by 11 years by choosing a trajectory supported by the gravitational field of Jupiter. The idea on which the following calculation is based rests on the assumption that an elastic collision takes place in the gravitational well of Jupiter, whereby an infinitesimal fraction of the planetary kinetic energy is transferred to the satellite (see following figures). We begin with the same heliocentric path (sun in the center of gravity) as in the preceding case, but choose the start time such that a meeting with Jupiter in its circulation orbit happens. The reaction of the satellite onto Jupiter and therefore onto its orbital velocity $\mathbf{V}_J$ are neglected because $M_J/m \gg 1$, and moreover the interaction time is small against the orbital period of the planet.

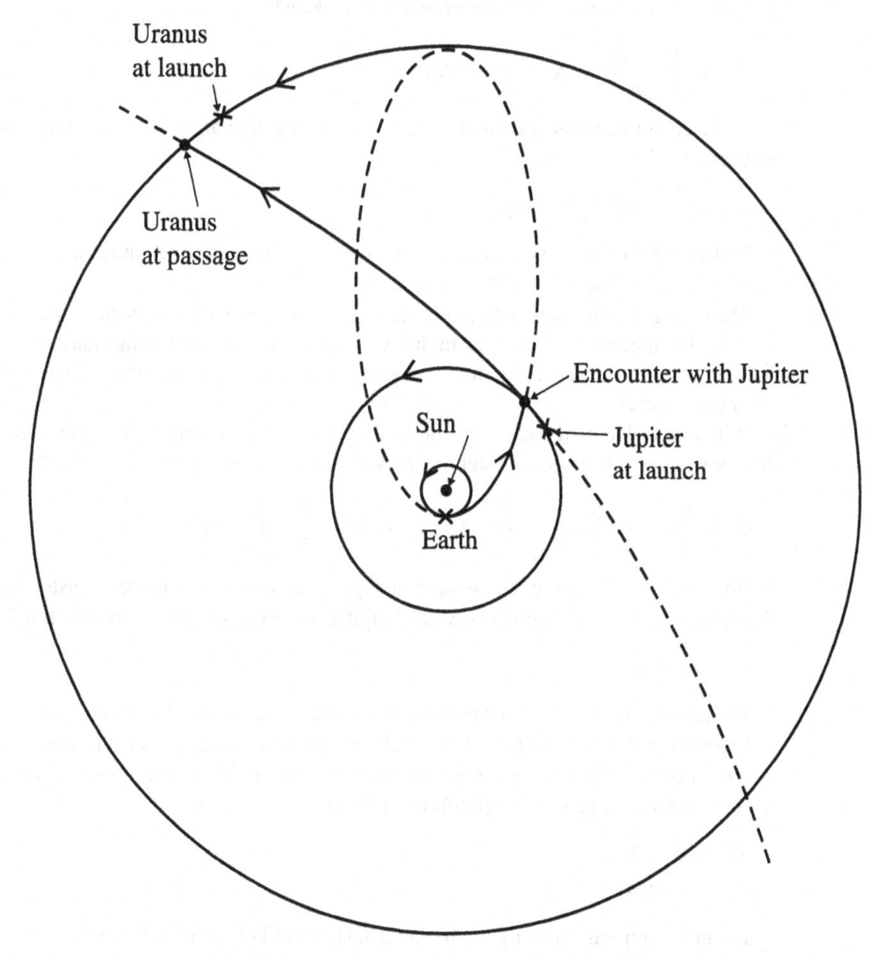

Using the gravitational potential well of Jupiter for an optimized voyage (shorter in time) of a space probe to Uranus.

The momenta of the spaceship before and after the meeting with Jupiter in the heliocentric system are denoted by $\mathbf{p}_i$ and $\mathbf{p}_f$, respectively. The following equations hold:

$$\mathbf{p}_i = \mathbf{p}'_i + m\mathbf{V}_J, \tag{28.33}$$

$$\mathbf{p}_f = \mathbf{p}'_f + m\mathbf{V}_J, \tag{28.34}$$

where $\mathbf{V}_J$ is the orbital velocity of Jupiter, and $\mathbf{p}'_i$, $\mathbf{p}'_f$ are the momenta of the spaceship in the center-of-mass system of the planet. The Galileo transformation 28.33, which is meaningful only for nonrelativistic velocities, yields for the momentum transfer

$$\Delta\mathbf{p} = \mathbf{p}_f - \mathbf{p}_i .$$

The momentum transfer is the same in both reference frames:

$$\Delta\mathbf{p} = \Delta\mathbf{p}'. \tag{28.35}$$

The change of kinetic energy, however, depends on the reference frame from which the spaceship is observed. In the heliocentric reference frame, we get

$$\Delta T = \frac{p_f^2 - p_i^2}{2m} = \Delta T' + \mathbf{V}_J \cdot \Delta\mathbf{p}'. \tag{28.36}$$

In the center-of-mass system of Jupiter, we had required an elastic scattering, such that $\Delta T' = 0$ and hence

$$\Delta T = \mathbf{V}_J \cdot \Delta\mathbf{p} = \mathbf{V}_J \cdot \Delta\mathbf{p}'. \tag{28.37}$$

In the center-of-mass system of the sun, however, the scattering causes an energy increase of the satellite which is supplied by the planet Jupiter.

The strong gravitational field of the sun almost exclusively governs the path curve of the satellite. Only in the immediate vicinity of Jupiter is the gravitational field of the sun relatively constant, and the trajectory of the satellite is then essentially determined by the gravitational field of Jupiter (see following figure).

If $\mathbf{u}_i$ and $\mathbf{u}_f$ denote the velocities of the probe when entering and leaving the range of attraction of Jupiter in its center-of-mass system, energy and momentum conservation lead to

$$E' = \frac{1}{2}mu_i^2 = \frac{1}{2}mu_f^2 + \frac{1}{2}M_\mathrm{J}\Delta v^2 = \frac{1}{2}mu_f^2 + \frac{1}{2}M_\mathrm{J}\left[\frac{m}{M_\mathrm{J}}(\mathbf{u}_f - \mathbf{u}_i)\right]^2 , \tag{28.38}$$

where Δv is the change in the velocity of Jupiter due to the momentum transfer $\Delta\mathbf{p} = m(\mathbf{u}_f - \mathbf{u}_i)$. Because $m \ll M_\mathrm{J}$, the recoil energy onto Jupiter may be neglected, from which it follows that

$$u_i \approx u_f \equiv u. \tag{28.39}$$

Because the energy E' is positive, we may conclude from the classification of conic sections on p. 261 that the path is a hyperbola. In the heliocentric frame, one gets for the velocity of the spaceship at the border of the attraction range of Jupiter (comparable to the gravitational field of the sun), neglecting the change of $\mathbf{V}_J$ (see following figure):

$$d\mathbf{v}_i = \mathbf{u}_i + \mathbf{V}_J, \tag{28.40}$$

$$\mathbf{v}_f = \mathbf{u}_f + \mathbf{V}_J. \tag{28.41}$$

For the asymptotic velocity on the hyperbola, it follows with 28.39 that

$$u = \sqrt{v_i^2 + V_J^2 - 2v_i V_J \cos\beta_i} . \tag{28.42}$$

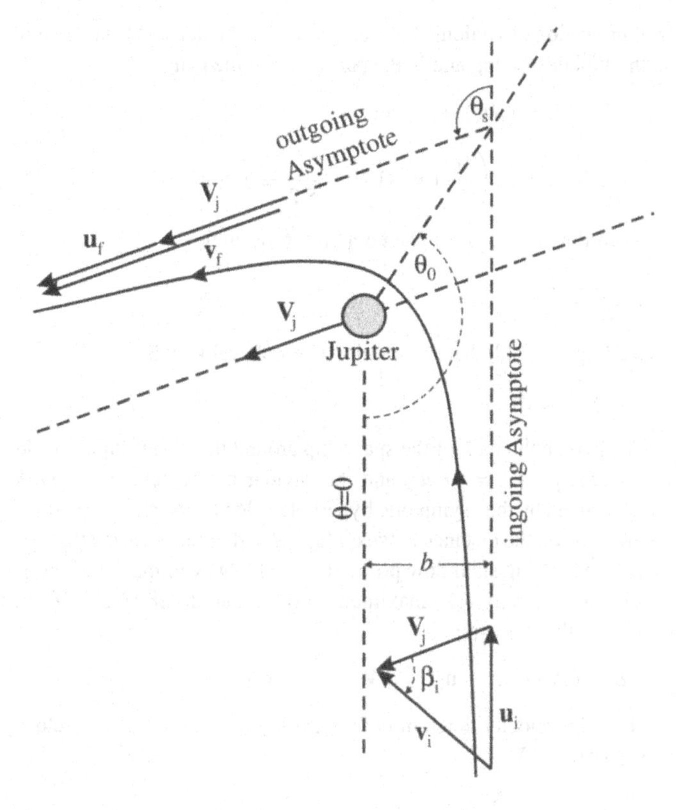

Geometry of the elastic scattering of the satellite at Jupiter ("swing-by").

With the circulation period of 11.9 years and the orbital radius $r_J = 5.2$ AU of Jupiter, the orbital velocity V_J is

$$V_J = \frac{2\pi}{\tau_J} r_J = \frac{2\pi \cdot 5.2 \cdot 1.5 \cdot 10^8\ \text{km}}{11.9 \cdot 365 \cdot 24 \cdot 3600\ \text{s}} = 13\ \text{km/s}. \tag{28.43}$$

The velocity of the spaceship when approaching Jupiter may be estimated from 28.30, with $r = r_J = 5.2$ AU:

$$v_i = 42\ \text{km/s} \sqrt{\frac{1}{5.2} - \frac{1}{2 \cdot 10.1}} = 16\ \text{km/s}. \tag{28.44}$$

We now calculate the angle β_i enclosed by the flight trajectory of the probe and the planetary orbit:

$$\cos \beta_i = \frac{\mathbf{v}_i \cdot \mathbf{V}_J}{v_i V_J} = \frac{(v_i)_\theta}{v_i}. \tag{28.45}$$

The projection of $\mathbf{v}_i$ along $\mathbf{V}_J$ (i.e., $(v_i)_\theta$) may be derived from the angular momentum conservation in the Jupiter meeting and in the perihelion of the path:

$$L = m(v_i)_\theta r_J = m v_p r_E \tag{28.46}$$

$$\Leftrightarrow \quad (v_i)_\theta = v_p \left(\frac{r_E}{r_J}\right) = 41\ \text{km/s}\ \frac{1}{5.2} \cong 8\ \text{km/s}. \tag{28.47}$$

According to 28.45, we then obtain for the angle

$$\cos\beta_i = \frac{1}{2}. \tag{28.48}$$

According to 28.42, the asymptotic hyperbola velocity is

$$u = 14.7\ \text{km/s}. \tag{28.49}$$

The hyperbolic path of the spaceship around the planet Jupiter is determined by the initial values for the energy, $E' = \frac{1}{2}mu^2$, and the angular momentum, $L' = mub$. Contrary to the energy E', which is fixed by the asymptotic hyperbola velocity, the angular momentum depends via the collision parameter b on the distance between Jupiter and satellite during the meeting and therefore on the start time. The meeting shall now proceed in such a way to make the energy transfer to the satellite, and thus its final velocity v_f, a maximum. From equations 28.36 and 28.40, we may calculate the energy transfer to the spaceship:

$$\Delta E = m\mathbf{V}_J \cdot (\mathbf{u}_f - \mathbf{u}_i) = m\mathbf{V}_J \cdot (\mathbf{v}_f - \mathbf{v}_i). \tag{28.50}$$

From the velocity diagram in the next figure, we see that the velocity $\mathbf{v}_f$ becomes a maximum if $\mathbf{v}_f$ is parallel to $\mathbf{V}_J$.

$$\mathbf{v}_f = (V_J + u)\frac{\mathbf{V}_J}{|\mathbf{V}_J|}. \tag{28.51}$$

From the available data, we obtain

$$v_f = 13\ \text{km/s} + 14.7\ \text{km/s} = 27.7\ \text{km/s}, \tag{28.52}$$

as compared to $v_i = 16$ km/s !

The scattering angle Θ_s between $\mathbf{u}_i$ and $\mathbf{u}_f$ is also determined from the following figure:

$$V_J = v_i \cos\beta_i + u_i \cos(\pi - \Theta_s) \tag{28.53}$$

$$\Leftrightarrow \quad \cos\Theta_s = \frac{v_i \cos\beta_i - V_J}{u_i} = -0.34.$$

Accordingly, the probe is deflected by $\Theta_s = 110°$.

We now shall investigate whether the minimum distance $r_{\min}$ of the hyperbolic orbit about Jupiter is indeed larger than its radius R_J. For this purpose we write the hyperbola path in the customary form:

$$r(\Theta) = \frac{k'}{1 + \varepsilon' \cos(\Theta - \Theta_0)}, \tag{28.54}$$

where r is the distance to Jupiter, and Θ_0 is the symmetry angle of the probe orbit with respect to the initial and final velocities $\mathbf{u}_i$ and $\mathbf{u}_f$. To calculate the eccentricity ε', we employ the initial condition that for $\Theta = 0$ $r \to \infty$. From that it follows that

$$\varepsilon' \cos(\Theta_0) = -1. \tag{28.55}$$

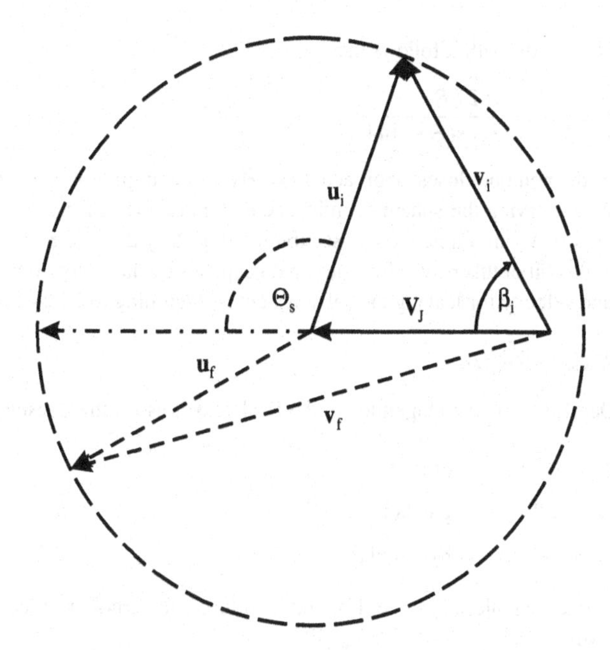

Velocity diagram for the passage of Jupiter in the general case. Note that in the special case used in the calculation (equation 28.53), $\mathbf{v}_f \parallel \mathbf{V}_J$. See also preceding figure.

We further see from the figure on p. 351 that $2\Theta_0 - \Theta_s = \pi$, and therefore

$$\varepsilon' = -\frac{1}{\cos(\pi/2 + \Theta_s/2)} = 1.23. \qquad \textbf{(28.56)}$$

To determine the angular momentum constant k', we make use of (26.37):

$$\varepsilon' = \sqrt{1 + \frac{2E'|L|^2}{mH^2}}; \qquad \textbf{(28.57)}$$

with $k' = |L|^2/(mH)$, it follows from 28.57 that

$$k' = \frac{H}{2E'}(\varepsilon'^2 - 1) \qquad \textbf{(28.58)}$$

$$\Leftrightarrow \quad k' = \frac{\gamma M_J}{u^2}(\varepsilon'^2 - 1), \qquad \textbf{(28.59)}$$

or in terms of the escape velocity from the planet Jupiter

$$v_{\mathrm{Fl}}^J = \sqrt{\frac{2\gamma M_J}{R_J}} = 60 \text{ km/s} \qquad \textbf{(28.60)}$$

$$\Rightarrow \quad k' = \frac{1}{2}R_J\left(\frac{v_{\mathrm{Fl}}^J}{u}\right)^2(\varepsilon'^2 - 1) = R_J \cdot \frac{1}{2}\left(\frac{60}{14.7}\right)^2(1.23^2 - 1) = 4.27\, R_J. \qquad \textbf{(28.61)}$$

Hence, for the path it follows that

$$r = \frac{4.27\,R_J}{1 + 1.23\cos(\Theta - 144^\circ)}. \tag{28.62}$$

At the point of closest approach, $\Theta = \Theta_0$, which implies $r = 1.9\,R_J$, ensuring a safe passage.

When leaving the sphere of influence of Jupiter, the spaceship switches with a final velocity $\mathbf{v}_f$ parallel to $\mathbf{V}_J$ into a new conic-sectionlike trajectory about the sun with perihelion at $r = r_J$. The type of orbit of this new heliocentric path depends on the energy transfer of Jupiter to the probe. The escape velocity for leaving the solar system is, according to 28.24,

$$v^{\odot}_{\text{escape}} = \sqrt{2}\,V_J. \tag{28.63}$$

Depending on the magnitude of the final velocities v_f, there result the following types of paths:

$$\begin{aligned} v_f &< \sqrt{2}\,V_J \qquad \text{ellipse,} \\ v_f &= \sqrt{2}\,V_J \qquad \text{parabola,} \\ v_f &> \sqrt{2}\,V_J \qquad \text{hyperbola.} \end{aligned} \tag{28.64}$$

In our example, $v_f/V_J = 1.5$, and therefore a hyperbola results, which again may be written in the usual form:

$$r = \frac{k''}{1 + \varepsilon''\cos\Theta}. \tag{28.65}$$

The distance of closest approach lies at $\Theta = 0, r = r_J$:

$$r_J = \frac{k''}{1 + \varepsilon''}. \tag{28.66}$$

Because the probe leaves Jupiter along its orbit about the sun, one has $L'' = mv_f r_J$, and with $k'' = L''^2/(mH)$ it follows that

$$k'' = \frac{v_f^2 r_J^2}{\gamma M^{\odot}} \quad\Rightarrow\quad \varepsilon'' = \frac{v_f^2}{\gamma M^{\odot}/r_J} - 1 = \left(\frac{v_f}{V_J}\right)^2 - 1 \tag{28.67}$$

$$\Rightarrow\quad \varepsilon'' = 3.54, \qquad k'' = 23.6\ \text{AU}. \tag{28.68}$$

The trajectory from Jupiter to Uranus is therefore completely given by

$$r(\Theta) = \frac{23.6\ \text{AU}}{1 + 3.54\cos\Theta}. \tag{28.69}$$

The start time for the path plotted in the figure on p. 349 must be chosen such that the planets are in a constellation enabling the gravitational-field-supported swing-by at the planet Jupiter and the passage flight at Uranus. The premise for such a Jupiter mission repeats every 14 years.

We still calculate the flight times for the path sections earth–Jupiter (equation 28.27) and Jupiter–Uranus (equation 28.69) plotted in the figure on p. 349 from the angular momentum conservation $|\mathbf{L}| = r^2\dot{\Theta}m$:

$$\Leftrightarrow\quad \Delta t = \int_{t_1}^{t_2} dt = \frac{m}{|\mathbf{L}|}\int_{\Theta_1}^{\Theta_2} r^2\,d\Theta; \tag{28.70}$$

with $r = k/(1+\varepsilon\cos\Theta)$ and $k = L^2/Hm$ it follows that

$$\Delta t = \frac{k^2 m}{L}\int_{\Theta_1}^{\Theta_2}\frac{d\Theta}{(1+\varepsilon\cos\Theta)^2} \tag{28.71}$$

$$\Leftrightarrow \quad \Delta t = \frac{k^{3/2}}{(\gamma M^{\odot})^{1/2}}\int_{\Theta_1}^{\Theta_2}\frac{d\Theta}{(1+\varepsilon\cos\Theta)^2}. \tag{28.72}$$

To work with convenient units, this is expressed by the orbital velocity of the earth (compare to 28.23):

$$\sqrt{\gamma M^{\odot}} = v_E\sqrt{r_E} = \frac{2\pi}{\tau_E}(r_E)^{3/2} \tag{28.73}$$

$$\Rightarrow \quad \Delta t = \left(\frac{\tau_E}{2\pi}\right)\left(\frac{k}{r_E}\right)^{3/2}\int_{\Theta_1}^{\Theta_2}\frac{d\Theta}{(1+\varepsilon\cos\Theta)^2}. \tag{28.74}$$

From the integral tables (e.g., Bronstein, # 350 and # 347), we find

$$\int\frac{d\Theta}{(1+\varepsilon\cos\Theta)^2} = \frac{\varepsilon\sin\Theta}{(\varepsilon^2-1)(1+\varepsilon\cos\Theta)} - \frac{1}{\varepsilon^2-1}\begin{cases}\dfrac{1}{\sqrt{\varepsilon^2-1}}\ln\left|\dfrac{(\varepsilon-1)\tan\frac{\Theta}{2}+\sqrt{\varepsilon^2-1}}{(\varepsilon-1)\tan\frac{\Theta}{2}-\sqrt{\varepsilon^2-1}}\right| & \text{for } \varepsilon^2>1,\\[2ex] \dfrac{2}{\sqrt{1-\varepsilon^2}}\arctan\left(\dfrac{(1-\varepsilon)\tan\frac{\Theta}{2}}{\sqrt{1-\varepsilon^2}}\right) & \text{for } \varepsilon^2<1.\end{cases} \tag{28.75}$$

For the elliptic path to Jupiter at the start $\Theta_1 = 0$, Θ_2 is determined from 28.27:

$$r = r_J = 5.2\ \text{AU} = \frac{1.9\ \text{AU}}{1+0.9\cos\Theta_2} \quad \Rightarrow \quad \Theta_2 \cong 135^\circ,$$

and with 28.75 and $\varepsilon^2 < 1$, we obtain

$$\Delta t \cong 1.21\ \text{years}. \tag{28.76}$$

For the hyperbola-type path from Jupiter to Uranus, we determine Θ_2 analogously from

$$r = r_U = \frac{23.6\ \text{AU}}{1+3.54\cos\Theta_2} \quad \Rightarrow \quad \Theta_2 \cong 86.3^\circ$$

and analogously from 28.75 with $\varepsilon^2 > 1$:

$$\Delta t \cong 3.74\ \text{years}. \tag{28.77}$$

The total flight time for an excursion from the earth to Uranus could be reduced from 16 to 5 years by a swing-by at Jupiter. The given data are, of course, approximate values, as we have assumed the gravitational forces of the planets and of the sun onto the spaceship to be independent of each other. By using numerical methods, one can drop this approximation and, for example, confirm the data in the figure on p. 345. For this goal, however, a number of refinements of the approximations made in our simple calculation are needed.

Problem 28.4: Perihelion motion

A planet of mass m moves in the gravitational potential of the sun:

$$U(r) = -\frac{\kappa}{r} - \frac{B}{r^3},$$

where the additional term is due to a polar flattening of the sun. Calculate the perihelion motion $\delta\Theta$ of the planetary orbit per revolution.

Hint

B shall be small such that the orbit may be assumed as a superposition of a fixed elliptic orbit and a perturbation:

$$u(\Theta) = u_0(\Theta) + \varepsilon v(\Theta) + O(\varepsilon^2).$$

Solution From the potential $U(r) = -\kappa/r - B/r^3$, the force $\mathbf{F}(r)$ follows:

$$\mathbf{F}(r) = -\nabla U(r) = -\left(\frac{\kappa}{r^2} + \frac{3B}{r^4}\right)\mathbf{e}_r \equiv F(r)\mathbf{e}_r,$$

or expressed in $u(\Theta) = r^{-1}(\Theta)$ (see (26.20) ff.)

$$F\left(\frac{1}{u}\right) = -\kappa u^2 - 3Bu^4. \tag{28.78}$$

We are dealing with a central force. The differential equation to be solved is therefore (see (26.22) of the lecture):

$$F\left(\frac{1}{u}\right) = -mh^2u^2\left(\frac{d^2u}{d\Theta^2} + u\right), \tag{28.79}$$

which explicitly reads

$$u''(\Theta) + u(\Theta) = \frac{\kappa}{mh^2} + \frac{3B}{mh^2}u^2 \tag{28.80}$$

$$= A + \frac{\varepsilon}{A}u^2. \tag{28.81}$$

Here we have set

$$A = \frac{\kappa}{mh^2} \quad \text{and} \quad \varepsilon = \frac{3\kappa B}{m^2h^4}.$$

One realizes that the differential equation without the term ε again leads to the Kepler problem with a fixed elliptic orbit.

Assuming that B and hence also ε is small, the following *ansatz* ("perturbation *ansatz*") is obvious:

$$u(\Theta) = u_0(\Theta) + \varepsilon v(\Theta) + O(\varepsilon^2). \tag{28.82}$$

We now demonstrate that u_0 yields the original Kepler ellipse, and v represents the perturbation leading to the perihelion motion. Insertion of 28.82 into 28.80 yields

$$u_0''(\Theta) + \varepsilon v''(\Theta) + u_0(\Theta) + \varepsilon v(\Theta)$$
$$= A + \frac{\varepsilon}{A}u_0^2(\Theta) + \frac{\varepsilon^3}{A}v^2(\Theta) + \frac{2\varepsilon^2}{A}u_0(\Theta)v(\Theta) + O\left(\varepsilon^4\right)$$

$$\Rightarrow u_0''(\Theta) + u_0(\Theta) + \varepsilon\left\{v''(\Theta) + v(\Theta)\right\}$$

$$= A + \varepsilon\left\{\frac{1}{A}u_0^2(\Theta)\right\} + O\left(\varepsilon^2\right) + O\left(\varepsilon^3\right) + O\left(\varepsilon^4\right). \tag{28.83}$$

Only terms without ε and terms linear in ε are considered; hence:
(a) Terms without ε:

$$u_0''(\Theta) + u_0(\Theta) = A. \tag{28.84}$$

This is the differential equation (26.25) of the Kepler motion known already from the lectures, which is solved by

$$u_0(\Theta) = A + C\sin\Theta + D\cos\Theta$$

or

$$u_0(\Theta) = A + E\cos(\Theta - \varphi).$$

Without restriction of generality, the coordinate frame may be selected such that $\varphi \equiv 0$, and one gets the trajectory

$$r(\Theta) = \frac{1}{A + E\cos\Theta} = \frac{A^{-1}}{1 + (E/A)\cos\Theta}. \tag{28.85}$$

(b) Terms linear in ε:

$$\begin{aligned} v''(\Theta) + v(\Theta) &= \frac{1}{A}u_0^2(\Theta) \\ &= \frac{1}{A}(A^2 + 2AE\cos\Theta + E^2\cos^2\Theta) \\ &= \left(A + \frac{E^2}{2A}\right) + 2E\cos\Theta + \frac{E^2}{2A}\cos 2\Theta, \end{aligned}$$

where $2\cos^2\varphi = 1 + \cos 2\varphi$ has been used. Because the differential equation is linear in v, we may write the solution as a superposition of three individual solutions:

$$v(\Theta) = v_1(\Theta) + v_2(\Theta) + v_3(\Theta),$$

with

$$v_1'' + v_1 = A + \frac{E^2}{2A}, \tag{28.86}$$

$$v_2'' + v_2 = 2E\cos\Theta,$$

$$v_3'' + v_3 = \frac{E^2}{2A}\cos 2\Theta. \tag{28.87}$$

The corresponding solutions are

$$v_1(\Theta) = A + \frac{E^2}{2A}, \tag{28.88}$$

$$v_2(\Theta) = E\Theta\sin\Theta,$$

$$v_3(\Theta) = -\frac{E^2}{6A}\cos 2\Theta. \tag{28.89}$$

The solution of the path equation up to first order in ε is then given by

$$\begin{aligned} u(\Theta) &= u_0(\Theta) + \varepsilon v(\Theta) \\ &= A + E\cos\Theta + \varepsilon\left(A + \frac{E^2}{2A}\right) + \varepsilon E\Theta\sin\Theta - \varepsilon\frac{E^2}{6A}\cos 2\Theta. \end{aligned} \tag{28.90}$$

The $\cos 2\Theta$-term is periodic in Θ; hence it cannot cause a perihelion motion. A perihelion motion must therefore originate from the $(\Theta\sin\Theta)$-term, which increases oscillatory with Θ.

We now employ the approximations

$$\begin{aligned} \cos\alpha &\approx 1 \qquad \text{for } \alpha \ll 1, \\ \sin\alpha &\approx \alpha \qquad \text{for } \alpha \ll 1, \end{aligned}$$

and the identity $\cos(\alpha - \beta) = \cos\alpha\cos\beta + \sin\alpha\sin\beta$:

$$\begin{aligned} \cos(\Theta - \varepsilon\Theta) &= \cos\Theta\cos(\varepsilon\Theta) + \sin\Theta\sin(\varepsilon\Theta) \\ &\approx \cos\Theta + \sin\Theta(\varepsilon\Theta) \qquad (\varepsilon\Theta \ll 1). \end{aligned}$$

Hence, $u(\Theta)$ may be written as

$$u(\Theta) = A + E\cos(\Theta - \varepsilon\Theta) + \varepsilon\left\{A + \frac{E^2}{2A} - \frac{E^2}{6A}\cos 2\Theta\right\}. \tag{28.91}$$

The last term oscillates with the period π between the values $\varepsilon(A + E^2/3A)$ and $\varepsilon(A + 2E^2/3A)$, that is, the radius shows, besides the variation due to the motion along the Kepler orbit, a slow periodical variation:

$$\begin{aligned} r(\Theta) &= \frac{1}{A + E\cos(\Theta - \varepsilon\Theta) + \varepsilon\,\Delta(2\Theta)}, \qquad \Delta(2\Theta) = A + \frac{E^2}{2A} - \frac{E^2}{6A}\cos 2\Theta \\ &= \frac{1}{A + E\cos(\Theta - \varepsilon\Theta)} \cdot \left[\frac{1}{1 + \dfrac{\varepsilon\,\Delta(2\Theta)}{A + E\cos(\Theta - \varepsilon\Theta)}}\right]; \end{aligned}$$

hence,

$$r(\Theta) \approx \frac{1}{A + E\cos(\Theta - \varepsilon\Theta)} \cdot \left[1 - \varepsilon\widetilde{\Delta}(\varepsilon, 2\Theta)\right], \qquad \widetilde{\Delta} = \frac{\Delta(2\Theta)}{A + E\cos(\Theta - \varepsilon\Theta)}.$$

The perihelion is defined as the minimum of $r(\Theta)$:

$$\begin{aligned} &\Rightarrow \quad \cos(\Theta - \varepsilon\Theta) = 1 \\ &\qquad \Rightarrow \quad \Theta - \varepsilon\Theta = 2\pi n, \qquad n = 0, 1, 2, \ldots. \end{aligned}$$

This yields

$$\Theta_{\min} = \frac{2\pi n}{1 - \varepsilon} = 2\pi n(1 + \varepsilon) + O(\varepsilon^2).$$

The perihelion thus moves for each circulation by the amount

$$\delta\Theta = 2\pi\varepsilon = \frac{6\pi\kappa B}{m^2 h^4}.$$

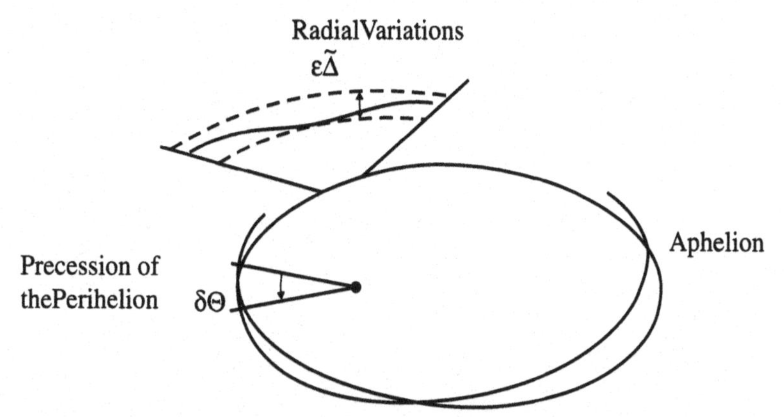

Perihel motion and variations of the radius due to a small, perturbative potential term.

PART III

THEORY OF RELATIVITY

29 Relativity Principle and Michelson–Morley Experiment

For the mathematical description of a mass point one specifies its relative motion with respect to a coordinate frame. It is convenient for this purpose to adopt a nonaccelerated reference frame (inertial system).

To an arbitrarily selected inertial system there are, however, arbitrarily many alternative ones that are moving uniformly against the first one. If one now changes from such an inertial system (K) into another one (K'), then the laws of Newtonian mechanics remain unchanged. As a consequence, one cannot decide from mechanical experiments whether an inertial system at absolute rest exists.

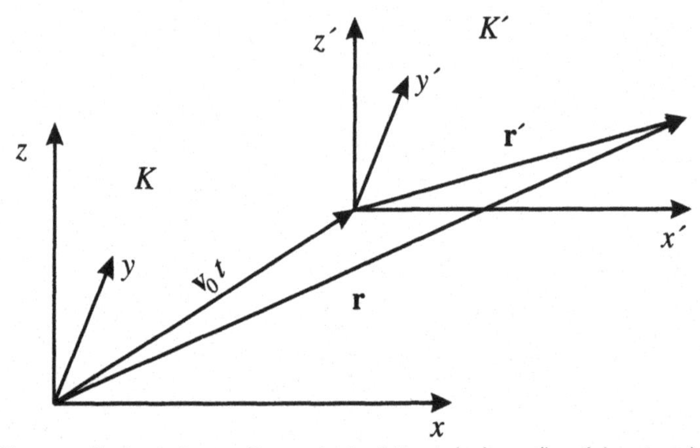

Two coordinate systems with constant relative velocity $\mathbf{v}_0$ (inertial systems).

The transformation specifying the transition from one reference frame to another frame moving with constant velocity $\mathbf{v}_0$ against the former one is the called *Galileo transformation.*[1] The corresponding transformation equations read (see figure)

$$x' = x - v_{0x}t,$$
$$y' = y - v_{0y}t,$$
$$z' = z - v_{0z}t.$$

[1]Named after *Galilei,* Galileo, Italian mathematician and philosopher, b. Feb. 15, 1564, Pisa—d. Jan. 8, 1642, Arcetri near Florence. He studied in Pisa. At the Florentine Accademia del Dissegno he got access to the writings of Archimedes. On recommendation of his patron Guidobaldo del Monte, in 1589 he got a professorship for mathematics in Pisa. Whether he performed fall experiments at the tilted tower is not proved incontestably; in any case they should confirm a false theory he proposed. In 1592 Galileo took the professorship of mathematics in Padua, not because of disagreement with colleagues but because of the better salary. He invented a proportional pair of compasses, furnished a precision mechanic workshop in his flat, found the laws for the string pendulum, and derived the fall laws in 1604 from false and in 1609 from correct assumptions. Galileo copied the telescope invented one year earlier in the Netherlands. He used it for astronomic observations and published the first results in 1610 in his *Nuncius Sidereus*, the *Star Message*. Galileo discovered the mountainous nature of the moon, the abundance of stars of the Milky Way, the phases of Venus, the moons of Jupiter (Jan. 7, 1610), and, in 1611 the sunspots, but on these Johannes Fabricius was before him.

Only since 1610 did Galileo, who returned to Florence as Court's mathematician and philosopher of the Grand Duke, publicly support the Copernican system. By his overkeenness in the following years he provoked, however, in 1614 the ban of this doctrine by the Pope. He was urged not to advocate it further by speech or writing. During a dispute on the nature of the comets of 1618, where Galileo was not in all points right, he wrote as one of his most profound treatises the *Saggiatore* (inspector with the gold balance, 1623), a paper dedicated to Pope Urban VIII. Because the former cardinal Maffeo Barberini had been well-disposed toward him, Galileo believed to win him as Pope for accepting the Copernican doctrine. He wrote his *Dialogo*, the *Talk on the two main world systems, the Ptolemyan and the Copernican,* gave the manuscript in Rome for examination, and published it in 1632 in Florence. Because he obviously had not included the agreed changes of the text thoroughly enough and had shown his siding with Copernicus too clearly, a trial set up against Galileo ended with his renouncement and condemnation on June 22, 1633. Galilieo was imprisoned in the building of inquisition for a few days. The statement "It (the earth) still moves" (Eppur si muove) is legendary. Galileo was sentenced to unrestricted arrest that he spent with short breaks in his country house at Arcetri near Florence. There he also wrote for the further development of physics his most important work: the *Discorsi e Dimonstrazioni Mathematiche*, the *Conversations and Proofs on Two New Branches of Science: The Mechanics (i.e. the Strength of Materials), and the Branches of Science Concerning Local Motions (Fall and Throw)* (Leiden, 1638).

In older representations of Galileo's life there are many exaggerations and mistakes. Galileo is not the creator of the experimental method, which he utilizes not more than many others of his contemporaries, although sometimes more critically than the competent Athanasius Kircher. Galileo was not an astronomer in the true sense, but a good observer, and as an excellent speaker and writer he won friends and patrons for a growing new science and its methods among the educated of his age, and he stimulated further research. Riccioli and Grimaldi in Bologna confirmed Galileo's laws of free fall by experiment. His scholars Torricelli and Viviani developed one of Galileo's experiments—for disproving the horror vacui—in 1643 to the barometric experiment. Chr. Huygens developed his pendulum clock based on Galileo's ideas, and he converted Galileo's kinematics to a real dynamics.

Galileo was one of the first Italians who also used their mother's language in their works for presenting scientific problems. He defended this attitude in his correspondence. His prose takes a special position within the Italian literature, since it distinguishes by its masterly clarity and simplicity from the prevailing baroque bombast Galileo had reproved also in his literary-critical essays on Tasso et al. In his works *Dialogo Sopra i due Massimi Sistemi* (Florence, 1632) and *I Dialoghi delle Nouve Scienze* (Leiden, 1638), he utilized the form of dialogue that came down from the Italian humanists, to be understood by a broad audience [BR].

In compact vector notation this simply becomes

$$\mathbf{r}' = \mathbf{r} - \mathbf{v}_0 t, \qquad \mathbf{v}_0 = \overrightarrow{\text{constant}}, \tag{29.1}$$

or *more general*:

$$\mathbf{r}' = \mathbf{r} - \mathbf{v}_0 t - \mathbf{R}_0 \qquad (\text{with } \mathbf{v}_0 = \text{constant}),$$

if the coordinate origins ($\mathbf{r} = \mathbf{r}' = 0$) at the time $t = 0$ differ by $\mathbf{R}_0$ in the x-frame.

Twofold differentiation yields

$$\mathbf{F} = m\ddot{\mathbf{r}} = m\ddot{\mathbf{r}}' = \mathbf{F}'. \tag{29.2}$$

From this equation (29.2), one immediately realizes that Newton's law, if it holds in one inertial system, also holds in any other inertial system, that is, the Newtonian mechanics remains unchanged. One says that the Newtonian mechanics is *Galileo-invariant*. In other words: The dynamic fundamental equation of mechanics is Galileo-invariant.

In the Galileo transformation it is assumed that in each inertial system the time t is the same, namely, when changing from one system to another one the time remains unchanged:

$$t = t'.$$

Hence, the time is an invariant; one speaks of an *absolute time*. In this premise one implicitly assumes that there is no upper limit for the velocity but that it is possible to transmit a message (comparison of clocks) with arbitrarily high velocity. Only then can one speak of an absolute time. We will come back to the problem of measuring times in the following chapters (see, for instance, Chapter 31).

The Michelson–Morley experiment

In the physics of the 19th century it was assumed that the light was bound to a material medium, the so-called *ether*. Just as sound propagates in air as a density oscillation, the light should propagate in the *world ether*.

It was obvious to declare the ether as "being at absolute rest" and then to try to find an "absolutely resting" inertial system, making use of electrodynamic experiments.

Imagine a spaceship moving in the ether. If this spaceship flies against the rays of light, then according to the ether theory the speed of light measured in the spaceship is larger; in the case of opposite direction of motion, it is lower. To check this theory Michelson[2]

[2] *Albert Abraham Michelson*, American physicist, b. Dec. 19, 1852, Strelno (Posen)—d. May 9, 1931, Pasadena (Calif.). From 1869–1881 he was a member of the Navy, taught at the Navy colleges in Annapolis, New York, and Washington, then was appointed professor in New York, Washington, Cleveland, Worchester, and Chicago. In 1880/81 Michelson performed an experiment in Potsdam, aimed at the proof of an absolute motion of earth. This attempt, as well as a repetition thereof performed commonly with the American chemist E.W. Morley (born Jan. 29, 1838, died Feb. 24, 1923), gave a negative result. Michelson further fixed the value of the normal meter to high accuracy, using interferometry. In 1925–1927 he performed precision measurements of the speed of light, and in 1923 he proposed an interference method to determine the absolute diameter of fixed stars. In 1907 he got the Nobel Prize of Physics for his "precision interferometer and the spectroscopic and meteorologic investigations performed with it." [BR]

adopted the earth as a spaceship moving with a speed of 30 km/s about the sun. If the ether theory applies, then the light must propagate along the direction of motion of the earth with higher speed than in any other direction.

To demonstrate these differences of speed, Michelson performed an experiment, which is sketched in the following.

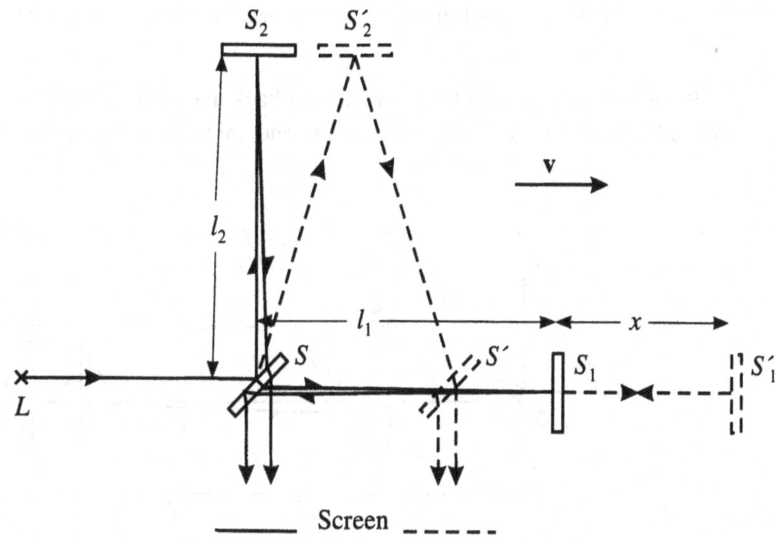

Scheme of the Michelson–Morley experiment.

The monochromatic light source L emits a light ray, which is split into two bundles by the semitransparent mirror S. After the distances l_1 and l_2, these hit the mirrors S_1 and S_2, respectively. Here they are reflected into themselves and finally again hit onto S, where the two bundles superpose. If the experiment is organized such that the two light bundles have different times of flight, one observes interference fringes on the screen.

The path difference between l_1 and l_2 in the frame at rest is

$$\Delta S = 2(l_1 - l_2).$$

In the frame ($v \,||\, l_1$) moving uniformly against the ether, the situation is as follows: The light ray passing the distance l_1 needs the time

$$t_L = \frac{\text{path} = l_1}{\text{velocity} = c}.$$

In the ether at rest, the speed of light is always equal to c. The path of the light ray is $l_1 + x$; x is the distance traversed by earth (or the mirror) during the time t_E.

$$t_E = \frac{x}{v}; \qquad t_L = \frac{l_1 + x}{c}.$$

Because $t_E = t_L$, it follows that

$$\frac{x}{v} = \frac{l_1 + x}{c} \quad \Rightarrow \quad x = \frac{l_1 v/c}{1 - v/c} = \frac{l_1 v}{c - v}. \tag{29.3}$$

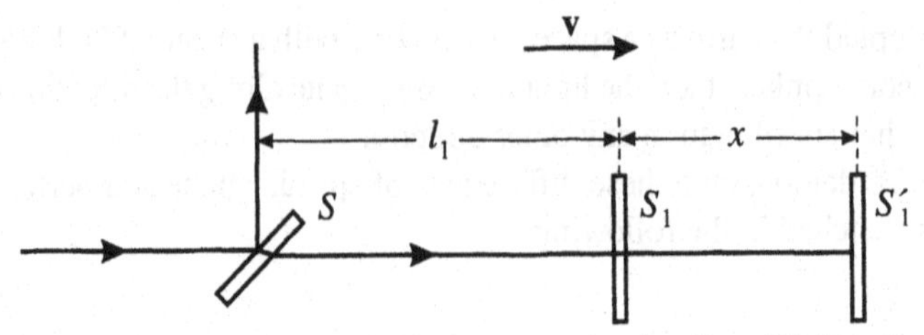

The path of the light moving ↑↑ with respect to the direction of flight is $l_1 + x$.

If we now consider the light ray moving back, the path traversed by the light ray equals $l_1 - x'$. x' is the distance traversed by the oncoming earth during the time $t_E' = x'/v$.

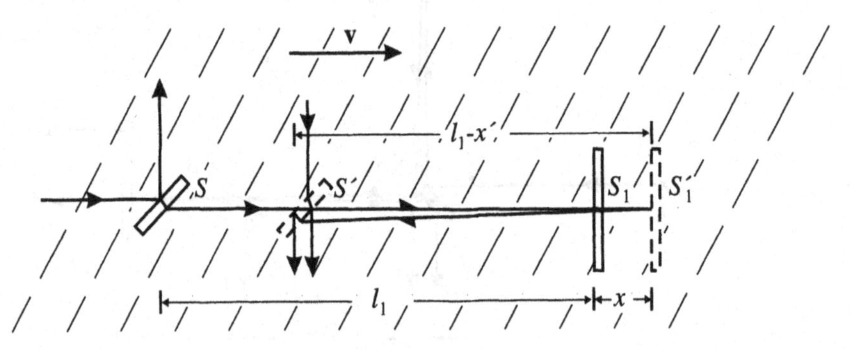

The path of the light moving ↑↓ with respect to the direction of flight is $l_1 - x$. The background should denote the ether, in which the light was supposed to propagate.

The backmoving light ray needs the time $t_L' = (l_1 - x')/c$. Because $t_E' = t_L'$, it follows that

$$\frac{x'}{v} = \frac{l_1 - x'}{c} \quad \Rightarrow \quad x' = \frac{l_1 v/c}{1 + v/c} = \frac{l_1 v}{c + v} . \tag{29.4}$$

The total distance passed by the light ray is

$$s_1 = l_1 + x + l_1 - x'. \tag{29.5}$$

By inserting x and x' (from (29.3) and (29.4)) into equation (29.5), we obtain

$$s_1 = 2l_1 + \frac{l_1 v}{c - v} - \frac{l_1 v}{c + v}$$

and after rewriting, we get

$$s_1 = \frac{2l_1}{1 - v^2/c^2} .$$

We now consider the path of rays of l_2: While the ray is running to S_2, the time

$$t = \frac{y}{v} = \frac{\sqrt{l_2^2 + y^2}}{c} \tag{29.6}$$

passes. On the way back the ray needs the same time, that is, it covers the same distance as on the way forward. From there it follows that

$$s_2 = 2\sqrt{l_2^2 + y^2}. \tag{29.7}$$

We first determine y^2 from (29.6):

$$\frac{y^2}{v^2} = \frac{l_2^2 + y^2}{c^2}.$$

Solving for y^2 yields

$$y^2 = \frac{(v^2/c^2)\, l_2^2}{1 - v^2/c^2}.$$

Inserting y^2 into (29.7) yields

$$s_2 = 2\sqrt{l_2^2 + \frac{(v^2/c^2)\, l_2^2}{1 - v^2/c^2}},$$

$$s_2 = 2l_2 \frac{1}{\sqrt{1 - v^2/c^2}}.$$

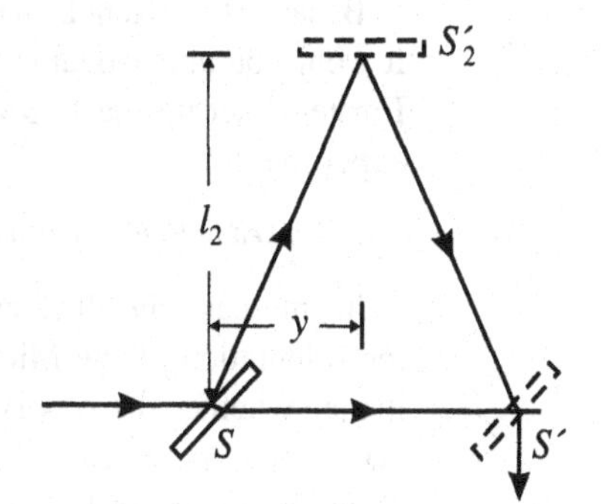

The path of the light moving ⊥ with respect to the direction of flight.

The path difference between s_1 and s_2 in the moving frame becomes

$$\Delta s = s_1 - s_2 = \frac{2l_1}{1 - (v/c)^2} - \frac{2l_2}{\sqrt{1 - (v/c)^2}}; \tag{29.8}$$

the difference of the transit times is correspondingly (the propagation speed of light in the ether is always equal to c)

$$\Delta t = \frac{\Delta s}{c} = \frac{2}{c}\left(\frac{l_1}{1 - (v/c)^2} - \frac{l_2}{\sqrt{1 - (v/c)^2}}\right). \tag{29.9}$$

If the experimental set-up is rotated by $-90°$, l_1 turns to the direction of l_2, and l_2 to the direction of l_1, that is, l_2 points along the direction of motion of earth. The light ray covers the distance l_2 faster than before the 90° rotation, and the distance l_1 is traversed slower. For $(v \,||\, l_2)$ an analogous expression results, namely

$$\Delta\widetilde{s} = s_1 - s_2 = \frac{2l_1}{\sqrt{1 - (v/c)^2}} - \frac{2l_2}{1 - (v/c)^2}. \tag{29.10}$$

This would cause a shift of the interference fringes, because

$$\Delta s - \Delta\widetilde{s} = (2l_1 + 2l_2) \cdot \left(\frac{1}{1 - (v/c)^2} - \frac{1}{\sqrt{1 - (v/c)^2}}\right) \approx (l_1 + l_2)\left(\frac{v}{c}\right)^2.$$

If, on the contrary, the frame is at rest ($v = 0$), then $\Delta s'$ in the rotated frame is equal to Δs in the nonrotated frame, namely the interference fringes are not shifted. Michelson observed, however, that also for $v \neq 0$ no shift of the interference fringes arises when rotating the apparatus.

This result might be understood if the speed of light were $c = \infty$. According to (29.8), one then would have $\Delta s = \Delta\widetilde{s} = 2(l_1 - l_2)$ for both orientations of the Michelson apparatus. Then there were no shift of the interference fringes. But it is known that $c = 300,000$ km/s. Hence this explanation is ruled out.

Because the Michelson experiment did not show any shift of the interference fringes and it would be unreasonable to assume the ether to follow the complicated motion of earth, Einstein[3] set up the following postulates to explain the result of the Michelson–Morley experiment:

The speed of light in vacuum has the same magnitude in all uniformly moving reference frames.

In this case the difference of the light paths $s_1 - s_2 = 2l_1 - 2l_2$ is independent of the orientation of the Michelson apparatus. If the speed of light is actually the same no matter whether the observer is moving toward the light source or away from it, always $\Delta s = s_1 - s_2 = 2l_1 - 2l_2$ and also $\Delta\widetilde{s} = s_1 - s_2 = 2l_1 - 2l_2$, hence $\Delta s - \Delta\widetilde{s} = 0$. Then there is no shift of the interference fringes, so to speak a priori. Moreover, Einstein postulated the *relativity principle*:

[3]*Albert Einstein*, physicist, b. March 14, 1879, Ulm—d. April 18, 1955, Princeton (N.J.). He grew up in Munich, then moved to Switzerland at the age of 15 years. As a "technical expert of third class" of the patent office at Bern he published in 1905 three highly important papers in Vol. 17 of the *Annalen der Physik*.

In his *Theory of the Brownian Motion* Einstein gave a direct and final proof of the atomistic structure of matter on a purely classical base. In the treatise "On the electrodynamics of moving bodies" he founded the *special theory of relativity*, based on a penetrating analysis of the concepts of space and time. From this theory he concluded a few months later on the general equivalence of mass and energy, expressed by the well-known formula $E = mc^2$. In the third paper Einstein extended the quantum theorem of M. Planck (1900) to the *hypothesis of light quanta* and thereby made the decisive second step in the development of quantum theory, which immediately implies the duality conception wave – particle. The idea of light quanta was considered by most physicists as being too radical and was accepted with much scepticism. The swing of opinion came only after the proposition of the theory of atoms by N. Bohr (1913).

In 1909 Einstein was appointed as professor at Zurich University. In 1911 he went to Prague; in 1912 again to Zurich to the Eidgenössische TH. In 1913 he was called to Berlin as full-time member of the Prussian Academy of Sciences and head of the Kaiser Wilhelm Institute for Physics. In 1914/15 he founded the *general theory of relativity*, starting from the strict proportionality of heavy and inert mass. The successful check of the theoretical prediction by the British expedition for observing the solar eclipse in 1919 made him publicly known far beyond the circle of experts. His political and scientific opponents tried to organize a campaign against him and the theory of relativity, which, however, remained meaningless. The Nobel Committee nevertheless considered it advisable to award the Nobel Prize for Physics of the year 1921 to Einstein not for the proposition of the theory of relativity but rather for his contributions to quantum theory.

As of 1920 Einstein tried to create a "unified theory of matter," which should comprise besides gravitation also electrodynamics. Even then when H. Yukawa had shown that besides the gravitation and electrodynamics there exist still other forces, Einstein continued his efforts that remained however without a final success. Although he had published in 1917 a landmark paper on the statistical interpretation of quantum theory, he later brought about serious objections against the "Copenhagen interpretation" by N. Bohr and W. Heisenberg, which originated in his philosophical world view.

Offenses because of his Jewish origin caused Einstein in 1933 to resign his academic positions in Germany. He found a new sphere of activity in the United States at the Institute for Advanced Studies in Princeton. Einstein's last life period was clouded by the fact that he—being all his life a convinced pacifist—had given the impetus to build the first American atomic bomb, initiated by a letter of August 2, 1939, commonly written with other scientists to President Roosevelt, that was motivated by fear of German aggression [BR].

In all uniformly moving frames there hold the same laws of nature.
(covariance of the laws of nature)

Henri Poincaré, the great French mathematician, statesman, and contemporary of Einstein, expressed it as follows: The relativity law states that the laws of physical phenomena shall be the same both for an observer being at rest, as well as for one put into uniform motion, that is, we have no, and cannot have a possibility to judge whether we are in such a kind of motion or not.

These requirements, first set up by Einstein, are not necessarily an implication of the Michelson–Morley experiment. On the contrary, many physicists tried to stick to the ether hypothesis by other possible explanations. An example of such an attempt is a hypothesis brought forward independently by Lorentz and Fitzgerald which is denoted in literature somewhat pathetically as "fatal cry of ether."

The basic idea is that the Maxwell equations hold in and only in the rest frame of the ether. Under this premise there result of course modifications of the electromagnetic interaction—additional electric and also magnetic fields—between those charged particles moving relative to the ether.

With this assumption Lorentz could prove that a system of charged particles moving in such a way against the "ether wind" is shortened by the modified electromagnetic forces.[4]

In order to apply this idea to the Michelson–Morley experiment, one has to get straight in mind that the surrounding matter, and in particular the measuring apparatus, consists of electric charges. In this way Lorentz could show that the arm of the device pointing along the motion of earth is shortened by just such an amount, that the actually longer transit time of light along this direction is compensated. As a consequence, one cannot observe changes of the interference patterns when rotating the measuring apparatus, just as is demonstrated experimentally.

Although this idea cannot be disproved in a simple way, it seems very unlikely that nature applies such complicated means to keep our absolute state of motion secret from us.

The goal is now, being confronted with the result of the Michelson–Morley experiment, to find transformation equations that mediate the transition between two inertial systems K and K'. These transformation equations are called *Lorentz transformations;* named after the Dutch theoretical physicist Hendrik Antoon Lorentz,[5] who for the first time derived the Lorentz transformation from the Michelson–Morley experiment, but did not realize its general validity, and hence also not the philosophically new element.

[4]H.A. Lorentz, *De Relative Bewegung van de AARDE en dem Aether*, Amsterdam (1892), Vers. 1, p. 74.

[5]*Hendrik Antoon Lorentz*, Dutch physicist, b. July 18, 1853, Arnheim—d. Feb. 4, 1928, Haarlem. He was professor in Leiden and since 1912 curator of the cabinet of natural sciences of the Teyler foundation in Haarlem. Lorentz joined the Maxwellian field theory with the electro-atomistic ideas: The most striking success of this theory is the explanation of the splitting of spectral lines in a magnetic field, detected in 1896 by P. Zeeman.

Lorentz thoroughly treated the relation between electric and optical phenomena in moving bodies, based on the electron theory, and he gave a first explanation of the result of the Michelson–Morley experiment by assuming a length contraction of the moving body in the direction of motion (Lorentz contraction). He contributed to the development of the theory of relativity and the quantum theory. After his retirement he was involved as leader in the scientific project to drain the Zuidersee. In 1902 he was awarded together with P. Zeeman with the Nobel Prize for Physics [BR].

30 The Lorentz Transformation

Let us consider two systems moving uniformly against each other with the relative velocity $\mathbf{v}$: The system (x, y, z, t) and the system (x', y', z', t'), and perform the following thought experiment.

At the time $t = t' = 0$, the origins of the two coordinate frames shall coincide. At this moment a light shall flash up at the origin of the two coordinate frames.

We follow Einstein's postulate—stated in order to explain the result of the Michelson–Morley experiment—that the speed of light has the same value c in each coordinate frame, that is, both an observer in the nonprimed system as well as an observer in the primed system see a *spherical wave* propagating with the same velocity c. That the wave is spherical in both systems and not an ellipsoidally deformed wave in the moving system (as one might expect at first) can be explained by the postulate of the *relativity principle.* This is an additional requirement set up by Einstein, according to which the state of motion (the velocity of the system) cannot be read off from any observation (equation) in any inertial system. An ellipsoidally deformed wave in the moving system or another propagation velocity would allow one to establish the state of motion, and thus would violate the relativity principle. Hence the light flash must be a spherical wave in both systems. Thus, the wave front obeys the equation

$$S: \qquad x^2 + y^2 + z^2 = c^2t^2 \tag{30.1}$$

in the nonprimed system and also

$$S': \qquad x'^2 + y'^2 + z'^2 = c^2t'^2 \tag{30.2}$$

in the primed system.

Because S specifies the spherical wave in K, the equation S' shall according to the relativity principle also specify a spherical wave in K'. We further postulate that also for a finite space-time distance $x^2 + y^2 + z^2 - c^2t^2 \neq 0$ in the system at rest, there shall be a corresponding finite space-time distance in the moving system, namely, $x'^2 + y'^2 + z'^2 - c^2t'^2 \neq 0$. One situation shall follow from the other, and vice versa. Therefore, S'

must follow from S; there must exist a corresponding functional relation between these equations; for example,

$$\begin{aligned} x'^2 + y'^2 + z'^2 - c^2t'^2 &= F\left(x^2 + y^2 + z^2 - c^2t^2, \mathbf{v}\right) \\ &= \widehat{F}(\mathbf{v})\left(x^2 + y^2 + z^2 - c^2t^2\right). \end{aligned}$$

In the last step we have written the functional connection as an operator equation. The operator $\widehat{F}$ thereby acts on the combination $(x^2 + y^2 + z^2 - c^2t^2)$. The function $F(x^2 + y^2 + z^2 - c^2t^2, \mathbf{v})$ still might explicitly depend on the space-time coordinates x, y, z, t (not only in the combination $x^2 + y^2 + z^2 - c^2t^2$), and on the relative velocity $\mathbf{v}$ of the inertial systems, namely, $F(x^2 + y^2 + z^2 - c^2t^2; x, y, z, t, \mathbf{v})$. In operator notation this reads $\widehat{F}(x, y, z, ct, \mathbf{v})(x^2 + y^2 + z^2 - c^2t^2)$. In this case the operator $\widehat{F}$ depends on the space-time point (x, y, z, ct) and on the velocity $\mathbf{v}$. But we require *homogeneity* of space and time. In other words: Each space-time point (x, y, z, t) shall have equal rights. The physical process then cannot depend on x, y, z, and t. This means that $\widehat{F}$ cannot explicitly depend on the space-time point x, y, z, t, and we obtain $F(x^2 + y^2 + z^2 - c^2t^2, \mathbf{v})$. Moreover, the space shall be *isotropic*, that is, the function F must not depend on the orientation of $\mathbf{v}$. In particular then it holds that

$$F\left(x^2 + y^2 + z^2 - c^2t^2, v\right) = F\left(x^2 + y^2 + z^2 - c^2t^2, -v\right),$$

or in operator form

$$\widehat{F}(v)\left(x^2 + y^2 + z^2 - c^2t^2\right) = \widehat{F}(-v)\left(x^2 + y^2 + z^2 - c^2t^2\right).$$

But because K is related to K' just as K' is to K, there must hold with the same function F:

$$\begin{aligned} x^2 + y^2 + z^2 - c^2t^2 &= \widehat{F}(-v)\left(x'^2 + y'^2 + z'^2 - c^2t'^2\right) \\ &= \widehat{F}(v)\left(x'^2 + y'^2 + z'^2 - c^2t'^2\right) \\ &= \widehat{F}(v)\widehat{F}(v)\left(x^2 + y^2 + z^2 - c^2t^2\right). \end{aligned} \tag{30.3}$$

This is only possible if the operator $\hat{F}$ means multiplication by ± 1. The negative sign is excluded because in the limit $v \to 0$ all primed quantities continuously turn into the nonprimed ones. The only remaining possibility is

$$x'^2 + y'^2 + z'^2 - c^2t'^2 = x^2 + y^2 + z^2 - c^2t^2. \tag{30.4}$$

This relation was derived for light waves, and for these it is actually trivial. It is now generalized in the following sense:

The transformation between the two systems K and K' shows similarities with a rotation of the coordinate frame in a three-dimensional space: Under a rotation the magnitude of the position vector $r^2 = x^2 + y^2 + z^2$ remains conserved; under a Lorentz transformation the quantity $s^2 \equiv x^2 + y^2 + z^2 - c^2t^2$ is conserved analogously.

We refer the reader to the subsequent considerations in the context of equation (30.43)! In other words: We now interprete the relation (30.4) in a more general sense, that is, we don't assume its validity to be restricted to light sources only, but require that the space-time length of the space-time vector $\{x, y, z, ct\}$ remains unchanged under Lorentz transformations.

In order to get more insight on the relation of the Lorentz transformation to a rotation, we first consider the rotation of a three-dimensional coordinate frame.

Rotation of a three-dimensional coordinate frame

To get a relation between the vector $\mathbf{r}'$ in the rotated frame S' and the vector $\mathbf{r}$ in the frame S, one adopts for simplicity orthogonal coordinate frames, where $\mathbf{r}$ and $\mathbf{r}'$ of course always describe the same physical points:

$$\mathbf{r} \to \mathbf{r}'.$$

A unit vector in the primed frame must be representable by a linear combination of the unit vectors in the nonprimed frame. One obtains the following system of equations:

$$\begin{aligned} \mathbf{e}_1' &= R_{11}\mathbf{e}_1 + R_{12}\mathbf{e}_2 + R_{13}\mathbf{e}_3, \\ \mathbf{e}_2' &= R_{21}\mathbf{e}_1 + R_{22}\mathbf{e}_2 + R_{23}\mathbf{e}_3, \\ \mathbf{e}_3' &= R_{31}\mathbf{e}_1 + R_{32}\mathbf{e}_2 + R_{33}\mathbf{e}_3. \end{aligned} \tag{30.5}$$

In matrix notation the three equations read

$$\begin{pmatrix} \mathbf{e}_1' \\ \mathbf{e}_2' \\ \mathbf{e}_3' \end{pmatrix} = \begin{pmatrix} R_{11} & R_{12} & R_{13} \\ R_{21} & R_{22} & R_{23} \\ R_{31} & R_{32} & R_{33} \end{pmatrix} \cdot \begin{pmatrix} \mathbf{e}_1 \\ \mathbf{e}_2 \\ \mathbf{e}_3 \end{pmatrix}. \tag{30.6}$$

For one of the equations one may also write

$$\mathbf{e}_i' = \sum_{k=1}^{3} R_{ik}\mathbf{e}_k, \qquad i = 1, 2, 3. \tag{30.7}$$

Let its inversion read as follows:

$$\mathbf{e}_i = \sum_{k=1}^{3} U_{ik}\mathbf{e}_k', \qquad i = 1, 2, 3. \tag{30.8}$$

To get an idea of what the coefficients R_{jk} are, we multiply equation (30.7) by $\mathbf{e}_m$:

$$\mathbf{e}_i' \cdot \mathbf{e}_m = \sum_{k=1}^{3} R_{ik}\mathbf{e}_k \cdot \mathbf{e}_m. \tag{30.9}$$

Because we have restricted ourselves to an orthogonal system, it holds that

$$\mathbf{e}_i \cdot \mathbf{e}_k = \delta_{ik}.$$

This means

$$\mathbf{e}_i' \cdot \mathbf{e}_m = R_{im} = \cos(\mathbf{e}_i', \mathbf{e}_m).$$

Just in the same way, (30.8) implies

$$\mathbf{e}_i \cdot \mathbf{e}_k' = U_{ik} = R_{ki}\,. \tag{30.10}$$

Thus, the inverse rotation matrix is the transposed (permuted indices) of the original matrix $\widehat{R}$ (compare to Example 6.6).

The coefficients represent the cosines of the angles between the corresponding primed and nonprimed coordinate axes. Such a cosine is also called *direction cosine.*

From $\mathbf{e}_i' \cdot \mathbf{e}_j' = \delta_{ij}$ it follows because of (30.7) that

$$\delta_{ij} = \sum_{k,k'=1}^{3} R_{ik} R_{jk'} \mathbf{e}_k \cdot \mathbf{e}_{k'} = \sum_{k=1}^{3} R_{ik} R_{jk}. \qquad \textbf{(30.11)}$$

This is the *row orthogonality* of the matrix R_{ij}. The *column orthogonality*

$$\sum_{k=1}^{3} R_{ki} R_{kj} = \delta_{ij} \qquad \textbf{(30.12)}$$

follows from the row orthogonality of the U_{ik} utilizing (30.10), namely, $U_{ik} = R_{ki}$.

For a vector $\mathbf{r}$ we have

$$\mathbf{r} = \sum_{i=1}^{3} x_i \mathbf{e}_i.$$

Because we have required that the vectors $\mathbf{r}$ and $\mathbf{r}'$ shall describe the same physical point, we have $\mathbf{r} = \mathbf{r}'$. The vector is kept fixed in space; the base frame rotates. Hence

$$\sum_{i=1}^{3} x_i' \mathbf{e}_i' = \sum_{i=1}^{3} x_i \mathbf{e}_i.$$

Multiplying this equation by $\mathbf{e}_k'$ yields

$$\sum_{i=1}^{3} x_i' \mathbf{e}_i' \cdot \mathbf{e}_k' = \sum_{i=1}^{3} x_i \mathbf{e}_i \cdot \mathbf{e}_k'.$$

We have $\mathbf{e}_i \cdot \mathbf{e}_k' = R_{ki}$ and $\mathbf{e}_i' \cdot \mathbf{e}_k' = \delta_{ik}$, from which it follows that

$$x_k' = \sum_{i=1}^{3} R_{ki} x_i,$$

or after renaming the indices

$$x_i' = \sum_{k=1}^{3} R_{ik} x_k, \qquad \textbf{(30.13)}$$

and analogously the inversion

$$x_i = \sum_{k=1}^{3} U_{ik} x_k' = \sum_{k=1}^{3} R_{ki} x_k'. \qquad \textbf{(30.14)}$$

Thus, the transformation equation for the components is completely analogous to the transformation equation for the unit vectors (equations (30.7), (30.8)).

Taking into account (30.13) and (30.14), for the normalization of $\mathbf{r}$ in both frames it results that

$$|\mathbf{r}|^2 = x^2 + y^2 + z^2 = x'^2 + y'^2 + z'^2,$$

or

$$\sum_{i=1}^{3} x_i^2 = \sum_{i=1}^{3} x_i'^2. \tag{30.15}$$

Inversely, from the invariance of the magnitude of a vector it follows according to (30.15) that the underlying transformation (30.13) and (30.14) must be an orthogonal transformation (i.e., (30.11) and (30.12) must hold). This will be proved in the following for four-dimensional vectors, starting from equation (30.17).

The Minkowski space[1]

In order to point out further analogies between a rotation in 3D space and the Lorentz transformation, we have to change to a four-dimensional space. This 4D space is called Minkowski space. We introduce the four coordinates

$$x_1 = x, \quad x_2 = y, \quad x_3 = z, \quad x_4 = ict.$$

A vector in the Minkowski space will be called *four-vector*. The position vector reads

$$\mathbf{r} = x_1\mathbf{e}_1 + x_2\mathbf{e}_2 + x_3\mathbf{e}_3 + x_4\mathbf{e}_4.$$

The Minkowski space is an orthogonal space. The orthogonality relations

$$\mathbf{e}_i \cdot \mathbf{e}_k = \delta_{ik}, \qquad i, k = 1, 2, 3, 4,$$

hold. By introducing these coordinates, the propagation of a flash of light as described by (30.1)

$$x^2 + y^2 + z^2 - c^2t^2 = 0$$

[1] *Hermann Minkowski*, b. June 22, 1864, Aleksotas (near Kaunas)—d. Jan. 12, 1909, Göttingen. Minkowski got his school leaving certificate in Königsberg (Kaliningrad) at the age of 15 years. Still during his college days in Königsberg and Berlin he won in 1883 the Great Prize of the mathematical sciences of the Academy at Paris with a paper on quadratic forms. In 1885 Minkowski did his doctorate in Königsberg, followed by his *Habilitation* in 1887 in Bonn, and since 1892 was appointed professor in Bonn, Königsberg, and Zurich, since 1902 in Göttingen. His most important achievement is the *"geometry of numbers,"* which he developed, allowing him to obtain number-theoretical results by means of geometric methods. These investigations naturally led him also to research on the *foundations of geometry.* He also contributed significantly to theoretical physics, in particular to electrodynamics, which deeply influenced the development of the special theory of relativity.

may be written in a simpler form, namely,

$$\sum_{j=1}^{4} x_j^2 = 0.$$

The expression

$$\sum_{i=1}^{4} x_i^2 \tag{30.16}$$

is the square of the magnitude (square of normalization) of the position vector in the Minkowski space. The particular feature is that the normalization of a four-vector may also be negative. We have seen by (30.4) that this normalization is conserved under a Lorentz transformation, that is, for Lorentz transformations

$$\sum_{i=1}^{4} x_i'^2 = \sum_{k=1}^{4} x_k^2. \tag{30.17}$$

holds. This important relation is *no* additional, intuitively found postulate. It may be concluded from the covariance of the light flashes (30.1) and (30.2). This will soon become evident: The starting points for determining the Lorentz transformation are equations (30.1) and (30.2). They express the covariance (equality of phenomena) of the spherical light wave in uniformly moving coordinate frames. Hence we look for a coordinate transformation between $x_i'(x', y', z', ict')$ and $x_k(x, y, z, ict)$ that converts (30.1) into (30.2), and vice versa. In analogy to the three-dimensional rotations, we try with a linear transformation

$$x_n' = \sum_{j=1}^{4} \alpha_{nj} x_j, \tag{30.18}$$

or written out

$$\begin{pmatrix} x_1' \\ x_2' \\ x_3' \\ x_4' \end{pmatrix} = \begin{pmatrix} \alpha_{11} & \alpha_{12} & \alpha_{13} & \alpha_{14} \\ \alpha_{21} & \alpha_{22} & \alpha_{23} & \alpha_{24} \\ \alpha_{31} & \alpha_{32} & \alpha_{33} & \alpha_{34} \\ \alpha_{41} & \alpha_{42} & \alpha_{43} & \alpha_{44} \end{pmatrix} \cdot \begin{pmatrix} x_1 \\ x_2 \\ x_3 \\ x_4 \end{pmatrix}, \tag{30.19}$$

where the α_{nj} constitute the transformation matrix. That the transformation must be linear may be understood as follows: Linear transformations are the only ones that map a straight line in one frame again on a straight line in other frames. In more general transformations it would happen that a uniform motion appears as accelerated motion in another inertial system. This would contradict the relativity principle. The matrix of a transformation that conserves the magnitude (30.16) of the position vector is an orthogonal matrix, that is, the row vectors or the column vectors are orthogonal to each other. The matrix α_{ik} is such an orthogonal matrix.

We may realize that by replacing in the relations

$$\sum_{k=1}^{4} x_k^2 = 0,$$

$$\sum_{i=1}^{4} x_i'^2 = 0$$

the primed coordinates by

$$x_i' = \sum_{k=1}^{4} \alpha_{ik} x_k, \qquad x_i' = \sum_{\nu=1}^{4} \alpha_{i\nu} x_\nu. \tag{30.20}$$

By introducing

$$x_i'^2 = \sum_{\nu=1}^{4} \sum_{k=1}^{4} \alpha_{ik} \alpha_{i\nu} x_k x_\nu$$

in (30.2), the requirement that (30.2) follows from (30.1) and vice versa implies conditions for the α_{ik}:

$$\begin{aligned} 0 = \sum_{i=1}^{4} x_i'^2 &= \sum_{i=1}^{4} \sum_{\nu=1}^{4} \sum_{k=1}^{4} \alpha_{ik} \alpha_{i\nu} x_k x_\nu \\ &= \sum_{\nu=1}^{4} \sum_{k=1}^{4} \left(\sum_{i=1}^{4} \alpha_{ik} \alpha_{i\nu} \right) x_k x_\nu \overset{!}{=} \sum_{k=1}^{4} x_k^2. \end{aligned}$$

That means that

$$\sum_{i=1}^{4} \alpha_{ik} \alpha_{i\nu} = 1 \qquad \text{for } k = \nu,$$

$$\sum_{i=1}^{4} \alpha_{ik} \alpha_{i\nu} = 0 \qquad \text{for } k \neq \nu.$$

must hold. This is written briefly as

$$\sum_{i=1}^{4} \alpha_{ik} \alpha_{i\nu} = \delta_{k\nu}. \tag{30.21}$$

Hence, the *column orthonormality* for the matrix (α_{ik}) holds. The row orthonormality also follows from equations (30.1) and (30.2) by starting from the transformation inverse to (30.20)

$$x_i = \sum_{k=1}^{4} b_{ik} x_k'. \tag{30.22}$$

From

$$0=\sum_{i=1}^{4}x_i^2=\sum_{i=1}^{4}\sum_{k,\nu=1}^{4}b_{ik}b_{i\nu}x'_kx'_\nu \stackrel{!}{=} \sum_{k=1}^{4}x_k'^2,$$

it then follows analogously that

$$\sum_{i=1}^{4}b_{ik}b_{i\nu}=\delta_{k\nu},$$

or after renaming the indices

$$\sum_{k=1}^{4}b_{ki}b_{k\nu}=\delta_{i\nu}. \tag{30.23}$$

Hence, the b_{ik}-matrix is orthonormal also with respect to the columns. But now the b_{ik} from (30.22) are related to the α_{ik} from (30.20), because from (30.20) and (30.22) it follows that

$$x'_i=\sum_{k=1}^{4}\alpha_{ik}x_k=\sum_{k=1}^{4}\sum_{\nu=1}^{4}\alpha_{ik}b_{k\nu}x'_\nu$$

$$\Rightarrow \quad \sum_{k=1}^{4}\alpha_{ik}b_{k\nu}=\delta_{i\nu}. \tag{30.24}$$

A comparison of (30.23) and (30.24) yields

$$\alpha_{ik}=b_{ki}\,, \tag{30.25}$$

namely, the matrix b_{ki} is the transposed α_{ik}-matrix. Insertion into (30.23) yields

$$\sum_{k=1}^{4}\alpha_{ik}\alpha_{\nu k}=\delta_{i\nu},$$

and a further renaming of indices yields

$$\sum_{i=1}^{4}\alpha_{ki}\alpha_{\nu i}=\delta_{k\nu}. \tag{30.26}$$

This is the *row orthonormality* of the matrix (α_{ik}). Although we have performed these considerations for the four-dimensional space, each individual step also holds in N dimensions. Hence, the relations (30.21) and (30.26) also hold in N dimensions.

With the column orthogonality (30.21) and the row orthogonality (30.26), it follows in general that always

$$\sum_i x_i'^2=\sum_i x_i^2.$$

This is the invariance of the "magnitude" of the space-time distance under Lorentz transformations (30.15). Hence, this relation (30.15) holds not only for zero vectors (light vectors)—that is, those for which $\sum x_i^2 = 0$—but for all vectors in the Minkowski space, hence also for those with $\sum_i x_i^2 \neq 0$. One says: It holds for all four-vectors. Later the concept of *four-vector* will be further specified.

We now turn to the explicit determination of the Lorentz transformation. In the following consideration the frames K and K' are moving against each other only in the x_1-direction. The x_1'-direction is chosen parallel to the x_1-direction; also the x_2- and x_2'- or x_3- and x_3'-directions are chosen parallel (see the figure). In this simple case it must be $y' = y$, $z' = z$. Moreover, because of the homogeneity of space, the values of x_1' and x_4' must not depend on x_2 and x_3 because the choice of the coordinate origin in the x_2, x_3-plane has no physical meaning. The Lorentz transformation therefore simplifies to

$$x_1' = \alpha_{11}x_1 + 0 + 0 + \alpha_{14}x_4, \tag{30.27}$$
$$x_2' = 0 + x_2 + 0 + 0, \tag{30.28}$$
$$x_3' = 0 + 0 + x_3 + 0, \tag{30.29}$$
$$x_4' = \alpha_{41}x_1 + 0 + 0 + \alpha_{44}x_4. \tag{30.30}$$

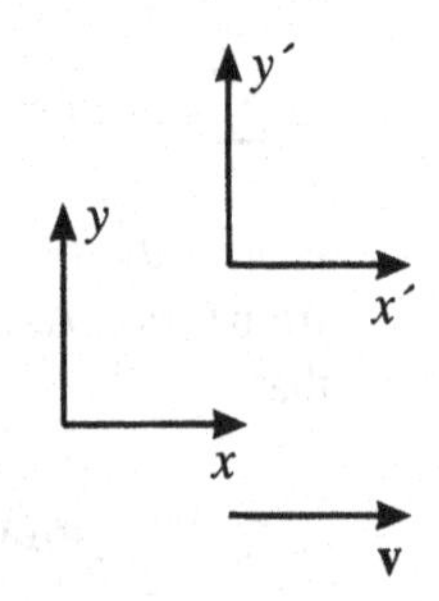

Two inertial systems of equal orientation move with relative velocity **v** along the z-axis.

The α_{jn} may now be determined by the already known orthonormality conditions (30.21) and (30.26). One has row orthonormality

$$\alpha_{11}^2 + \alpha_{14}^2 = 1, \tag{30.31}$$
$$\alpha_{41}^2 + \alpha_{44}^2 = 1, \tag{30.32}$$
$$\alpha_{11}\alpha_{41} + \alpha_{14}\alpha_{44} = 0, \tag{30.33}$$

and column orthonormality

$$\alpha_{11}^2 + \alpha_{41}^2 = 1, \tag{30.34}$$
$$\alpha_{14}^2 + \alpha_{44}^2 = 1, \tag{30.35}$$
$$\alpha_{11}\alpha_{14} + \alpha_{41}\alpha_{44} = 0. \tag{30.36}$$

From (30.27), it results that

$$x_1' = \alpha_{11}x_1 + \alpha_{14}x_4 = \alpha_{11}\left(x_1 + \frac{\alpha_{14}}{\alpha_{11}}x_4\right) = \alpha_{11}\left(x_1 + \frac{\alpha_{14}}{\alpha_{11}}ict\right).$$

Now we consider the coordinate origin of K'. There $x_1' = 0$; thus

$$0 = \alpha_{11}\left(x_1 + \frac{\alpha_{14}}{\alpha_{11}}ict\right) \quad \Rightarrow \quad x_1 = -\frac{\alpha_{14}}{\alpha_{11}}ict.$$

For the velocity it holds that

$$v = \dot{x}_1 = -\frac{\alpha_{14}}{\alpha_{11}}ic \quad \Rightarrow \quad \frac{\alpha_{14}}{\alpha_{11}} = i\frac{v}{c} \equiv i\beta \qquad \text{with} \quad \beta = \frac{v}{c}. \tag{30.37}$$

According to (30.31),

$$\alpha_{11}^2 + \alpha_{14}^2 = 1,$$

$$\alpha_{11}^2 \left(1 + \frac{\alpha_{14}^2}{\alpha_{11}^2}\right) = 1,$$
$$\alpha_{11}^2 \left(1 - \beta^2\right) = 1,$$
$$\alpha_{11} = \frac{1}{\pm\sqrt{1-\beta^2}}.$$

For low velocities the relativistic mechanics must turn into the Newtonian mechanics. But there $x_1' = x_1$,; hence $\alpha_{11} = 1$. Therefore, we find the limit

$$\beta \to 0: \quad \frac{1}{\pm\sqrt{1-\beta^2}} \to 1.$$

From there it follows that only the positive sign holds. We therefore conclude

$$\alpha_{11} = \frac{1}{\sqrt{1-\beta^2}} \quad \Rightarrow \quad \alpha_{14} = \frac{i\beta}{\sqrt{1-\beta^2}}.$$

From (30.32) and (30.35), we obtain

$$\alpha_{14}^2 = \alpha_{41}^2,$$
$$\alpha_{14} = \pm\alpha_{41}, \tag{30.38}$$

and from (30.33) it follows that

$$\alpha_{44} = -\frac{\alpha_{11}\alpha_{41}}{\alpha_{14}} = \mp\alpha_{11} = \mp\frac{1}{\sqrt{1-\beta^2}}. \tag{30.39}$$

The sign may be fixed by a similar consideration as that above. We have

$$x_4' = \alpha_{41}x_1 + \alpha_{44}x_4$$

or

$$ict' = \alpha_{41}x + \alpha_{44}ict.$$

For

$$v \to 0 \quad \Rightarrow \quad t' \to t$$

and for

$$\beta \to 0 \quad \Rightarrow \quad \frac{1}{\sqrt{1-\beta^2}} \to 1.$$

Again only the positive sign holds. This implies

$$\alpha_{44} = \frac{1}{\sqrt{1-\beta^2}}.$$

From the relation (30.39)

$$\alpha_{44} = +\alpha_{11}$$

and equation (30.38), we obtain

$$\alpha_{14} = -\alpha_{41};$$

hence

$$\alpha_{41} = \frac{-i\beta}{\sqrt{1-\beta^2}}.$$

The sign again may be determined directly, just as above. A compilation (relative motion of the two frames only in x-direction) leads to the transformation matrix:

$$(\alpha_{ik}) = \begin{pmatrix} \frac{1}{\sqrt{1-\beta^2}} & 0 & 0 & \frac{i\beta}{\sqrt{1-\beta^2}} \\ 0 & 1 & 0 & 0 \\ 0 & 0 & 1 & 0 \\ \frac{-i\beta}{\sqrt{1-\beta^2}} & 0 & 0 & \frac{1}{\sqrt{1-\beta^2}} \end{pmatrix}. \tag{30.40}$$

The Lorentz transformation equations (30.27) to (30.30) therefore read

$$\begin{aligned} x' &= \frac{x}{\sqrt{1-\beta^2}} - \frac{v}{\sqrt{1-\beta^2}}t, \\ y' &= y, \\ z' &= z, \\ t' &= \frac{t}{\sqrt{1-\beta^2}} - \frac{v/c^2}{\sqrt{1-\beta^2}}x. \end{aligned} \tag{30.41}$$

A quick glance at these equations shows that for $v \ll c$, that is, $v \to 0$ and/or $c \to \infty$, the Lorentz transformation (30.41) turns into the Galileo transformation (29.1). Actually, for $v \to 0$ both coordinate frames become identical ($x' = x, y' = y, z' = z, t' = t$), and for $c \to \infty$ the Lorentz transformations (30.41) turn into the known Galileo transformations

$$\begin{aligned} x' &= x - vt, \qquad z' = z, \\ y' &= y, \qquad t' = t \end{aligned} \tag{30.42}$$

(compare Chapter 17, section on inertial systems).

Definition of the four-vector

Four numbers $\{x_1, x_2, x_3, x_4 = ict\}$ that, with the base vectors $\mathbf{e}_1, \mathbf{e}_2, \mathbf{e}_3, \mathbf{e}_4$ of the Minkowski space, form a four-vector according to

$$\hat{r} = x_1\mathbf{e}_1 + x_2\mathbf{e}_2 + x_3\mathbf{e}_3 + x_4\mathbf{e}_4$$

are called components of the four-vector. They transform under Lorentz transformations according to (30.41). Conversely, if four numbers x_ν ($\nu = 1, 2, 3, 4$) transform according to (30.41), namely, under Lorentz transformations in a transition from an inertial system K to another one K', then these numbers form the components of a four-vector. They form—briefly spoken—a *four-vector.* This is similar to the vectors of the three-dimensional space, the components of which must transform under space rotations according to the rotational matrix.

We still note that the magnitude of a four-vector remains unchanged under Lorentz transformations (30.41). This is evident from equation (30.17) implied by the row and column orthonormality, but may also be verified explicitly by calculation. Actually, from (30.41) it follows that

$$
\begin{aligned}
x_1'^2 + x_2'^2 + x_3'^2 + x_4'^2 &= x'^2 + y'^2 + z'^2 - c^2t'^2 \\
&= \frac{1}{1-\beta^2}(x - vt)^2 + y^2 + z^2 - \frac{c^2}{1-\beta^2}\left(t - \frac{v}{c^2}x\right)^2 \\
&= \left[\frac{1}{1-\beta^2} - \frac{v^2/c^2}{1-\beta^2}\right]x^2 + y^2 + z^2 - c^2t^2\left[\frac{1}{1-\beta^2} - \frac{v^2/c^2}{1-\beta^2}\right] \\
&\quad - tx\left[\frac{2v}{1-\beta^2} - \frac{2v}{1-\beta^2}\right] \\
&= x^2 + y^2 + z^2 - c^2t^2 \\
&= x_1^2 + x_2^2 + x_3^2 + x_4^2 .
\end{aligned}
\tag{30.43}
$$

Thus, the invariance of the magnitude (30.16, 30.17) of a four-vector generally holds for arbitrary vectors in Minkowski space. Analogously to the rotations of the three-dimensional space for which according to (30.15) the magnitude of a vector remains invariant, Lorentz transformations are also denoted as *rotations in the Minkowski space.* In equation (30.4) the invariance (30.43) served as the starting point of our derivation of the Lorentz transformation. To be more precise, we have postulated the invariance of (30.43) because of the covariance of the spherical flash of light, (30.1) and (30.2), and constructed the Lorentz transformation from this postulate. We have confirmed this once again for the particular transformation (30.40).

It is important to note once more that we inferred the Lorentz transformations from the covariance (and invariance) of the expression

$$
x^2 + y^2 + z^2 - c^2t^2 = 0 = x'^2 + y'^2 + z'^2 - c^2t'^2 .
$$

Vectors of this kind are called *light vectors* or better *zero vectors.* The propagation of light is described by such a zero vector. But now we point out that these Lorentz transformations keep also arbitrary (i.e., not only zero-type) four-vectors invariant in magnitude.

Problem 30.1: Lorentz invariance of the wave equation

Show that the wave equation $\Delta\psi - (1/c^2)(\partial^2\psi/\partial t^2) = 0$ is invariant under Lorentz transformation, but is not invariant under Galileo transformation.

To simplify the problem, only the time and one space component shall be considered, that is, $\psi(x, y, z, t)$ shall be restricted to $\psi(x, t)$ or $\psi(x', t')$.

Solution The equation then reads

$$\frac{\partial^2\psi}{\partial x^2} - \frac{1}{c^2}\frac{\partial^2\psi}{\partial t^2} = 0.$$

The Lorentz transformation for the position and time coordinate reads

$$x' = \frac{x - vt}{\sqrt{1-\beta^2}}, \qquad t' = \frac{t - vx/c^2}{\sqrt{1-\beta^2}}.$$

The partial derivatives with respect to the nonprimed coordinates must be replaced by derivatives with respect to the primed coordinates. For $\partial/\partial x$ one has as complete partial derivative $\partial/\partial x_i = \sum_j(\partial x_j'/\partial x_i)(\partial/\partial x_j')$, and hence

$$\frac{\partial}{\partial x} = \frac{\partial x'}{\partial x}\frac{\partial}{\partial x'} + \frac{\partial t'}{\partial x}\frac{\partial}{\partial t'},$$

$$\frac{\partial}{\partial x} = \frac{1}{\sqrt{1-\beta^2}}\frac{\partial}{\partial x'} - \frac{v/c^2}{\sqrt{1-\beta^2}}\frac{\partial}{\partial t'}.$$

According to the same scheme we obtain the second derivative:

$$\frac{\partial^2}{\partial x^2} = \frac{1}{1-\beta^2}\frac{\partial^2}{\partial x'^2} - \frac{2v}{c^2}\frac{1}{1-\beta^2}\frac{\partial}{\partial x'}\frac{\partial}{\partial t'} + \frac{v^2}{c^4(1-\beta^2)}\frac{\partial^2}{\partial t'^2}.$$

$\partial/\partial t$ may be written as

$$\frac{\partial}{\partial t} = \frac{\partial t'}{\partial t}\frac{\partial}{\partial t'} + \frac{\partial x'}{\partial t}\frac{\partial}{\partial x'},$$

$$\frac{\partial}{\partial t} = \frac{1}{\sqrt{1-\beta^2}}\frac{\partial}{\partial t'} - \frac{v}{\sqrt{1-\beta^2}}\frac{\partial}{\partial x'},$$

$$\frac{\partial^2}{\partial t^2} = \frac{1}{1-\beta^2}\frac{\partial^2}{\partial t'^2} - \frac{2v}{1-\beta^2}\frac{\partial}{\partial x'}\frac{\partial}{\partial t'} + \frac{v^2}{1-\beta^2}\frac{\partial^2}{\partial x'^2}.$$

By insertion into the wave equation, we obtain

$$\begin{aligned}\frac{\partial^2\psi}{\partial x^2} - \frac{1}{c^2}\frac{\partial^2\psi}{\partial t^2} &= \frac{1}{1-\beta^2}\left(\frac{\partial^2\psi}{\partial x'^2} - \frac{2v}{c^2}\frac{\partial^2\psi}{\partial x'\partial t'} + \frac{v^2\partial^2\psi}{c^4\partial t'^2} - \frac{1}{c^2}\frac{\partial^2\psi}{\partial t'^2} + \frac{2v}{c^2}\frac{\partial^2\psi}{\partial x'\partial t'} - \frac{v^2}{c^2}\frac{\partial^2\psi}{\partial x'^2}\right)\\ &= \frac{1}{1-\beta^2}\left[\frac{\partial^2\psi}{\partial x'^2} - \frac{1}{c^2}\frac{\partial^2\psi}{\partial t'^2}\right] - \frac{v^2/c^2}{1-\beta^2}\left[\frac{\partial^2\psi}{\partial x'^2} - \frac{1}{c^2}\frac{\partial^2\psi}{\partial t'^2}\right]\\ &= \frac{\partial^2\psi}{\partial x'^2} - \frac{1}{c^2}\frac{\partial^2\psi}{\partial t'^2} = 0.\end{aligned}$$

Hence, the invariance under the Lorentz transformation is proved. This result may be obtained more quickly by noting that the four-gradient

$$\widehat{\nabla} = \frac{\partial}{\partial x_1}\mathbf{e}_1 + \frac{\partial}{\partial x_2}\mathbf{e}_2 + \frac{\partial}{\partial x_3}\mathbf{e}_3 + \frac{\partial}{\partial x_4}\mathbf{e}_4$$

is a four-vector, and therefore the four-scalar product

$$\begin{aligned} \widehat{\nabla}\cdot\widehat{\nabla} &= \frac{\partial^2}{\partial x_1^2} + \frac{\partial^2}{\partial x_2^2} + \frac{\partial^2}{\partial x_3^2} + \frac{\partial^2}{\partial x_4^2} \\ &= \frac{\partial^2}{\partial x^2} + \frac{\partial^2}{\partial y^2} + \frac{\partial^2}{\partial z^2} - \frac{1}{c^2}\frac{\partial^2}{\partial t^2}. \end{aligned}$$

must be a Lorentz invariant.

We still investigate the wave equation with respect to the Galileo transformation. The Galileo transformation reads

$$x' = x - vt, \qquad t' = t.$$

The partial derivatives are related by

$$\begin{aligned} \frac{\partial}{\partial x} &= \frac{\partial x'}{\partial x}\frac{\partial}{\partial x'} + \frac{\partial t'}{\partial x}\frac{\partial}{\partial t'}, & \frac{\partial}{\partial t} &= \frac{\partial t'}{\partial t}\frac{\partial}{\partial t'} + \frac{\partial x'}{\partial t}\frac{\partial}{\partial x'}, \\ \frac{\partial}{\partial x} &= \frac{\partial}{\partial x'}, & \frac{\partial}{\partial t} &= \frac{\partial}{\partial t'} - v\frac{\partial}{\partial x'}, \\ \frac{\partial^2}{\partial x^2} &= \frac{\partial^2}{\partial x'^2}, & \frac{\partial^2}{\partial t^2} &= \frac{\partial^2}{\partial t'^2} + v^2\frac{\partial^2}{\partial x'^2} - 2v\frac{\partial}{\partial t'}\frac{\partial}{\partial x'}. \end{aligned}$$

Insertion into the equation yields

$$\begin{aligned} \frac{\partial^2\psi}{\partial x^2} - \frac{1}{c^2}\frac{\partial^2\psi}{\partial t^2} &= \frac{\partial^2\psi}{\partial x'^2} - \frac{1}{c^2}\frac{\partial^2\psi}{\partial t'^2} - \frac{v^2}{c^2}\frac{\partial^2\psi}{\partial x'^2} + \frac{2v}{c^2}\frac{\partial^2\psi}{\partial t'\partial x'} \\ &= 0 = \frac{\partial^2\psi}{\partial x'^2} - \frac{1}{c^2}\left(\frac{\partial}{\partial t'} - v\frac{\partial}{\partial x'}\right)^2\psi. \end{aligned}$$

Obviously the wave equation is not invariant under the Galileo transformation. It is noteworthy, and we may be surprised in retrospect that the Lorentz transformations, as those coordinate transformations which keep the wave equation invariant, were not detected long before Einstein. After all, the wave equation was known since Maxwell. Obviously, basic discoveries mostly are not made in a straightforward way.

Group property of the Lorentz transformation

A nonvoid set G of elements $G = \{g_0, g_1, g_2, \ldots\}$ with $g_i, g_k, g_j \in G$ and a combination law ($\otimes$) are called a group if they have the following properties:

1. The combination ($\otimes$) is an inner combination that to each pair of elements $g_i, g_k \in G$ assigns a uniquely determined element $g_j = g_i \otimes g_k$ out of G.
2. The associative law $(g_i \otimes g_j) \otimes g_k = g_i \otimes (g_j \otimes g_k)$ holds.

3. There exists a unit element g_0 in G with the property

$$g_0 \otimes g_i = g_i \otimes g_0 = g_i \qquad \text{for all } g_i \in G.$$

4. To each $g_i \in G$ there exists an inverse element g_i^{-1} also belonging to G and satisfying $g_i \otimes g_i^{-1} = g_0$.

The set G has now the Lorentz transformations as elements (set of operations—as an operation one considers here the transition from one coordinate frame to a second coordinate frame moving with uniform relative velocity v with respect to the first frame); the combination means a successive application of the Lorentz transformations. As far as condition (1) is concerned, this means that the Lorentz transformation from K to K'' is equivalent to the successive application of two transformations from K to K' and from K' to K''. For sake of simplicity, here we again consider only particular Lorentz transformations in x_1-direction with parallel axes of K, K', and K''.

Transformation from K to K':

$$x'_\sigma = \sum_{\mu=1}^{4} \alpha_{\sigma\mu}(\beta_1)x_\mu,$$

with

$$\alpha_{\sigma\mu}(\beta_1) = \begin{pmatrix} \dfrac{1}{\sqrt{1-\beta_1^2}} & 0 & 0 & \dfrac{i\beta_1}{\sqrt{1-\beta_1^2}} \\ 0 & 1 & 0 & 0 \\ 0 & 0 & 1 & 0 \\ \dfrac{-i\beta_1}{\sqrt{1-\beta_1^2}} & 0 & 0 & \dfrac{1}{\sqrt{1-\beta_1^2}} \end{pmatrix}.$$

The transformation matrix from K' to K'' is

$$x''_\nu = \sum_{\sigma=1}^{4} \alpha_{\nu\sigma}(\beta_2)x'_\sigma .$$

$\alpha_{\nu\sigma}(\beta_2)$ is composed just as $\alpha_{\sigma\mu}(\beta_1)$, with the only difference that v_1 and β_1 are to be substituted by v_2 and β_2, respectively. For the transformation from K to K'' it now results that

$$\begin{aligned} x''_\nu &= \sum_\sigma \alpha_{\nu\sigma}(\beta_2) \sum_\mu \alpha_{\sigma\mu}(\beta_1)x_\mu \qquad (\nu, \sigma, \mu = 1, 2, 3, 4) \\ &= \sum_{\sigma,\mu} \alpha_{\nu\sigma}(\beta_2)\alpha_{\sigma\mu}(\beta_1)x_\mu \\ &= \sum_\mu \alpha_{\nu\mu}(\beta)x_\mu, \end{aligned}$$

where we have set $\alpha_{\nu\mu}(\beta) = \sum_\sigma \alpha_{\nu\sigma}(\beta_2)\alpha_{\sigma\mu}(\beta_1)$.

The expression $\sum_\sigma \alpha_{\nu\sigma}(\beta_2)\alpha_{\sigma\mu}(\beta_1)$ simply means a matrix multiplication. In order to calculate $\alpha_{\nu\mu}(\beta)$, we will determine the individual coefficients of this matrix. For example:

$$\alpha_{11}(\beta) = \alpha_{11}(\beta_2)\alpha_{11}(\beta_1) + \alpha_{12}(\beta_2)\alpha_{21}(\beta_1) + \alpha_{13}(\beta_2)\alpha_{31}(\beta_1) + \alpha_{14}(\beta_2)\alpha_{41}(\beta_1),$$

$$\begin{aligned} \alpha_{11}(\beta) &= \frac{1}{\sqrt{1-\beta_2^2}}\frac{1}{\sqrt{1-\beta_1^2}} + \frac{\beta_2\cdot\beta_1}{\sqrt{1-\beta_2^2}\sqrt{1-\beta_1^2}} \\ &= \frac{1}{\sqrt{(1-\beta_2^2-\beta_1^2+\beta_1^2\beta_2^2)/(1+\beta_1\beta_2)^2}} \\ &= \frac{1}{\sqrt{1-[(\beta_1+\beta_2)/(1+\beta_1\beta_2)]^2}} \\ &= \frac{1}{\sqrt{1-\beta^2}}, \end{aligned}$$

with

$$\beta = \frac{\beta_1+\beta_2}{1+\beta_1\beta_2}. \tag{30.44}$$

From here we already get a prescription for the *addition of velocities*, namely

$$v = \frac{v_1+v_2}{1+v_1v_2/c^2}. \tag{30.45}$$

In the subsequent text this "addition theorem" of velocities still shall be derived directly by another method. The other coefficients of the matrix $\alpha_{ik}(\beta)$ may be determined in the same manner. We finally obtain

$$\alpha_{\nu\mu}(\beta) = \begin{pmatrix} \dfrac{1}{\sqrt{1-\beta^2}} & 0 & 0 & \dfrac{i\beta}{\sqrt{1-\beta^2}} \\ 0 & 1 & 0 & 0 \\ 0 & 0 & 1 & 0 \\ \dfrac{-i\beta}{\sqrt{1-\beta^2}} & 0 & 0 & \dfrac{1}{\sqrt{1-\beta^2}} \end{pmatrix}, \tag{30.46}$$

where v and β are determined according to (30.45).

From there it follows that the velocity of K'' against K is equal to the addition of the velocities of K' against K, and of K'' against K', according to the addition law (30.45) for relativistic velocities. At the same time we see: Two Lorentz transformations applied successively again yield a Lorentz transformation. This is nothing else but the closure property of the set of Lorentz transformations performed successively (condition (1)).

Because of the principal equality of inertial frames K, there is no difference whether one performs at first a transformation from K to K' and subsequently from K' to K'' or vice versa, that is, the combination *"Lorentz transformation" is even commutative*; the successive application of Lorentz transformations is arbitrary with respect to the sequence. This can also be seen immediately from (30.44). But be careful! This holds only for Lorentz transformations with the same direction of the velocities, thus for inertial frames moving in the same direction.

The second group property, the associativity of the Lorentz transformation, is also fulfilled. This follows by repeated application of equation (30.45), because for the Lorentz transformation with the velocities $\beta_1, \beta_2, \beta_3$, we get

$$(L(\beta_1) \otimes L(\beta_2)) \otimes L(\beta_3) = L(\beta),$$

with

$$\beta = \frac{(\beta_1 + \beta_2)/(1 + \beta_1\beta_2) + \beta_3}{1 + (\beta_1 + \beta_2)\beta_3/(1 + \beta_1\beta_2)} = \frac{\beta_1 + \beta_2 + \beta_3 + \beta_1\beta_2\beta_3}{1 + \beta_1\beta_2 + \beta_1\beta_3 + \beta_2\beta_3}$$

and

$$L(\beta_1) \otimes (L(\beta_2) \otimes L(\beta_3)) = L(\beta'),$$

with

$$\beta' = \frac{\beta_1 + (\beta_2 + \beta_3)/(1 + \beta_2\beta_3)}{1 + \beta_1(\beta_2 + \beta_3)/(1 + \beta_2\beta_3)} = \frac{\beta_1 + \beta_2 + \beta_3 + \beta_1\beta_2\beta_3}{1 + \beta_1\beta_2 + \beta_1\beta_3 + \beta_2\beta_3}.$$

Obviously, $\beta' = \beta$ and, therefore, $L(\beta) = L(\beta')$.
But this means

$$(L(\beta_1) \otimes L(\beta_2)) \otimes L(\beta_3) = L(\beta_1) \otimes (L(\beta_2) \otimes L(\beta_3)), \qquad \text{q.e.d.}$$

The unit element has the form

$$g_0 = \begin{pmatrix} 1 & 0 & 0 & 0 \\ 0 & 1 & 0 & 0 \\ 0 & 0 & 1 & 0 \\ 0 & 0 & 0 & 1 \end{pmatrix}.$$

It corresponds to the Lorentz transformation from a system onto itself, namely, no change of the inertial frame. As required, the combination with the unit matrix is commutative (condition (3)).

To any Lorentz transformation there exists an inverse one of the form

$$v \to -v \qquad \text{or} \qquad \beta \to -\beta,$$

$$g_{ij}^{-1} = \alpha_{ij}(-\beta) = \begin{pmatrix} \dfrac{1}{\sqrt{1-\beta^2}} & 0 & 0 & \dfrac{-i\beta}{\sqrt{1-\beta^2}} \\ 0 & 1 & 0 & 0 \\ 0 & 0 & 1 & 0 \\ \dfrac{i\beta}{\sqrt{1-\beta^2}} & 0 & 0 & \dfrac{1}{\sqrt{1-\beta^2}} \end{pmatrix}.$$

Due to the orthogonality of the Lorentz transformation, the inverse element is obtained by permuting columns and rows in the transformation matrix; this simply means a reflection at the main diagonal of the matrix. One may easily verify that $\sum_j \alpha_{ij}(-\beta)\alpha_{jk}(+\beta) = \delta_{ik}$

(condition (4)). Hence, the initially imposed four conditions for the group properties of a set are fulfilled for the set of Lorentz transformations, that is, the Lorentz transformations form an infinite, continuous group (the number of elements of the set is not restricted).

Problem 30.2: Rapidity

The Lorentz transformation relating the coordinates t, z and t', z' of two coordinate systems S and S' in uniform relative motion along the z-axis with velocity $v = \beta c$ is given by

$$t = +t'\gamma + z'\beta\gamma, \qquad z = -t'\beta\gamma + z'\gamma\,. \tag{30.47}$$

This transformation is similar in its structure to a rotation in the t, z-plane,

$$t = +t'\cos\varphi + z'\sin\varphi, \qquad z = -t'\sin\varphi + z'\cos\varphi\,. \tag{30.48}$$

However, the factors γ and $\beta\gamma$ in (30.47) are greater than 1, which can be achieved with the sine and cosine functions only for an imaginary argument φ. One can make instead an *ansatz* for the transformation using the hyperbolic functions sinh and cosh with a real argument y:

$$t = +t'\cosh y + z'\sinh y, \qquad z = -t'\sinh y + z'\cosh y. \tag{30.49}$$

The argument y in these transformation equations is called the *rapidity*.

(a) Calculate the dependence of y from γ and β.

(b) When applying two consecutive rotations, the two rotation angles can simply be added. Check whether this relation also holds for the rapidity in the consecutive application of two Lorentz transformations.

Solution (a) Comparision of (30.47) and (30.49) yields

$$\gamma = \cosh y \qquad \text{and} \qquad \gamma\beta = \sinh y\,.$$

The factor β can thus be obtained as

$$\beta = \tanh y\,.$$

(b) If t' and z' are written as

$$t' = +t''\cosh y' + z''\sinh y', \qquad z' = -t''\sinh y' + z''\cosh y', \tag{30.50}$$

then inserting this in (30.49) yields

$$t = +t''(\cosh y\cosh y' + \sinh y\sinh y') + z''(\sinh y\cosh y' + \cosh y\sinh y'),$$
$$z = -t''(\sinh y\cosh y' + \cosh y\sinh y') + z''(\cosh y\cosh y' + \sinh y\sinh y'). \tag{30.51}$$

Using the addition theorems of the hyperbolic functions or reducing them to the exponential functions gives the resulting equations

$$\begin{aligned} t &= +t'' \cosh(y + y') + z'' \sinh(y + y'), \\ z &= -t'' \sinh(y + y') + z'' \cosh(y + y') . \end{aligned} \tag{30.52}$$

This corresponds to a single Lorentz transformation with rapidity

$$y'' = y + y' .$$

Thus, the rapidity variable is additive for two consecutive Lorentz transformations along the same direction, in the same way as the rotation angle is additive for two consecutive rotations around the same axis.

31 Properties of the Lorentz transformation

Time dilatation

We first note that clocks at distinct positions $x_1, x_2, \ldots$ in an inertial frame may always be mutually synchronized, that is, made to show equal times. This may be achieved, for example, by emission of light signals in second intervals from clock 1 (time t_1) to clock 2 (time t_2). At the moment of arrival at x_2, the time $(x_2 - x_1)/c$ passed, such that

$$t_2 = t_1 + \frac{x_2 - x_1}{c}.$$

We now consider the following example:

A light ray is emitted by the light source Q in the system K and after reflection by the mirror S is received at E. The measured time interval is $\Delta t = 2l/c$.

In the system K' flying by, one measures a longer time interval for the same process, as in this system the light has to traverse a longer path to reach the receiver.

Vice versa, an observer in the system K would also see such a time interval in the system K' as dilated, as the path now appears longer.

In the K-system it holds that

$$\Delta t = t_2 - t_1 = 2 \cdot \frac{l}{c}.$$

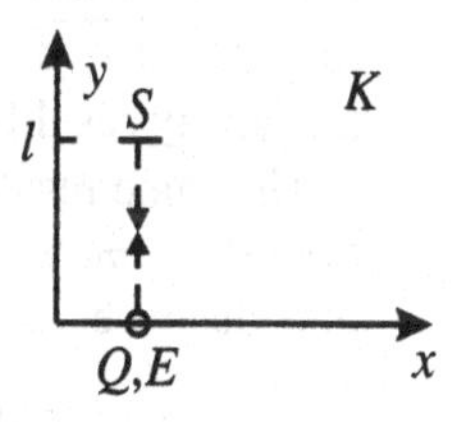

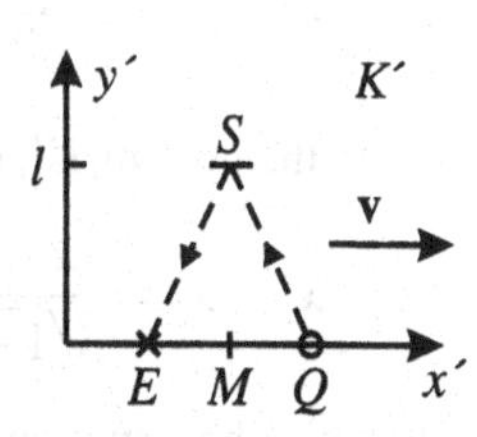

The path of a light ray on the way from sender Q to detector E in the inertial system at rest (K) and moving with respect to the source (K'). An observer at rest in K' sees the light ray emerging from point Q, hitting the mirror (at rest in K) at M and reaching the x'-axis again at E.

After the Lorentz transformation in the system K', it holds that

$$\Delta t' = t_2' - t_1',$$

where

$$t_\nu' = \frac{t_\nu - (v/c^2)\,x_\nu}{\sqrt{1-\beta^2}}$$

for $\nu = 1, 2$. We now have $x_1 = x_2$ because the light pulse is emitted and received at the same position in the K-system. Hence the interval is given by

$$\Delta t' = \frac{\Delta t}{\sqrt{1-\beta^2}}.$$

The time interval Δt in the system at rest corresponds to the time interval $\Delta t'$ in the moving system. For our example it results that

$$\Delta t' = 2\frac{l}{c}\frac{1}{\sqrt{1-\beta^2}}.$$

The dilatation of the time intervals by the Lorentz transformation is of course independent of the special definition of the time interval adopted here. If in one system the time T passed, an observer moving relative to the system finds that his clock displays the longer time $T/\sqrt{1-\beta^2}$. An observer will consider time intervals in systems moving relative to him always as dilated. This fact led to the concept of *time dilatation.*

The same result is obtained in a somewhat modified experiment: If signals are emitted *from the same position x in K* at the times t_1 and t_2, they will be received in K' with the time distance

$$t_2' - t_1' = \frac{t_2 - (v/c^2)\,x}{\sqrt{1-\beta^2}} - \frac{t_1 - (v/c^2)\,x}{\sqrt{1-\beta^2}} = \frac{t_2 - t_1}{\sqrt{1-\beta^2}}.$$

In the system K', the signals are emitted at distinct positions x_1' and x_2'. We have

$$x_1' - x_2' = \frac{x - vt_1}{\sqrt{1-\beta^2}} - \frac{x - vt_2}{\sqrt{1-\beta^2}} = \frac{v(t_2 - t_1)}{\sqrt{1-\beta^2}} = v(t_2' - t_1').$$

This phenomenon will be elucidated further by the following example 31.1. It is important that the clock in the system at rest (in our case, the system K) always ticks at the same position ($x_1 = x_2$) while, on the contrary, in the moving system (in our case, the system K') these signals are emitted at distinct positions ($x_1' \neq x_2'$). This type of measuring process is the reason for the different values of the observation times in both systems.

One may construct, although somewhat artificially, a measurement of the time intervals in such a way that the moving observer faces a shortening: At the times t_1 and t_2 in the system K at rest there occur two events at all points of a distance that is parallel to the x-axis (flashing of various lamps connected in coincidence – note that this cannot be a fluorescent tube). The time distance $t_2' - t_1'$ of the events is measured by means of a moving clock from the moving coordinate frame.

We then have

$$t_2' - t_1' = \frac{t_2 - t_1 - (v/c^2)(x_2 - x_1)}{\sqrt{1-\beta^2}}.$$

Let the measurement be performed always at the same position in the moving system; hence we have

$$x_2' - x_1' = 0 = \frac{x_2 - x_1 - v(t_2 - t_1)}{\sqrt{1-\beta^2}},$$

and by elimination of $x_2 - x_1$

$$t_2' - t_1' = (t_2 - t_1)\sqrt{1-\beta^2}.$$

It is evident that this kind of measuring time intervals, for example, for the decaying muon in the following example, does not apply.

Example 31.1: Decay of the muons

The time dilatation may be proved by means of a cosmic process: The earth is surrounded by an atmosphere of about 30-km thickness screening us off from influences from the universe. If a proton from the cosmic radiation hits the atmosphere, π-mesons are produced; several of them decay further into a muon (a "heavy electron") and a neutrino each. Now one establishes the following: The muon has a mean lifetime of $\Delta t = 2 \cdot 10^{-6}$ s in its rest system. Classically, according to $s = v \cdot \Delta t$, it might traverse even with the speed of light only a distance of 600 m. Nevertheless the particle has been recorded at the earth's surface.

In the relativistic approach, however, this contradiction is resolved: Muons at rest have a mass of $m_0 c^2 = 10^8$ eV. The "cosmic" muons are created at an altitude of ca. 10 km with a total energy of $E = 5 \cdot 10^9$ eV.

Hence we have

$$s' = v\Delta t' = \frac{v\Delta t}{\sqrt{1-\beta^2}} = \Delta x' :$$

$$s' = \frac{v m_0 c^2}{m_0 c^2\sqrt{1-\beta^2}}\Delta t = \frac{v}{m_0 c^2} E\Delta t.$$

The expression for the relativistic energy $E = m_0 c^2/\sqrt{1-\beta^2}$ used here will be derived later on in Chapter 33.

Δt is the lifetime of the muons in their rest system. $\Delta x'$ is the path of the muon during its lifetime $\Delta t' = \Delta t/\sqrt{1-\beta^2}$ in the moving system (i.e., we and the detector, fixed to the earth). $\Delta t'$ is determined by emission of two signals: The first one indicates the creation, the second one the decay of the muon in the moving system K'.

To get an upper estimate, we replace v by the speed of light; thus we find

$$s' \approx \frac{3 \cdot 10^{10}}{10^8} \cdot 5 \cdot 10^9 \cdot 2 \cdot 10^{-6}\ \text{cm} = 30\ \text{km}.$$

More precise measurements[1] actually gave a value of 38 km.

Problem 31.2: On time dilatation

We consider a spaceship that moves away from earth with the velocity $v = 0.866c$. It emits two light signals to earth spaced by $\Delta t' = 4$ s (spaceship time).

(a) What is the time distance ΔT (earth time) between the two signals arriving on earth?

(b) What distance, measured from earth, did the spaceship cover between emitting the two signals?

(c) A body at rest in the spaceship has the mass $m_0 = 1$ kg. What is its kinetic energy measured from earth?

Solution (a) We denote the *emission* of the first and second light flashes as events A and B, respectively. In the spaceship frame S' they have the space-time coordinates (x'_A, t'_A) and $(x'_B = x'_A, t'_B = t'_A + \Delta t')$; in the earth-fixed frame S the coordinates (x_A, t_A) and $(x_B = x_A + \Delta x, t_B = t_A + \Delta t)$. The relation between the two coordinate frames is given by

$$x = \gamma(x' + vt'), \qquad t = \gamma\left(t' + \frac{v}{c^2}x'\right),$$

with $\gamma = (1 - (v/c)^2)^{-1/2} = 2$ for $x = x_A$ or x_B, etc. Therefore,

$$x_A = \gamma(x'_A + vt'_A), \qquad t_A = \gamma\left(t'_A + \frac{v}{c^2}x'_A\right),$$

$$x_B = \gamma(x'_B + vt'_B), \qquad t_B = \gamma\left(t'_B + \frac{v}{c^2}x'_B\right),$$

hence, using $x'_B - x'_A = 0$,

$$x_B - x_A = \Delta x = \gamma v \Delta t', \qquad t_B - t_A = \Delta t = \gamma \Delta t'.$$

In S the two signals are emitted at the distance $\Delta t = \gamma \Delta t'$. During this time the spaceship moved forward by the distance Δx. The two light signals arrive at the earth-fixed point x_0 at the time T_A and $T_B = T_A + \Delta T$, respectively. T_A and T_B are calculated as

$$T_A = t_A + \frac{x_A - x_0}{c}, \qquad T_B = t_B + \frac{x_B - x_0}{c},$$

where $(x_A - x_0)/c$ and $(x_B - x_0)/c$ represent the transit times of the signals in S from point x_A and x_B, respectively, to the point x_0. Hence

$$T_B - T_A = \Delta T = t_B - t_A + \frac{1}{c}(x_B - x_A) = \Delta t + \frac{1}{c}\Delta x,$$

[1] The first experiments of this kind were carried out in the late 1930s and early 1940s; see, e.g., Bruno Rossi, Norman Hilberry, J. Barton Hoag: "The variation of the hard component of cosmic rays with height and the disintegration of mesotrons," *Phys. Rev.* **57** (1940) 461–469, and Bruno Rossi, David B. Hall: "Variation of the rate of decay of mesotrons with momentum," *Phys. Rev.* **59** (1941) 223–228. The "mesotron" in these papers is the lepton, which is today known as the muon.

that is, the measured time difference between the received signals is composed of the time difference Δt in S (the emissions), and a transit time difference. With the equations derived above, we have

$$\Delta T = \gamma \left(1 + \frac{v}{c}\right) \Delta t' = \frac{(1 + v/c)\Delta t'}{\sqrt{(1 + v/c)(1 - v/c)}} = \sqrt{\frac{1 + v/c}{1 - v/c}}\,\Delta t'.$$

With the data of our example, we obtain $\Delta T = 15$ s.

(b) The path Δx covered by the spaceship between the two emissions as seen from earth is $\Delta x = \gamma v \Delta t' = 2.1 \cdot 10^9$ m.

(c) The body has a total mass of $m = m_0\gamma = 2$ kg and a kinetic energy of $E_{\text{kin}} = (m - m_0)c^2 = 9 \cdot 10^{16}$ J, which corresponds to about 0.7% of the total electric energy produced in the United States in 1999. The expression for the kinetic energy used here will be substantiated in detail in Chapter 33.

Problem 31.3: Relativity of simultaneity

We observe that in a remote galaxy two events A and B happen at the same position within the galaxy. In galaxy time the event B happens by 4 s later than the event A. Further let the distance between earth and galaxy be practically constant for our problem, that is, the galaxy shall move with a constant velocity $\mathbf{v}$ perpendicular to the visual line earth–galaxy (see figure).

On earth the event B is recorded by 6 s later than the event A. Find the velocity $|\mathbf{v}|$ of the galaxy relative to earth.

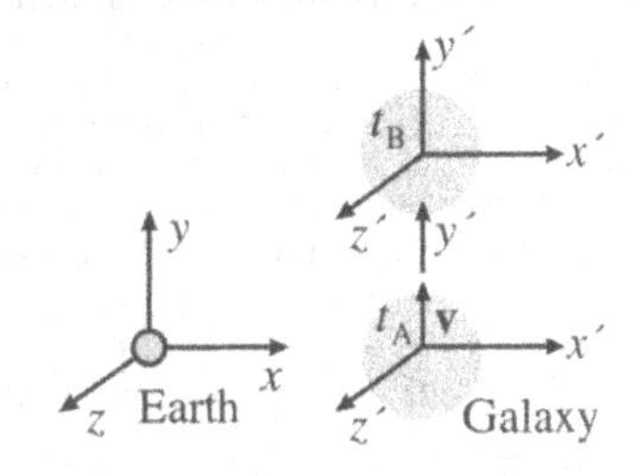

A galaxy moving with velocity **v** perpendicular to the distance from earth.

Solution The coordinate frame in the galaxy is denoted by primed quantities (S'), the earth-fixed frame by nonprimed quantities (S). The event A [B] takes place at galaxy time t'_A [t'_B], and the signal originating there is received on earth at the time t_A [t_B]. According to the condition the signals emitted by the galaxy from the events A and B traverse the same way to earth, such that the time difference between $\Delta t' = t'_B - t'_A = 4$ s and $\Delta t = t_B - t_A = 6$ s is caused by the time dilatation only. Hence:

$$t_B - t_A = \gamma(t'_B - t'_A)$$

or

$$\gamma = \frac{t_B - t_A}{t'_B - t'_A} = \frac{6}{4} = 1.5.$$

From γ immediately follows the velocity $v = |\mathbf{v}|$ of the galaxy relative to earth:

$$\gamma = \frac{1}{\sqrt{1 - v^2/c^2}}, \qquad \gamma^2 = \frac{1}{1 - v^2/c^2}.$$

$1 - v^2/c^2 = 1/\gamma^2$, and v is thus obtained as

$$v = c\sqrt{\frac{\gamma^2 - 1}{\gamma^2}} = c\sqrt{\frac{1,5^2 - 1}{1,5^2}} = 0.75\,c.$$

Lorentz–Fitzgerald length contraction

A further property of the Lorentz transformation is the length contraction measured under a relative motion of observer and object. Let us consider a rod of length l resting in the nonprimed frame K, and an observer in the moving frame K'; the frame K' moves with a relative velocity v parallel to the rod axis.

The measurement of the length is performed in such a way that the coordinates of the rod ends are determined in the observer's system at the same time ($\Delta t' = 0$) and the difference is formed, $l' = x_2' - x_1'$.

According to the Lorentz transformation,

$$x' = \frac{x - vt}{\sqrt{1-\beta^2}}.$$

The rod length is then

$$x_2' - x_1' = \frac{x_2 - x_1 - v(t_2 - t_1)}{\sqrt{1-\beta^2}}. \tag{31.1}$$

Simultaneity of the reading-off for the observer means $t_2' - t_1' = 0$; that means

$$t_2' - t_1' = \frac{(t_2 - t_1) - (v/c^2)(x_2 - x_1)}{\sqrt{1-\beta^2}} = 0.$$

From that we may determine the time interval $t_2 - t_1$. If we still set $x_2 - x_1 = l$ and investigate equation (31.1), there results

$$l' = x_2' - x_1' = l\sqrt{1-\beta^2}. \tag{31.2}$$

For the moving observer, the rod resting in K appears to be shortened by the factor $\sqrt{1-\beta^2}$z.

The cause of the length contraction is again the finiteness of the speed of light. Among the light rays from the rod ends—exploited in the measurement—which arrive simultaneously at the observer's position, the first one leaves the rod at the time t_1; then a time interval $t_2 - t_1 = vl/c^2$ passes until the second light ray leaves the other rod end.

Because the rod (or the observer's frame) is moving farther during this interval, a contraction of the rod is seen by the observer. Because only the relative motion of observer and rod matters, we always get a length contraction, no matter whether the frame of the observer or that of the rod is considered as being at rest (or moving).

Let the volume of the cube in its rest frame be $V = \Delta x\, \Delta y\, \Delta z$; in the moving frame the volume is

$$V' = \Delta x'\, \Delta y'\, \Delta z' = \Delta x\sqrt{1-\beta^2}\,\Delta y\, \Delta z = V\sqrt{1-\beta^2}. \tag{31.3}$$

Thus, the moving observer measures a smaller volume. This measurement proceeds in such a way that one measures from the moving frame perpendicularly to the direction of motion the distances $\Delta y' = \Delta y$ and $\Delta z' = \Delta z$, and parallel to the direction of motion the distance $\Delta x' = \Delta x\sqrt{1-\beta^2}$.

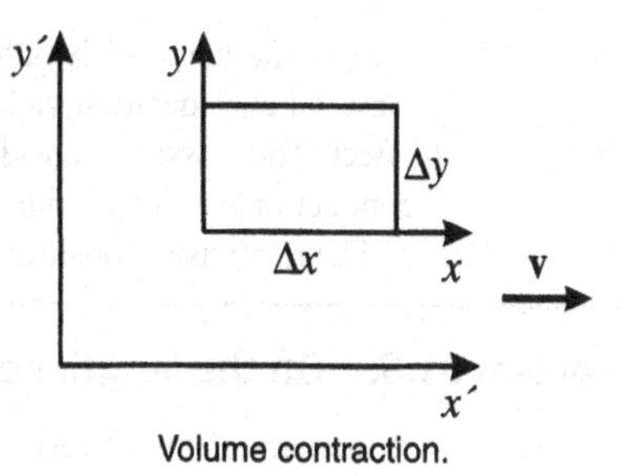

Volume contraction.

The phenomenon that time intervals appear as extended for the moving observer, while space distances appear as shortened, is due to the distinct nature of the measuring process in these cases (in the case of time measurement, we have already met two possibilities leading to dilatation or shortening, respectively).

If the measurement of length were performed by emitting signals at the ends of the distance that are simultaneous in the resting system, and by determining the position of the signals with the moving rule, then $t_1 = t_2$ and

$$x_2' - x_1' = \frac{x_2 - x_1}{\sqrt{1-\beta^2}} .$$

In this measurement the moving observer would find no contraction but rather a dilatation of the distance. The difference as compared with the earlier prescription of measurement lies in the fact that the two measured values are now recorded simultaneously in the resting frame, but formerly simultaneously in the moving frame.

Problem 31.4: Classical length contraction

Let a rod of length l_0 move with constant velocity v along the z-axis of a coordinate frame. Show that an observer at rest in this frame sees this rod as contracted also "fully without the theory of relativity" if the light propagates with finite velocity (classical length contraction). *Hint*: One should think about how the observer will define the length of the rod.

Solution The observer will conclude on the length of the rod from the light emitted by the beginning and by the end of the rod, and arriving simultaneously at his position. For simplicity we assume the observer to be at one end of the rod at some instant.

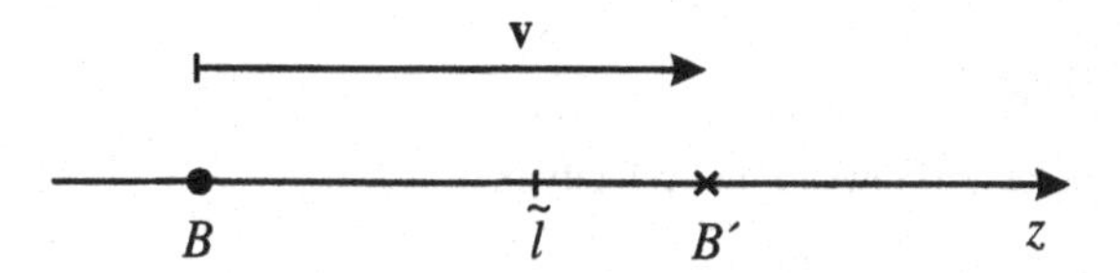

Because of the finite speed of light c, the observer B sees the end of the rod at an earlier instant $\tau = \tilde{l}/c$ at which the rod still was shifted left by the amount $v\tau$. Thus, he finds for the length of the rod

$$\tilde{l} = l - v\tau = l - v\frac{\tilde{l}}{c}$$

$$\Rightarrow \quad \tilde{l} = \frac{l}{1 + v/c} .$$

This is the classical length contraction. However, if the rod is moving toward the observer, this classical consideration yields a length dilatation—similar to the situation with the classical Doppler effect. The classical consideration thus results—depending on the case at hand—in either a length contraction or a length dilatation.

The relativistic consideration, however, yields a length contraction in all cases.

Problem 31.5: On the length contraction

A measuring rule of rest length l moves relative to an observer with the velocity v. The observer measures the length of the rule to be $\frac{2}{3}l$. Find the velocity v.

Solution We first derive the equation for the Lorentz contraction; according to the Lorentz transformations, it holds that

$$x' = \frac{x - vt}{\sqrt{1-\beta^2}}. \tag{31.4}$$

The length of the rule as measured by the observer is then

$$x_2' - x_1' = \frac{x_2 - x_1 - v(t_2 - t_1)}{\sqrt{1-\beta^2}}. \tag{31.5}$$

Simultaneity of the reading-off for the observer means that $t_2' - t_1' = 0$; that is,

$$t_2' - t_1' = \frac{(t_2 - t_1) - (v/c^2)(x_2 - x_1)}{\sqrt{1-\beta^2}} = 0$$

$$\Rightarrow \quad t_2 - t_1 = \frac{v}{c^2}(x_2 - x_1).$$

With $l' = x_2' - x_1'$ and $l = x_2 - x_1$, it follows after insertion in (31.5) that

$$l' = l\sqrt{1-\beta^2}\,. \tag{31.6}$$

According to the data, $l' = \frac{2}{3}\,l$. Equation (31.6) then implies

$$\sqrt{1 - \frac{v^2}{c^2}} = \frac{2}{3}.$$

From there it follows for the velocity that

$$\left(\frac{v}{c}\right)^2 = 1 - \frac{4}{9} = \frac{5}{9} \quad \Rightarrow \quad v = 0.745\,c.$$

Note on the invisibility of the Lorentz–Fitzgerald length contraction

From the result of the length contraction it has been concluded that an observer would see a cube moving relative to him as a cuboid, and a sphere as an ellipsoid. However, this is not the case, as we will see now.

This fallacy has been noted by the Austrian physicist Anton Lampa[2] in 1924, but his short paper (in German) remained virtually unnoticed. Only in 1959 were Lampa's main ideas independently rediscovered by James Terrell[3].

It turns out that the length contraction of spatial distances along the direction of motion under distinct methods of observation has distinct consequences. For elucidation we consider the optical image of a moving cube produced on a photographic plate parallel to a lateral face of the cube.

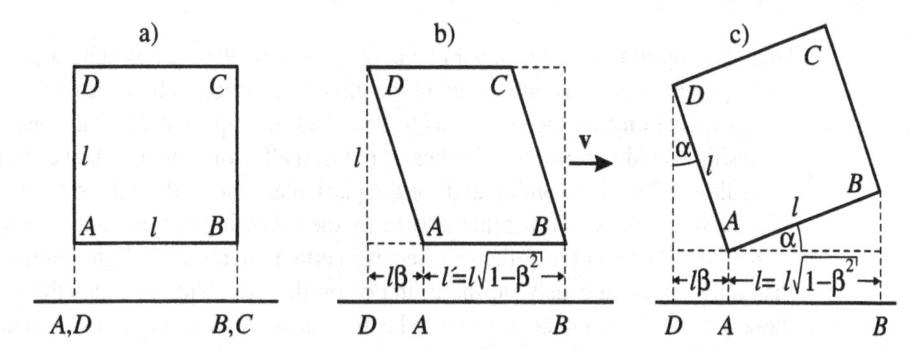

Optical appearance of a cube at rest (a) and in uniform motion (b). Figure (c) shows the apparent rotation of the cube by the angle α.

The condition of recording is again the simultaneous arrival of all light in the frame of the photographic plate. If the relative velocity v equals zero, we see (for the appropriate arrangement) only the face AB; the lateral faces AD and BC are not visible.

If the cube is moving then, due to the finite speed of light, the light arriving simultaneously on the plate has been emitted by the cube at distinct times. Although the record is made under the same conditions as in the first case, this implies that the side face AD now becomes visible.

A light ray from the point D travels by the time $T = l/c$ longer, that is, it was emitted earlier by this time, than a ray from the point A arriving simultaneously with the first ray. The same holds for the other points of the face AD. During the time l/c, the cube moved

[2] *Anton Lampa*, "Wie erscheint nach der Relativitätstheorie ein bewegter Stab einem ruhenden Beobachter?" (How does a moving rod appear to an observer at rest according to the theory of relativity?), *Z. Physik* **27** (1924) 138–148. Anton Lampa, b. Jan. 17 1868, Budapest, Hungary—d. Jan. 27 1938, Vienna. Lampa was a distiguished experimentalist working in the field of electrodynamics and electromagnetic properties of matter, and a talented teacher. The eminent nuclear physicist Lise Meitner was among his students in Vienna. From 1909–1919, Lampa was professor for experimental physics and head of the physics institute at the German University in Prague (now Czechia). He was one of the first German physicists fully grasping the importance of Einstein's new special theory of relativity and managed to get Einstein on his first full professorship, the chair for theoretical physics at the German University in Prague, in 1909. After the Great War, Lampa had to resign from his post in Prague and returned to Vienna. He did hardly any physics research anymore, but committed himself to adult education. In fact, his note on the appearence of a moving rod is his sole physics paper after 1919.

[3] J. Terrell, "Invisibility of the Lorentz contraction," *Phys. Rev.* **116** (1959) 1041–1045.

farther by the distance $s = v \cdot l/c = l \cdot \beta$, namely, the face AD is recorded on the plate as shortened by the factor β. According to the Lorentz–Fitzgerald contraction, the face AB is recorded as shortened by the factor $\sqrt{1-\beta^2}$. From the two-dimensional photoplate one thus gets the impression that the cube is rotated by the angle α ($\tan\alpha = \gamma\beta = \beta/\sqrt{1-\beta^2}$) and that the body apparently retained its shape.

The visible appearance of quickly moving bodies[4]

Until the appearance of the paper of James Terrell in 1959, it was generally believed that a moving body seems to contract along the direction of motion by a factor $(1-(v/c)^2)^{1/2}$—Lampa's note on this subject seems to have had no impact at all. The passenger of a fast spaceship would see spherical bodies as reduced ellipsoids, which, however, is impossible according to Terrell's opinion and for the particular case of the sphere has been proved by R. Penrose.[5] The reason for that is seen by the following consideration: If we see or take a photograph of an object, then we receive certain quantities of light emitted by the body that arrive simultaneously on the retina or on the film. This includes the possibility that these quantities of light are not emitted simultaneously by all points of the body. The eye or the photographic device therefore perceives a distorted image of the moving object. In the special theory of relativity, this distortion has the remarkable consequence to compensate the Lorentz contraction such that the object appears not as distorted but only as rotated. This, however, holds exactly only for bodies that lie within a small solid angle—only then the image consists mainly of parallel light pulses.

Optical appearance of a quickly moving cube

To elucidate the situation we consider the image distortion under *nonrelativistic conditions*, that is, the light propagates with the velocity c in a frame at rest against the observer, and that the motion of the object does not cause a Lorentz contraction. In the frame of the object moving with velocity v, the speed of light along the direction of motion would be $c - v$, and in the opposite direction it would be equal to $c + v$.

We first consider a cube of edge length l that moves parallel to an edge and is observed from a direction perpendicular to the direction of motion (it is observed from a large distance to keep the solid angle covered by the cube possibly small). The square $ABCD$ opposite to the observer is perceived as nondistorted as all points of the face have the same distance to the observer. The situation is different for the square $ABEF$ being perpendicular to the direction of motion. If the cube is moving, the face $ABEF$ becomes visible: Due to the time shift of the light signals from the points E and F, which are emitted by (l/c) seconds

[4] We follow a paper by V.F. Weisskopf, *Physics Today* Sept. 1960.

[5] R. Penrose, *Cambridge Phil. Soc.* **55** (1959) 137.

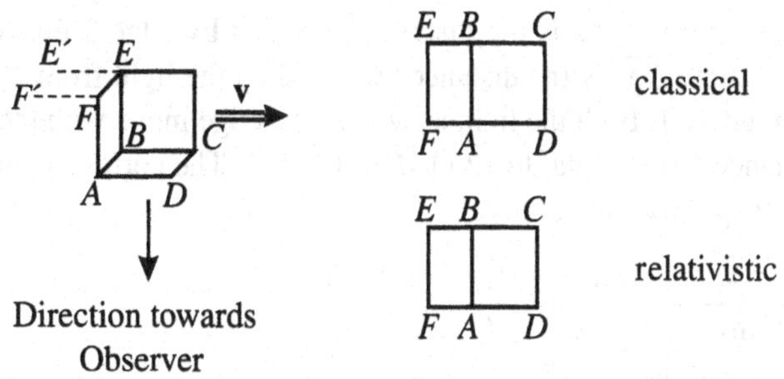

Optical appearance of a cube (classical and relativistic).

earlier than the signals from the points A and B, the points E and F are observed with a spatial displacement of $(l/c)v$ at the positions E' and F'.

The face $ABEF$ is thus seen as a rectangle with a height l and a width $(v/c)\,l$. This means that the image of the cube is distorted. In a nondistorted image of a rotated cube, both faces would appear as shortened; if the face $ABEF$ were shortened by the factor (v/c), the face $ABCD$ should be shortened by the factor $(1 - (v/c)^2)^{1/2}$, while nevertheless the face $ABCD$ appears as a square. Therefore, in the classical consideration, the image of the cube appears as extended in the direction of motion. A similar consideration for a moving sphere shows that it would appear as an ellipsoid extended along the direction of motion by the factor $(1 + (v/c)^2)^{1/2}$. One gets still considerably more paradox results if the image of a moving cube in a nonrelativistic world is not considered under an angle of 90° relative to the direction of motion but under an angle of $180° - \alpha$, where α is a very small angle. We now look at the object to the left while it moves from the left toward our position. In order to simplify the consideration we assume $v/c = 1$. The figure illustrates the new situation. The edges $\overline{AB}$, $\overline{CD}$, $\overline{EF}$ are denoted by the numbers 1, 3, 2. We assume that the edge 1 emits its light quantum at the moment $t = 0$. One sees that edge 2 must emit its light much earlier and edge 3 much later to get a simultaneous arrival at the observer's position.

Actually edge 2 must emit its light if it reaches the position of the point $2'$, which is defined by the equality of the distances $\overline{2'2}$ and $\overline{2'M}$ (the velocity v was assumed to coincide with the speed of light c!).

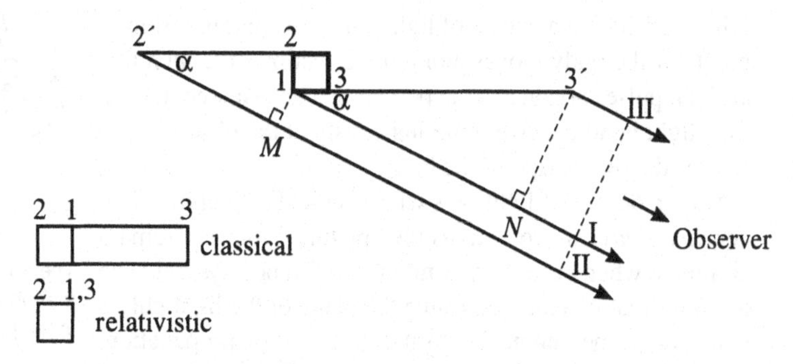

Optical appearance of a cube moving (nearly) toward the observer.

The interval $\overline{2'2}$ is the distance traversed by edge 2 between the emission of 1 and 2. The length $\overline{2'M}$ is the distance traversed by the light from 2′ to be "in line" with the light emitted by 1. Both the light as well as the edge move with the speed c; one can see that the distance $\overline{1M}$ is equal to $\overline{12}$ ($\overline{1M} = \overline{12} = l$). The corresponding also holds for edge 3. The intercept theorems imply

$$\frac{\overline{3'N}}{l\sin\alpha} = \frac{\overline{1N}}{l\cos\alpha} = \frac{\overline{13'}}{l} = \frac{l+\overline{1N}}{l},$$

and therefore,

$$\overline{3'N} = l\sin\alpha\,(1-\cos\alpha)^{-1}.$$

Note that $\overline{33'} = \overline{1N}$!

The image of the cube is indicated in the figure by the points I, II, III. We see a strongly deformed cube, with edge 2 to the left of 1, such as if the cube were viewed from backward, and edge 3 far to the right of 1. In the direction of flight there results an extended image; the area between 1 and 2 appears as a square.

The *theory of relativity* simplifies the situation. It removes the image distortion such that there results a nondistorted but rotated image of the object. This may be seen directly from the quoted examples. Let us assume that the cube is observed perpendicularly to its direction of motion; the Lorentz contraction reduces the spacing between edges $\overline{AB}$ and $\overline{CD}$ by the factor $(1-(v/c)^2)^{1/2}$ and keeps the spacing between $\overline{AB}$ and $\overline{EF}$ invariant. The image of the face $ABCD$ is thus represented as shortened exactly by the amount needed to yield a nondistorted image of the cube rotated by the angle $\arcsin(v/c)$. If the cube moves with the speed of light toward us ($\alpha = 0$), then the Lorentz contraction reduces the spacing between edges 1 and 3 to zero. The resulting image is a regular square that is identical with the lateral face $ABEF$ of the cube. In the general case (finite α), the cube is observed as nondistorted, but rotated by an angle of $(180° - \alpha)$.

Basing on the following consideration, we may show that this result is generally valid for any object.

Optical appearance of bodies moving with almost the speed of light

It is assumed that a bundle of light pulses originating from N points of the body moves along the direction of $\mathbf{k}$ such that all light pulses are on a plane perpendicular to $\mathbf{k}$ (see figure). This light bundle arrives simultaneously at the observer and creates the seen shape of the body.

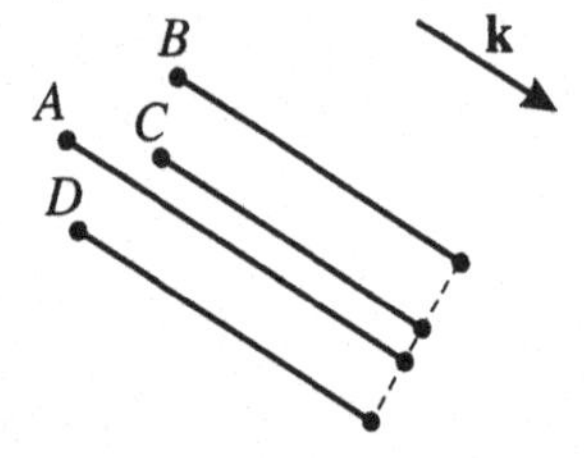

The points A, B, ... emit light that arrives at the same time in the plane of the observer.

Such a bundle of light rays will be called a "picture." Under *nonrelativistic* conditions the "picture" does not remain an image when seen from a moving reference system. The reason is that in a moving frame the plane of the light pulses is no longer perpendicular to the direction of propagation. In a relativistic world the situation is different. There the "picture" remains an image in any reference frame. The light pulses arrive simultaneously at the observer in any reference frame.

This fact may be proved in the following manner. The light pulses are visible, that is, one may imagine them as being embedded in an electromagnetic wave just there where this wave has a peak (wave group). It is known that *electromagnetic waves are transverse in all reference systems*, namely, that the front side of the wave or the plane of the wave peak is perpendicular to the propagation direction in any system (the vectors of the electric and magnetic field oscillate $\perp$ to the propagation direction $\mathbf{k}$). It may also be shown that the spacing between the light pulses is an invariant quantity. One only has to introduce a coordinate frame, the x-axis of which is parallel to the propagation direction.

The only variable quantity is the direction of propagation—the vector $\mathbf{k}$. The change of the propagation direction is given by the *aberration relation* to be derived in the following.

A light ray that encloses the angle $\overline{\theta}$ with the x-axis is observed under the angle $\overline{\theta}'$ in a frame moving with the velocity v along the x-axis. The angle $\overline{\theta}'$ is the angle under which the observer sees the incident light coming in (see figure). As may be seen from the figure, in the resting frame the light needs the time $t = \overline{P_0P_1}/c = \overline{P_0P_3}/(c \cdot \sin\overline{\theta})$ for traversing the distance $\overline{P_0P_1}$.

During this time, point P_2 moves to P_1. The distance $\overline{P_2P_1}$ is

$$\overline{P_2P_1} = v \cdot t = \frac{v}{c}\frac{\overline{P_0P_3}}{\sin\overline{\theta}}.$$

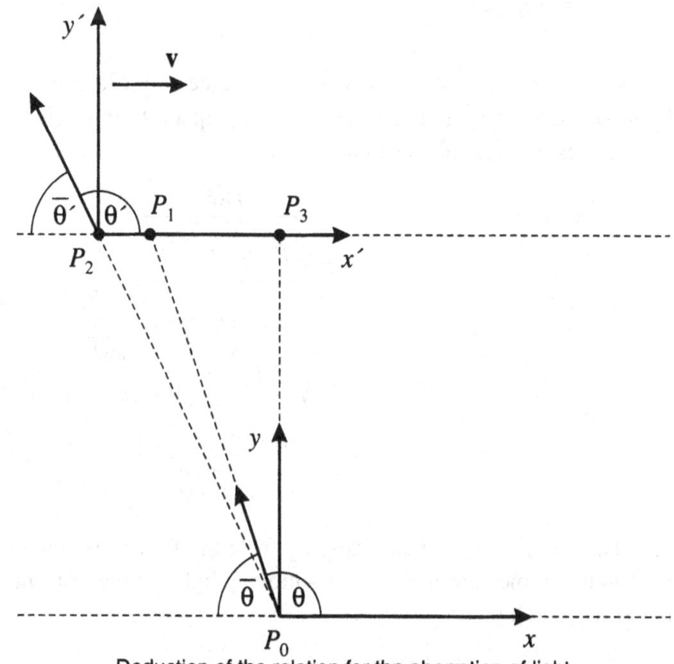

Deduction of the relation for the aberration of light.

Taking into account the relation

$$\overline{P_1P_3} = c \cdot \cos\overline{\theta} \cdot t = \overline{P_0P_3} \cdot \cot\overline{\theta}$$

for the distance $\overline{P_2P_3}$ results in the expression

$$\overline{P_2P_3} = \overline{P_2P_1} + \overline{P_1P_3} = \frac{\overline{P_0P_3}}{\sin\overline{\theta}}\left(\frac{v}{c} + \cos\overline{\theta}\right).$$

The nonrelativistic aberration results from the fact that the light is observed under the angle $(\theta')_{\text{n.r.}}$ given by

$$\sin(\theta')_{\text{n.r.}} = \sin(\pi - \overline{\theta}'_{\text{n.r.}}) = \sin(\overline{\theta}')_{\text{n.r.}}$$
$$= \frac{\overline{P_0P_3}}{\sqrt{\overline{P_0P_3}^2 + \overline{P_2P_3}^2}} = \frac{\sin\overline{\theta}}{\sqrt{1 + 2(v/c)\cos\overline{\theta} + v^2/c^2}}.$$

Because $\overline{\theta} = \pi - \theta$, we thereby obtained the relation between θ' and θ in the nonrelativistic case. To get the functional dependence of the observer angle θ' on the angle θ in the relativistic case, one has to take into account that the determined distance $\overline{P_2P_3}$ in the rest frame of the telescope (observer) because of the length contraction has the value $\overline{P_2P_3}'$, which is calculated from the relations

$$\overline{P_2P_3}'\sqrt{1 - v^2/c^2} = \overline{P_2P_3}$$

or

$$\overline{P_2P_3}' = \frac{\overline{P_2P_3}}{\sqrt{1 - v^2/c^2}} \tag{31.7}$$

(see (31.2)). The rest length here is the distance $\overline{P_2P_3'}$. It appears in the frame of the resting light source as $\overline{P_2P_3}$ and is related with that quantity through (31.2) or (31.7). From there now results the wanted aberration relation:

$$\sin\theta' = \sin(\pi - \overline{\theta}') = \sin\overline{\theta}' = \frac{\overline{P_0P_3}}{\sqrt{\overline{P_0P_3}^2 + \overline{P_2P_3}'^2}}$$
$$= \frac{\overline{P_0P_3}}{\sqrt{\overline{P_0P_3}^2 + \dfrac{\overline{P_0P_3}^2}{(1 - (v/c)^2)\sin^2\overline{\theta}}(v/c + \cos\overline{\theta})^2}}$$
$$= \frac{\sqrt{1 - (v/c)^2}\sin\overline{\theta}}{1 + (v/c)\cos\overline{\theta}} = \frac{\sqrt{1 - (v/c)^2}\sin\theta}{1 - (v/c)\cos\theta}, \tag{31.8}$$

because $\overline{\theta} = \pi - \theta$. When changing from the frame "moving observer – light source at rest" to the frame "observer at rest – moving light source," one has only to replace $v \to -v$ in (31.8), which yields

$$\sin\theta' = \frac{\sqrt{1 - (v/c)^2}\sin\theta}{1 + (v/c)\cos\theta}. \tag{31.9}$$

Expressed by the angles (compare the figure)

$$\overline{\theta'} = \pi - \theta', \qquad \overline{\theta} = \pi - \theta, \tag{31.10}$$

(31.9) finally reads

$$\sin\overline{\theta'} = \frac{\sqrt{1-(v/c)^2}\sin\overline{\theta}}{1-(v/c)\cos\overline{\theta}}. \tag{31.11}$$

Formally this is the same relation as (31.8), but—and this is important—the angles changed their meaning: According to (31.10), they are the supplement angles for θ', θ to 180°.

By the way, the inversion of equation (31.11) reads

$$\sin\overline{\theta} = \frac{\sqrt{1-(v^2/c^2)}\sin\overline{\theta'}}{1-(v/c)\cos\overline{\theta'}}, \tag{31.12}$$

which is symmetric to (31.11), that is, only $\overline{\theta}$ and $\overline{\theta'}$ are interchanged, as one would expect.

From the invariance of the image of a point, we may draw the following conclusions: The image of a moving point observed under the angle θ' is identical with the image of the same point at rest and observed under the angle θ. We therefore see a nondistorted image of a moving object (point set) that is virtually rotated by the angle $\theta' - \theta$. A spherical object therefore continues to appear as a sphere.

This should not be interpreted as nonexistence of Lorentz contraction. Of course, the Lorentz contraction happens, but it only compensates for the extension of the image caused by the finite propagation speed of light (see (31.7)). The classically expected image extension is just balanced by the Lorentz contraction!

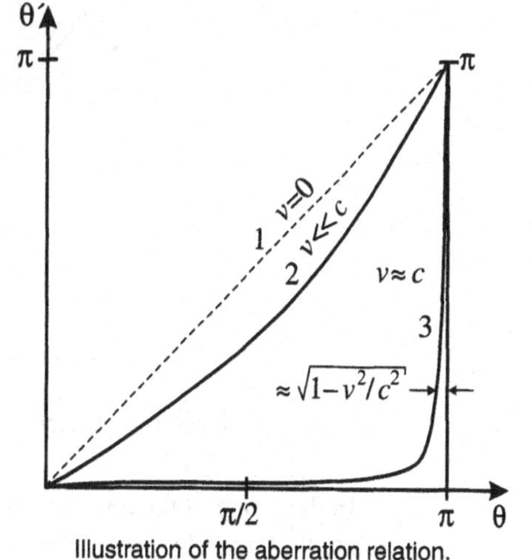

Illustration of the aberration relation.

It is appropriate to plot the angle θ' according to equation (31.11) as a function of θ. The figure shows this graph for $v = 0$ (1), for a small value of v/c (2), and also for $v/c \approx 1$ (3). We see that the virtual rotation is always negative. This means that one sees also that side of an object that points opposite to the direction of motion. In the extremum case $v \approx c$, the angle θ' is extraordinarily small for all values of θ, except for those where the angle $180° - \theta$ corresponds to the value $(1-(v/c)^2)^{1/2}$.

Because for an object moving past the angle θ ranges from 180° to 0°, we find that for $v \approx c$ the front face of the object is visible only in the very beginning. During the oncoming the object rotates; hence also its face pointing opposite to the direction of motion becomes visible to us. This state continues until the object leaves us. From that moment one sees the object from the back. This paradox situation is possibly not so surprising, if we remember the fact that the aberration angle amounts to almost 180° for $v \approx c$. If the object is moving toward our position, we see the light from it coming up to us.

Light intensity distribution of a moving isotropic emitter

The situation becomes more transparent if we consider the light distribution as seen from the observer in more detail. Let us assume the moving object to emit rays that are isotropic in their own reference system, namely, their intensity is independent of the emission angle θ. This radiation does not at all appear as isotropic in the frame of an observer at rest (laboratory system): Here it seems to be concentrated in the forward direction. For $v \approx c$, most of the emitted light seems to form a very small angle θ' with the direction of motion. This effect implies that an isotropic radiation appears as if almost the entire radiation were emitted into a spotlight cone.

The connection between the angular distribution $I(\theta)$ of the radiation intensity in the rest frame of the light source and the angular distribution I (θ') in the rest frame of the observer (in which the light source is moving) is obtained as follows: We consider a light beam that in the frame of the light source is emitted with the intensity $I(\theta)$ under the angle θ and passes an infinitesimal area element $dF = r^2 \sin\theta d\theta d\varphi$ (compare the figure).

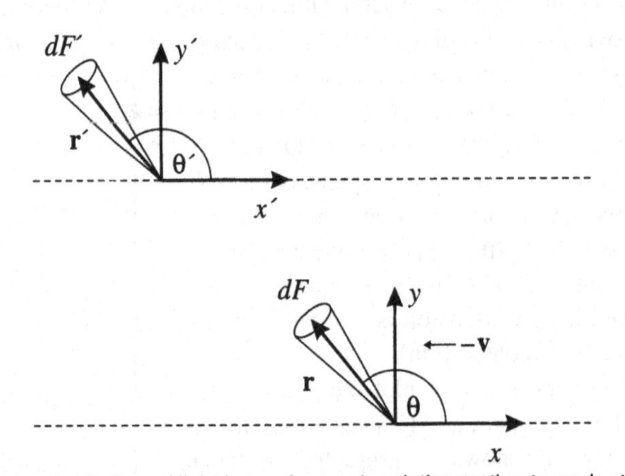

Angular distribution of light in two frames in relative motion to each other.

In the frame of the observer, this light beam is detected with the intensity $I(\theta')$ under the angle θ'. It thereby passes through the infinitesimal area element $dF' = r'^2 \sin\theta' d\theta' d\varphi'$. It is clear that the quantity of light passing through dF' in the frame of the observer must be the same as that passing through dF in the frame of the light source,

$$I(\theta)dF = I(\theta')dF'.$$

It is also clear that the increment $d\varphi' = d\varphi$, because here $d\varphi$ is perpendicular to the image plane (compare the figure) and therefore is not affected by the transformation between the moving and the resting frame.

Moreover, we may choose $r = r' = 1$: The factor $r^2(r'^2)$ in the equation for $dF(dF')$ describes only the geometric widening of the light beam if we don't let it pass at a defined distance (here $r = r' = 1$) from the coordinate origin through the test area. We therefore obtain

$$I(\theta)\sin\theta d\theta = I(\theta')\sin\theta' d\theta'.$$

The aberration formula (31.9) thereby provides the relation between θ and θ'. We use (31.9) because the light source shall move relative to the observer (see the following figure).

$$\frac{d\theta'}{d\theta} = \frac{\sqrt{1-v^2/c^2}}{1+(v/c)\cos\theta}, \qquad \frac{d\theta}{d\theta'} = \frac{\sqrt{1-v^2/c^2}}{1+(v/c)\cos\theta'}.$$

Thus, we obtain as ratio of the radiation intensities

$$\frac{I(\theta)}{I(\theta')} = \frac{\sin\theta' \, d\theta'}{\sin\theta \, d\theta} = K(\theta) = \frac{1-v^2/c^2}{(1+(v/c)\cos\theta)^2} \tag{31.13}$$

or

$$\frac{I(\theta')}{I(\theta)} = \frac{\sin\theta \, d\theta}{\sin\theta' \, d\theta'} = K(\theta') = \frac{1-v^2/c^2}{(1+(v/c)\cos\theta')^2}. \tag{31.14}$$

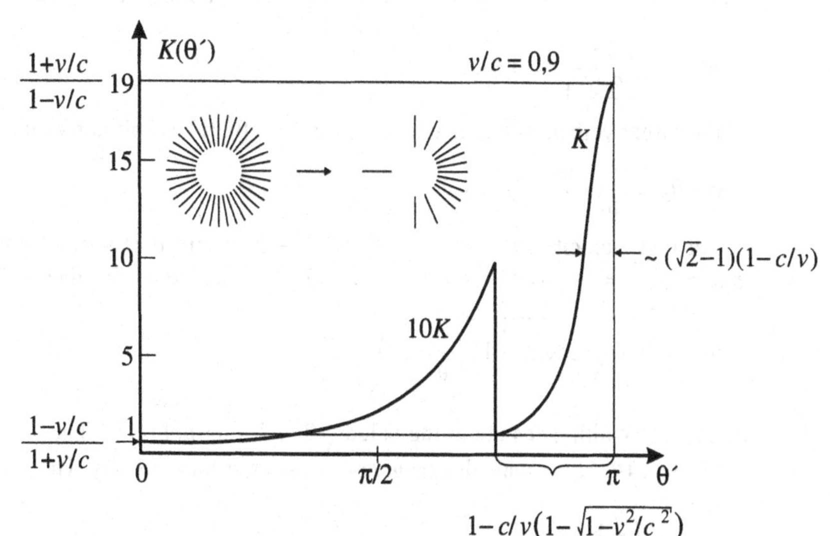

The relation of intensities $K(\theta')$ as function of the angle of observation θ'.

The last formula (31.14) is the really interesting one because it expresses the intensity $I(\theta')$ in the frame of the resting observer as function of his observation angle θ'. $I(\theta)$ is the intensity distribution of the light source in the frame of the resting source. As was already stated above, we will assume the source as isotropic, namely, we set $I(\theta) = \text{constant}$.

Here θ' is the observation angle with forward direction at $\theta' = \pi$. The function $K(\theta')$ is plotted versus θ' in the figure. We see the maximum in the forward direction ($\theta' = \pi$) and the minimum in the backward direction ($\theta' = 0$).

At high velocities $v/c = 1$, the maximum becomes extremely sharp, such that the main fraction of the radiation is emitted within a small angle about $\theta' = \pi$. The beam width is obtained from

$$\frac{1-(v/c)^2}{(1+(v/c)\cos\theta_B)^2} = \frac{1}{2}\,\frac{1+v/c}{1-v/c} = \frac{1}{2}K(\pi).$$

Here, θ_B is the so-called half-maximum angle: At the angle θ_B, the intensity of the forward-directed "spotlight cone" is reduced to half of the maximal intensity, which is reached at $\theta_{\max} = \pi$. Rewriting of the first equation yields

$$\sqrt{2}\left(1-\frac{v}{c}\right) = 1 + \frac{v}{c}\cos\theta_B.$$

One immediately sees that not every value of v is allowed, for example, $v = 0$ leads to the contradiction $\sqrt{2} = 1$. The reason is that for $v = 0$ there is no change of the intensity $I(\theta) =$ constant due to aberration. Hence there is no forward-directed "spotlight cone" that reaches half of its maximum intensity ratio at θ_B.

It is evident that the light source must move at least with such a speed that θ_B may take at least the value 0 (at backward angles the intensity ratio reaches only half of the maximum value). With $\cos\theta = \cos 0 = 1$, it then follows that v must take at least the value

$$v = c \cdot \frac{\sqrt{2}-1}{\sqrt{2}+1} \approx 0.172c.$$

For larger velocities one may evaluate a $\theta_B \in [0, \pi]$ as solution of the equation

$$\cos\theta_B = \frac{c}{v}(\sqrt{2}-1) - \sqrt{2}.$$

For high velocities $\theta_B \approx \pi$, or $\theta_B = \pi - \vartheta$, where ϑ is a small positive angle. With $\cos(\pi - \vartheta) = -\cos\vartheta \approx -1 + \frac{1}{2}\theta^2 = -1\pi + \frac{1}{2}(\pi - \theta_B)^2$, it follows that

$$\theta_B \approx \pi - \sqrt{2(\sqrt{2}-1)\left(\frac{c}{v}-1\right)},$$

that is, the width takes about the value $\sqrt{2(\sqrt{2}-1)(c/v-1)}$. The value of θ' for which $K(\theta') = 1$ ($\theta' = \theta_1$) may also be given in a straightforward way. Then

$$\frac{1-\left(\frac{v}{c}\right)^2}{\left(1+\frac{v}{c}\cos\theta_1\right)^2} = K(\theta_1) = 1$$

must hold, that is,

$$\cos\theta_1 = \frac{c}{v}\left(\sqrt{1-\left(\frac{v}{c}\right)^2} - 1\right).$$

For high velocities again $\theta_1 \approx \pi$, namely, by means of an argument analogous to that above, one may determine θ_1 as

$$\theta_1 \approx \pi - \sqrt{1 - \frac{c}{v}\left(1 - \sqrt{1 - \left(\frac{v}{c}\right)^2}\right)}\,.$$

Doppler shift of quickly moving bodies[6]

An observer moving with the velocity v observes in the moving frame (i.e., in his rest frame) light of frequency $\omega' = 2\pi\nu'$ emitted by a resting light source with frequency $\omega = 2\pi\nu$.

We are working in the frame of the light source K.

The light source emits light of frequency ω (period T) under an angle θ against the x-axis. In the figure each bar perpendicular to **k** indicates a "wave peak" of the light wave. What is the situation for the moving observer?

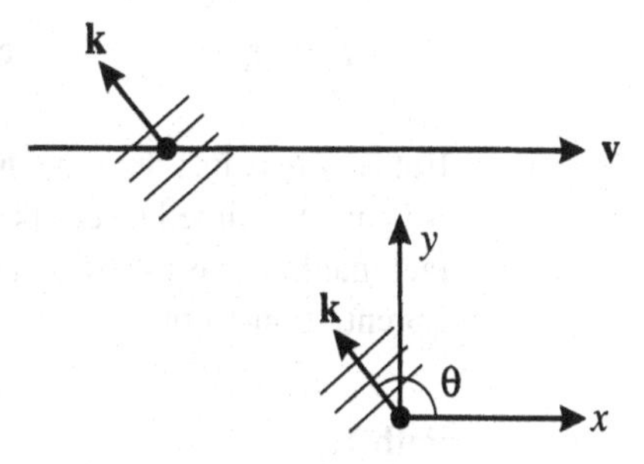

The light source is at rest, while the observer moves by with velocity v.

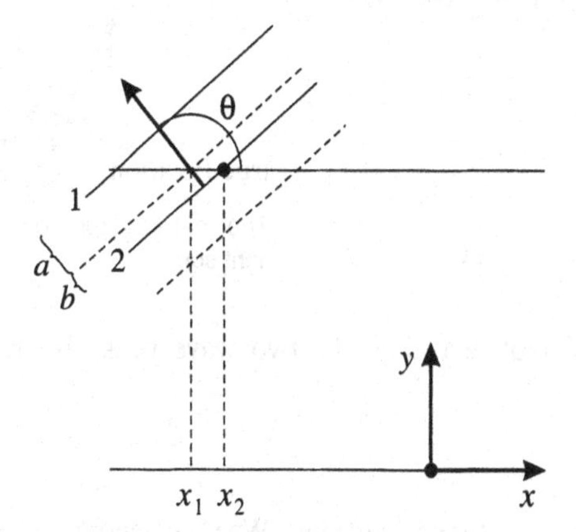

Instant t_1: The first wave peak arrives and is detected.

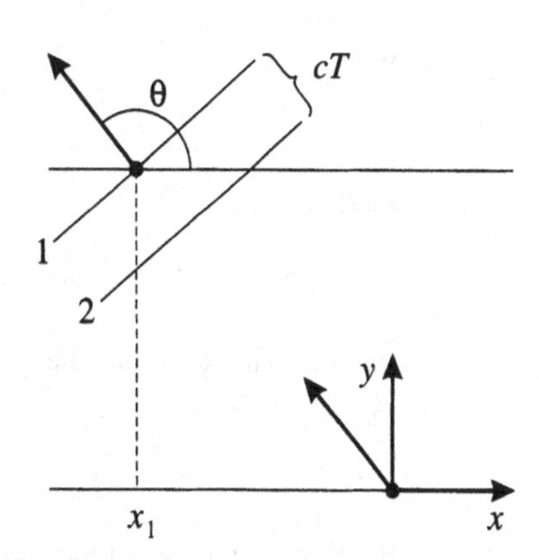

Instant t_2: The second wave peak arrives and is detected.

[6]We also refer to the detailed work of Hasselkamp, Mandry, and Scharmann, *Zeitschrift für Physik* **A289** (1979) 151.

However, the observer moved with v toward the wave peak (therefore, he will measure a shorter spacing between the wave peaks, and, consequently, measure the ultraviolet shift of the Doppler effect).

There holds (sure): $x_2 - x_1 = v(t_2 - t_1)$. (The observer meanwhile moved with v in the x-direction.) Moreover (compare figure): $\lambda = cT = a + b = c(t_2 - t_1) + (x_2 - x_1)\cos(\pi - \theta)$. a is the distance covered by the wave peak 1 during the time $t_2 - t_1$; b follows from the geometry of the right-angled triangle.

Therefore,

$$cT = c(t_2 - t_1) + v(t_2 - t_1)\cos(\pi - \theta)$$

$$\Leftrightarrow T = (t_2 - t_1)\left(1 - \frac{v}{c}\cos\theta\right).$$

But $t_2 - t_1$ is just the time difference that would be measured by the observer as a period: It is just the time he sees passing between the arrival of two wave peaks—apart from the fact that he measures with a clock that is at rest in his moving frame. Thus one still has to Lorentz-transform:

Path 1:

$$t_2' - t_1' = \gamma\left(t_2 - t_1 - \frac{v}{c^2}(x_2 - x_1)\right)$$

$$= \gamma(t_2 - t_1)\left(1 - \frac{v^2}{c^2}\right) = \frac{1}{\gamma}(t_2 - t_1),$$

thus

$$t_2 - t_1 = \gamma(t_2' - t_1').$$

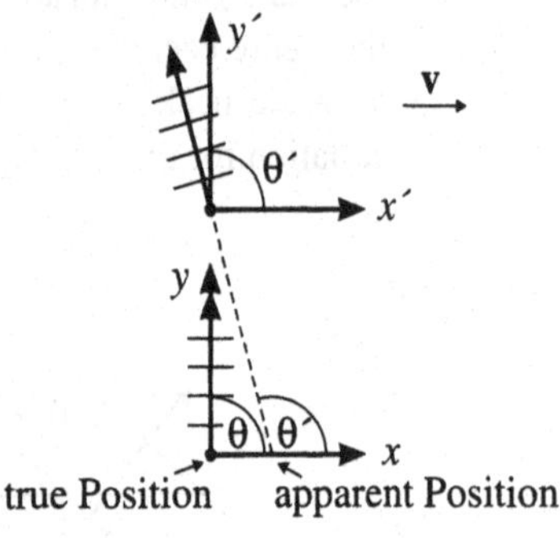

The real and apparent position of the light source.

Path 2:

$$t_2 - t_1 = \gamma\left(t_2' - t_1' + \frac{v}{c^2}(x_2' - x_1')\right)$$

(in the primed frame the point of arrival of the two wave peaks is always the origin $x_2' = x_1' = 0$)

$$t_2 - t_1 = \gamma\left(t_2' - t_1'\right).$$

Both arguments yield the same result (as it must be). We now denote $t_2' - t_1' = T'$, that is, the period measured by the moving observer in his frame.

We get

$$T = \gamma T'\left(1 - \frac{v}{c}\cos\theta\right)$$

or

$$T' = \frac{1}{\gamma}T\frac{1}{1 - \dfrac{v}{c}\cos\theta}$$

or

$$\omega' = \gamma\omega\left(1 - \frac{v}{c}\cos\theta\right)$$

This is the Doppler formula!

Actually ω' is larger if the observer moves toward the light source, because then $\theta \in [\pi/2, \pi] \Leftrightarrow \cos\theta \in [-1, 0]$. The additional factor γ then provides the Doppler shift caused by the aberration: Even if the light source emits its radiation under $\theta = 90°$, the observer measures a *higher* frequency. The reason is: The observer still must tilt his telescope *against* the direction of motion because of the relativistic aberration. He therefore *virtually* sees the light source coming up to him (although it just passes him; see figure). This implies the (relativistic) Doppler effect!

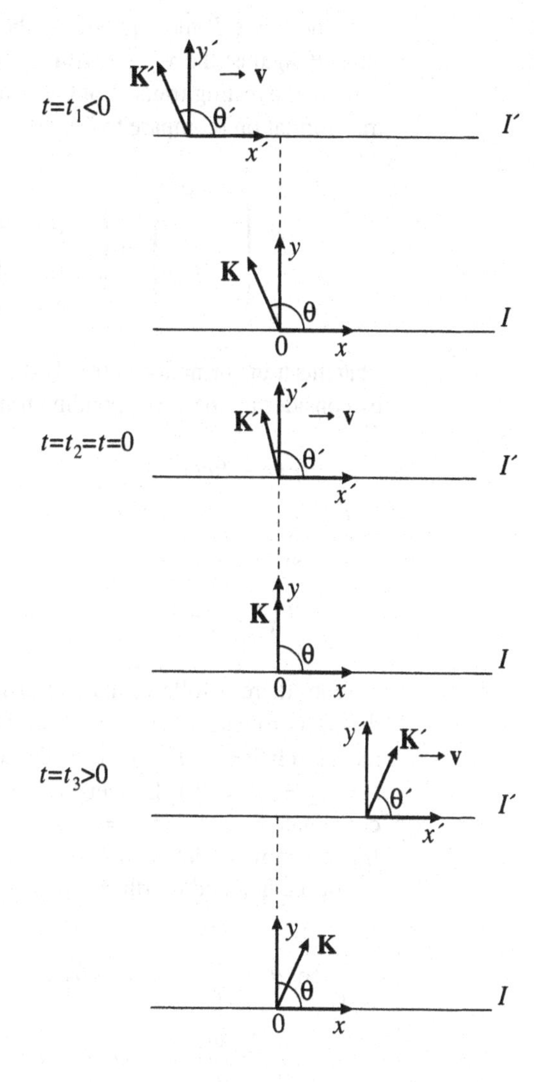

This important relation still may be understood in another way: If we consider the light as a plane wave[7]

$$\psi = \psi_0 e^{i(\mathbf{K}\cdot\mathbf{r}-\omega t)},$$

and generalize the wave number vector to a four-vector (compare to Chapter 33)

$$K_\mu = \left(\mathbf{K}, i\frac{\omega}{c}\right),$$

$$\psi = \psi_0 e^{i\sum_\mu K_\mu x_\mu},$$

we may investigate the behavior of the four-wave number vector under Lorentz transformations and also calculate the Doppler effect. K_μ must be a four-vector. The phase $\sum_{\mu=1}^{4} K_\mu x_\mu$ in the plane wave must be a scalar, because otherwise the interference properties in distinct Lorentz systems would be different. But this cannot be true. Because now x_μ is a four-vector, K_μ also must be a four-vector.

[7]Plane waves can be described by functions of the type $\Psi_0\cos(\mathbf{K}\cdot\mathbf{r}-\omega t)$. The planes of constant phase $\phi = \mathbf{K}\cdot\mathbf{r}-\omega t$ are planes that are moving in the direction of $\mathbf{K}$ with the velocity $v = \omega/|\mathbf{K}| = \omega/K$. We use the complex exponential in our description of the plane wave for technical reasons only. The imaginary part $i\Psi_0\sin(\mathbf{K}\cdot\mathbf{r}-\omega t)$ is also a plane wave. It is carried along, but not used.

In the frame I' moving with v, the plane wave is observed in the x', y'-plane under the angle θ' against the x'-axis with a frequency ω'. The wave number vector K_μ of the plane wave in the resting frame I of the light source is related to the four-vector K'_μ via a Lorentz transformation (compare to (30.40)):

$$K'_\mu = \frac{\omega'}{c}\begin{pmatrix}\cos\theta' \\ \sin\theta' \\ 0 \\ i\end{pmatrix} = \begin{pmatrix}\gamma & 0 & 0 & i\beta\gamma \\ 0 & 1 & 0 & 0 \\ 0 & 0 & 1 & 0 \\ -i\beta\gamma & 0 & 0 & \gamma\end{pmatrix}\frac{\omega}{c}\begin{pmatrix}\cos\theta \\ \sin\theta \\ 0 \\ i\end{pmatrix}.$$

That this transformation correctly describes the situation of the figure may easily be realized by considering the corresponding transformation of the position vector,

$$\begin{aligned} x' &= \gamma(x - \beta ct), \\ y' &= y, \\ z' &= z, \\ t' &= \gamma\left(t - \frac{\beta}{c}x\right). \end{aligned}$$

From there it follows that the origin of the coordinate frame $I'(x' = y' = z' = 0)$ has the x-coordinate $x = \beta ct = vt$ in the frame I, as it must be because the frame I' moves relative to I with v in the x-direction (and we have synchronized the times at $t = t_2 = t' = 0$). Conversely, the origin of the frame $I(x = y = z = 0)$ has the x'-coordinate $x' = -\gamma\beta ct = -\beta ct' = -vt'$ in the frame I', which is evident because the frame I moves relative to I' with v in the $(-x')$-direction.

For the first and fourth components of the K'_μ-vector, we obtain

$$\begin{aligned} \frac{\omega'}{c}\cos\theta' &= \frac{\omega}{c}(\gamma\cos\theta - \beta\gamma), \\ i\frac{\omega'}{c} &= \frac{\omega}{c}(-i\beta\gamma\cos\theta + i\gamma). \end{aligned}$$

Solving the system of equations for ω' and $\cos\theta'$ yields

$$\cos\theta' = \frac{-\beta + \cos\theta}{1 - \beta\cos\theta}, \qquad \cos\theta = \frac{\beta + \cos\theta'}{1 + \beta\cos\theta'},$$

$$\omega' = \frac{\sqrt{1-\beta^2}}{1 + \beta\cos\theta'}\omega, \qquad \omega' = \sqrt{K(\theta')}\omega.$$

Here $K(\theta')$ is the quantity already defined in equation (31.14). As is easily checked by using the relation $\sin\theta' = \sqrt{1 - \cos^2\theta'}$, the first line is equivalent to equation (31.8) (see above). The dependence of the frequency ω' of the received light on the observation angle θ' coincides with the relation obtained by geometric consideration. This is the wanted *aberration relation.*

The aberration of the light emitted by fixed stars was first discovered and explained by James Bradley[8] (1728).

To ensure that the light from a far remote star hits the eye of the observer moving with the earth, the observer must tilt his telescope according to the aberration relation.

We shall get this phenomenon straight to our mind by a particular case of the aberration relation. Let us assume that the **k**-vector in the resting frame I of the light source just takes the angle $\theta = \pi/2$ against the x-axis, that is, that the light is emitted just perpendicular to the x-axis and parallel to the y-axis. This corresponds to the case $t = t_2 = t' = 0$ in the above figure. We now ask under which angle θ' the observer in the moving frame I' receives the light. According to the aberration relation, one then has ($\cos\theta = \cos\pi/2 = 0$)

$$\cos\theta' = -\beta,$$

namely, $\theta' > \pi/2$, as is also indicated in the figure. But this means that the observer has to tilt his telescope against the direction of motion to get the $\mathbf{k}'$-vector pointing along the telescope axis (see the following figure).

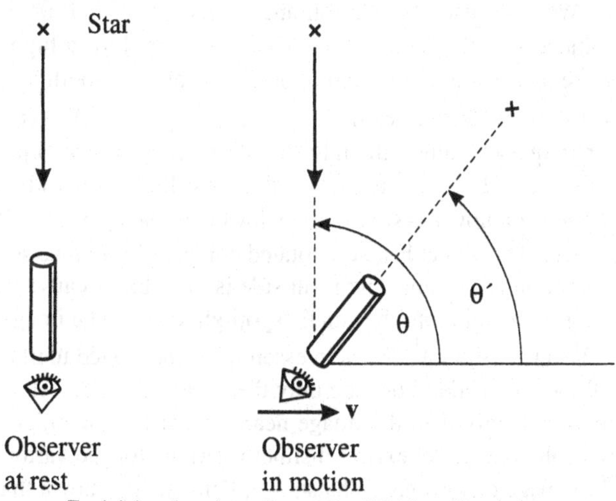

Explaining the aberration of the light from fixed stars.

We still discuss the Doppler shift. If the observer moves from a wide distance directly toward the light source, the light must be emitted under $\theta = \pi$ for him to receive it. According to the above formula, then also $\theta' = \pi$ and

$$\omega' = \sqrt{\frac{1+\beta}{1-\beta}}\,\omega > \omega,$$

[8] *James Bradley*, British astronomer, b. end of March 1693, Sherborne (near Dorchester)—d. July 13, 1762, Chalford (near Gloucester). Bradley was professor of astronomy at Oxford, then, following Edmund Halley on this post, astronomer royal and professor at Greenwich Observatory. In 1728, while searching for the parallax of stars, Bradley discovered the aberration of star light and used this observation to calculate the speed of light. In 1748, he confirmed the nutation of the earth's axis, which had been predicted by I. Newton. Bradley compiled at Greenwich a catalog with the precise locations of more than 3200 stars.

that is, the received frequency ω' is larger than the genuine light frequency ω. If the observer at far distance moves off the light source, then $\theta = \theta' = 0$ and

$$\omega' = \sqrt{\frac{1-\beta}{1+\beta}}\,\omega < \omega,$$

that is, the received frequency ω' is smaller than the genuine light frequency. The particular case $\theta = \pi/2$ is also of interest. We have seen that then $\cos\theta' = -\beta$ (aberration formula), and therefore

$$\omega' = \frac{1}{\sqrt{1-\beta^2}}\omega = \gamma\omega > \omega. \tag{31.15}$$

Although the light was emitted under $\theta = \pi/2$, it is received in the observer's frame under $\theta' > \pi/2$, that is, such as if the observer moved toward the light source. This is accompanied with the usual Doppler shift to higher frequencies!

We now describe what will be seen if an object moves away with nearly the speed of light: We first observe under an angle close to $\theta' = 180°$. Here we see the front side of the object whereby, due to the strong Doppler shift, a high intensity and a shift to very high frequencies are observed. One looks into the spotlight beam of the radiation. If the observation angle reaches the magnitude $\theta' = \pi - (1-(v/c)^2)^{1/2}$, the color changes to lower-frequency values, the intensity decreases, and the object seems to rotate.

If $\theta \approx \pi - 2^{1/4}(1-(v/c)^2)^{1/2}$, thus is still close to 180°, we are beyond the "spotlight ray"; the color now has significantly lower-frequency values than in a frame convected with the object. The object has now rotated completely, and we see its side pointing opposite to the direction of motion. The front side is invisible because all rays emitted forward in the moving frame join into the small "spotlight cone." The images seen at angles smaller than $\pi - 2^{1/4}(1-(v/c)^2)^{1/2}$ remain essentially unchanged until the object disappears.

All these considerations are only then exact if the object is confined within a very small solid angle. Only then the image nearly consists of parallel light pulses. At larger values of the solid angle, we expect distinct rotations for the various fractions of the image that lead to image distortions. Whatsoever, Penrose has shown that the image of a sphere has a circular circumference also at large observation angles.

Relativistic space-time structure—space-time events

In a four-dimensional coordinate frame, as was introduced for the mathematical description of the Minkowski space, we may no longer operate with the concept "position" as in three-dimensional space. We therefore introduce the concept "event" to stress the equality of spatial and time coordinates. The four-dimensional space of three position coordinates and one time coordinate is frequently denoted simply as *space-time*.

A mass point that moves or is at rest relative to its inertial frame is described as a function of time and space, thus as a curve in space-time. This curve in the Minkowski space is called a *world line*.

The time behavior of a point being at rest (A) (representation in the two-dimensional subspace of the Minkowski space), as well as that of a mass point moving relative to an inertial frame (B) may be described geometrically as is shown in the graph. The reciprocal slope of the curve specifies the velocity of a moving mass point.

At an angle of 45° against the x-axis one has the line of light; it holds that

$$\tan\alpha = \frac{ct}{x} = 1 \quad \Rightarrow \quad c = \frac{x}{t}.$$

A curve bent to the right represents a mass point getting faster; a curve bent to the left represents a permanently decelerated point. Because the speed of light cannot be exceeded, the smallest possible slope equals 1.

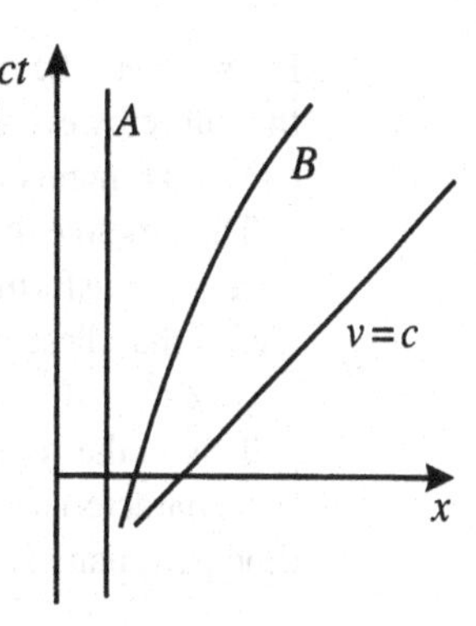

World line of a point at rest (a), in accelerated motion (b), and moving at the speed of light (c).

Relativistic past, present, future

In the Minkowski space the length element $ds^2 = dx^2 + dy^2 + dz^2 - c^2dt^2$ is invariant against Lorentz transformations. Because of the coordinate ict in the four-dimensional space-time compound, the length element is no longer positive-definite. The following cases may be distinguished:

(a) $ds^2 > 0$

This distance is called *spacelike* since the "spatial" part of the length element is larger than the time part, that is,

$$\underbrace{dx^2 + dy^2 + dz^2}_{\text{spatial part of } ds^2} > \underbrace{c^2dt^2}_{\text{time part of } ds^2}$$

If, for example, two events happen at the same time but at distinct positions, then $dx^2 + dy^2 + dz^2 \neq 0$ and $c^2dt^2 = 0$.

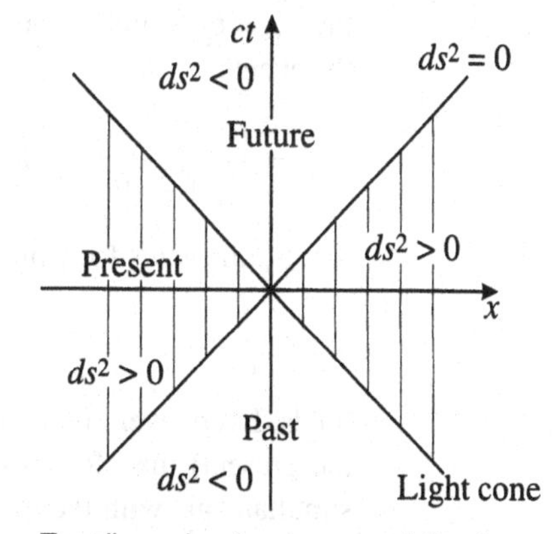

Two-dimensional subspace of the four-dimensional space-time.

For an observer at the origin of the above frame, those events that have a space-like distance to him cannot be found out because of the finite speed of light. No information may be obtained from this region. The speed of the information transfer would have to be larger than the speed of light. The spacelike distance remains spacelike under any Lorentz transformation.

(b) $ds^2 < 0$

Such distances are called *timelike,* because the time part of ds^2 dominates, that is,

$$c^2\,dt^2 > dx^2 + dy^2 + dz^2.$$

One is dealing with events that happened already or will happen, thus events that we "have seen" or "shall see" if we understand ourselves again as observers at the origin.

Events from the past may be identified by their aftereffects; conversely, we may influence the future by events that have lasting effects. The region of the Minkowski space for which $ds^2 > 0$ remains inaccessable to us.

The timelike distance specifies that $dx^2 + dy^2 + dz^2 < c^2\,dt^2$. In this case there exists a Lorentz transformation for which $ds^2 = ds'^2 = -c^2\,dt'^2$ and $dx'^2 + dy'^2 + dz'^2 = 0$. This means that these events may be observed.

(c) $ds^2 = 0$

This is the region of the *light cone;* the region of the greatest possible signal velocity that characterizes the zero elements. The spatial part of the length element is equal to the time part, namely

$$dx^2 + dy^2 + dz^2 = c^2\,dt^2. \tag{31.16}$$

Vectors $d\hat{r}$ with $ds^2 = d\hat{r}\cdot d\hat{r} = 0$ are also called *zerolike* or *lightlike*. They lie on a cone in four dimensions, because we would have to draw four coordinate axes for a complete description of this hypersurface.

For a resting observer at the position $x = 0$ at the instant $t = 0$ all those events constitute the *present* which also happen at the time $t = 0$. The *past* corresponds to the events with $t < 0$, the *future* to all events with $t > 0$. This convention is independent of the position. The observer has access only to those events that for him are in the timelike region.

As *simultaneous* one declares all those events for which in any moving frame it holds that $t_1' = t_2'$. Simultaneous with the event $x' = 0$, $t' = 0$ for an observer moving with v are the events

$$t' = \frac{t - (v/c^2)x}{\sqrt{1-\beta^2}} = 0,$$

that is, all events for which in a resting frame

$$t = \frac{v}{c^2}x.$$

holds. Every event in the interval $-x/c < t < x/c$ (the hatched region in the last figure—the present) may for an observer moving with the appropriate velocity between $\pm c$ be simultaneous with the event at $x' = 0$, $ct' = 0$. In other words: Two events that lie in a spacelike distance to each other can be made simultaneous. For this purpose one has only to describe these events in an inertial system with the appropriate velocity.

The causality principle

The causality principle of classical mechanics states that an event cannot happen earlier than its cause, that is, the triggering event must have taken place earlier than the resulting one.

If this principle shall continue to hold in the theory of relativity, there must not exist an inertial system in which the causal relation of the events is inverted.

As an example of an appropriate course, one may take the blackening of a photographic plate following a light flash. If the causing event happens in the system K at the time t_1 at the position x_1, the resulting event at the later time $t_2 > t_1$ at the position x_2, then any transformation to a K'-system must satisfy

$$t_2' - t_1' \geq 0.$$

As the speed of light represents the greatest possible signal velocity, for the causal relation in the frame K

$$c \geq \frac{x_2 - x_1}{t_2 - t_1}$$

holds; that means

$$c(t_2 - t_1) \geq (x_2 - x_1).$$

For the time difference in the frame K' moving with v relative to K,

$$t_2' - t_1' = \frac{c(t_2 - t_1) - (v/c)(x_2 - x_1)}{c\sqrt{1-\beta^2}}.$$

holds. Because now $c(t_2 - t_1) \geq (x_2 - x_1)$ and $v/c \leq 1$, there follows that for all inertial frames

$$t_2' - t_1' \geq 0.$$

The order of sequence of causally related events is therefore independent of the reference frame; the causality principle remains valid in relativistic mechanics.

The Lorentz transformation in the two-dimensional subspace of the Minkowski space

The length contraction and the time dilatation may well be visualized in this subspace. We distinguish between the real coordinates $x(x')$ and $ct(ct')$ on the one hand, and the Minkowski coordinates $x(x')$ and $ict(ict')$ on the other hand. At first the representation is in real coordinates:

The relation between two systems moving relative to each other is given by

$$x' = \frac{x - (v/c) \cdot ct}{\sqrt{1-\beta^2}}, \qquad ct' = \frac{ct - (v/c) \cdot x}{\sqrt{1-\beta^2}}. \tag{31.17}$$

To get the position of the primed coordinate axes, we set

$$x' = 0 = x - \frac{v}{c}ct \qquad (t'\text{-axis})$$

and

$$ct' = 0 = ct - \frac{v}{c}x \qquad (x'\text{-axis}).$$

The inclination angle α of the ct'-axis against the ct-axis is determined by $\tan\alpha = x/ct = v/c$. The inclination angle β of the x'-axis against the x-axis is given by $\tan\beta = ct/x = v/c$. Hence $\alpha = \beta$, that is, both axes are inclined by the same angles against the corresponding coordinate axes of the resting system (x, ct) (compare the figure).

For a complete representation of Lorentz contraction and time dilatation we consider the behavior of the unit scales on the two axes. Because $s^2 = s'^2 = x^2 - c^2t^2 = x'^2 - c^2t'^2$ is invariant under Lorentz transformation, $x^2 - c^2t^2 = 1$ represents the invariant unit scale in all Lorentz systems. The associated world lines are equilateral hyperbolas with light cone as asymptote (compare the figure).

These hyperbolas cut out the unit scales on the axes. The unit scale in the (x, ct)-frame (K) is OA. An observer at rest in the frame (x', ct') (K') sees it with the length OB', that is, shorter than his own scale OA'. The measuring signals are namely emitted at the points $x = 0$ and $x = 1$ in K; the endpoints of the distance 01 in K are represented by the world lines $x = 0$ and $x = 1$ (parallels to the t-axis). This corresponds to the unit distance at rest in K. In the frame K' at the same time $(t' = 0)$ a picture is taken, that is, the intersection point of the x'-axis with the world lines of the points 0 and 1 resting in K is determined.

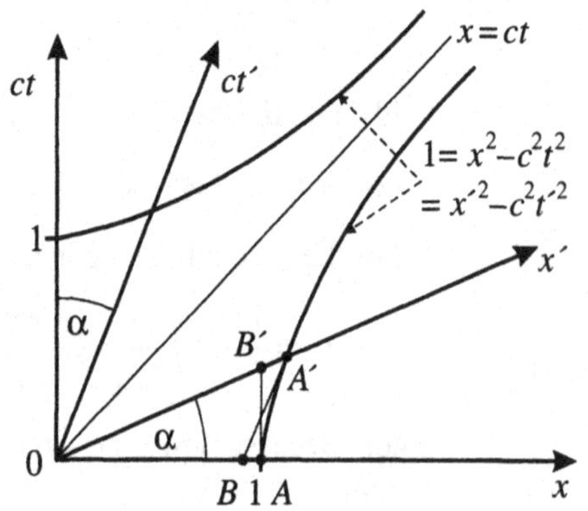

Graphical representation of the Lorentz transformation in real coordinates.

Conversely, an observer at rest in K sees the scale OA' as OB, that is, shorter than his own scale OA. Hence the Lorentz contraction is a mutual effect. The mutual control of the clocks proceeds in the corresponding manner.

A more convenient geometrical representation of time dilatation and Lorentz contraction that makes the comparison with unit scales unnecessary is obtained by using the coordinate $x_4 = ict$ instead of the time coordinate ct. Equations (31.17) then turn into

$$x_1' = \frac{(x_1 + i\beta x_4)}{\sqrt{1-\beta^2}}\,, \qquad x_2' = x_2\,, \qquad x_3' = x_3\,, \qquad x_4' = \frac{(x_4 - i\beta x_1)}{\sqrt{1-\beta^2}}\,. \tag{31.18}$$

The associated Lorentz transformation is

$$\alpha_{\mu\nu} = \begin{pmatrix} \gamma & 0 & 0 & i\gamma\beta \\ 0 & 1 & 0 & 0 \\ 0 & 0 & 1 & 0 \\ -i\gamma\beta & 0 & 0 & \gamma \end{pmatrix}. \tag{31.19}$$

Here $1/\sqrt{1-\beta^2} = \gamma$ has been abbreviated. $\alpha_{\mu\nu}$ is an orthogonal transformation and may therefore be represented as

$$x_1' = \cos\varphi\, x_1 + \sin\varphi\, x_4\,, \qquad x_4' = -\sin\varphi\, x_1 + \cos\varphi\, x_4\,. \tag{31.20}$$

By comparison of coefficients of (31.18) and (31.20), we get

$$\cos\varphi = \gamma \geq 1 \quad \text{and} \quad \sin\varphi = i\beta\gamma \quad \text{or} \quad \tan\varphi = i\beta. \tag{31.21}$$

Because $\cos\varphi = \gamma \geq 1$, φ must be an *imaginary angle*. The trigonometric functions $\cos\varphi$, $\sin\varphi$, $\tan\varphi$, $\cot\varphi$ for imaginary arguments $\varphi = i\alpha$ (α real) are defined by the corresponding series expansions. For example, $\cos\varphi = \cos(i\alpha) = 1-(i\alpha)^2/2!+(i\alpha)^4/4!-\cdots = 1+\alpha^2/2!+\alpha^4/4!+\cdots > 1$. Hence, $\cos i\alpha$ is larger than 1 and may even diverge to infinity in the limit $\alpha \to \infty$. Correspondingly, $\sin i\alpha = i\alpha/1! - (i\alpha)^3/3! + \cdots = i(\alpha/1! + \alpha^3/3! + \cdots)$, namely, purely imaginary. In fact, the series expansions of the trigonometric functions sin and cos for imaginary argument $i\alpha$ yield the hyperbolic functions, sinh and cosh. If we compare with Example 30.2, we see that α is just the rapidity introduced there.

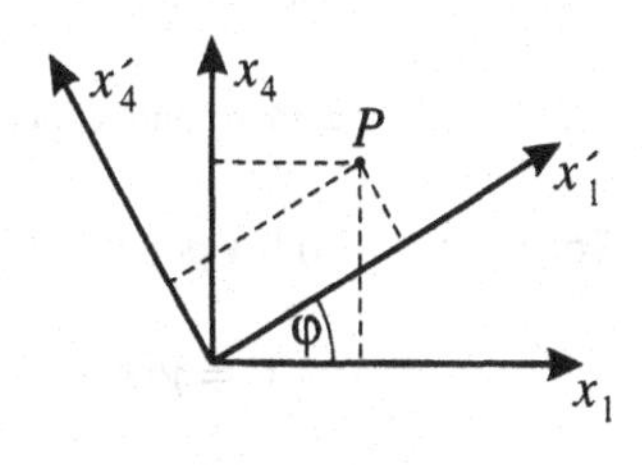

Graphical representation of the Lorentz transformation in Minkowski coordinates.

Because orthogonal transformations are angle-conserving, (31.20) may be represented as a simple rotation of the axes (compare figure).

Lorentz contraction and time dilatation become evident from the figure by geometrical considerations:

One has

$$L = \frac{L_0}{\cos\varphi} = \frac{L_0}{\gamma}$$

and

$$T = T_0 \cos\varphi = T_0\gamma. \tag{31.22}$$

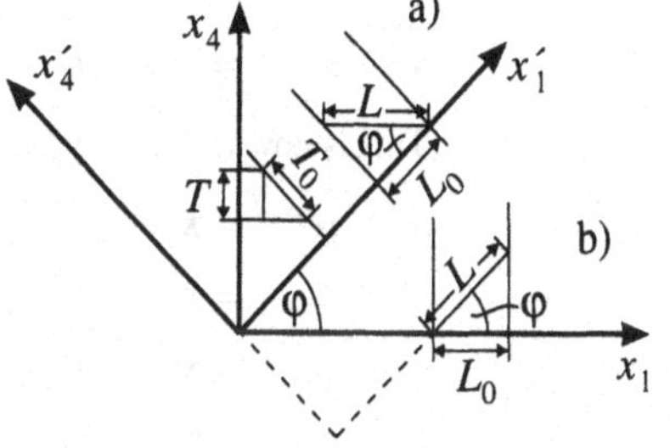

Lorentz transformation in Minkowski coordinates.

When using $x_4 = ict$, the geometric relations for the Lorentz contraction and time dilatation may be read off directly from the diagram. It is not necessary to investigate the behavior of the unit scales! But be careful! *Only the geometry* is correctly reproduced by the drawing: For Example, T in the diagram is smaller than T_0, but actually it holds that

$$T = \frac{T_0}{\sqrt{1-\beta^2}}; \qquad \text{thus} \quad T > T_0.$$

The relation for the length contraction and time dilatation is mutual for both inertial frames. This fact has been explained for the length contraction by the cases (a) L_0 rests in K', and (b) L_0 rests in K in the above figure: In both cases one always measures in the correspondingly other (moving) system

$$L = \frac{L_0}{\cos\varphi} = L_0\sqrt{1-\beta^2}.$$

The length measurement in a coordinate frame is always performed at the same instant; for example, in case (a) at fixed x_4, and in case (b) at fixed x_4'.

Problem 31.6: Lorentz transformation for arbitrarily oriented relative velocity

Let S be an inertial frame. Let a frame S' move with uniform velocity $\mathbf{v}$ against S. Show that the Lorentz transformation from S to S' looks as follows:

$$\mathbf{x}' = \mathbf{x}_\perp + \gamma\left[\mathbf{x}_{||} - \boldsymbol{\beta}(ct)\right], \qquad \gamma = \frac{1}{\sqrt{1-\beta^2}},$$

$$ct' = \gamma\left[ct - \boldsymbol{\beta}\cdot\mathbf{x}\right], \qquad \boldsymbol{\beta} = \frac{\mathbf{v}}{c},$$

with $\mathbf{x}_\perp$ and $\mathbf{x}_{||}$ denoting the components of $\mathbf{x}$ perpendicular and parallel to $\boldsymbol{\beta}$, respectively.

Solution (a) If $\mathbf{v}$ points in the x-direction, then one gets the well-known relation

$$x' = \gamma[x - \beta(ct)], \qquad y' = y, \qquad \gamma = \frac{1}{\sqrt{1-\beta^2}},$$

$$ct' = \gamma[ct - \beta x], \qquad z' = z, \qquad \beta = \frac{v}{c}.$$

(b) The general Lorentz transformation is then determined by the condition

$$\mathbf{x}'^2 - (ct')^2 = \mathbf{x}^2 - (ct)^2.$$

This is now fulfilled by the above relations:

$$\begin{aligned}\mathbf{x}'^2 - (ct')^2 &= \mathbf{x}_\perp^2 + \gamma^2\left[\mathbf{x}_{||}^2 - 2\boldsymbol{\beta}\,\mathbf{x}_{||}ct + \boldsymbol{\beta}^2(ct)^2 - (ct)^2 + 2\boldsymbol{\beta}\,\mathbf{x}\,ct - (\boldsymbol{\beta}\,\mathbf{x})^2\right]\\ &= \mathbf{x}_\perp^2 + \gamma^2 = \left[(1-\boldsymbol{\beta}^2)\mathbf{x}_{||}^2 - (ct)^2(1-\boldsymbol{\beta}^2)\right]\\ &= \mathbf{x}^2 - (ct)^2.\end{aligned}$$

Remark: The generalization of the Lorentz transformation for an arbitrarily oriented relative velocity may also be performed by writing down the formulas analogous to (31.18) for a translation parallel to the y- and z-axis, respectively, and then performing these three special Lorentz transformations successively. But one has to be careful, because Lorentz transformations do generally not commute, that is, the order of the transformations is important.

The second generalization for arbitrarily oriented axes may be made based on the remark that the rotations of the ordinary three-dimensional space, for unchanged time, also belong to the general Lorentz transformation. It then suffices to add such rotations to the special Lorentz transformations and to suspend the parallelity of the axes.

32 Addition Theorem of the Velocities

In this chapter we investigate the behavior of the velocities under a Lorentz transformation. For this purpose we consider a particle with the velocity $\mathbf{w}$ in the coordinate frame K. What is the velocity of the particle in the frame K' moving against K with the relative velocity $\mathbf{v} = (v_x, 0, 0)$?

We first restrict ourselves to the x-components of the velocity. According to the Lorentz transformation, we have

$$x' = \frac{x - vt}{\sqrt{1-\beta^2}}, \qquad t' = \frac{t - (v/c^2)\,x}{\sqrt{1-\beta^2}},$$

or for the differentials:

$$dx' = \frac{dx - v\,dt}{\sqrt{1-\beta^2}}, \qquad dt' = \frac{dt - (v/c^2)\,dx}{\sqrt{1-\beta^2}}.$$

In the frame K we have $dx = w_x dt$, $dy = w_y dt$, $dz = w_z dt$, with $\mathbf{w} = (w_x, w_y, w_z)$ being the velocity in the frame K. By inserting $dx = w_x dt$ in dx' and dt', we get

$$dx' = \frac{(w_x - v)\,dt}{\sqrt{1-\beta^2}}, \qquad dt' = \frac{(1 - (v/c^2)w_x)\,dt}{\sqrt{1-\beta^2}}. \tag{32.1}$$

The x-component of the velocity in the primed system is given by $w'_x = dx'/dt'$. By forming the quotient of the differentials (32.1), we find

$$\frac{dx'}{dt'} = w'_x = \frac{w_x - v}{1 - (v/c^2)w_x}.$$

w'_y is obtained in a similar way from (32.1) with $y' = y$, $dy' = dy = w_y dt$, and dt':

$$w'_y = \frac{w_y\sqrt{1-\beta^2}}{1 - (v/c^2)w_x}.$$

w'_z is obtained in the same manner as w'_y:

$$w'_z = \frac{w_z\sqrt{1-\beta^2}}{1-(v/c^2)w_x}.$$

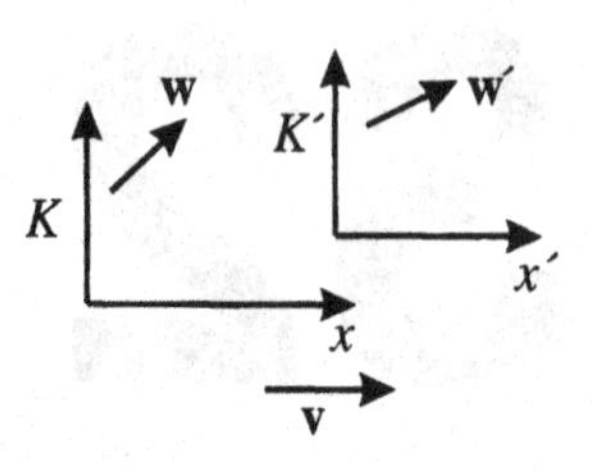

Illustration of the velocity vectors **w** (in K) and **w**′ (in K'). The relative velocity of both systems is $\mathbf{v} = v\mathbf{e}_z$.

Therefore, the velocity $\mathbf{w}'$ of the particle (with the velocity $\mathbf{w}$ in K) as seen from the frame K' moving relative to K is completely determined by the transformation equation for the three components w'_x, w'_y, w'_z:

$$\mathbf{w}' = \frac{1}{1-(v/c^2)w_x}\left(w_x - v,\, w_y\sqrt{1-\beta^2},\, w_z\sqrt{1-\beta^2}\right). \tag{32.2}$$

The first component of this result is identical with our earlier one, equation (30.45), when replacing $v \to -v$.

If one assumes that a massless particle propagates in K with the speed of light $|\mathbf{w}| = c$ and that the relative velocity of K' with respect to K again equals $\mathbf{v} = (v_0, 0, 0)$ the question arises which velocity $\mathbf{w}'$ is observed in K'.

We insert in $|\mathbf{w}'|^2 = w'^2 = w'^2_x + w'^2_y + w'^2_z$ the nonprimed quantities from (32.2):

$$w'^2 = \frac{(w_x - v)^2 + (w_y^2 + w_z^2)(1-\beta^2)}{(1 - v\,w_x/c^2)^2},$$

$$= c^4\left[\frac{w_x^2 - 2w_x v + (v^2/c^2)w_x^2 + v^2 + w_y^2 + w_z^2 - (v^2/c^2)(w_x^2 + w_y^2 + w_z^2)}{(c^2 - vw_x)^2}\right].$$

Because the particle is moving in K with the speed of light, we have $w_x^2 + w_y^2 + w_z^2 = c^2$. Hence we obtain

$$w'^2 = c^4\left[\frac{c^2 - 2w_x v + (v^2/c^2)w_x^2}{(c^2 - vw_x)^2}\right] = c^2\frac{(c^2 - vw_x)^2}{(c^2 - vw_x)^2} = c^2.$$

It is evident that also in K' no larger velocity than the speed of light c can be measured, independent of the magnitude of the velocity $\mathbf{v}$ of the relative motion of the two coordinate frames against each other. If we set

$$\mathbf{v} = (-c, 0, 0),$$
$$\mathbf{w} = (c, 0, 0),$$

the particle moves in K with the speed of light, and K' also moves with the speed of light relative to K in the opposite direction.

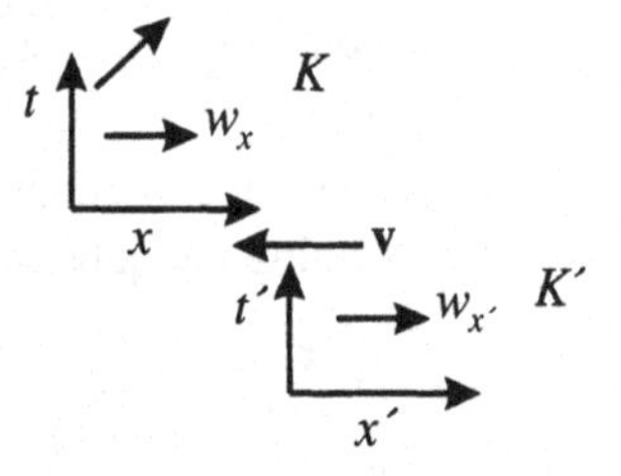

In the inertial frame K, light is moving with the velocity c along the x-axis. The frame K' is moving with velocity $\mathbf{v} = -c\mathbf{e}_x$ onto the frame K.

This interesting case shall be discussed here in brief. Naively one might expect to get "twice as fast light." But this is not true: For the x-component, according to (32.2), it holds that

$$w'_x = \frac{w_x - v}{1 - (v/c^2)\, w_x}.$$

After insertion we get

$$w'_x = \frac{2c}{1 + c^2/c^2} = c, \quad \text{i.e.,} \quad w'_x = c.$$

One might also try to generate "light resting" in K' by setting $\mathbf{v} = (c, 0, 0)$. The K'-frame moves so to speak with the speed of light parallel to the light beam. The transformation (32.2) yields in this case

$$w'_x = \frac{w_x - v}{1 - (v/c^2)\, w_x} = \frac{w_x - c}{(c - w_x)/c} = -c,$$

also in the limit $w_x \to c$. The observer in the system K' thus sees the light as propagating with the speed of light along the negative x'-direction. One again realizes the meaning of the speed of light c as limiting velocity for any motions. For $v \ll c$, (32.2) turns into the Galileo transformation:

$$\mathbf{w}' = (w_x - v, w_y, w_z),$$

as expected.

Supervelocity of light, phase, and group velocity

The addition theorem of velocities discussed in the preceding sections implies that the speed of light must be considered as upper limiting velocity for the propagation of physical phenomena.

But nevertheless, one may quote physical phenomena or experiments where a supervelocity of light may be reached:

1. The light ray emitted by a rotating light source (compare figure) shall hit on a far remote screen. If the screen is sufficiently far away from the light source, then the luminous spot caused by the light ray on the screen moves with supervelocity of light.

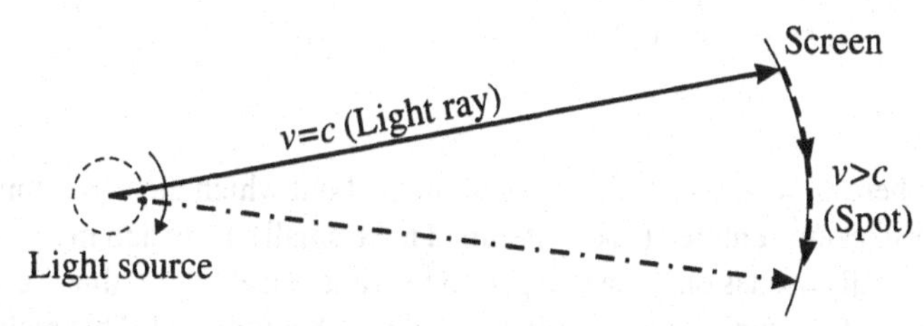

The luminous spot on the screen may move with superluminous velocity.

2. In optics the speed of light in a dispersive medium is calculated from the law of refraction

$$c_0/c = n, \tag{32.3}$$

where c_0 is the vacuum speed of light, n is the refractive index, and c is the wanted propagation velocity of light in the corresponding medium. There are substances (e.g., metals) with a refractive index $n < 1$, such that because $c = c_0/n$ one has $c > c_0$, that is, supervelocity of light in media with $n < 1$.

One has to distinguish between the *phase* and the *group velocity*:

The phase velocity is the traveling velocity of the phase of a propagating wave. Visually, the phase is the instantaneous state of motion of a vibration. For example: $\psi = A\cos(kx - \omega t)$ is a wave (more strictly: a plane harmonic wave). Its maximum amplitude $\psi = A$ is reached, for example, for values of the argument (the phase) $kx - \omega t = 0$. This maximum amplitude obviously moves with the velocity $dx/dt = x/t = \omega/k$. For the traveling velocity of the other maximum amplitudes at $kx - \omega t = n\pi$, one obtains the same result. This is the *phase velocity*

$$v_{\mathrm{ph}} = \frac{\omega}{k}.$$

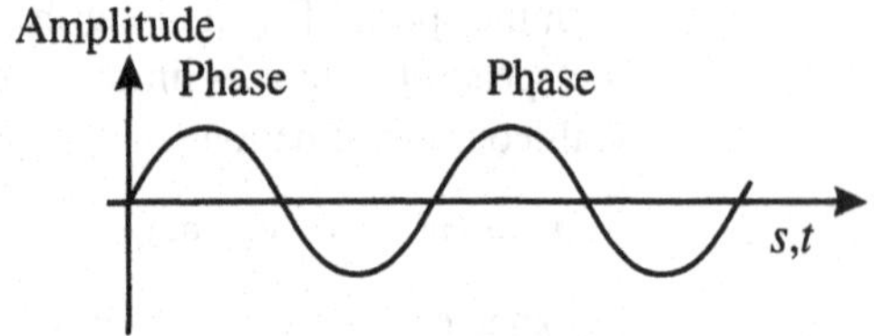

Illustration of a plane wave.

It is important to understand that such a plane wave extending from $-\infty$ to $+\infty$ cannot transfer information. In order to transfer information, the uniformity and "monotony" of the wave must be destroyed, that is, one must create a wave peak (wave group) and see how it propagates. Only this perturbation is visible (recordable).

The group velocity, on the contrary, is the propagation velocity of a wave packet (pulse of waves), that is, the superposition of several individual waves.

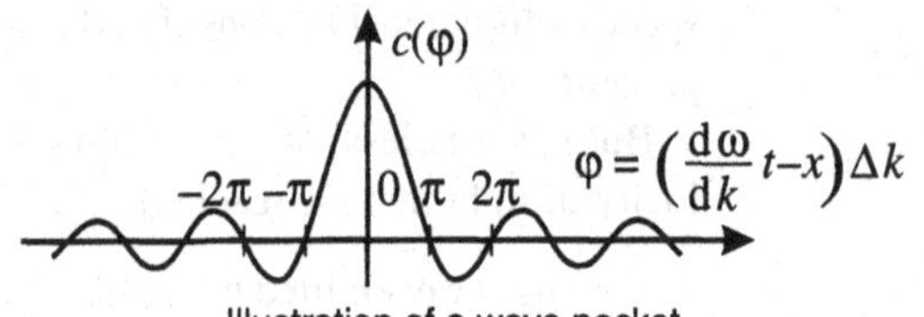

Illustration of a wave packet.

According to the given definition of the wave group, a wave packet $\psi(x, t)$ may be represented by the expression

$$\psi(x, t) = \int\limits_{k_0-\Delta k}^{k_0+\Delta k} c(k)e^{i(\omega t - kx)}\, dk, \tag{32.4}$$

where $k_0 = 2\pi/\lambda_0$ is the wave number about which the wave numbers involved in the wave packet are centered (Δk is assumed to be small). Here and in the following discussion, what is really used is only the real part of the function $e^{i(\omega t - kx)}$, thus $e\cos(\omega t - kx)$. The imaginary part of the function is simply taken along, but not used. This makes the calcultations easier.

Because Δk is small, we may expand the frequency ω, which in general is a function of k, into powers of $(k - k_0)$:

$$\omega = \omega_0 + \left(\frac{d\omega}{dk}\right)_0 (k - k_0) + \cdots ;$$
$$k = k_0 + (k - k_0),$$

and set $k - k_0 = \xi$. Taking $\xi = k - k_0$ as the new integration variable and assuming the amplitude $c(k)$ to be a slowly varying function of k, $\psi(x, t)$ may be represented in the form

$$\psi(x, t) = c(k_0) e^{i(\omega_0 t - k_0 x)} \int_{-\Delta k}^{\Delta k} e^{i((d\omega/dk)_0 t - x)\xi} \, d\xi .$$

Performing the simple integration with respect to ξ, we find

$$\begin{aligned} \psi(x, t) &= 2c(k_0) \frac{\sin\{[(d\omega/dk)_0 t - x]\, \Delta k\}}{[(d\omega/dk)_0 t - x]} e^{i(\omega_0 t - k_0 x)} \\ &= c(x, t) \cdot e^{i(\omega_0 t - k_0 x)}. \end{aligned} \qquad (32.5)$$

Because the argument of the sine involves the small quantity Δk, the quantity $c(x, t)$ as a function of the time t and the coordinate x will vary only slowly. Hence, $c(x, t)$ may be considered as the amplitude of an almost monochromatic wave, and $(\omega_0 t - k_0 x)$ as the phase. We now evaluate the point x where the amplitude $c(x, t)$ takes its maximum. This point shall be denoted as the *center* of the wave group. Obviously the desired maximum occurs at the point

$$x = \left(\frac{d\omega}{dk}\right)_0 t .$$

This implies that the group center will move with a velocity v that is obtained by differentiating the preceding equation with respect to t; this velocity is the *group velocity*

$$v_{\text{gr}} = \left(\frac{d\omega}{dk}\right)_0 . \qquad (32.6)$$

The theory of relativity makes only a statement on the speed of light as an upper limit for the propagation of particles and the transport of energy (signals), that is, on the group velocity. For the phase velocity, on the contrary, which is not capable of transmitting signals, namely, cannot transport energy and therefore cannot mediate causal relations, such a restriction (as expressed by an upper limiting velocity) does not exist.

In the first example, this means that the observer at the screen cannot use the luminous spot moving with $v > c$ to transmit signals with supervelocity of light. He would have to "radio" to the light source after passage of the luminous spot, in order to control the further course of the spot on the screen.

Also in the second example (32.2) the phase velocities c_0, c determine the refractive index. We shall see in the lectures on electrodynamics that also in media with $n < 1$ the signal velocity of light is always $< c_0$ (compare the volume of the lectures about Classical Electrodynamics, Chapter 19).

This distinction between the two velocities removes the seeming contradiction occuring in the two examples: Supervelocities of light may occur only for the phase velocity, that is, the phase of a wave may actually propagate with a velocity $v > c$. Physical information may, however, be transferred only by a wave group. The group velocity of signals (signal velocity) is always smaller than the speed of light in vacuum for all physical situations studied so far.

33 The Basic Quantities of Mechanics in Minkowski Space

A vector in R^3 is characterized by specifying three quantities, for example, the position vector

$$\mathbf{r} = (x, y, z)$$

by the three spatial coordinates. They transform under rotations of the coordinate frame according to the three-dimensional rotation group (see equations (30.13), (30.14)). Correspondingly, a *four-vector* is characterized by four quantities that transform according to the Lorentz transformation (compare the discussion in Chapter 30).

The analog to the position vector is in the four-dimensional Minkowski space the vector

$$\vec{\vec{r}} = (x_1, x_2, x_3, x_4) = (x, y, z, ict),$$

which is denoted as *world vector* (four-vector). It includes, besides the three space coordinates, an imaginary additional component being proportional to the time. Four-vectors shall be identified by a double-arrow, such as $\vec{\vec{r}}$.

A four-vector transforms under the Lorentz transformation similarly as a vector in R^3 transforms under a rotation. This will become clearer by interpreting the Lorentz transformation as a rotation in the Minkowski space with an imaginary rotational angle φ, compare (31.20) and (31.21), yielding

$$\cos\varphi = \frac{1}{\sqrt{1-\beta^2}} > 1.$$

Lorentz scalars

Scalar quantities, both in R^3 as well as in R^4, are characterized by their *invariance* against a corresponding rotation. Let us consider once again the square of the distance. By using the orthonormality relations, we obtain

$$\begin{aligned} s'^2 &= \sum_n x_n'^2 = \sum_n x_n' x_n' = \sum_n \left(\sum_j R_{nj} x_j \sum_k R_{nk} x_k \right) \\ &= \sum_n \sum_k \sum_j R_{nj} R_{nk} x_j x_k = \sum_k \sum_j \left(\sum_n R_{nj} R_{nk} \right) x_j x_k \\ &= \sum_k \sum_j \delta_{jk} x_j x_k = \sum_j x_j x_j = \sum_j x_j^2 = s^2, \end{aligned}$$

with $n, k, j = 1, 2, 3$ in R^3, and $n, k, j = 1, 2, 3, 4$ in R^4. The orthonormality of the transformation matrices R_{ni} reads

$$\delta_{jk} = \sum_n R_{nj} R_{nk}$$

with $n, j, k = 1, 2, 3$ in R^3, and $n, j, k = 1, 2, 3, 4$ in R^4.

Such an invariant (*scalar*) against *Lorentz transformations* is also the infinitesimal square of distance in the Minkowski space

$$ds^2 = ds'^2 = dx^2 + dy^2 + dz^2 - c^2 dt^2 = dx'^2 + dy'^2 + dz'^2 - c^2 dt'^2,$$

because it is the four-scalar product $d\,\vec{\vec{r}} \cdot d\,\vec{\vec{r}}$, where $d\,\vec{\vec{r}} = \{dx, dy, dz, ic\,dt\}$ is the infinitesimal world vector. One then also speaks of a *Lorentz invariant* or of a *Lorentz scalar.* The time t by which one differentiates in Newtonian mechanics, for example, when calculating the velocity or the acceleration, is not transformation-invariant, because "ict" is the fourth component of the world vector, and hence is no scalar. But now we have to find a Lorentz-invariant time, mainly for the reason to obtain again a four-vector when differentiating a four-vector. In other words: We want to establish clear relations concerning the transformation behavior of the various quantities (velocity, acceleration).

To get a Lorentz-invariant time unit, we start from

$$-ds^2 = c^2 dt^2 - (dx^2 + dy^2 + dz^2)$$

and define

$$\begin{aligned} d\tau &\equiv +\sqrt{\frac{-ds^2}{c^2}} = +\sqrt{dt^2 - \frac{dx^2 + dy^2 + dz^2}{c^2}} \\ &= dt\sqrt{1 - \frac{1}{c^2}\frac{dx^2 + dy^2 + dz^2}{dt^2}} \\ &= dt\sqrt{1 - \beta^2}. \end{aligned}$$

The quantity $d\tau$ has the dimension of a time. $d\tau$ is denoted as the *proper time of the system*, as in the rest system (proper system) it is identical with the coordinate time dt measured there, because there $v = 0$ and therefore $\beta = 0$. Depending on whether $d\tau$ is real or imaginary, one distinguishes timelike- and spacelike-related domains of the Minkowski space.

As already stated: *In the rest frame of a body its proper time τ is equal to the coordinate time t*; from there also originates the name "proper time."

We consider in the following how the three-dimensional quantities of the Newtonian mechanics are modified in the four-dimensional Minkowski space. We thereby follow the idea that *the natural laws are Lorentz-covariant, namely, must be formulated as four-dimensional laws (expressed by four-scalars, four-vectors, etc.)*. This is basically the principle of relativity: In all inertial frames there hold (formally) equal natural laws.

Four-velocity in Minkowski space

To get the *four-velocity*, one must differentiate the world vector

$$\vec{\vec{r}} = (x_1, x_2, x_3, x_4)$$

with respect to the Lorentz-invariant proper time $d\tau$:

$$\begin{aligned}\vec{\vec{v}} = \frac{d\vec{\vec{r}}}{d\tau} &= \left(\frac{\dot{x}_1}{\sqrt{1-\beta^2}}, \frac{\dot{x}_2}{\sqrt{1-\beta^2}}, \frac{\dot{x}_3}{\sqrt{1-\beta^2}}, \frac{\dot{x}_4}{\sqrt{1-\beta^2}}\right)\\ &= \frac{1}{\sqrt{1-\beta^2}}(\mathbf{v}, ic).\end{aligned} \tag{33.1}$$

Obviously,

$$\vec{\vec{v}} \cdot \vec{\vec{v}} = \sum_{i=1}^{4} v_i v_i = \frac{1}{1 - v^2/c^2}(v^2 - c^2) = -c^2. \tag{33.2}$$

The expression

$$\vec{\vec{v}} = \frac{1}{\sqrt{1-\beta^2}}(\mathbf{v}, ic) \tag{33.3}$$

represents the *four-velocity* and reflects the relation with the "ordinary" three-dimensional velocity $\mathbf{v}$. The fourth component at first sight has no particular meaning. The components of $\vec{\vec{v}} = \{v_1, v_2, v_3, v_4\}$ transform under Lorentz transformations (30.40) according to

$$v_i' = \sum_k \alpha_{ik} v_k.$$

One should be clear that if we had differentiated in (33.1) with respect to the ordinary coordinate-time t (and not with respect to the proper time τ), then we would have obtained the four-component quantity $\{\dot{x}_1, \dot{x}_2, \dot{x}_3, ic\}$. But this quantity is no four-vector. Its transformation behavior against Lorentz transformations is not clear (complicated). Only the pre-factor $1/\sqrt{1-\beta^2}$ in (33.1) converts this four-component quantity into a four-vector.

Momentum in Minkowski space

In R^3 the momentum is defined as

$$\mathbf{p} = m_0\mathbf{v}. \tag{33.4}$$

The question arises as how to generalize this momentum into the four-dimensional. The nonrelativistic relation (33.4) must be generalized in such a way that (33.4) is always obtained as the nonrelativistic limit. We are looking for a four-momentum vector.

Analogously to (33.4), we therefore define the momentum in R^4 by

$$\begin{aligned}\stackrel{\Rightarrow}{p} = m_0 \stackrel{\Rightarrow}{v} &= \left(\frac{m_0}{\sqrt{1-\beta^2}}v_x, \frac{m_0}{\sqrt{1-\beta^2}}v_y, \frac{m_0}{\sqrt{1-\beta^2}}v_z, \frac{icm_0}{\sqrt{1-\beta^2}}\right)\\ &= (m\mathbf{v}, icm) = (\mathbf{p}, icm).\end{aligned} \tag{33.5}$$

The first three components, as it must be, convert in the nonrelativistic limit into the Newtonian momentum (33.4). The fourth component will be interpreted later on. $\stackrel{\Rightarrow}{p}$ obviously is a four-vector because the rest mass m_0 shall be a scalar, and $\stackrel{\Rightarrow}{v}$ is a four-vector, as we just have seen.

Note that the mass m is no longer a constant but varies according to the equation

$$m = \frac{m_0}{\sqrt{1-\beta^2}}, \tag{33.6}$$

with m_0 being the *rest mass* of the particle in the state of rest ($m = m_0$ for $\mathbf{v} = 0$). The rest mass is a Lorentz scalar, that is, it is the same in any inertial frame. The mass m, on the contrary, is *no* Lorentz scalar but, as is seen, up to the factor ic is the fourth component of the four-momentum vector. The mass m thus varies with the velocity. For $v \to c$, the mass becomes infinitely large. Therefore, one must expend more and more energy in particle accelerators to further increase the velocity of highly relativistic particles ($v \approx c$).

Minkowski force (four-force)

In R^3 the force is defined by Newton's force law as

$$\mathbf{K} = \frac{d}{dt}(m\mathbf{v}) = \frac{d}{dt}(\mathbf{p}), \tag{33.7}$$

the Newtonian force. This relation must also be generalized to four dimensions, namely such that the four-force becomes a four-vector and that the Newtonian force (33.7) results as the nonrelativistic limit. Analogously to (33.7), we define the force in R^4 by

$$\overset{\Rightarrow}{F} = \frac{d}{d\tau}(\overset{\Rightarrow}{p}) = \frac{1}{\sqrt{1-\beta^2}}\frac{d}{dt}(\overset{\Rightarrow}{p}). \tag{33.8}$$

This is also the Lorentz-covariant basic equation of relativistic mechanics. There occur four-vectors to the left and right, similarly as in Newtonian mechanics expressed by the basic law (33.7) with three-vectors on both sides of the equation. This dynamic basic law (33.8) has been guessed. The principle of relativity (Lorentz covariance of the equations), the simplicity and the analogy to the nonrelativistic basic law (33.7), as well as the fact that the latter one must be contained in the new law (33.8) as a particular case served as guide for setting up (finding, guessing) equation (33.8). Similar to the nonrelativistic case, the basic law (33.8) has not only statutory character but also the character of a definition. Equation (33.8) defines the special form of the four-force and its relation to the three-force, which reads in detail

$$\overset{\Rightarrow}{F} = \frac{1}{\sqrt{1-\beta^2}}\frac{d}{dt}(m\mathbf{v}, icm). \tag{33.9}$$

Because in (33.8) the four-vector $\overset{\Rightarrow}{p}$ is differentiated with respect to the (Lorentz-scalar) proper time τ, the four-force formed this way is again a four-vector. From there result as components of the Minkowski force or four-force:

$$F_1 = \frac{K_x}{\sqrt{1-\beta^2}} \quad \text{with} \quad K_x = \frac{d}{dt}(mv_x) = m_0\frac{d}{dt}\left(\frac{v_x}{\sqrt{1-\beta^2}}\right);$$

$$F_2 = \frac{K_y}{\sqrt{1-\beta^2}} \quad \text{with} \quad K_y = \frac{d}{dt}(mv_y) = m_0\frac{d}{dt}\left(\frac{v_y}{\sqrt{1-\beta^2}}\right);$$

$$F_3 = \frac{K_z}{\sqrt{1-\beta^2}} \quad \text{with} \quad K_z = \frac{d}{dt}(mv_z) = m_0\frac{d}{dt}\left(\frac{v_z}{\sqrt{1-\beta^2}}\right);$$

$$F_4 = \frac{1}{\sqrt{1-\beta^2}}\frac{d}{dt}\left(\frac{icm_0}{\sqrt{1-\beta^2}}\right) = \frac{icm_0}{\sqrt{1-\beta^2}}\left[\frac{\beta\cdot\dot{\beta}}{(1-\beta^2)^{3/2}}\right]$$
$$= icm_0\frac{\beta\cdot\dot{\beta}}{(1-\beta^2)^2}. \tag{33.10}$$

Here K_x, K_y, K_z are the components of the ordinary three-dimensional force. The fourth component F_4 has for the time being no meaning in the three-dimensional case. But one should note that the relativistic mass (33.6) has already been included in K_x, K_y, K_z. This is also an important point of relativistic mechanics. The velocity-dependent mass is no fiction but manifests itself directly in the basic law. This has been proved experimentally by many experiments; such as by the experiments of Kaufmann, who demonstrated that electrons of high velocity are deflected in magnetic fields actually according to the relativistic mass (compare Example 33.2). In the rest frame of the particle ($\beta = 0$) the four-force

$$(F_1, F_2, F_3, F_4) = (K_x, K_y, K_z, 0) \tag{33.11}$$

in its first three components is identical with the ordinary (three-) force. We may in principle construct the four-force also by starting from the rest frame, namely, from the right side of (33.11), and derive the four-force in an inertial frame in which the particle is moving by a Lorentz transformation. This idea will be pursued in the following example.

Example 33.1: Construction of the four-force by Lorentz transformation

We will derive the four-force

$$F^\mu = \{\gamma \mathbf{K}, i\gamma \frac{\mathbf{v}}{c} \cdot \mathbf{K}\} \tag{33.12}$$

from the Lorentz transformation properties of F^μ. Here $\mathbf{K} = d/dt\,(m\mathbf{v})$ is the three-force, and $\mathbf{v} = c\boldsymbol{\beta}$ is the velocity of the particle. In the rest frame of the particle it shall hold that

$$F_0^\mu = \{\mathbf{K}_0, 0\}, \tag{33.13}$$

that is, the relativistic four-force is in its first three components identical with the three-force in this system. This equation is consistent with (33.12) to be proved. In a frame in which the particle moves with $\mathbf{v}$, (because we obtain this frame from the rest frame of the particle by a boost in $(-\mathbf{v})$-direction) it holds that

$$F^\mu = \alpha^\mu_\nu(-\mathbf{v}) F_0^\nu$$

or

$$\begin{aligned} F_\| &= \gamma\left(F_{0\|} - i\frac{v}{c}F_0^4\right) = \gamma F_{0\|} = \gamma K_{0\|}, \\ \mathbf{F}_\perp &= \mathbf{F}_{0\perp} = \mathbf{K}_{0\perp}, \\ F^4 &= \gamma\left(F_0^4 + i\frac{v}{c}F_{0\|}\right) = i\gamma\frac{v}{c}F_{0\|} = i\gamma\frac{v}{c}K_{0\|}. \end{aligned} \tag{33.14}$$

Here $F_\|$ and $F_\perp$ denote the spatial components of the four-force parallel and perpendicular to the direction of motion, respectively. The similar holds for $K_{0\|}$ and $K_{0\perp}$. In order to prove equation (33.12), we still have only to find out how $\mathbf{K}$ is related to $\mathbf{K}_0$. Then $\mathbf{K}_0$ may be substituted on the right sides of (33.14), and we shall obtain (33.12). The relation between the three-forces $\mathbf{K}$ and $\mathbf{K}_0$ may be derived as follows:

We consider the force acting on a particle of velocity $\mathbf{v}$ and mass $m = m_0\gamma_v = m_0/\sqrt{1 - \frac{v^2}{c^2}}$ in the inertial frame S:

$$\mathbf{K} = \frac{d}{dt}(m\mathbf{v}). \tag{33.15}$$

In another inertial frame S′ moving relative to S with $\mathbf{V} = (V, 0, 0)$, the force on this particle is given by

$$\mathbf{K}' = \frac{d}{dt'}(m'\mathbf{v}'), \tag{33.16}$$

with

$$
\begin{aligned}
t' &= \gamma_V (t - (V/c^2)x), \\
m' &= m_0 \gamma_{v'} = m \frac{\gamma_{v'}}{\gamma_v}, \\
\mathbf{v}' &= \left(\frac{v_x - V}{1 - (v_x V)/c^2}, \frac{\sqrt{1 - V^2/c^2} v_y}{1 - (v_x V)/c^2}, \frac{\sqrt{1 - V^2/c^2} v_z}{1 - (v_x V)/c^2} \right), \\
\gamma_V &= \frac{1}{\sqrt{1 - V^2/c^2}}, \qquad \gamma_{v'} = \frac{1}{\sqrt{1 - (v'^2)/c^2}}, \qquad \gamma_v = \frac{1}{\sqrt{1 - v^2/c^2}}.
\end{aligned}
\tag{33.17}
$$

Obviously (addition theorem of velocities),

$$
\begin{aligned}
\frac{1}{1 - v'^2/c^2} &= \frac{1}{1 - \dfrac{v_x^2 + V^2 - 2 v_x V + \left(1 - V^2/c^2\right) v_y^2 + \left(1 - V^2/c^2\right) v_z^2}{\left(1 - v_x V/c^2\right)^2 c^2}} \\
&= \frac{\left(1 - v_x V/c^2\right)^2 c^2}{c^2 - 2 v_x V + v_x^2 V^2/c^2 - v_x^2 - V^2 + 2 v_x V - v_y^2 - v_z^2 + \left(V^2/c^2\right)\left(v_y^2 + v_z^2\right)} \\
&= \frac{\left(1 - v_x V/c^2\right)^2 c^2}{c^2 - v^2 - V^2 + v^2 V^2/c^2} = \frac{\left(1 - v_x V/c^2\right)^2}{\left(1 - v^2/c^2\right)\left(1 - V^2/c^2\right)},
\end{aligned}
$$

and therefore,

$$
m' = m \frac{1 - v_x V/c^2}{\sqrt{1 - v^2/c^2}\sqrt{1 - V^2/c^2}} \sqrt{1 - v^2/c^2} = m \frac{1 - v_x V/c^2}{\sqrt{1 - V^2/c^2}}.
\tag{33.18}
$$

Thus we have

$$
\begin{aligned}
K'_x &= \frac{dt}{dt'} \frac{d}{dt} \left(m \frac{1 - v_x V/c^2}{\sqrt{1 - V^2/c^2}} \frac{v_x - V}{1 - v_x V/c^2} \right) \\
&= \frac{1}{\dfrac{dt'}{dt}} \frac{d}{dt} \left(m \frac{v_x - V}{\sqrt{1 - V^2/c^2}} \right) \\
&= \frac{1}{\gamma_V \left(1 - \dfrac{V}{c^2} v_x\right)} \frac{d}{dt} \left(m \frac{v_x - V}{\sqrt{1 - \dfrac{V^2}{c^2}}} \right)
\end{aligned}
\tag{33.19}
$$

because $V = \text{constant}$ and $dx/dt = v_x$.

It further follows (because $\gamma_V = 1/\sqrt{1 - V^2/c^2} = \text{constant}$) that

$$
K'_x = \frac{1}{1 - V v_x/c^2} \left(\frac{d}{dt} m v_x - V \frac{dm}{dt} \right).
\tag{33.20}
$$

Now

$$
\frac{d}{dt} m v_x = K_x ,
$$

and

$$\begin{aligned}\frac{dm}{dt} &= m_0 \frac{d}{dt}\gamma_v = m_0\gamma_v^3 \frac{\mathbf{v}\cdot\dot{\mathbf{v}}}{c^2} = m_0\gamma_v\left(1+\frac{v^2}{c^2}\gamma_v^2\right)\frac{\mathbf{v}\cdot\dot{\mathbf{v}}}{c^2}\\ &= \frac{\mathbf{v}}{c}\cdot\left(m_0\gamma_v\frac{\dot{\mathbf{v}}}{c}\right)+\frac{v^2}{c^2}m_0\dot{\gamma}_v = \frac{\mathbf{v}}{c}\cdot\left(m_0\gamma_v\frac{\dot{\mathbf{v}}}{c}\right)+\frac{\mathbf{v}}{c}\left(\frac{\mathbf{v}}{c}m_0\dot{\gamma}_v\right)\\ &= \frac{\mathbf{v}}{c}\frac{d}{dt}\left(m_0\gamma_v\frac{\mathbf{v}}{c}\right) = \mathbf{v}\cdot\mathbf{K}\frac{1}{c^2},\end{aligned} \tag{33.21}$$

yielding for K_x'

$$\begin{aligned}K_x' &= \frac{1}{1-\dfrac{v_x V}{c^2}}\left(K_x - \frac{V}{c^2}v_x K_x - \frac{V}{c^2}v_y K_y - \frac{V}{c^2}v_z K_z\right)\\ &= K_x - \frac{\dfrac{V}{c^2}}{1-\dfrac{v_x V}{c^2}}(v_y K_y + v_z K_z)\,.\end{aligned} \tag{33.22}$$

If the particle was at rest in the frame S ($v_x = v_y = v_z = 0$), then

$$K_x' = K_x. \tag{33.23}$$

For the other components of the force $\mathbf{K}'$ it holds that

$$\begin{aligned}K_y' &= \frac{1}{\gamma_V\left(1-\dfrac{Vv_x}{c^2}\right)}\frac{d}{dt}\left(m\frac{1-v_xV/c^2}{\sqrt{1-V^2/c^2}}\,\frac{\sqrt{1-V^2/c^2}\,v_y}{1-v_xV/c^2}\right)\\ &= \frac{1}{\gamma_V\left(1-\dfrac{Vv_x}{c^2}\right)}\frac{d}{dt}(mv_y) = \frac{\sqrt{1-V^2/c^2}}{1-\dfrac{Vv_x}{c^2}}K_y,\end{aligned} \tag{33.24}$$

and analogously:

$$K_z' = \frac{\sqrt{1-V^2/c^2}}{1-\dfrac{Vv_x}{c^2}}K_z. \tag{33.25}$$

If the particle was at rest in the frame S, then

$$\begin{aligned}K_y' &= \sqrt{1-V^2/c^2}\,K_y,\\ K_z' &= \sqrt{1-V^2/c^2}\,K_z.\end{aligned} \tag{33.26}$$

Equations (33.26) have been derived with the assumption that the frame S' moves relative to S with $\mathbf{V} = (V, 0, 0)$. For an arbitrary direction of motion we introduce the notations $||$ (for components parallel to $\mathbf{V}$) and $\perp$ (for components perpendicular to $\mathbf{V}$) and obtain

$$\begin{aligned}K_{||}' &= K_{||},\\ \mathbf{K}_\perp' &= \frac{1}{\gamma_V}\,\mathbf{K}_\perp.\end{aligned} \tag{33.27}$$

The frame S shall now be that frame in which the particle is at rest, the frame S' that one in which the particle moves with $\mathbf{v}$. Then obviously $\gamma_V = \frac{1}{\sqrt{1-V^2/c^2}} = \frac{1}{\sqrt{1-v^2/c^2}}$, and therefore (with the notations of (33.14)),

$$K_{\|} = K_{0\|},$$
$$\mathbf{K}_{\perp} = \frac{1}{\gamma}\mathbf{K}_{0\perp}. \tag{33.28}$$

By inserting this in (33.14), it follows that

$$F_{\|} = \gamma K_{\|},$$
$$\mathbf{F}_{\perp} = \gamma \mathbf{K}_{\perp},$$
$$F^4 = i\gamma\frac{v}{c}K_{\|} = i\gamma\frac{\mathbf{v}}{c}\cdot\mathbf{K}, \tag{33.29}$$

because $\mathbf{v}\cdot\mathbf{K}_{\perp} = 0$ by definition, that is,

$$F^{\mu} = \left\{\gamma\mathbf{K}, i\gamma\frac{\mathbf{v}}{c}\cdot\mathbf{K}\right\}, \qquad \text{q.e.d.} \tag{33.30}$$

Because of (33.21) $(\mathbf{v}\cdot\mathbf{K})/c^2 = dm/dt$, one immediately sees that

$$F^4 = ci\gamma\frac{dm}{dt}. \tag{33.31}$$

Hence one obtains the expression for F^4 already known from (33.10).

Kinetic energy

The kinetic energy in Newtonian mechanics is calculated as follows:

$$T(t) = \int_{t_0}^{t} \mathbf{K}\cdot\frac{d\mathbf{r}}{dt'}\,dt'.$$

Differentiation with respect to the time yields

$$\frac{dT}{dt} = \mathbf{K}\cdot\mathbf{v} = \frac{d\mathbf{p}}{dt}\cdot\mathbf{v}. \tag{33.32}$$

By inserting here for $\mathbf{K} = d\mathbf{p}/dt = m_0\,d\mathbf{v}/dt$, that is, the relation according to Newtonian mechanics, we find

$$dT = m_0\,\mathbf{v}\cdot d\mathbf{v}$$

or after integration

$$T_2 - T_1 = \frac{m_0}{2}v_2^2 - \frac{m_0}{2}v_1^2.$$

This is the well-known expression for the kinetic energy in classical (Newtonian) mechanics.

On the contrary, when inserting for $\mathbf{p} = m\mathbf{v} = (m_0/\sqrt{1-\beta^2})\,\mathbf{v}$, that is, the relativistic (three-dimensional) momentum (see equation (33.5)), into the relation (33.32), we obtain

$$\frac{dT}{dt} = \mathbf{v}\cdot\frac{d}{dt}\left(\frac{m_0}{\sqrt{1-\beta^2}}\mathbf{v}\right),$$

and with $\mathbf{v} = v\mathbf{e}$, we get

$$\frac{dT}{dt} = v\frac{d}{dt}\left(\frac{m_0 v}{\sqrt{1-\beta^2}}\right)\mathbf{e}\cdot\mathbf{e} + v\left(\frac{m_0 v}{\sqrt{1-\beta^2}}\right)\mathbf{e}\cdot\dot{\mathbf{e}},$$

and because $\mathbf{e}\cdot\mathbf{e} = 1, \quad \mathbf{e}\cdot\dot{\mathbf{e}} = 0$:

$$\begin{aligned}\frac{dT}{dt} &= v\frac{d}{dt}\left(\frac{m_0 v}{\sqrt{1-\beta^2}}\right) = c^2\beta\frac{d}{dt}\left(m_0\frac{\beta}{\sqrt{1-\beta^2}}\right)\\ &= m_0c^2\frac{d}{dt}\left(\frac{1}{\sqrt{1-\beta^2}}\right),\end{aligned}$$

because

$$\beta\frac{d}{dt}\left(\frac{\beta}{\sqrt{1-\beta^2}}\right) = \frac{d}{dt}\left(\frac{1}{\sqrt{1-\beta^2}}\right),$$

as one may prove by differentiation. Integration with respect to the time yields

$$\begin{aligned}T &= m_0c^2\int_{t_0}^{t}\frac{d}{dt}\left(\frac{1}{\sqrt{1-\beta^2}}\right)dt = \left.\frac{m_0c^2}{\sqrt{1-\beta^2}}\right|_{t_0}^{t}\\ &= m_0c^2\left[\frac{1}{\sqrt{1-v^2(t)/c^2}} - \frac{1}{\sqrt{1-v^2(0)/c^2}}\right].\end{aligned}$$

If for $t_0 = 0$, $v = 0$, or $\beta = 0$, one finally obtains

$$T = \frac{m_0c^2}{\sqrt{1-\beta^2}} - m_0c^2 = (m - m_0)c^2. \tag{33.33}$$

The expression m_0c^2 is practically denoted as *rest energy*. By rearranging the terms, we get the relation

$$T + m_0c^2 = mc^2 = E. \tag{33.34}$$

The famous equation

$$E = mc^2 \tag{33.35}$$

is one of the most important statements of the theory of relativity: *Energy and mass are equivalent.* E is called total energy: That is the entire energy of a *free particle*. For free particles it is composed of the rest energy $(m_0\,c^2)$ and the kinetic energy $((m - m_0)c^2)$.

For particles in a force field, the total energy includes also the potential energy (compare the text later on): The interpretation of the rest energy $m_0 c^2$ as a new independent fraction of energy must ultimately be verified by questioning of nature (experiment). Examples in this context will be presented in the following. But we may already now provide an argument for the physical reality of the rest energy, by considering a fission process of a particle of mass m_0 into two daughter particles m_1 and m_2. In general, $m_0 \neq m_1 + m_2$. The rest energy therefore contributes to the energy balance in the decomposition of a particle. This possibility would get lost if we would consider in (33.34) the rest energy as always being constant and would absorb it into the constant E on the right-hand side. For $v \ll c$, thus $\beta \ll 1$, the relativistic kinetic energy must turn into the kinetic energy of Newtonian mechanics. From

$$T = \frac{m_0 c^2}{\sqrt{1-\beta^2}} - m_0 c^2, \tag{33.36}$$

one obtains by expanding the square root

$$T = m_0 c^2 \left(1 + \frac{1}{2}\beta^2 + \frac{1 \cdot 3}{2 \cdot 4}\beta^4 + \cdots \right) - m_0 c^2$$

or

$$T = m_0 c^2 + \frac{1}{2} m_0 v^2 + \cdots - m_0 c^2 \approx \frac{1}{2} m_0 v^2 + \cdots .$$

At low velocity ($v \ll c$) one has to a very good approximation $T = \frac{1}{2} m_0 v^2$, which corresponds to the nonrelativistic expression for the kinetic energy.

The equivalence between mass and energy (33.34) has been confirmed in nuclear physics in a variety of cases; for example, in nuclear fission an atomic nucleus of mass M splits into two nuclei of about the same size with the masses M_1 and M_2. One finds $M > M_1 + M_2$. The mass defect corresponds to the energy difference

$$\Delta E = (M - M_1 - M_2)c^2,$$

which is released as kinetic energy in the fission process. [1]

Example 33.2: Einstein's box

In the following thought experiment invented by A. Einstein in 1906[2] we shall consider the relation between the inertia of matter and radiation energy. We will investigate which amount of inert mass (quotient of momentum and velocity) is equivalent to a given energy. For this purpose we assume that at the left end of a box of mass M and length L being initially at rest (displayed in the figure (a)) a cloud of photons of energy E is emitted.

[1] A detailed discussion of masses and energy relations may be found in J.M. Eisenberg and W. Greiner, *Nuclear Theory Vol 1: Nuclear Models*, 3rd ed., North-Holland, Amsterdam, 1987.

[2] A. Einstein, *Annalen der Physik* **20** (1906) 627–633.

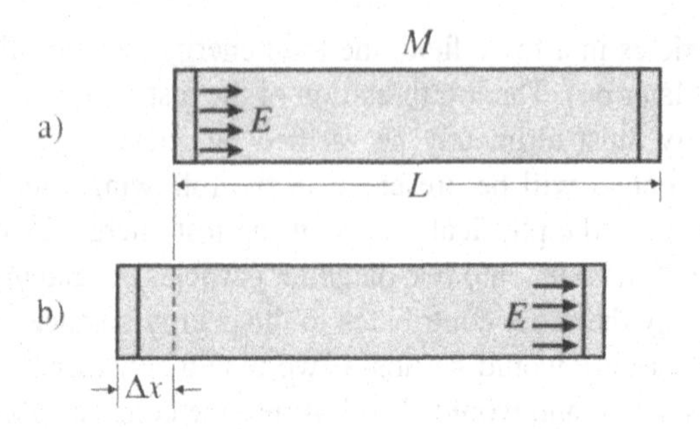

Einsteins box: The emission of a bunch of photons (a) at the left border of the box creates a recoil. This results in a motion of the box towards the left by the distance Δx, until the light is reabsorbed at the right border of the box (b).

The photon cloud or radiation carries a momentum of $p = E/c$; see (33.61). Because the total momentum of the system must vanish as before the emission act, the box gets a momentum transfer of $p = -E/c$. Because of this recoil, the box moves with the velocity v:

$$v = -\frac{E}{Mc}. \tag{33.37}$$

After the time Δt the radiation hits on the opposite wall of the box, which thereby again comes to rest, because the momentum transferred by the stopping equals the negative initial momentum. Therefore, the box is displaced by a distance Δx given by

$$\Delta x = v\Delta t = -\frac{EL}{Mc^2}. \tag{33.38}$$

If we put the center of mass R_s of the system into the coordinate origin, then its position must remain unchanged also after termination of the experiment. This is only then possible if we attribute a mass m to the photon cloud, such that

$$R_s = \frac{\Delta x M + mL}{m + M} = 0. \tag{33.39}$$

Together with (33.38), we thereby obtain

$$-\frac{mL}{M} = -\frac{EL}{Mc^2} \quad \Leftrightarrow \quad E = mc^2. \tag{33.40}$$

Verbally expressed: Equation (33.40) describes the inertia of the energy, that is, any change ΔE of the energy of a body causes a corresponding change Δm of its inert mass.

In our example this means: At that end of the box where the photon cloud is emitted, the inert mass reduces by E/c^2. Correspondingly, the inert mass of the box increases again by the same amount when the photon cloud is stopped or thermalized at the other end of the box. We still note that by taking into account this circumstance as well as the transit time changed by the recoil of the box, the result of (33.40) remains unchanged.

Example 33.3: On the increase of mass with the velocity

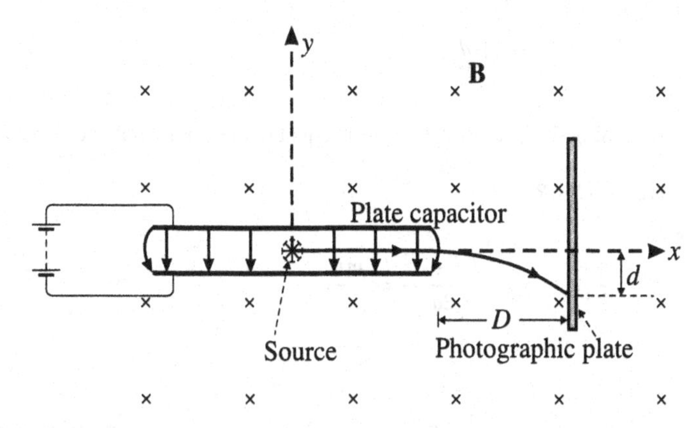

Simplified, schematic view of Bucherer's experiment, which uses a capacitor as a velocity filter. After leaving the capacitor, the β rays (electrons) are deviated by a magnetic field and detected on a photographic plate. The magnetic field **B** is oriented into the plane of the drawing. It is denoted by the crosses ($\times$).

Already in 1897 Thomson could measure the ratio of e/m for electrons by using cathode rays. In 1901 W. Kaufmann[3] demonstrated, utilizing the parabola method, that the value e/m depends on the velocity of the β-rays. In 1908 A.H. Bucherer[4] from Bonn performed an improved experiment to determine e/m using β-rays. The experimental set-up is shown in the figure: β-rays from a radium source were emitted between the plates of a large capacitor. The potential difference between the plates creates an **E**-field in negative y-direction, whereby an electron experiences the force $\mathbf{F}_E = -e\mathbf{E}$ along the y-direction ($e > 0$!). Due to the applied magnetic field an electron moving in the x-direction undergoes the Lorentz force $\mathbf{F}_B = -e\mathbf{v}/c \times \mathbf{B}$ along the negative y-axis (compare Volume III of the lectures: *Classical Electrodynamics*). Because the plate diameter of the capacitor is large against the spacing of the plates, only such electrons may escape for which $|\mathbf{F}_E| = |\mathbf{F}_B|$, hence,

$$e\frac{v}{c}B = eE\,,$$

$$\frac{v}{c} = \frac{E}{B}\,. \tag{33.41}$$

(Remark: This relation holds also in the relativistic case although we have calculated with the nonrelativistic expressions for $\mathbf{F}_E$ and $\mathbf{F}_B$. The reason for that is that a factor $1/\sqrt{1-v^2/c^2}$ cancels on both sides of (33.41), compare vol. III of the lectures.) Hence the capacitor acts as a *velocity filter (crossed fields)*.

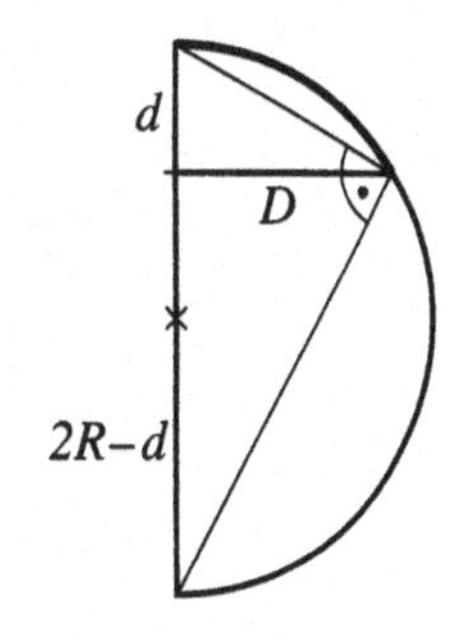

Height D in the right-angled triangle.

[3] W. Kaufmann, *Gött. math.-nat. Klasse* 143 (1901); *Phys. Zeitschr.* **4** (1902) 55.

[4] A. H. Bucherer, *Verh. d. Deutschen Phys. Ges.* **6** (1908) 688.

After leaving the **E**-field the electron moves on a circular path of radius R due to the **B**-field (this also holds in the relativistic case; compare the volume of the lectures about electrodynamics). From the geometry of the figure one reads off, using the well-known theorem for right-angled triangles,

$$d(2R-d) = D^2,$$
$$R = \frac{D^2+d^2}{2d}. \qquad \textbf{(33.42)}$$

By setting the Lorentz acceleration equal to the centripetal acceleration, one gets

$$mv = \frac{e}{c}B \cdot R$$

or

$$R = \frac{mv}{Be}c \quad \Rightarrow \quad \frac{D^2+d^2}{2d} = \frac{mv}{Be}c.$$

Bucherer's results for e/m of β-rays (electrons)

v/c	$e/m = \dfrac{e\sqrt{1-v^2/c^2}}{m_0}$	e/m_0
0.3173	$1.661 \cdot 10^{11}$ C/kg	$1.752 \cdot 10^{11}$ C/kg
0.3787	1.630	1.761
0.4281	1.590	1.759
0.5154	1.511	1.763
0.6870	1.283	1.766

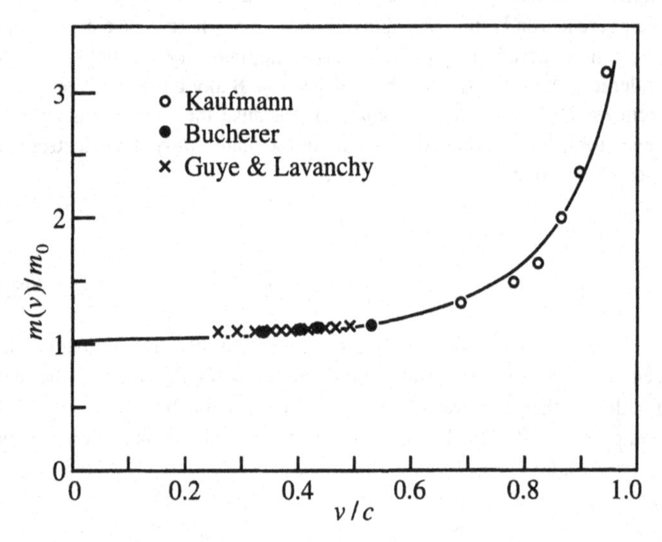

The inertial mass of the electron as a function of its velocity. The measurements are by Kaufmann (*Phys. Zeitschr.* **4** (1902) 55), Bucherer (*Verh. DPG* **6** (1908) 688), and Guye and Lavenchy (*Arch. de Genève* **41** (1916) 286, 353, 441).

With equation (33.41), it results that

$$\frac{e}{m} = \frac{2d}{(D^2 + d^2)} \frac{E}{B^2} c^2 . \tag{33.43}$$

Bucherer reversed the polarity of the **E**- and **B**-fields, yielding a second luminous spot on the photographic plate, and he determined d as the half-distance between the two luminous spots. The experiment was performed for various **B**- and **E**-field intensities or electron velocities. The results are listed in the table.

The value for e/m_0 is calculated from the measured values for e/m and v/c. The following figure summarizes the experiments of Kaufmann, Bucherer, and Guye and Lavanchy,[5] which impressively demonstrate the velocity dependence of the electron mass.

Problem 33.4: Relativistic mass increase

Calculate the velocity and the path of a relativistic particle of rest mass m_0 in the gravitational field of the earth for the initial condition $\mathbf{r}(t=0) = \mathbf{0}$ and $\mathbf{v}(t=0) = v_0 \mathbf{e}_z$.

Solution Insertion of the velocity-dependent mass

$$m(v) = \frac{m_0\, c}{\sqrt{c^2 - v^2}} \tag{33.44}$$

in the equation of motion yields

$$\begin{aligned} \frac{d}{dt}(m(v)\mathbf{v}) &= m(v)\dot{\mathbf{v}} + \frac{m(v)}{c^2 - v^2}(\dot{\mathbf{v}} \cdot \mathbf{v})\,\mathbf{v} \\ &= m(v) g \mathbf{e}_z . \end{aligned} \tag{33.45}$$

The velocity components in x- and y-directions vanish because of the initial condition. From (33.45) it then follows that

$$\dot{v}_z + \frac{1}{c^2 - v^2} v_z^2\, \dot{v}_z = g$$

$$\Rightarrow \quad \dot{v}_z \left(1 + \frac{v_z^2}{c^2 - v_z^2}\right) = g$$

$$\Rightarrow \quad \dot{v}_z = g \frac{1}{1 + v_z^2/(c^2 - v_z^2)} = g \frac{c^2 - v_z^2}{c^2 - v_z^2 + v_z^2} = g\left(1 - \left(\frac{v_z}{c}\right)^2\right) . \tag{33.46}$$

The solution of (33.46) results from

$$\int_{v_0}^{v_z} dv_z' (c^2 - v_z'^2)^{-1} = \frac{1}{c}\left(\text{Artanh}\, \frac{v_z}{c} - \text{Artanh}\, \frac{v_0}{c}\right)$$

$$= \frac{1}{c}\, \text{Artanh}\left(\frac{v_z - v_0}{c - v_z\, v_0/c}\right) = \frac{g}{c^2} t . \tag{33.47}$$

[5] Ch.E. Guye and Ch. Lavanchy, *Arch. de Genève* **41** (1916) 286, 353, 441.

We thereby have used the relation

$$\operatorname{Artanh} x - \operatorname{Artanh} y = \operatorname{Artanh} \frac{x-y}{1-xy}. \tag{33.48}$$

The velocity of the relativistic particle is

$$\mathbf{v}(t) = \left(v_0 + c\tanh\left(\frac{g}{c}t\right)\right)\left(c + v_0\tanh\left(\frac{g}{c}t\right)\right)^{-1} c\,\mathbf{e}_z. \tag{33.49}$$

For $t \to \infty$ (or $\tanh gt/c \to 1$) it approaches the limit velocity c.
The function $z(t)$ is obtained by integration of (33.49):

$$\begin{aligned} z(t) &= c\int_0^t dt' \frac{v_0\cosh\left(\frac{g}{c}t'\right) + c\sinh\left(\frac{g}{c}t'\right)}{c\cosh\left(\frac{g}{c}t'\right) + v_0\sinh\left(\frac{g}{c}t'\right)} \\ &= \frac{c^2}{g}\ln\left[\cosh\left(\frac{g}{c}t\right) + \frac{v_0}{c}\sinh\left(\frac{g}{c}t\right)\right]. \end{aligned} \tag{33.50}$$

With $\cosh x \simeq 1 + x^2/2$, $\sinh x \simeq x$ and $\ln(x+1) \simeq x$ for $x \ll 1$, we obtain

$$z(t) \simeq \frac{1}{2}gt^2 + v_0 t \qquad \text{for } t \ll \frac{c}{g},$$

namely, the normal free falling.
For $t \to \infty$, one has $z \simeq ct$ if $v_0 \ll c$.

Problem 33.5: Deflection of light in the gravitational field

Einstein speculated in 1911 whether the relation $m = E/c^2$ for the inert mass of radiation energy may be inserted in the gravitational field to describe the deflection of light rays from remote stars by the sun. The deflection causes that an observer supposes the position of the star to be along the extension of the straight line a (dashed line). Thus, the direction of the star seems to be displaced (see figure, in particular figure (b)).

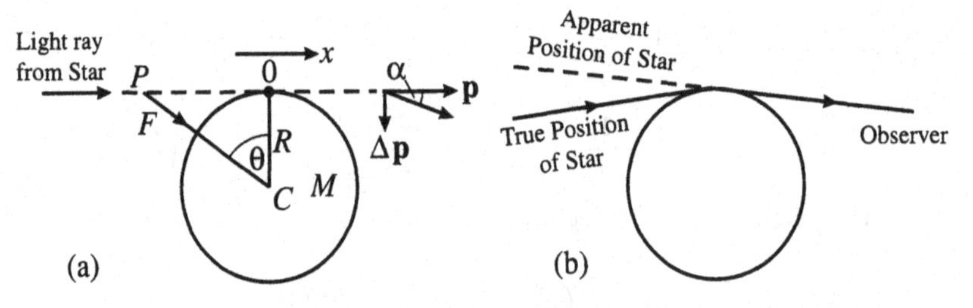

Classical sketch (a) of the deflection of a photon grazing the rim of the sun at O, and the real deflection behavior (b).

Already in 1901 the German astronomer J. Soldner had made a similar calculation in which he described the light as a Newtonian particle with the velocity c. Calculate the deflection angle α of a photon grazing the border of the sun (see figure) with the assumption that the photon passes the sun with the velocity c on a straight line. Let the component of the gravitational force perpendicular to the path of flight ($F\cos\theta$) integrated over the entire flight orbit provide the transverse momentum component.

Solution The transverse momentum component $\Delta p = \int F\cos\theta\, dt$ represented in the figure is calculated between the limits $\pm\infty$ whereby the origin of the path x is put into the point of contact 0. The momentum is $p = E/c$—see equation (33.61)—and $dt = dx/c$. For the distance $\overline{CP}$, one may take from the figure

$$(\overline{CP})^2 = x^2 + R^2$$

$$\Rightarrow \quad \Delta p = \int_{-\infty}^{\infty} F\cos\theta \frac{dx}{c} = \frac{1}{c}\int_{-\infty}^{\infty} F\frac{R}{\sqrt{x^2+R^2}}\,dx \tag{33.51}$$

$$= \frac{\gamma m M R}{c}\int_{-\infty}^{\infty}(x^2+R^2)^{-3/2}dx$$

$$= \frac{\gamma m M R}{c}\frac{x}{R^2\sqrt{x^2+R^2}}\Bigg|_{-\infty}^{\infty} = \frac{2\gamma m M}{Rc}.$$

We thus obtain for the deflection angle $\alpha \approx \tan\alpha = \Delta p/p$

$$\alpha = \frac{2\gamma m M}{Rcmc} = \frac{2\gamma M}{Rc^2}. \tag{33.52}$$

Insertion of the numerical values $M_\odot = 1.99\cdot 10^{30}$ kg, $R_\odot = 6.96\cdot 10^8\,m$, $\gamma = 6.67\cdot 10^{-11}\,\mathrm{m^3/(kg\,s^2)}$, $c = 2.998\cdot 10^8$ m/s yields a deflection angle of $\alpha = 0.875''$, a result that at first is believed as quantitatively only conditionally correct. Surprisingly, in the general theory of relativity the calculation of the deflection of a light ray in the Schwarzschild field yields the same value except for a factor of 2, thus $\alpha = 4\gamma M/Rc^2 = 1.75''$. Experimental investigations between 1919 and 1954 yielded values between $1.5''$ and about $3''$ (Finlay-Freundlich, 1955; von Kluber, 1960). These measurements on the average seem to yield $2.2''$, which would be too large by 25 %. In 1952 van Biesbroeck found in a precision experiment the value $1.7''\pm 0.1''$. More recent measurements from 1970 (Hill, 1971; Sramek, 1971) at Mullard Radio Astronomy Observatory of Cambridge University and at the National Radio Observatory (USA) essentially confirm the value obtained by van Biesbroeck, which agrees well with the theoretical prediction. The most accurate measurements of the deflection of radio waves grazing the sun using state-of-the-art long-baseline interferometry[6] yield a confirmation of the general relativistic prediction for the deflection of 0.9998±0.0008. It should be noted that the most sophisticated optical observations[7] during solar eclipses can give no better confirmation than 0.95±0.11 of the Einstein prediction.

---

[6] D. E. Lebach et al., "Measurement of the solar gravitational deflection of radio waves using very-long-baseline interferometry," *Phys. Rev. Lett.* **75** (1995) 1439–1442.

[7] R. A. Brune, Jr. et al., "Gravitation deflection of light: solar eclipse of 30 June 1973. I. Description of procedures and final results," *Astron. J.* **81** (1976) 452.

The Tachyon hypothesis

We have seen that the speed of light is an *upper* limiting velocity. But a limit has two sides. With this hint the hypothesis has been set up that there might exist particles with a *lower* limiting velocity that equals the speed of light.

These hypothetical particles are called *tachyons* (Greek: *tachys* = fast). Their existence does not contradict the theory of relativity. If the relativistic energy $E = m_0 c^2/\sqrt{1-\beta^2}$ (compare to (33.33)) is plotted as function of the velocity, then the pole at $\beta = 1$ separates the velocity range into two regions. The range $v < c$ is the (so far) accessible one, the range $v > c$ is that of the tachyons. However, it must be assumed that the rest mass M_0 of the tachyons is purely imaginary ($M_0 = im_0$) to ensure that their energy

$$E = \frac{M_0 c^2}{\sqrt{1-\beta^2}} = \frac{im_0 c^2}{i\sqrt{\beta^2 - 1}} = \frac{m_0 c^2}{\sqrt{\beta^2 - 1}}$$

for $\beta > 1$ (which characterizes the tachyons) remains real. Thus one drops the reality of the rest mass but sticks to the requirement for always real energy. Finally, it is the energy of a particle that is being measured. Its mass is – more or less – a proportionality factor (e.g., in the basic law of dynamics).

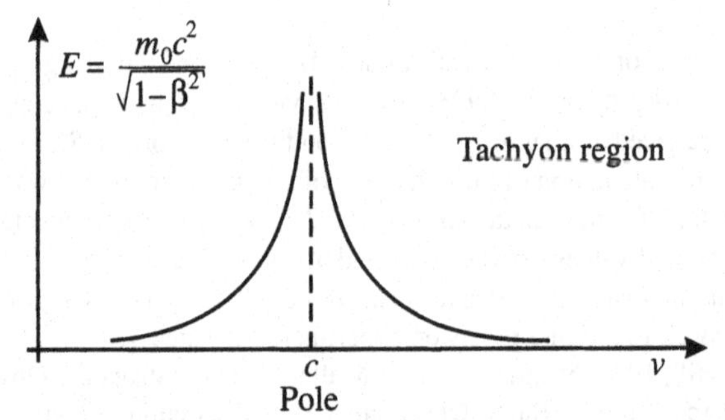

The tachyon hypothesis: What is the energy at $v > c$?

From the preceding sketch one may immediately read off several further properties of tachyons.

1. Tachyons have a lower limit of velocity = c. There is no upper velocity limit ($c \leq |\mathbf{v}_{\text{tach.}}| < \infty$).
2. For real energy the rest mass M_0 of tachyons has an imaginary value.
3. If a tachyon has the speed of light, then its energy and momentum become infinitely large.
4. If a tachyon loses energy, then its velocity increases. At $E = 0$ one has $|\mathbf{v}_{\text{tach.}}| = \infty$.

Further properties shall be

5. Tachyons may in any energy state emit massless particles (photons, neutrinos). They therefore must carry additional quantum numbers, such as electric charge.
6. The number of tachyons at a given instant in a given space is not uniquely determined. It depends on the position of the observer.
7. Assumption: Tachyons are electrically charged particles. This is necessary, as already noted in (5), to enable them to radiate light waves (photons).

This latter property significantly increases the chance of detection (if these particles should exist at all). According to the theory an electrically charged tachyon should release a Tscherenkov radiation[8] (these are electromagnetic head shock waves (more precisely: Mach shock waves), which always arise if ordinary charged particles are passing a medium with a higher speed than the speed of light in this medium).[9]

A useful comparison for elucidation is the Mach cone.[10] It arises, for example, by an aeroplane flying with supersonic speed.

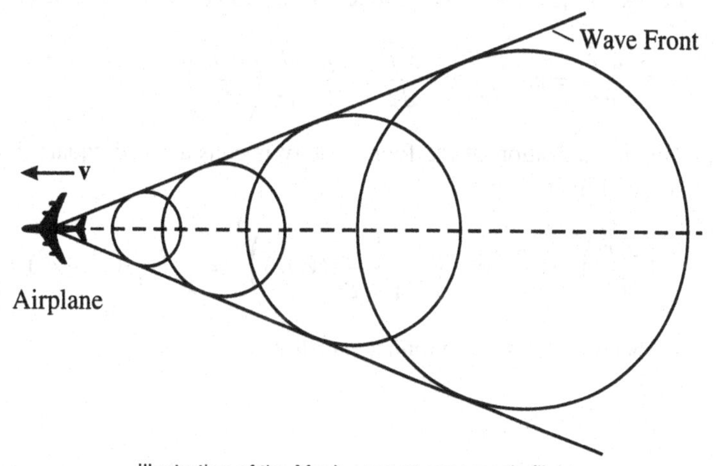

Illustration of the Mach cone at supersonic flight.

[8] *Ravel Aleksejevich Tscherenkov*, Soviet physicist, b. July 28, 1904, near Voronesh, since 1959 professor and since 1964 member of the Academy in Moscow. In 1934 he discovered the Tscherenkov radiation. In 1958 he was awarded with the Nobel Prize, together with I.M. Frank and I. Tamm. After the war Tscherenkov was involved in the construction of an electron synchrotron at the Lebedev Institute.

[9] The principle of Tscherenkov radiation finds application in experimental high-energy physics and nuclear physics in the so-called Tscherenkov counters. They consist essentially of a medium of high refractive index, such that the velocity of fast charged particles entering the counter exceeds the speed of light in this medium and these particles therefore emit Tscherenkov radiation. The radiation may be observed and thus indirectly serves for detecting the particles.

[10] *Ernst Mach*, physicist and philosopher, b. Feb. 18, 1838, Turas (Moravia)—d. Feb. 19, 1916, Haar near Munich. In 1864 he was appointed professor of physics in Graz, in 1867 in Prague, in 1895–1901 professor of philosophy in Vienna. As physicist he investigated in particular acoustic and optical problems. He improved the stroboscopic method and successfully applied Toepler's schlieren method for investigating flying missiles. He especially studied the motion of solids with supersonic speed.

The Tscherenkov radiation is the analog to Mach's shock wave that arises during the motion of a body with a speed above the phase velocity of the elastic wave in the surrounding medium. [11]

Because a tachyon always moves with superspeed of light, it should permanently emit visible Tscherenkov radiation in vacuum. The disadvantage is that the tachyon loses energy, and therefore its velocity increases to infinity during the radiation process. This difficulty shall be circumvented by a permanent supply of energy to the tachyon mediated by an electric field. Thereby the velocity would decrease again, and the Tscherenkov radiation should be recordable.

Finally, it should be pointed out that the tachyons so far are nothing else but a "possibility of theory." An experimental proof is still missing.

Derivation of the energy law in the Minkowski space

Let us consider the scalar product of four-force and four-velocity:

$$\overset{\Rightarrow}{F}\cdot\frac{d\overset{\Rightarrow}{r}}{d\tau}=m_0\frac{d^2\overset{\Rightarrow}{r}}{d\tau^2}q\cdot\frac{d\overset{\Rightarrow}{r}}{d\tau}=\frac{m_0}{2}\frac{d}{d\tau}\left(\frac{d\overset{\Rightarrow}{r}}{d\tau}\right)^2. \tag{33.53}$$

The normalization of the four-velocity is constant and equals the negative square of the speed of light

$$\left(\frac{d\overset{\Rightarrow}{r}}{d\tau}\right)^2=\overset{\Rightarrow}{v}\cdot\overset{\Rightarrow}{v}=\left(\frac{1}{\sqrt{1-\beta^2}}(\mathbf{v},ic)\right)^2=\frac{1}{1-\beta^2}(v^2-c^2)=-c^2. \tag{33.54}$$

Therefore the scalar product vanishes,

$$\overset{\Rightarrow}{F}\cdot\frac{d\overset{\Rightarrow}{r}}{d\tau}=0.$$

Evaluation of the scalar product component by component yields the relation

$$\left(\frac{K_x}{\sqrt{1-\beta^2}},\frac{K_y}{\sqrt{1-\beta^2}},\frac{K_z}{\sqrt{1-\beta^2}},\frac{1}{\sqrt{1-\beta^2}}\frac{d}{dt}\left(\frac{icm_0}{\sqrt{1-\beta^2}}\right)\right)\cdot\left(\frac{1}{\sqrt{1-\beta^2}}\left(\frac{dx}{dt},\frac{dy}{dt},\frac{dz}{dt},ic\right)\right)=0$$

[11] In nuclear physics Mach shock waves were detected by H. Gutbrod et al. in a fast, "supersonic" collision of a small nucleus through a large one. They were predicted almost 15 years earlier by Scheid, Müller, and Greiner (W. Scheid, H. Müller, and W. Greiner, *Phys. Rev. Lett.* **32** (1974) 741). These nuclear compression waves represent the key mechanism for the compression of nuclear matter. This phenomenon is exploited for studying the equation of state of nuclear matter, see, e.g., W. Greiner and H. Stöcker, *Scientific American*, Jan. 1985, and H. Gutbrod and H. Stöcker, *Scientific American*, Nov. 1991. Furthermore, the shock-induced compression of nuclear matter can compress and at the same time heat the elementary matter so strongly that the nucleons are decomposed to their constituents, quarks and gluons. This reaction is supposed to create a quark-gluon plasma, a state of matter that existed in the very first instants of the world shortly after the Big Bang.

or

$$\frac{1}{1-\beta^2}\left(K_x\frac{dx}{dt}+K_y\frac{dy}{dt}+K_z\frac{dz}{dt}+\left(\frac{d}{dt}(imc)\right)ic\right)=\frac{1}{1-\beta^2}\left(\mathbf{K}\cdot\frac{d\mathbf{r}}{dt}-\frac{d}{dt}(mc^2)\right)$$
$$=0.$$

From that follows

$$(\mathbf{K}\cdot d\mathbf{r}-d(mc^2))=-(dV+d(mc^2))=0, \tag{33.55}$$

where the relation $V(\mathbf{r})=V(x,y,z)=-\int_{\mathbf{r}_0}^{\mathbf{r}}\mathbf{K}\cdot d\mathbf{r}$ has been used, which means a restriction to conservative force fields. The integration of (33.55) yields the *relativistic energy law*

$$V(x,y,z)+mc^2=\text{constant}=E. \tag{33.56}$$

Also here we see, in a different way than earlier, that mc^2 must be interpreted as the total energy (rest energy m_0c^2 + kinetic energy $(mc^2-m_0c^2)$) of the mass m.

The fourth momentum component

So far we could not yet interpret the fourth component of the four-momentum (33.5). We will now express the momentum by the energy. For this purpose we first calculate the fourth momentum component. By insertion of $E=mc^2$ into the fourth momentum component $p_4=imc$, it follows that

$$p_4=imc=i\frac{mc^2}{c}=\frac{iE}{c}.$$

The components of the four-momentum then read

$$p_1=mv_1,\qquad p_2=mv_2,\qquad p_3=mv_3,\qquad p_4=\frac{iE}{c}, \tag{33.57}$$

with

$$m=\frac{m_0}{\sqrt{1-\beta^2}}.$$

Hence, the fourth momentum component essentially represents the energy of the mass point.

Conservation of momentum and energy for a free particle

From (33.8) and (33.9), it follows (see also (33.10)) that

$$\frac{d\overset{\Rightarrow}{p}}{d\tau} = \frac{1}{\sqrt{1-\beta^2}}\frac{d}{dt}\left(m\mathbf{v}, i\frac{E}{c}\right) = \overset{\Rightarrow}{F}$$

$$= \left(\frac{\mathbf{K}}{\sqrt{1-\beta^2}}, \frac{icm_0\beta\dot{\beta}}{(1-\beta^2)^2}\right) . \qquad (33.58)$$

If no three-forces are acting (i.e., $\mathbf{K} = 0$), from the first three components it obviously follows that

$$\frac{d}{dt}(m\mathbf{v}) = 0;$$

hence,

$$m\mathbf{v} = \frac{m_0}{\sqrt{1-\beta^2}}\mathbf{v} = \overrightarrow{\text{constant}} .$$

This is the relativistic form of the momentum conservation law for a free particle. This vector equation immediately implies that the direction of $\mathbf{v}$ is constant. If we now consider the magnitude of the vector equation and employ m_0 = constant, then it follows that $(v \equiv |\mathbf{v}|)$

$$\frac{v}{\sqrt{1-(v/c)^2}} = \text{constant},$$

that is, the magnitude of $\mathbf{v}$ also must be constant. Therefore, also $\beta = v/c$ = constant, i.e., $\dot{\beta} = 0$. Hence, from the fourth component of (33.58) follows:

$$\frac{dE}{dt} = 0 \quad \text{or} \quad E = mc^2 = \text{constant}$$

This is the energy law for a free particle.

Relativistic energy for free particles

The scalar multiplication of two four-momenta yields

$$\overset{\Rightarrow}{p} \cdot \overset{\Rightarrow}{p} = (\mathbf{p}, imc) \cdot (\mathbf{p}, imc) = \mathbf{p}^2 - m^2c^2 = p^2 - m^2c^2$$

and also

$$\overset{\Rightarrow}{p} \cdot \overset{\Rightarrow}{p} = m_0^2 \overset{\Rightarrow}{v} \cdot \overset{\Rightarrow}{v} = -m_0^2c^2,$$

because $\overset{\Rightarrow}{v} \cdot \overset{\Rightarrow}{v} = -c^2$ (see equation (33.54) above). From there we get

$$p^2 - m^2c^2 = -m_0^2c^2,$$

and with $mc = \dfrac{E}{c}$:

$$p^2 - \frac{E^2}{c^2} = -m_0^2 c^2,$$

$$E^2 = p^2c^2 + m_0^2 c^4,$$

or

$$E^2 = (\mathbf{p}c)^2 + (m_0c^2)^2 = (mc^2)^2. \tag{33.59}$$

This is the *relativistic energy–momentum relation for a free particle* since no additional potential occurs. Note that formally also negative energies are possible.

$$E_1 = +\sqrt{(\mathbf{p}c)^2 + (m_0c^2)^2}\,, \qquad E_2 = -\sqrt{(\mathbf{p}c)^2 + (m_0c^2)^2}. \tag{33.60}$$

If a particle has the rest mass zero (photon, neutrino), then

$$E = p \cdot c. \tag{33.61}$$

For photons the quantum theory states that their energy is proportional to the frequency, i.e., $E = \hbar\omega$. Here $\hbar$ denotes Planck's elementary quantum of action. According to equation (33.61), for the momentum p of the photon it immediately follows that

$$p = \frac{E}{c} = \hbar\frac{\omega}{c} = \hbar\frac{2\pi\nu}{c} = \hbar 2\pi\frac{1}{Tc} = \hbar\frac{2\pi}{\lambda} = \hbar k, \tag{33.62}$$

with k as wave number. This is the *de Broglie relation* between momentum p and wave number k. It plays an important role in the discovery of quantum mechanics. Since now the momentum direction $\mathbf{p}$ surely must coincide with the propagation direction $\mathbf{k}$ of the light wave (only this is physically meaningful), this equation may also be written in vector form, namely

$$\mathbf{p} = \hbar\mathbf{k}. \tag{33.63}$$

The relativistic energy spectrum (33.60) is illustrated in the figure. This spectrum later on follows also in the relativistic quantum mechanics from the Dirac equation, the relativistic form of the Schrödinger equation. It is valid for fermions with spin 1/2, hence, for example, for electrons. An electron being in a state of positive energy might "spontaneously" switch to arbitrary lower states and thereby radiate off energy. This process would never terminate since there always exist further, lower states for electron transitions.

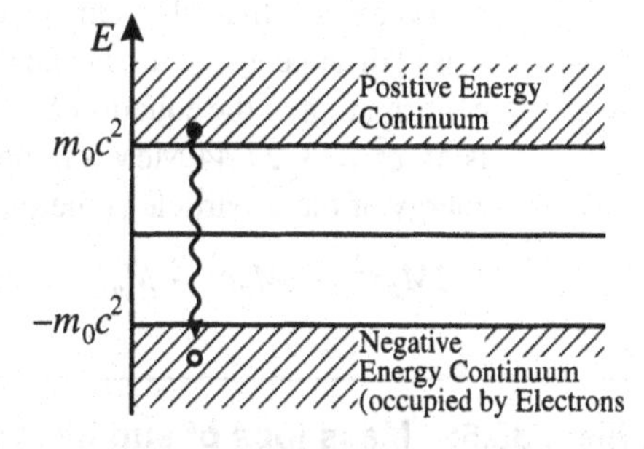

The relativistic energy spectrum of a free particle. • means an electron, o means a hole (positron).

A radiation catastrophe that of course never has been observed would be unavoidable. In order to avoid this difficulty, one must assume that the states of negative energy are completely occupied: An electron then cannot change to the negative energy states since this is forbidden according to the Pauli principle. The energy continuum occupied with electrons (the "Dirac sea") is homogeneously and isotropically distributed over the entire space. The *Dirac sea so to speak represents the vacuum.* It shall carry neither charge nor mass. A hole (unoccupied electron state in the sea) behaves like a positive electron, which is a *positron.* A light quantum (photon) with sufficient energy $\hbar\omega > 2m_0c^2$ may lift an electron from the negative sea into the positive energy continuum and thereby leave behind a hole (positron). This is the base of the electron–positron pair production or, more generally, the particle–antiparticle production, which thus is founded by the theory of relativity.

The theory sketched here also meets with difficulties; first of all the infinitely large mass and the infinitely large charge of the vacuum (occupied negative energy continuum) must be eliminated ("renormalized"). This concept will be formulated and realized in quantum electrodynamics.[12]

Examples on the equivalence of mass and energy

(a) An example on the equivalence of mass and energy is the positron–electron annihilation. The positron is the antiparticle of the electron. Antiparticles are in general elementary particles that may arise in reactions with very large energy conversions together with ordinary particles, and in essential properties (electric charge, magnetic moment) appear so to speak as their mirror image. An interpretation of their appearance is given by quantum mechanics in its relativistic generalization of the *Dirac wave equation.* According to this theory the particles may have both positive and negative energy states. Particles and antiparticles disappear (annihilate) in common just as they appear in common (pair annihilation and pair production, respectively).

(b) The mass defect: If one adds the individual masses of the protons and neutrons forming an atomic nucleus, and compares the sum with the result of measuring the mass of that nucleus in the mass spectrograph, one realizes that the composite nucleus has a lower mass than the sum of the individual masses of its nucleons. A fraction of the mass "disappeared"; it has been converted into energy (binding energy). This is a further confirmation of the equation $E = mc^2$. For example: The mass of an He nucleus (α particle) is $M_\alpha c^2 = 3727.44$ MeV; on the contrary $2M_pc^2 + 2M_nc^2 = 3755.44$ MeV. The binding energy of the α particle is therefore

$$2M_pc^2 + 2M_nc^2 - M_\alpha c^2 = 28\ \text{MeV}.$$

Problem 33.6: Mass loss of sun by radiation

The mean sun energy density irradiated onto the earth's surface is

$$\varepsilon = 1.4 \cdot 10^6\ \text{erg} \cdot \text{cm}^{-2} \cdot \text{s}^{-1}\,.$$

[12] We refer to Volume 7 of the lectures: W. Greiner and J. Reinhardt: *Quantum Electrodynamics*, Spinger Verlag New York 2003.

How much of mass is lost per second by the sun when recalculating this energy loss to mass loss? What would be the lifetime of the sun if this rate of loss remained constant ($m_s = 1.99 \cdot 10^{33}$ g)?

Solution The sun shall radiate energy uniformly and isotropically. A spherical surface about the sun at the distance sun–earth ($r_e = 1.5 \cdot 10^{13}$ cm) has the area

$$F = 4\pi r_e^2 = 2.83 \cdot 10^{27}\ \mathrm{cm}^2.$$

The energy release in the interval $\Delta t = 1$ s is therefore

$$\begin{aligned}\Delta E &= \varepsilon \cdot F \cdot \Delta t \\ &= 1.4 \cdot 10^6 \cdot 2.82 \cdot 10^{27} \cdot \mathrm{erg} \cdot \mathrm{cm}^{-2}\,\mathrm{s}^{-1} \cdot \mathrm{cm}^2 \cdot \mathrm{s} \\ &= 3.96 \cdot 10^{33}\ \mathrm{erg}.\end{aligned}$$

This corresponds to a loss of mass per second of

$$\Delta m = \frac{\Delta E}{c^2} = 4.4 \cdot 10^{12}\ \mathrm{g} \qquad (c = 3 \cdot 10^{10}\ \mathrm{cm} \cdot \mathrm{s}^{-1}).$$

For the lifetime of the sun, we then get

$$T = \Delta t \frac{m_s}{\Delta m} = \frac{1.99 \cdot 10^{33}\ \mathrm{g} \cdot 1\ \mathrm{s}}{4.4 \cdot 10^{12}\ \mathrm{g}} = 4.53 \cdot 10^{20}\ \mathrm{s} = 1.43 \cdot 10^{13}\ \text{years}.$$

This problem is, however, unrealistic because due to energy conservation laws for the elementary particles only a fraction of the mass may annihilate at all. If one assumes that about 1/1000 of the sun mass may annihilate, there remains a lifetime of the sun of about 10^{10} years, which compares with the estimated age of the world.

Problem 33.7: Velocity dependence of the proton mass

The rest mass of the proton is $m_0(p) = 1.66 \cdot 10^{-27}$ kg. Calculate the mass of the proton moving with (a) $3 \cdot 10^7$ m/s and (b) $2.7 \cdot 10^8$ m/s.

Compare the kinetic energy of the proton in both cases according to the classical and the relativistic calculation. (1 Joule = 1 kg m^2/s^2 = $0.62 \cdot 10^{13}$ MeV.)

Solution For the given velocities one evaluates the following values for

$$\beta^2 = \frac{v^2}{c^2} \qquad \text{and} \qquad \gamma = \frac{1}{\sqrt{1-\beta^2}}$$

(a) $\beta = 0.1, \gamma = 1.005$,

(b) $\beta = 0.91, \gamma = 2.3$.

For the proton mass it follows from the relation that

$$m = \frac{m_0}{\sqrt{1-\beta^2}} = \gamma m_0$$

(a) $m = 1.005 m_0 \cong 1.67 \cdot 10^{-27}$ kg,

(b) $m = 2.300 m_0 \cong 3.82 \cdot 10^{-27}$ kg.

The relativistic kinetic energy

$$T = E - E_0 = m_0 c^2 (\gamma - 1)$$

is then

(a)

$$T = m_0 \cdot (3 \cdot 10^8)^2 \frac{\text{m}^2}{\text{s}^2} \cdot 0.005 = (1.5 \cdot 10^{-10}) \text{kg} \frac{\text{m}^2}{\text{s}^2} \cdot 0.005$$
$$= 7.5 \cdot 10^{-13} \text{ Joule,}$$

(b)

$$T = m_0 \cdot (3 \cdot 10^8)^2 \frac{\text{m}^2}{\text{s}^2} \cdot 1.3 = 1.3 \cdot 10^{-10} \text{ Joule.}$$

A comparison of the velocities and the kinetic energies shows that cases (a) and (b) differ by a factor 9 in the velocity, but by a factor 260 in the energy.

The classical calculation of the kinetic energy

$$T = \frac{1}{2} m_0 v^2 = \frac{1}{2} m_0 c^2 \beta^2$$

yields

(a) $T = 7.5 \cdot 10^{-13}$ J and (b) $T = 6.1 \cdot 10^{-11}$ J.

For case (a) the classical and relativistic energy are roughly equal. For case (b) the relativistic value is by a factor 3.2 higher than the classical result, which is also expected from the calculated β-values.

Problem 33.8: Efficiency of a working fusion reactor

In 1970 the total energy consumption of the world amounted to $5.5 \cdot 10^{13}$ kWh (kilowatt hour). A fusion reactor could produce energy by the reaction ${}^2\text{D} + {}^2\text{D} \rightarrow {}^4\text{He} + \text{energy}$. (^{2}D—deuterium with $m_0({}^2\text{D}) = 2.0147$ amu, ^{4}He—helium with $m_0({}^4\text{He}) = 4.0039$ amu, where 1 amu = 1 atomic mass unit = 1/12 (rest mass of ${}^{12}_{6}\text{C}$) = $1.685 \cdot 10^{-27}$ kg.) How many kg deuterium would be needed to generate the world energy consumption of 1970?

Solution The rest mass of two deuterium nuclei before the reaction is

$$m_0(\text{before}) = 2m_0({}^2\text{D}) = 4.0294 \text{ amu},$$

while the rest mass after the reaction is

$$m_0(\text{after}) = m_0({}^4\text{He}) = 4.0039 \text{ amu}.$$

Hence, the mass loss during the reaction is

$$\Delta m = m_0(\text{before}) - m_0(\text{after}) = 0.0255 \text{ amu}.$$

The released energy ΔE is calculated according to the relation $E = mc^2$ as $\Delta E = (\Delta m)c^2$.

This means that per deuterium–mass the energy

$$\frac{\Delta E}{2m_0({}^2\text{D})} = \frac{\Delta m}{2m_0({}^2\text{D})} c^2 = (0.00633)c^2$$

is produced. Inversely, the quantity of deuterium (mass M) needed for a definite quantity of energy E to be produced is

$$M({}^2\text{D}) = \frac{E}{c^2} \cdot \frac{1}{0.00633}.$$

The factor 0.00633 is a measure of the efficiency of the reaction $^2\mathrm{D} + ^2\mathrm{D} \rightarrow ^4\mathrm{He} + \text{energy}$. For an annual energy consumption of ($1\,\mathrm{kWh} = 3.6 \cdot 10^6$ Joule)

$$E = 5.5 \cdot 10^{13} \cdot 3.6 \cdot 10^6 \text{ Joule} \approx 2 \cdot 10^{20} \text{ Joule},$$

one therefore would need (1 Joule = 1 kg m^2 s^{-2})

$$M(^2\mathrm{D}) = \frac{2 \cdot 10^{20}}{(3 \cdot 10^5)^2} \cdot \frac{1}{0.00635} \frac{\text{Joule s}^2}{\text{km}^2} \approx 3.5 \cdot 10^5 \text{ kg} = 350 \text{ t}$$

of deuterium.

As the earth's oceans contain ca. $0.2\,{}^0/_{00}$ deuterium, mankind would get rid of any energy problem for 1 million years—if fusion reactors were available.

Problem 33.9: Decay of the π^+-meson

The rest mass of the π^+-meson is $m_\pi = 139.6\,\mathrm{MeV/c^2}$. The π^+-meson decays into a anti-muon (μ^+ lepton, a "heavy positron") with the rest mass $m_\mu = 105.7\,\mathrm{MeV/c^2}$ and a neutrino ν_μ with the rest mass $m_\nu = 0$. Find the momentum and the energy of the arising muon μ^+.

Solution In the figure below the quoted decay is sketched, (a) in the rest frame of the π^+, and (b) in the laboratory system, as bubble-chamber record, with the subsequent decay $\mu^+ \rightarrow e^+ \nu_e \bar{\nu}_\mu$ in a positron and two neutrinos.

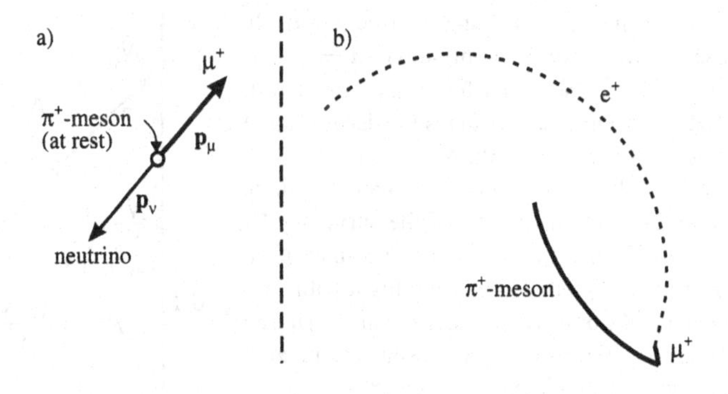

The requirement for conservation of the four-momentum

$$\vec{\vec{p}}_\pi = \vec{\vec{p}}_\mu + \vec{\vec{p}}_\nu$$

implies the conservation of both the momentum as well as the energy. In the following we employ the relation for the magnitude of the three-momentum $p = |\, \mathbf{p}_\nu \,| = |\, \mathbf{p}_\mu \,|$. The total energy of the anti-muon μ^+ is then

$$E_\mu^2 = c^2 p^2 + m_\mu^2 c^4,$$

and the energy of the neutrino is

$$E_\nu^2 = c^2 p^2 \quad \text{because } m_\nu = 0.$$

It further holds that

$$E_\pi = E_\mu + E_\nu$$

or

$$m_\pi c^2 = \sqrt{c^2p^2 + c^4m_\mu^2} + cp.$$

Forming the square and rearranging yields

$$cp = \frac{c^2}{2}\left(\frac{m_\pi^2 - m_\mu^2}{m_\pi}\right) = \frac{1}{2}\left[m_\pi c^2 - \frac{(m_\mu c^2)^2}{m_\pi c^2}\right],$$

and by inserting the rest masses $m_\pi = 139.6\ \mathrm{MeV/c^2}$ and $m_\mu = 105.7\ \mathrm{MeV/c^2}$ we get

$$cp = \frac{1}{2}\left[139.6 - \frac{105.7}{139.6}\cdot 105.7\right]\ \mathrm{MeV} = 29.8\ \mathrm{MeV}.$$

For the kinetic energy of the anti-muon μ^+, it then follows that

$$\begin{aligned} T_\mu &= E - m_\mu c^2 = \sqrt{c^2p^2 + m_\mu^2c^4} - m_\mu c^2 \\ &= \left[\sqrt{(29.8)^2 + (105.7)^2} - 105.7\right]\ \mathrm{MeV} = 4.1\ \mathrm{MeV}. \end{aligned} \tag{33.64}$$

Problem 33.10: Lifetime of the K^+-mesons

The lifetime of a K^+-meson (the positively charged variety of the K-mesons) is $\tau = 1.235 \cdot 10^{-8}$ s when measured for K-mesons at rest. The following figure displays the decay data of a K-meson emitter with a momentum of 1.6 GeV/c and 2.0 GeV/c in the laboratory system. Here the fraction (N/N_0) of the surviving K-mesons (N_0 is the total number of K-mesons in the beam) is plotted versus the flight path covered. Let the origin of the length scale be chosen arbitrarily. Because the K-mesons are moving practically with the speed of light, the scaling on the abscissa may also be understood as a time scale as adopted in the laboratory system.

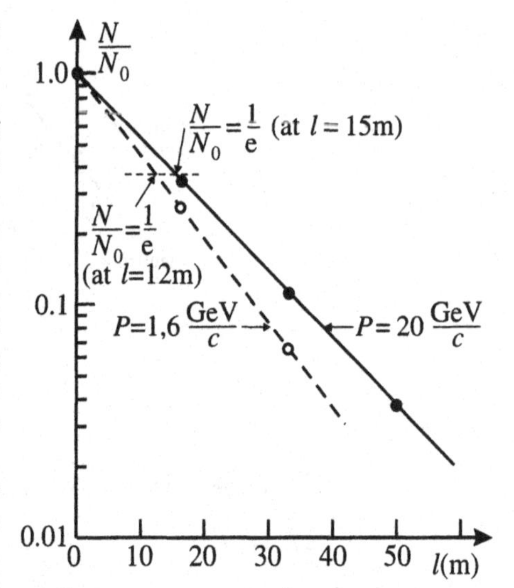

Decay of Kaons from emitters with different velocities, as seen in the laboratory frame.

One sees, however, from the figure that the K-mesons with a larger laboratory momentum are virtually (in the laboratory) longer-lived. However, to a certain time interval in the laboratory system there corresponds a shortened time interval in the rest frame of the K-meson, due to the time dilatation. The latter interval becomes shorter, the larger the momentum of the K-meson in the laboratory system is. Show that the data given in the figure are consistent with the lifetime of a K^+-meson quoted above, if the phenomenon of time dilatation, as is required by the special theory of relativity, is taken into account. (The rest energy of the K-meson is $m_0c^2 = 0.494$ GeV.)

Solution The figure shows the quantity N/N_0 versus the flight path l in semilogarithmic scale. The two decay curves are straight lines in this representation, i.e. the decay data obey the equation

$$N = N_0 e^{-l/\lambda}, \tag{33.65}$$

because then $\ln(N/N_0)$ depends linearly on l, corresponding to the scale on the ordinate. But then it holds that

$$\ln\left(\frac{N}{N_0}\right) = -\frac{l}{\lambda}, \tag{33.66}$$

and we see right now that λ denotes that flight length after which the surviving rate of K-mesons dropped to the value $1/e$. From the figure one therefore extracts

$$\lambda = 12\ \text{m} \qquad \text{for momentum } p = 1.6\ \text{GeV/c},$$
$$\lambda = 15\ \text{m} \qquad \text{for momentum } p = 2.0\ \text{GeV/c}.$$

Denoting the velocity of a K-meson by $v = \beta c$, a meson needs the time Δt to cover the distance l:

$$\Delta t = \frac{l}{\beta c}.$$

We therefore may write equation (33.65) in the form

$$N = N_0 e^{-\Delta t \beta c/\lambda}. \tag{33.67}$$

We compare this result with the known form of the decay law:

$$N = N_0 e^{-\Delta t'/\tau}, \tag{33.68}$$

where $\Delta t'$ denotes a time interval measured in the *rest frame* of the K-meson. Such an interval $\Delta t'$ undergoes a dilatation in the laboratory frame and is measured there as $\Delta t = \gamma \Delta t'$. A comparison of (33.67) with

$$N = N_0 e^{-\Delta t'/\tau} = N_0 e^{-\Delta t/\gamma\tau} \tag{33.69}$$

therefore leads to

$$\frac{\beta c}{\lambda} = \frac{1}{\gamma\tau} \qquad \text{or} \qquad \tau = \frac{\lambda}{\beta\gamma c}. \tag{33.70}$$

Because of the relation (33.68) τ is the lifetime of the K-meson as is measured in its rest frame. For K-mesons with momenta in the given order of magnitude the velocity v practically equals the speed of light, i.e., $\beta \approx 1$. The momentum of the particles may, however, be calculated exactly; according to

$$pc = m_o\gamma\beta c^2 \qquad (p = mv = m_0\gamma v = m_0\gamma\beta c), \tag{33.71}$$

we obtain

$$\beta\gamma = \frac{pc}{m_0c^2} = \frac{1.6}{0.494} = 3.239 \qquad \text{for } p = 1.6\ \text{GeV/c},$$
$$= \frac{2.0}{0.494} = 4.049 \qquad \text{for } p = 2.0\ \text{GeV/c}.$$

For the mean lifetime τ in the rest frame of the K-meson, one thus obtains from relation (33.70)

$$\tau = \left(\frac{12\ \text{m}}{3\cdot 10^8\ \text{m/s}}\right)\cdot\frac{1}{3.239} = 1.235\cdot 10^{-8}\ \text{s} \qquad \text{for } p = 1.6\ \text{GeV/c},$$

and

$$\tau = \left(\frac{15\ \text{m}}{3\cdot 10^8\ \text{m/s}}\right)\cdot\frac{1}{4.049} = 1.235\cdot 10^{-8}\ \text{s} \qquad \text{for } p = 2.0\ \text{GeV/c},$$

which agrees with the value given in the formulation of the problem. However, the lifetime measured in the laboratory frame is obviously

$$\tau' = \gamma\tau = \frac{\lambda}{\beta c} \tag{33.72}$$

because of relation (33.69), such that the lifetime in the laboratory frame appears as extended by a factor of ≈ 3 or ≈ 4 against the lifetime in the rest frame of the K-meson. The faster the K-meson moves, the larger is the time dilatation and the longer is its "lifetime" in the laboratory frame. To determine the velocity of the K-mesons, we write $\alpha = \beta\gamma$ and obtain

$$\alpha^2 = \beta^2\gamma^2 = \frac{\beta^2}{1-\beta^2} \qquad \text{or} \qquad \beta^2 = \alpha^2(1-\beta^2)$$

or rewritten

$$\left(1+\alpha^2 = 1 + \frac{\beta^2}{1-\beta^2} = \frac{1}{1-\beta^2}\right)$$

$$\beta^2 = \frac{\alpha^2}{1+\alpha^2} \qquad \text{or} \qquad \beta = \frac{\alpha}{\sqrt{1+\alpha^2}} = \frac{\beta\gamma}{\sqrt{1+\beta^2\gamma^2}}.$$

For a K-meson of momentum 1.6 GeV/c, we found $\beta\gamma = 3.239$; thus for β

$$\beta = \frac{3.239}{\sqrt{1+10.49}} = 0.955,$$

results, that is, $v = 0.955c$. The K-mesons practically move with the speed of light, as was assumed above.

Problem 33.11: On nuclear fission

One of the basic reactions in nuclear fission is

$$\text{n} + {}^{235}_{92}\text{U} \rightarrow {}^{236}_{92}\text{U} \rightarrow {}^{92}_{38}\text{Sr} + {}^{140}_{54}\text{Xe} + 4\text{n}\,.$$

The masses of the essential reaction partners are:

$$\begin{aligned} m_0(^{235}\text{U}) &= 235.175\ \text{amu},\\ m_0(^{92}\text{Sr}) &= 91.937\ \text{amu},\\ m_0(^{140}\text{Xe}) &= 139.947\ \text{amu},\\ m_0(\text{n}) &= 1.009\ \text{amu} \end{aligned}$$

(amu = "atomic mass unit," 1 amu = $1.6585\cdot 10^{-27}$ kg). Calculate the energy released per reaction. How many kg of uranium are needed to produce the worldwide total electric energy consumed in 1970 ($5.5\cdot 10^{12}$ kWh) with an efficiency of $\eta = 0.5$?

Solution

$$m_0(n) + m_0(^{235}\mathrm{U}) = 236.184 \text{ amu} = 391.711 \cdot 10^{-27} \text{ kg}.$$
$$m_0(\mathrm{Sr}^{92}) + m_0(\mathrm{Xe}^{140}) + 4m_0(n) = 235.92 \text{ amu} = 391.273 \cdot 10^{-27} \text{ kg}.$$

Thus, the mass defect of the reaction is

$$\Delta m = 0.438 \cdot 10^{-27} \text{ kg}$$

or

$$\Delta E = 3.94 \cdot 10^{-11} \text{ J}.$$

To release $5.5 \cdot 10^{12}$ kWh of electric energy (1 kWh $= 3.6 \cdot 10^6$ J), one needs about 3920 t of uranium ^{235}U. With a density of 18.7 g/cm^3, this would correspond to a cube with an edge length of 5.94 m.

Problem 33.12: Mass–energy equivalence in the example of the π^0-meson

The π^0-meson is an electrically neutral particle that decays into two high-energetic photons. The rest energy of the π^0-meson is $m_0c^2 = 135$ MeV.

(a) Find the energy of the photons if a π^0 decays at rest.

(b) Find the maximum and minimum energy of the γ-rays in the laboratory frame if the π^0 there has a total energy of $E_{\text{tot}} = 426$ MeV.

Solution (a) Let the two emitted photons have the energies E_1, E_2 and the momenta $\mathbf{p}_1$ and $\mathbf{p}_2$, respectively. Because of the energy–momentum conservation law then

$$E_1 + E_2 = E = m_0c^2, \qquad \mathbf{p}_1 + \mathbf{p}_2 = 0,$$

thus $|\mathbf{p}_1| = |\mathbf{p}_2|$ (see figure).

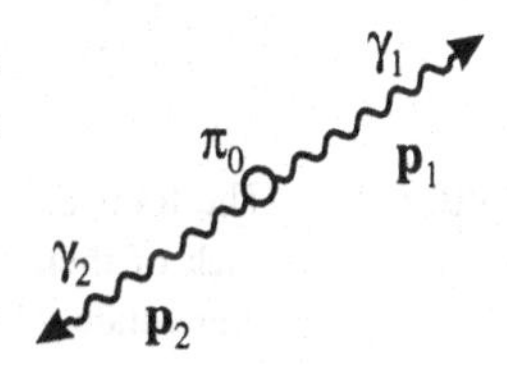

Decay of the π^0-meson into two photons.

Moreover,

$$|\mathbf{p}_i| = \frac{E_i}{c} \qquad (i = 1, 2),$$

and therefore,

$$E_1 = E_2 = \frac{E}{2} = 67.5 \text{ MeV}.$$

(b) Because $E_{\text{tot}} = mc^2 = 3.16\, m_0c^2 = \sqrt{10}\, m_0c^2$ for the velocity of the π^0-meson in the laboratory frame, it follows that $\gamma^2 = 10$ or $\beta = 0.9486$. The π^0-meson thus moves with a velocity of $|\mathbf{v}| = 0,9486\, c$ in the laboratory frame. In the rest frame of the meson it now decays into two photons as was described in (a). In the laboratory frame the two $\gamma's$ may now be emitted under arbitrary angles against the beam axis (direction of $\mathbf{v}$) and there also appear as more or less red- or ultraviolet-shifted (see below).

$E_{\gamma(\text{max})}$ is obtained if an emitted γ moves along the direction of $\mathbf{v}$ (see figure), $E_{\gamma(\text{min})}$ is obtained if a γ moves against the direction of $\mathbf{v}$. We denote the rest frame S of π^0 by nonprimed quantities, the laboratory frame S' by primed quantities. The energy $E_0(\gamma) = 67.5$ MeV (measured in S) of an emitted photon transforms as the timelike component of a four-vector, thus:

$$E' = \gamma[E_0 - \boldsymbol{\beta} \cdot (c\mathbf{p})].$$

The situation before (left) and after (right) the decay of the π^0-meson.

In S one has $|\mathbf{p}| = E_0/c$, and the maximum (minimum) γ-energy is obtained if $\mathbf{p}$ points in negative (positive) $\boldsymbol{\beta}$-direction (as seen from the S-frame). Hence:

$$E_{\gamma(\max)} = \gamma \cdot E_0(1+\beta) = \sqrt{\frac{1+\beta}{1-\beta}} \cdot E_0 = 416\,\text{MeV},$$

and

$$E_{\gamma(\min)} = \gamma \cdot E_0(1-\beta) = \sqrt{\frac{1-\beta}{1+\beta}} \cdot E_0 = 10.9\,\text{MeV}.$$

Problem 33.13: On pair annihilation

Let an isolated system contain $6 \cdot 10^{27}$ protons and the same number of antiprotons at rest. ($m_0(p) = m_0(\overline{p}) = 1.7 \cdot 10^{-27}$ kg.) Let all protons and antiprotons annihilate each other and produce $30 \cdot 10^{27}\,\pi$-mesons. What is the mean kinetic energy of the π-mesons? ($m_0(\pi)/m_0(p) = 0.15$.)

Solution The total mass of the system is $M_{\text{total}} = 12 \cdot 10^{27}\, m(p)$. Because $30 \cdot 10^{27}\pi$-mesons are created, each of them on the average has a total energy of $m_{\text{total}}(\pi) = \frac{12}{30}\, m_0(p) = 0.4\, m_0(p)$. From there immediately follows a mean kinetic energy of

$$\begin{aligned} E_{\text{kin}}(\pi) &= [m_{\text{total}}(\pi) - m_0(\pi)]c^2 = \left[0.4 - \frac{m_0(\pi)}{m_0(p)}\right] m_p c^2 \\ &= (0.4 - 0.15) \cdot 0.937\,\text{GeV} = 234\,\text{MeV}. \end{aligned}$$

Problem 33.14: Kinetic energy of the photon

A certain radioactive nucleus emits photons of energy $E = h\nu$ and momentum $p = h\nu/c$. A precision technique based on the "Mössbauer effect" allows frequency measurements up to an accuracy of $d\nu/\nu = 10^{-15}$. The photons of frequency ν are absorbed by a detector. If the emitter and the detector are at equal altitude above the earth's surface, the detector receives a photon of frequency $\nu' = \nu$. This is no longer true if the emitter is at an altitude L above the detector (see figure).

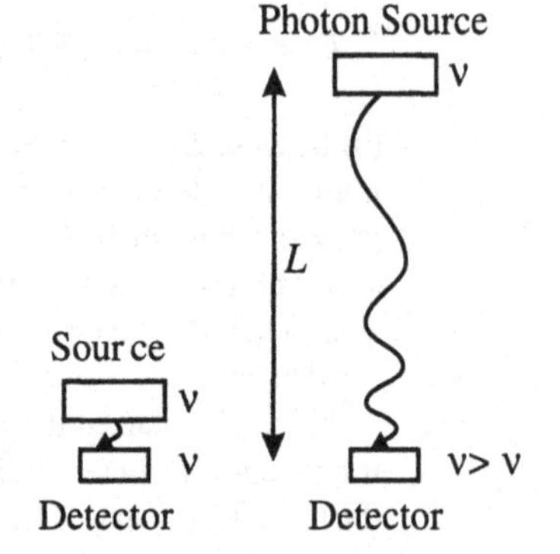

Emission and detection of photons in the gravitational field of the earth.

(a) The rest mass of a photon is $m_0 = 0$. What is the actual mass of a photon of energy $E = h\nu$?

(b) If a photon falls through an altitude L in the earth's gravitational field, then its potential energy decreases.

As a result, the photon gains "kinetic" energy. How large is the photon energy E' when the photon hits the detector?

(c) What frequency ν' is measured by the detector?

(d) Assume that the photon falls from an altitude $L = 10$ m. Could one measure the frequency shift in the gravitational field of earth by means of the Mössbauer effect?

(e) Could this effect have a bearing on the light emission of very heavy stars?

Solution (a) Because $E = h\nu = mc^2$, it follows for the mass of a photon that

$$m = \frac{h\nu}{c^2}.$$

(b) The potential energy of a photon of mass m at the altitude L above ground follows from

$$\frac{dE}{dx} = mg = \frac{E}{c^2}g$$

by integration; thus

$$E' = E\,\mathrm{e}^{gl/c^2} \approx E\left(1 + \frac{gL}{c^2}\right) = h\nu\left(1 + \frac{gL}{c^2}\right).$$

If the photon "falls down" by the altitude L, it gains this energy according to

$$E' = E + mgL = \left(1 + \frac{gL}{c^2}\right)h\nu.$$

(c) The new frequency immediately results from there as

$$\nu' = \left(1 + \frac{gL}{c^2}\right)\nu.$$

(d) As a relative frequency shift, we consider the quantity

$$\frac{\Delta\nu}{\nu} = \frac{gL}{c^2} = 10^{-15}, \qquad \text{for } L = 10\text{ m}.$$

That is, with a fall distance of 10 m on earth, the effect is measurable by means of the Mössbauer effect. The experiment was performed first in 1960 by Pound[13] and Rebka with a fall distance of 72 ft (about 22 m). They obtained an experimental value of $(5.13 \pm 0.51) \cdot 10^{-15}$ as compared with the theoretical prediction of $4.92 \cdot 10^{-15}$.

(e) If light leaves the gravitational field of the emitting body, it gains potential energy and therefore appears as red-shifted. We therefore see very heavy stars "colder" than they actually are at their surface.

---

[13] *Robert Vivian Pound*, b. 1919, Ridgeway, Ontario, Canada. He held research positions at MIT from 1942–1946, and at Harvard from 1946–1989. Pound did pioneering work in the areas of nuclear magnetic resonance (NMR), radar, and experimental tests of Einstein's general theory of relativity. The experiment of "weighting photons" was done in collaboration with his graduate student Glen Rebka.

Problem 33.15: The so-called twin paradox

On the earth are living triplets A, B, and C. At earth time $t = 0$, B and C each board a spaceship and go away from earth on straight lines. A observes the travels of the brothers from the earth and realizes the following by his clocks and scales: B for one year experiences a uniform acceleration, such that he comes from velocity zero to a velocity of $v = 0.8\,c$. He then flies for another year with this constant velocity. During a further year B reduces his velocity and reverses it to $-0.8\,c$. He again flies for one year with this velocity, and in the course of a further year he reduces his velocity to zero to land again on earth. C makes a similar trip as B, during which he, as stated by A, in one year uniformly accelerates to a velocity of $0.8\,c$, but then flies 11 years with this velocity, within a further year returns in the same way as B, with constant velocity flies back 11 years, and then reduces his velocity during one year to zero and lands again at A on earth. Let B and C have determined the duration of their trip with the same kinds of clocks as A did.

(a) Sketch the states of motion of the three brothers in a space-time (t, x) diagram.

(b) The two brothers B and C compare the duration of their trips after the landing of C on earth. What difference exists between the duration of C's trip determined by C's clock and the duration of B's trip determined by B's clock?

(c) For the observer A on earth, the time difference between the duration of C's trip and that of B amounted to 20 years. Compare this statement with the result from (b). Doesn't this lead to a contradiction to the postulate of relativity of the special theory of relativity, which states that all inertial frames are on equal rights?

(d) Let us assume that a further observer D at the moment $t = 0$ was accelerated *instantaneously* to its velocity $v = 0.8\,c$. According to A on earth, D moves away from earth for 10 years with this constant velocity; then he instantaneously turns his velocity around by a strong acceleration and flies back to earth with $v = 0.8\,c$. After 10 more years of flight with constant speed, he again reaches the earth, where he instantaneously reduces his speed to zero and lands at A. We shall assume that the number of heartbeats of A and D measures their corresponding proper time that has passed. How does the space-time diagram of A and D look like? How much did A and D age during the flight of D?

Solution (a) In the space-time diagram, the motions of A, B, and C look as plotted in the figure:

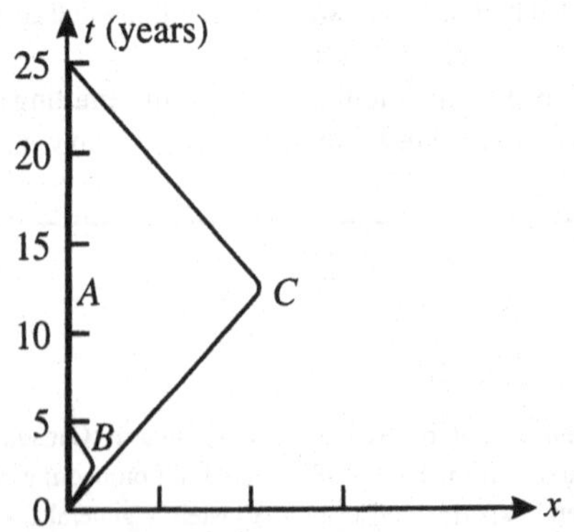

Space-time map of the journeys of the triplets A, B, and C.

(b) Since B and C have passed acceleration periods, we cannot calculate the proper times τ_B and τ_C of the trips of B and C, respectively. However, the acceleration periods for B and C were identical, hence we may calculate the difference of the proper times $\tau_C - \tau_B$ by considering only the intervals of constant velocity. As seen from earth these are for the motion of B in total 2 years, for C in total 22 years. B and C flew during these periods with the same velocity of $v = 0.8\,c$ relative to earth. We then may calculate $\Delta\tau \equiv \tau_C - \tau_B$ by means of the time dilatation factor:

$$\Delta t = 20\,\mathrm{y} = \gamma\Delta\tau$$

or

$$\Delta\tau = \gamma^{-1}\cdot 20\,\mathrm{y} = \sqrt{1-0.8^2}\cdot 20\,\mathrm{y} = 12\text{ years.}$$

Thus, if B and C compare the trip durations measured by their own clocks, they realize that C was 12 years longer on the way than B. If they directly compare the display of their clocks, it turns out that C's clock shows 8 years less than the clock of B, who after his return to earth still had to wait 20 earth years for C's return.

(c) Although after landing of C on earth all three brothers may compare their clocks at one position on earth, the time difference of the trips of C and B according to their statements amounts to 12 years, although A states 20 years for that. One might now argue as follows: During the phases of constant speed A moves with a speed of $|v| = 0.8\,c$ relative to B or C, such that both B and C see the time evolutions at A slowed down. For reasons of symmetry the relativity principle should then imply that the above difference on their statements was not permissible. This reasoning is, however, wrong. There is no symmetry between the motion of A and the motion of B or C, because the latter both were accelerated (absolutely) and thus did not always stay in an inertial frame.

(d) The space-time diagram of A and D looks as follows:

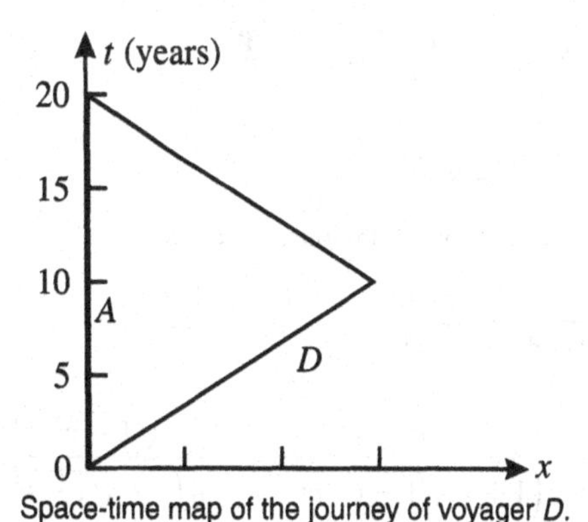

Space-time map of the journey of voyager D.

During the flight of D 20 years of proper time passed for A, while D ages by only $\Delta\tau_D = \Delta t\cdot\gamma^{-1} = 12$ years. Thus D aged by 8 years less than A did. The proper time that passed between two space-time points x and y thus depends on the trajectory T of the observer between x and y. It is given by the arc length $\tau(T) = \frac{1}{c}\int_x^y ds$ of the trajectory between x and y. That in our example $\tau_D < \tau_A$, although the world line of D in the above figure is larger than that of A, is a consequence of the indefinite metric of the space-time continuum.

Problem 33.16: Kinetic energy of a relativistic particle

The kinetic energy of a nonrelativistic particle reads

$$T = \frac{1}{2} m_0 \mathbf{v}^2 = \frac{\mathbf{p}^2}{2m_0},$$

where $\mathbf{p} = m_0 \mathbf{v}$ is the momentum of the particle. Find a formally similar expression for the relativistic kinetic energy.

Solution One has

$$E = mc^2 = \frac{m_0 c^2}{\sqrt{1 - v^2/c^2}} \tag{33.73}$$

for the relativistic total energy of a free particle and

$$\mathbf{p} = m\mathbf{v} = \frac{m_0 \mathbf{v}}{\sqrt{1 - v^2/c^2}} \tag{33.74}$$

for the relativistic momentum. According to (33.59) for (33.73) the following form still exists:

$$\begin{aligned} E^2 &= c^2 \mathbf{p}^2 + \left(m_0 c^2\right)^2 \\ &= c^2 \mathbf{p}^2 + E_0^2, \end{aligned} \tag{33.75}$$

with $E_0 = m_0 c^2$ the rest energy. Hence

$$c^2 \mathbf{p}^2 = E^2 - E_0^2 = (E - E_0)(E + E_0),$$

and therefore it follows for the relativistic kinetic energy that

$$\begin{aligned} T = E - E_0 &= \frac{c^2 \mathbf{p}^2}{(E + E_0)} = \frac{c^2 \mathbf{p}^2}{(m + m_0)c^2} \\ &= \frac{p^2}{m + m_0}. \end{aligned} \tag{33.76}$$

This is the desired form. Obviously one has

$$\lim_{v \to 0} T = \frac{\mathbf{p}^2}{m_0 + m_0} = \frac{\mathbf{p}^2}{2m_0} \tag{33.77}$$

and

$$\begin{aligned} T &= \frac{m^2 \mathbf{v}^2}{m_0 \left(1 + 1/\sqrt{1 - v^2/c^2}\right)} = \frac{m^2 \sqrt{1 - v^2/c^2}\, \mathbf{v}^2}{m_0 \left(1 + \sqrt{1 - v^2/c^2}\right)} \\ &= \frac{m \mathbf{v}^2}{1 + \sqrt{1 - v^2/c^2}} \underbrace{=}_{v \to c} m \mathbf{v}^2 = \mathbf{p} \cdot \mathbf{v}. \end{aligned}$$

These relations were pointed out first by W.G. Holladay (Vanderbilt University, Nashville, Tennessee.).

34 Applications of the Special Theory of Relativity

The elastic collision

In the general collision problem we are interested in the changes of momenta and energies of the colliding particles. The only assumption about the interaction adopted here is that it shall act only at very small distances between the particles. The problem may be solved by means of the conservation laws of momentum and energy. We denote the four-momenta of the two particles before the collision by $\vec{\vec{p}}$ and $\vec{\vec{P}}$, and after the collision by $\vec{\vec{p}}'$ and $\vec{\vec{P}}'$. The four-momentum conservation law then reads

$$\vec{\vec{p}} + \vec{\vec{P}} = \vec{\vec{p}}' + \vec{\vec{P}}' . \tag{34.1}$$

The four-vector equation comprises two conservation laws, namely that for the usual three-momentum

$$\mathbf{p} + \mathbf{P} = \mathbf{p}' + \mathbf{P}' \tag{34.2}$$

(see equation (33.57)), and that for the energy

$$e + E = e' + E' = \overline{E}, \tag{34.3}$$

where with the rest masses of the particles m_0 and M_0

$$\frac{e^2}{c^2} = m_0^2 c^2 + \mathbf{p}^2, \qquad \frac{E^2}{c^2} = M_0^2 c^2 + \mathbf{P}^2 \tag{34.4}$$

are the energies of the particles before the collision. Because the collision was supposed as elastic, the rest masses m_0 and M_0 remain unchanged in the collision process. Therefore, the energies of the particles after the collision are given by e' and E', where

$$\frac{e'^2}{c^2} = m_0^2c^2 + \mathbf{p}'^2, \qquad \frac{E'^2}{c^2} = M_0^2c^2 + \mathbf{P}'^2. \tag{34.5}$$

The total energy of the considered system has been denoted by $\overline{E}$. In (34.2) and (34.3), the rest masses and the components of the initial momenta are to be considered as given. We are looking for the components of the final momenta $\mathbf{p}'$ and $\mathbf{P}'$. Hence, we have four equations for six unknown quantities, such that the general solution will contain two undetermined parameters.

The collision problem becomes most simple in that coordinate frame where the initial momenta $\mathbf{p}$ and $\mathbf{P}$ are oppositely equal. This is the frame in which the total momentum vanishes and therefore the center of mass of the two particles is at rest. This is the rest frame of the center of mass, which is often denoted more briefly as the center-of-mass system. In this frame because of (34.1) also the final values $\mathbf{p}'$ and $\mathbf{P}'$ of the momenta must be oppositely equal.

On the other hand, the energy law (34.3) requires that the magnitudes $|\mathbf{p}| = |\mathbf{P}|$ and $|\mathbf{p}'| = |\mathbf{P}'|$ of the momenta remain unchanged in the collision:

$$|\mathbf{p}| = |\mathbf{p}'|$$

or simply

$$p = p', \tag{34.6}$$

with the abbreviations $p = |\mathbf{p}|$, $P = |\mathbf{P}'|$, etc. In the collision process only the straight line along the initial direction of the two momenta is arbitrarily rotated in space (see figure).

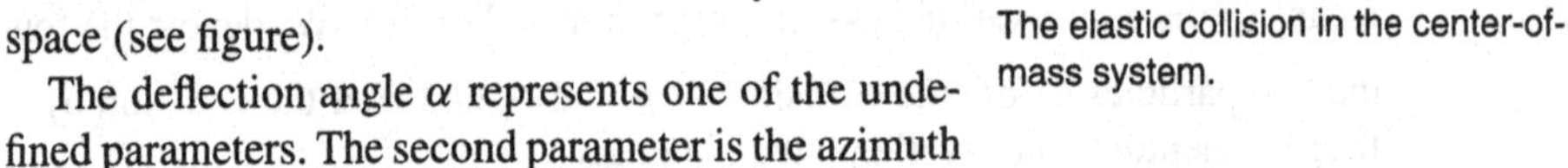

The elastic collision in the center-of-mass system.

The deflection angle α represents one of the undefined parameters. The second parameter is the azimuth angle specifying the position of the plane defined by $\mathbf{p}$ and $\mathbf{p}'$, which evidently may be arbitrarily rotated about the direction of $\mathbf{p}$.

Of particular interest for physical applications is the case with one of the particles, for example, the second one, being at rest before the collision:

$$\mathbf{P} = 0. \tag{34.7}$$

Formula (34.2) then simplifies to

$$\mathbf{p} = \mathbf{p}' + \mathbf{P}'. \tag{34.8}$$

The energy equation retains its form (34.3), but now from (34.4) it follows that

$$\frac{E}{c} = M_0 c. \tag{34.9}$$

In this case one may choose the angle θ or α as the first undetermined parameter (figure). The second parameter is again the azimuth angle, which determines the position of the drawing plane of the figure that may be arbitrarily rotated about the direction of $\mathbf{p}$; it has no meaning for the following calculation.

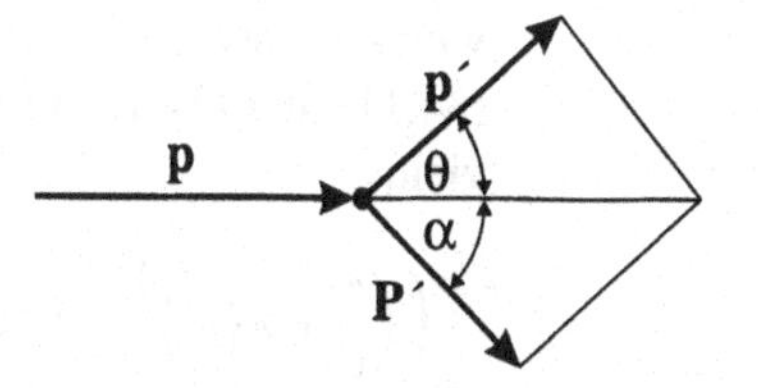

Momentum balance of a colliding particle and a particle at rest.

The solution of this particular collision problem might be found from the solution in the center-of-mass system by Lorentz-transforming to the rest frame of the second particle. Here we shall derive the final formula from (34.8), (34.3), and (34.9) by straightforward calculation.

We choose the angle θ as parameter and write down the trigonometric formula that follows immediately from the figure (see also equation (34.8))

$$P'^2 = p'^2 + p^2 - 2p\,p'\cos\theta. \tag{34.10}$$

By forming the square of the relation following from (34.5) and (34.3) ($\varepsilon = \overline{E}/c$)

$$\varepsilon - \sqrt{M_0^2c^2 + P'^2} = \sqrt{m_0^2c^2 + p'^2}\,,$$

there results

$$p'^2 + m_0^2c^2 = P'^2 + M_0^2c^2 + \varepsilon^2 - 2\varepsilon\sqrt{M_0^2c^2 + P'^2}\,.$$

When inserting in this relation the value (34.10) of P'^2, the resulting equation contains besides the term with the square root only a part linear in p'. Thus, one may eliminate the square root by one more squaring and obtain an equation of second degree in p':

$$(m_0^2c^2 - M_0^2c^2 - p^2 - \varepsilon^2 + 2p\,p'\cos\theta)^2 - 4\varepsilon^2(M_0^2c^2 + p^2 + p'^2 - 2p\,p'\cos\theta) = 0\,.$$

By explicitly calculating the square and rearranging the terms, one gets

$$\begin{aligned} &4\left(p^2\cos\theta - \varepsilon^2\right)p'^2 + 4p\,p'\cos\theta\left(m_0^2c^2 - M_0^2c^2 - p^2 + \varepsilon^2\right) \\ &\quad + \left(m_0^2c^2 - M_0^2c^2\right)^2 - 2\left(p^2 + \varepsilon^2\right)m_0^2c^2 + \left(\varepsilon^2 - p^2\right)M_0^2c^2 + \left(p^2 - \varepsilon^2\right)^2 = 0\,. \end{aligned} \tag{34.11}$$

This equation can be brought to a simpler form by making use of the relations

$$\varepsilon = \sqrt{p^2 + m_0^2c^2} + M_0c, \tag{34.12}$$

$$\varepsilon = \frac{e}{c} + M_0c \tag{34.13}$$

which follow from (34.3), (34.4), and (34.7). By inserting (34.13) in the first summand of (34.11) and (34.12) in the second and third summands, a further muliplication by $-1/4$ yields

$$\left(\left(\frac{e}{c}+M_0c\right)^2 - p^2\cos\theta\right)p'^2$$
$$- 2pp'\cos\theta\left(m_0^2c^2 + M_0c\sqrt{p^2+m_0^2c^2}\right) - p^2c^2\left(M_0^2 - m_0^2\right) = 0\,. \tag{34.14}$$

In the second summand, one now replaces $\sqrt{p^2+m_0^2c^2}$ by e/c according to (34.4), and finally one ends up with a quadratic equation for p':

$$\left(\left(\frac{e}{c}+M_0c\right)^2 - p^2\cos\theta\right)p'^2$$
$$- 2pp'\cos\theta\left(m_0^2c^2 + M_0c\frac{e}{c}\right) - p^2c^2\left(M_0^2 - m_0^2\right) = 0\,. \tag{34.15}$$

Solving for p' yields the final result

$$p' = \frac{p}{(M_0c+e/c)^2 - p^2\cos^2\theta}\left\{\cos\theta\left(m_0^2c^2 + M_0c\frac{e}{c}\right)\right.$$
$$\left.\pm\left(\frac{e}{c}c + M_0c^2\right)\sqrt{M_0^2 - m_0^2\sin^2\theta}\right\}. \tag{34.16}$$

We still note that for $M_0 > m_0$ only the positive sign before the square root in (34.16) must be admitted. According to (34.12), $\varepsilon > p$ and therefore $\varepsilon^2 - p^2\cos^2\theta > 0$. The explanation for this behavior is provided by (34.16). In the case $M_0 < m_0$, the angle θ passes twice through the range $0 \leq \theta \leq \theta_{\max}$, whereby $\theta_{\max}$ follows from

$$M_0 = m_0 \sin\theta_{\max}\,.$$

Therefore to any angle θ in this range will correspond two solutions of the collision problem. In the case $M_0 > m_0$, however, θ passes the range $0 \leq \theta \leq \pi$ once, hence for any value of θ there is only one solution.

By inserting the value (34.16) of p' in the first of equations (34.5), there results after an elementary calculation the value of e'/c. The final formula reads

$$\frac{e'}{c} = \frac{1}{(M_0c+e/c)^2 - p^2\cos^2\theta}\left\{\left(\frac{e}{c}+M_0c\right)\left(M_0c\frac{e}{c}+m_0^2c^2\right)\right.$$
$$\left.\pm\, cp^2\cos\theta\sqrt{M_0^2 - m_0^2\sin^2\theta}\right\}. \tag{34.17}$$

To finish the calculation, we still have to give formulas for $\mathbf{P}'$ and E'/c. $\mathbf{P}'$ may be calulated immediately from (34.7), namely

$$\mathbf{P}' = \mathbf{p} - \mathbf{p}'\,, \tag{34.18}$$

since $\mathbf{p}'$ is determined by p' and the angle θ. The energy E'/c is most simply calculated from (34.3) and (34.9):

$$\frac{E'}{c} = M_0 c + \frac{e}{c} - \frac{e'}{c}. \tag{34.19}$$

The formulas (34.16) to (34.19) represent the complete solution of the given collision problem.

Compton scattering

We will apply these formulas to the case of the collision of a photon with an electron at rest. The special feature of the photon is that its rest mass is extremely small against the rest mass of the electron and possibly even strictly equals zero. This simplifies considerably the algebra we had needed in the previous section.

In the course on quantum mechanics, we will see in detail how the photon energy e is related to the frequency and the wavelength λ of the radiation. We will find

$$e = \hbar\omega = h\nu = \frac{\hbar c \cdot 2\pi}{\lambda} = \frac{hc}{\lambda}. \tag{34.20}$$

Here $\hbar = h/2\pi$ is the Planck quantum of action ($\hbar = 1.054571 \times 10^{-34}$ Js), and $\omega = 2\pi\nu$ is the angular frequency of the photon oscillation. The momentum of the photon is given by

$$\mathbf{p} = \hbar\mathbf{k} \tag{34.21}$$

which defines the so-called wave vector $\mathbf{k}$ of the photon.

When photons of X-rays are scattered by electrons, a frequency shift can be observed, the amount of this shift depending on the scattering angle. This effect was discovered by Compton in 1923 and explained on the basis of the photon picture simultanously by Compton himself and Debye.

The figure illustrates again the kinematical situation. We assume the electron is unbound and at rest before the collision. Then the conservation of energy and momentum reads

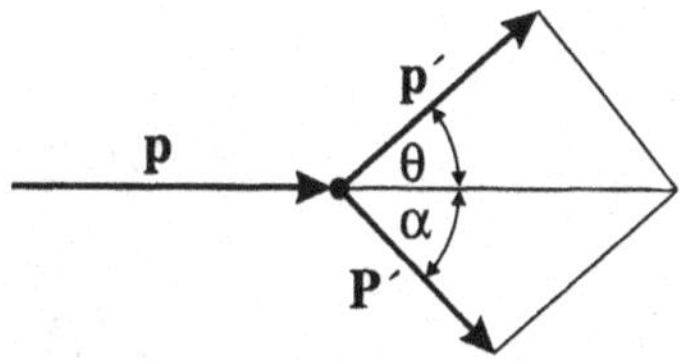

$$\hbar\omega + m_0c^2 = \hbar\omega' + \frac{m_0c^2}{\sqrt{1-\beta^2}}, \tag{34.22}$$

$$\hbar\mathbf{k} = \hbar\mathbf{k}' + \frac{m_0\mathbf{v}}{\sqrt{1-\beta^2}} \tag{34.23}$$

Conservation of momentum in Compton scattering.

To obtain a relation between the scattering angle θ and the frequency shift, we split up (34.23) into components parallel and vertical to the direction of incidence. This yields, with $k = \omega/c$,

$$\frac{\hbar\omega}{c} = \frac{\hbar\omega'}{c}\cos\theta + \frac{m_0 v}{\sqrt{1-\beta^2}}\cos\alpha \tag{34.24}$$

and

$$\frac{\hbar\omega'}{c}\sin\theta = +\frac{m_0 v}{\sqrt{1-\beta^2}}\sin\alpha\,. \tag{34.25}$$

From these two component equations, we can first eliminate α and then, by (34.22), the electron velocity v ($\beta = v/c$). To this end, we bring the $\cos\theta$ term from equation (34.24) to the left hand side, sqare and add the square of equation (34.25):

$$\left(\frac{\hbar\omega}{c} - \frac{\hbar\omega'}{c}\cos\theta\right)^2 + \left(\frac{\hbar\omega'}{c}\sin\theta\right)^2 = \left(\frac{m_0 v}{\sqrt{1-\beta^2}}\right)^2 \left(\cos^2\alpha + \sin^2\alpha\right) \tag{34.26}$$

or

$$\left(\frac{\hbar\omega}{c}\right)^2 - 2\frac{\hbar\omega}{c}\frac{\hbar\omega'}{c}\cos\theta + \left(\frac{\hbar\omega'}{c}\right)^2 = \left(\frac{m_0 v}{\sqrt{1-\beta^2}}\right)^2 = \frac{m_0^2 v^2}{1-\frac{v}{c}^2}\,. \tag{34.27}$$

From (34.22) we get

$$\left(\hbar\omega - \hbar\omega' + m_0c^2\right)^2 = \frac{\left(m_0c^2\right)^2}{1-\frac{v^2}{c^2}}\,, \tag{34.28}$$

or

$$1-\frac{v^2}{c^2} = \frac{\left(m_0c^2\right)^2}{\left(\hbar\omega - \hbar\omega' + m_0c^2\right)^2}\,, \qquad v^2 = c^2\left(1 - \frac{\left(m_0c^2\right)^2}{\left(\hbar\omega - \hbar\omega' + m_0c^2\right)^2}\right). \tag{34.29}$$

Hence,

$$\begin{aligned}
\frac{m_0^2 v^2}{1-\frac{v}{c}^2} &= m_0^2c^2\left(1 - \frac{\left(m_0c^2\right)^2}{\left(\hbar\omega - \hbar\omega' + m_0c^2\right)^2}\right)\cdot\frac{\left(\hbar\omega - \hbar\omega' + m_0c^2\right)^2}{\left(m_0c^2\right)^2} \\
&= m_0^2c^2\left(\frac{\left(\hbar\omega - \hbar\omega' + m_0c^2\right)^2}{\left(m_0c^2\right)^2} - 1\right)T \\
&= \frac{1}{c^2}\left(\left(\hbar\omega - \hbar\omega' + m_0c^2\right)^2 - \left(m_0c^2\right)^2\right) \\
&= \frac{1}{c^2}\left(\left(\hbar\omega - \hbar\omega'\right)^2 + 2\left(\hbar\omega - \hbar\omega'\right)m_0c^2\right) \\
&= \left(\frac{\hbar\omega}{c}\right)^2 + \left(\frac{\hbar\omega'}{c}\right)^2 - 2\frac{\hbar\omega}{c}\frac{\hbar\omega'}{c} + 2\hbar\left(\omega - \omega'\right)m_0\,.
\end{aligned}$$

Inserting this in (34.23), we obtain

$$2\frac{\hbar\omega}{c}\frac{\hbar\omega'}{c}\left(1-\cos\theta\right) = 2\hbar\left(\omega - \omega'\right)m_0\,. \tag{34.30}$$

Using the trigonometric identity

$$1 - \cos\theta = 2\sin^2\frac{\theta}{2}, \tag{34.31}$$

we finally obtain the following result for the frequency difference:

$$\omega - \omega' = \frac{2\hbar}{m_0 c^2}\omega\omega' \sin^2\frac{\theta}{2}. \tag{34.32}$$

If we put $\omega = 2\pi c/\lambda$, we obtain the *Compton scattering formula* in the usual form with the difference in wavelength as a function of the scattering angle θ,

$$\lambda' - \lambda = 4\pi\frac{\hbar}{m_0 c}\sin^2\frac{\theta}{2} = 2\lambda_c \sin^2\frac{\theta}{2}. \tag{34.33}$$

The scattering formula shows that the change in wavelength depends only on the scattering angle θ. During the collision the photon loses a part of its energy, and the wavelength increases ($\lambda' > \lambda$).

The factor $2\pi\hbar/m_0 c$ is called the *Compton wavelength* λ_c of a particle with rest mass m_0 (here, an electron). The Compton wavelength can be used as a measure of the size of a particle. The electron has the Compton wavelength $\lambda_c = 2.426 \times 10^{-12}$ m.

The kinetic energy of the scattered electron is

$$T = \hbar\omega - \hbar\omega' = \hbar 2\pi c\left(\frac{1}{\lambda} - \frac{1}{\lambda'}\right), \tag{34.34}$$

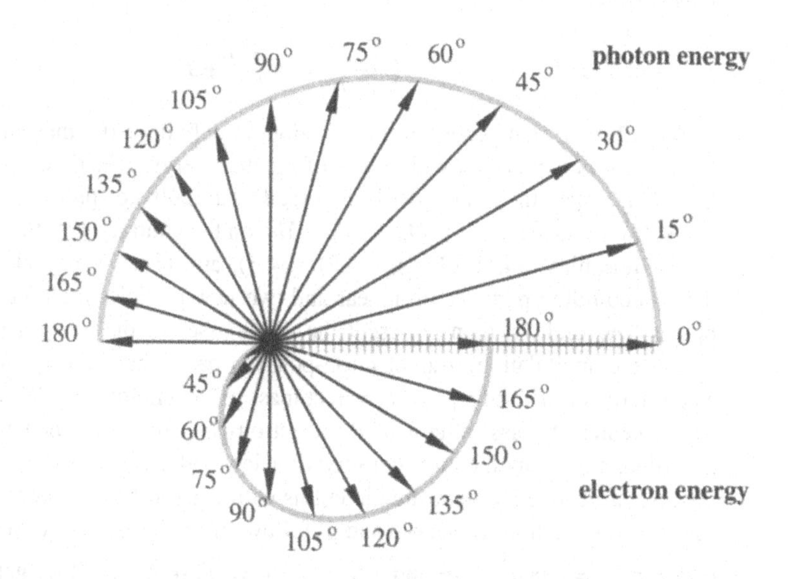

The Compton effect energy distribution of photons and electrons, showing the dependence on the scattering angle. The total available photon energy is the dashed grey line, corresponding to no scattering of the photon, $\theta = 0°$.

or, using

$$\frac{1}{\lambda} - \frac{1}{\lambda'} = \frac{\lambda' - \lambda}{\lambda'\lambda} = \frac{1}{\lambda}\frac{2\lambda_C \sin^2\frac{\theta}{2}}{\lambda + 2\lambda_C \sin^2\frac{\theta}{2}} \tag{34.35}$$

and, again with $2\pi c/\lambda = \omega$,

$$T = \hbar\omega \frac{2\lambda_C \sin^2\frac{\theta}{2}}{\lambda + 2\lambda_C \sin^2\frac{\theta}{2}} . \tag{34.36}$$

Thus the energy of the scattered electron is directly proportional to the energy of the photon. Therefore the Compton effect can only be observed in the domain of short wavelengths, such as X-rays and γ-rays. To appreciate this observation fully, we mention that in classical electrodynamics, no alteration in frequency is permitted in the scattering of electromagnetic waves – this change in frequency is only possible if scattering occurs at light quanta with momentum $\mathbf{p} = \hbar\mathbf{k}$ end energy $e = \hbar\omega$. Thus the idea of light quanta has been experimentally confirmed by the Compton effect.

The inelastic collision

In an inelastic collision, kinetic energy is lost. By definition, also the rest masses of both or at least one of the colliding particles are changing. In this case the values of the rest masses before the collision shall be denoted by m_0 and M_0, the values after the collision by m_0' and M_0'. Equations (34.1), (34.4), and (34.2, 34.3) hold also here without any modification. On the contrary, (34.5) changes to

$$\left(\frac{e'}{c}\right)^2 = m_0'^2 c^2 + p'^2, \qquad \left(\frac{E'}{c}\right)^2 = M_0'^2 c^2 + P'^2. \tag{34.37}$$

We still note that according to the adopted definition the inelastic collision must not necessarily be connected with a loss of kinetic energy. Kinetic energy is consumed only then if the sum of the rest masses is increased by the collision process, namely, if $m_0' + M_0' > m_0 + M_0$. In the case $m_0' + M_0' < m_0 + M_0$, on the contrary, kinetic energy is created.

The formulas (34.1), (34.4), (34.2), (34.3), and (34.37) may also then be applied if the two colliding particles disappear and two new particles are produced in the collision process. m_0' and M_0' then represent the rest masses of the new arising particles. Such a case is the annihilation of an electron-positron pair, where $m_0 = M_0$ = rest mass of the electron (positron) and $m_0' = M_0'$ = rest mass of the photon = 0. We consider this process in the center-of-mass system of the electron-positron pair. Then $\mathbf{p} = -\mathbf{P}$. Therefore, according to the momentum conservation law (34.2), also $\mathbf{p}' = -\mathbf{P}'$, that is, the two photons are emitted in opposite directions with equal momenta, hence also equal energies (figure). The relation between p and p' follows from the energy conservation law. Because $e/c = E/c = \sqrt{m_0^2c^2 + p^2}$ and $e'/c = E'/c = p'$, from (34.3) we get

$$p' = \sqrt{m_0^2 c^2 + p^2}. \tag{34.38}$$

According to (34.20), the energy of every photon is

$$h\nu = cp' = \sqrt{m_0^2c^4 + p^2c^2}. \tag{34.39}$$

The smallest photon energy therefore corresponds to the case $p = 0$ and is equal to the rest energy m_0c^2 of the electron.

The inverse process of annihilation—the creation of an electron-positron pair—occurs in the interaction of a sufficiently energetic photon with an atomic nucleus. This is a process that differs from the collision considered so far. Before the interaction two particles are present also here: the photon and the atomic nucleus. After the collision, however, three particles are involved: the atomic nucleus and the electron-positron pair. The most important feature of this process, which is significant for the experimental verification of the theorem on the inertia of energy, can however be derived immediately from the energy conservation law.

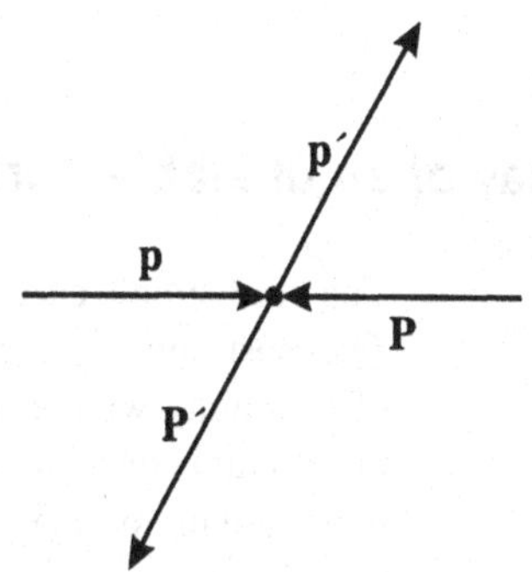

Momentum balance of a binary collision in the center-of-mass frame.

If we denote the photon energy by $\hbar\omega$, the rest mass of the atomic nucleus and the electron (positron) by M_0 and m_0, and the final values of the momenta of the three particles by $\mathbf{P}$, $\mathbf{p}_-$, and $\mathbf{p}_+$, we have

$$\hbar\omega + M_0c^2 = \sqrt{M_0^2c^4 + P^2c^2} + \sqrt{m_0^2c^4 + p_-^2c^2} + \sqrt{m_0^2c^4 + p_+^2c^2}. \tag{34.40}$$

Here we assumed that the atomic nucleus was at rest before the interaction. From this relation follows immediately

$$\hbar\omega > 2m_0c^2. \tag{34.41}$$

This means that the process is possible only with photons of an energy exceeding the sum of the rest energies of two electrons. This has been explained already in the preceding chapter.

As another example of inelastic collisions we mention the *nuclear reactions*. If thereby only two particles are present also after the collision then the process is again described by equations (34.1), (34.4), (34.2), (34.3), and (34.37). But in this case the energy equation may be simplified since, as a rule, the velocities of all particles are small against the speed of light. For such particles, which are usually denoted as nonrelativistic, there holds to good approximation the relation following from (34.4) and (33.34):

$$e = m_0c^2 + \frac{1}{2}m_0q^2.$$

Therefore, the energy law (34.3) reduces to

$$(m_0 + M_0)c^2 + E_{\text{kin}} = (m_0' + M_0')c^2 + E_{\text{kin}}'. \tag{34.42}$$

Here E_{kin} and E'_{kin} denote the sum of the kinetic energies before and after the collision calculated by means of (33.34). When denoting the velocities of the particles before and after the collision by q, Q and q', Q', respectively, this means

$$E_{\text{kin}} = \frac{1}{2}m_0 q^2 + \frac{1}{2}M_0 Q^2, \qquad E'_{\text{kin}} = \frac{1}{2}m'_0 q'^2 + \frac{1}{2}M'_0 Q'^2. \tag{34.43}$$

Decay of an unstable particle

The most simple case is obviously that the unstable particle decays into two new particles. The most important results may also be derived here from the conservation laws. For sake of simplicity we consider the process in the rest frame of the original particle. From the momentum conservation then follows that the momenta $\mathbf{p}'$ and $\mathbf{p}''$ of the two new particles must sum up to zero,

$$\mathbf{p}' = -\mathbf{p}''. \tag{34.44}$$

When denoting the rest mass of the original particle by M_0, and the rest masses of the new particles by m'_0, m''_0, the energy conservation law reads

$$M_0 c = \sqrt{m_0'^2 c^2 + p'^2} + \sqrt{m_0''^2 c^2 + p'^2}. \tag{34.45}$$

We thereby have taken into account the relation $p'' = p'$ following from (34.44). If the rest masses M_0, m'_0 and m''_0 are known, one may determine the value of p' from (34.45). In this decay mode thus in the decay of particles at rest the new particles are always emitted with the momentum p' following from (34.45), and therefore with the uniquely defined energy values

$$e' = \sqrt{m_0'^2 c^4 + p'^2 c^2}, \qquad e'' = \sqrt{m_0''^2 c^4 + p'^2 c^2}. \tag{34.46}$$

The decay into more than two new particles may also be treated in a similar way. If one deals, for example, with a decay into three new particles, the conservation laws in the rest frame of the original particle read

$$0 = \mathbf{p}' + \mathbf{p}'' + \mathbf{p}''', \tag{34.47}$$

$$M_0 c = \sqrt{m_0'^2 c^2 + p'^2} + \sqrt{m_0''^2 c^2 + p''^2} + \sqrt{m_0'''^2 c^2 + p'''^2}. \tag{34.48}$$

We shall not discuss these equations in detail but mention only an important qualitative conclusion. Contrary to the preceding case, the momentum equation now can no longer reduce the momentum values arising after the decay p', p'', p''' to a single quantity. Therefore, from the energy equation determined values of these quantities no longer follow uniquely. In a decay with more than two product particles, the new particles are no longer—as in the case of two product particles—emitted with uniquely determined energy values but shall show a continuous energy spectrum. This fact, by the way, led W. Pauli[1] in 1930 to postulate the neutrino as a hypothetical decay product in the β-decay of the neutron.[2] Later on these particles were actually experimentally detected.

Problem 34.1: The relativistic rocket

If rockets reach velocity ranges comparable to the speed of light, the equations of motions must be based on relativistic mechanics. Formulate the general equation of motion for this problem and discuss it for the one-dimensional case.

Solution We denote by p^μ the four-momentum of the spaceship, and by $dq^\mu = \delta m \cdot \omega^\mu$ the four-momentum of the mass δm expelled by the ship per unit time, as seen from an inertial frame. Energy conservation requires

$$p^\mu = dq^\mu + (p^\mu + dp^\mu) \tag{34.49}$$

with the new four-momentum of the ship $p^\mu + dp^\mu$. We insert dq^μ and $dp^\mu = d(mu^\mu)$ and divide by the proper time $d\tau$:

$$0 = \frac{\delta m}{d\tau}\omega^\mu + \frac{dm}{d\tau}u^\mu + m\frac{du^\mu}{d\tau}. \tag{34.50}$$

$\delta m/d\tau$ und $dm/d\tau$ are the rates for the expelled masses and for the related decreasing mass of the ship. The relation $\delta m/d\tau = -dm/d\tau$ now no longer holds! We define $\lambda = \delta m/d\tau$:

$$\dot{m}u^\mu + m\dot{u}^\mu = -\lambda\omega^\mu. \tag{34.51}$$

We multiply (34.51) by u_μ and employ $u_\mu u^\mu = -c^2$ and $u_\mu \dot{u}^\mu = 0$ (here the Einstein sum convention is used, which means an automatic sum over pairs of equal indices), because

$$\frac{d}{d\tau}(u_\mu u^\mu) = \dot{u}_\mu u^\mu + u_\mu \dot{u}^\mu = 0.$$

[1] *Wolfgang Pauli*, Swiss physicist of Austrian descent, b. April 25, 1900, Vienna—d. Dec. 15, 1958, Zurich. Pauli was a student of Arnold Sommerfeld in Munich, where he also became acquainted with Werner Heisenberg. At a student in his third year, he wrote a review article on the theory of relativity for the *Enzyklopädie der mathematischen Wissenschaften*. With his doctoral thesis from 1921 he for the first time cast doubt on the then-prevailing quantum theory (model of the atom of Bohr and Sommerfeld). His discussions with Heisenberg, Max Born, and Niels Bohr did contribute eminently to the development of matrix mechanics, the algebraic formulation of quantum mechanics. In 1926 he successfully applied the new theory on the hydrogen atom. Already in 1924 he had discovered the exclusion principle (Pauli priciple), for which he was honored with the 1945 Nobel Price (awarded in 1946). Also in 1924 Pauli postulated the existence of the spin of the atomic nucleus in oder to explain the hyperfine structure of atomic spectra. In 1927 he formulated a field equation for the electron, taking into account the spin in a non-relativistic manner. Pauli was a professor in Hamburg from 1926–1928, and at ETH Zurich from 1928 on. In 1930, he formulated the neutrino hypothesis, which was later on corroborated by experiment. From 1940–1945, he stayed in the United States, working mainly on the theory of mesons. His later works back in Zurich in 1946 centered mainly around particle physics and the quantum theory of fields. Pauli made a lasting impression on modern physics and its way of thinking. With his profound analysis of the epistemological foundations of this science and his harsh criticism of obscure thinking, he was known as the "conscience of physics."

[2] see e. g. W. Greiner and B. Müller, *Gauge Theories of Weak Interactions*, Springer Verlag New York, 2000

From there it follows that

$$\lambda = \dot{m}\frac{c^2}{u_\mu \omega^\mu}. \tag{34.52}$$

Hence we have the equation of motion of a body with variable mass, as is represented by the rocket.

$$\frac{d}{d\tau}(mu^\mu) = -\frac{\dot{m}c^2}{u_\nu \omega^\nu}\omega^\mu. \tag{34.53}$$

Solving this equation for a one-dimensional problem causes no trouble. We write for the two velocities ω and u

$$\frac{\omega}{c} \equiv \tanh\phi, \qquad \frac{u}{c} \equiv \tanh\theta, \tag{34.54}$$

and using $h = c\tan\alpha$ and $\alpha = \theta - \phi$ we may express the velocity of the expelled matter relative to the ship.

$$u_\nu \omega^\mu = c^2(\sinh\theta\sinh\phi - \cosh\theta\cosh\phi) = -c^2\cosh(\theta - \phi) = -c^2\cosh\alpha. \tag{34.55}$$

Equation (34.53) for $\mu = 1$ then takes the following form:

$$\frac{d}{d\tau}(m\,c\sinh\theta) = \frac{\dot{m}c^2}{c^2\cosh\alpha}c\sinh\phi, \tag{34.56}$$

and finally reduces to the simple differential equation

$$\begin{aligned}
\dot{m}\sinh\theta + m\dot{\theta}\cosh\theta &= \dot{m}\frac{\sinh\phi}{\cosh\alpha}, \\
\dot{m}\sinh\theta\cosh\alpha + m\dot{\theta}\cosh\theta\cosh\alpha &= \dot{m}\sinh\phi, \\
\dot{m}(\sinh\theta\cosh\alpha - \sinh\phi) + m\dot{\theta}\cosh\theta\cosh\alpha &= 0, \\
\dot{m}\sinh\alpha + m\dot{\theta}\cosh\alpha &= 0 \\
\Rightarrow \quad m\dot{\theta} + \dot{m}\frac{h}{c} &= 0.
\end{aligned} \tag{34.57}$$

Here we have made use of the relation

$$\sinh\phi = \sinh(\theta - \alpha) = \sinh\theta\cosh\alpha - \cosh\theta\sinh\alpha\,.$$

If with h the relative expulsion velocity of the mass is constant, θ may be given as a function of the mass:

$$\theta = \log\left(\frac{m}{M}\right)^{-h/c}. \tag{34.58}$$

M is the integration constant, which here plays the role of the start mass of the spaceship.

$$\frac{u}{c} = \tanh\theta = \frac{1 - e^{-2\theta}}{1 + e^{-2\theta}} = \frac{1 - (m/M)^{2h/c}}{1 - (m/M)^{2h/c}}.$$

If we assume that the relative expulsion velocity of the expelled mass is about $h \approx c$ and half of the start mass is released, then the final velocity is

$$\frac{u}{c} = \frac{1 - (0.5)^2}{1 + (0.5)^2} = \frac{3}{5}.$$

Problem 34.2: The photon rocket

The emission of electromagnetic radiation is considered as an option for driving spaceships in future.
Start from the equation (34.51) of the problem of the relativistic rocket and compare the two propulsion systems.

Solution The equation of motion reads

$$\frac{d}{d\tau}(mu^\mu) = -\lambda P^\mu. \tag{34.59}$$

P^μ is the four-momentum vector of the emitted radiation. We again multiply by u_μ and get, using again the Einstein sum convention, which means an automatic sum over pairs of equal indices,

$$m u_\mu \dot{u}^\mu + \dot{m} u_\mu u^\mu = -\lambda u_\mu P^\mu \tag{34.60}$$

$$\Rightarrow \quad \lambda = \frac{\dot{m}^2 c^2}{u_\mu P^\mu} \qquad \text{because of } u_\mu u^\mu = -c^2 \text{ and } u_\mu \dot{u}^\mu = 0.$$

Thus, equation (34.59) reads

$$\frac{d}{d\tau}(mu^\mu) = -\left(\frac{\dot{m}c^2}{u_\nu P^\nu}\right) P^\mu. \tag{34.61}$$

We are already familiar with this result from the relativistic rocket. The difference is that the photon four-momentum vector is a zero vector, because of the vanishing mass:

$$P_\mu P^\mu = 0. \tag{34.62}$$

We again consider the one-dimensional case. Equation (34.62) then reduces to

$$\left(P^1\right)^2 - \left(P^4\right)^2 = 0 \qquad \Leftrightarrow \qquad P^1 = \pm P^4. \tag{34.63}$$

If the spaceship flies in positive x-direction, the photons necessarily should have a negative momentum:

$$P^1 = -P^4. \tag{34.64}$$

We now write down the discrete photon energy, using the de Broglie relation, thus

$$W = h\nu \qquad \Rightarrow \qquad P^4 = \frac{h\nu}{c}, \quad P^1 = -\frac{h\nu}{c}, \tag{34.65}$$

$$u_\nu P^\nu = u^1 P^1 - u^4 P^4 = \left(u^1 + u^4\right) P^1 = -\left(u^1 + u^4\right) P^4. \tag{34.66}$$

Equation (34.61) for $\mu = 1$ and $\mu = 4$ then becomes

$$\left(u^1 + u^4\right) \frac{d}{d\tau}\left(mu^1\right) = -\dot{m}c^2, \tag{34.67}$$

$$\left(u^1 + u^4\right) \frac{d}{d\tau}\left(mu^4\right) = \dot{m}c^2. \tag{34.68}$$

The sum of these two equations yields

$$\frac{d}{d\tau}\left(mu^1 + mu^4\right) = 0 \qquad \Rightarrow \qquad mu^1 + mu^4 = Mc, \tag{34.69}$$

where with M the start mass of the rocket was introduced via the integration constant; this is the situation with $u^1 = 0$ and $u^4 = c$.

Rewriting again the equation by means of the definitions

$$u^1 = c \sinh\theta \qquad u^4 = c\cosh\theta \tag{34.70}$$

there results

$$m(\sinh\theta + \cosh\theta) = M \quad \Leftrightarrow \quad m = Me^{-\theta}$$
$$\Rightarrow \quad \frac{u}{c} = \tanh\theta = \frac{1-(m/M)^2}{1+(m/M)^2}. \tag{34.71}$$

Thus, the final velocity does not depend on the frequency of the radiation. Nevertheless, it will presumably take some time to overcome the enormous difficulties in developing photon engines. Such an engine should of course provide a sufficient thrust. We also don't see an advantage of such a type of engine over engines emitting massive particles near the speed of light.

Problem 34.3: The relativistic central force problem

Solve the central force problem relativistically for a particle of mass m with the charge q and a central charge Q that is tightly fixed to the origin of the coordinate frame. You should take into consideration only the electrostatic interaction $\mathbf{K} = (Qq/r^2)\,\mathbf{e}_r$.

Solution The relativistic form of the second Newtonian axiom is the four-vector equation

$$F^\mu = m_0 \frac{d}{d\tau} u^\mu , \tag{34.72}$$

with F^μ being the four-force (33.8), p. 429, and u^μ the four-velocity (33.3):

$$F^\mu = \left(\gamma\mathbf{K}, i\gamma\frac{\mathbf{v}}{c}\cdot\mathbf{K}\right), \tag{34.73}$$

$$u^\mu = (\gamma\mathbf{v}, i\gamma c), \qquad \gamma = \frac{1}{\sqrt{1-\dfrac{v^2}{c^2}}}. \tag{34.74}$$

$\mathbf{K}$ and $\mathbf{u}$ are the force and velocity according to Newtonian mechanics, respectively. The expression used here for the fourth component F^4 of the four-force F^μ,

$$F^4 = i\gamma\frac{\mathbf{v}}{c}\cdot\mathbf{K} = i\frac{\gamma}{c}\frac{d}{dt}mc^2,$$

follows immediately from (**??**), which states

$$\frac{d}{dt}mc^2 = \mathbf{v}\cdot\mathbf{K}.$$

But we must use as the only force $\mathbf{K}$ in the relativistic expression only the Lorentz force acting on a charged particle. The other important interaction, the gravitation, *cannot* be treated without further ado in this calculus, since it depends on the masses involved. These problems will be treated in the general theory of relativity. Our central force problem is based on the electrostatic interaction of the two charges Q and q.

$$\mathbf{K} = \frac{Qq}{r^2}\mathbf{e}_r , \tag{34.75}$$

$$\Rightarrow \quad F^\mu = \left(\gamma \frac{Qq}{r^2}\mathbf{e}_r, i\gamma \frac{Qq}{r^2}\frac{\mathbf{v}}{c}\mathbf{e}_r \frac{Qq}{r^2}\mathbf{u}\cdot\mathbf{e}_r\right). \tag{34.76}$$

We shall use cylindrical coordinates for this problem. But we have to take into account the dependence of the unit vectors on the time. To do this, we recall that the four-velocity $u^\mu(\mu = 1, 2, 3, 4)$ is the derivative of the world vector $x^\mu = (x^1, x^2, x^3, ict)$ with respect to proper time,

$$u^\mu = \frac{dx^\mu}{d\tau} = \left(\frac{1}{\sqrt{1-\beta^2}}\frac{d\mathbf{x}}{dt}, \frac{ic}{\sqrt{1-\beta^2}}\right) = (\gamma\mathbf{v}, ic\gamma),$$

where $\gamma = 1/\sqrt{1-\beta^2}$, $\beta = v/c$, and $\mathbf{v} = d\mathbf{x}/dt$ the usual three-velocity. In cylindrical coordinates, we have

$$\begin{aligned} x^\mu &= (r\mathbf{e}_r, z, ict), \\ u^\mu &= \frac{dx^\mu}{d\tau} = \left(\frac{d}{d\tau}(r\mathbf{e}_r), \frac{dz}{d\tau}, ic\frac{dt}{d\tau}\right) \\ &= ((r\mathbf{e}_r)^\bullet, \dot{z}, ic\gamma). \end{aligned} \tag{34.77}$$

Here, the dot • means the derivative with respact to proper time τ, thus

$$(\ldots)^\bullet = \frac{d}{d\tau} = \frac{1}{\sqrt{1-\beta^2}}\frac{d}{dt}.$$

In planar cylindrical coordiantes ($z = 0$), the world vector reduces to

$$x^\mu = (r\mathbf{e}_r, 0, ict),$$

where $\mathbf{r} = r\mathbf{e}_r$ and the unit vectors in radial and in φ-direction are

$$\begin{aligned} \mathbf{e}_r &= (\cos\varphi, \sin\varphi), \\ \mathbf{e}_\varphi &= (-\sin\varphi, \cos\varphi). \end{aligned}$$

The four-velocity hence is

$$\frac{dx^\mu}{d\tau} = ((r\mathbf{e}_r)^\bullet, 0, ic\gamma) = (\dot{r}\mathbf{e}_r + r\dot{\mathbf{e}}_r, 0, ic\gamma).$$

Because

$$\dot{\mathbf{e}}_r = \dot{\varphi}\mathbf{e}_\varphi, \qquad \dot{\mathbf{e}}_\varphi = -\dot{\varphi}\mathbf{e}_r,$$

we get

$$\overset{\Rightarrow}{u} = \frac{dx^\mu}{d\tau} = \left(\dot{r}\mathbf{e}_r + r\dot{\varphi}\mathbf{e}_\varphi, 0, ic\gamma\right), \tag{34.78}$$

$$\begin{aligned} \frac{d^2x^\mu}{d\tau^2} &= \left(\ddot{r}\mathbf{e}_r + \dot{r}\dot{\mathbf{e}}_r + \dot{r}\dot{\varphi}\mathbf{e}_\varphi + r\ddot{\varphi}\mathbf{e}_\varphi + r\dot{\varphi}\dot{\mathbf{e}}_\varphi, 0, ic\frac{d\gamma}{d\tau}\right), \\ &= \left(\left(\ddot{r} - r\dot{\varphi}^2\right)\mathbf{e}_r + (2\dot{r}\dot{\varphi} + r\ddot{\varphi})\,\mathbf{e}_\varphi, 0, ic\dot{\gamma}\right). \end{aligned} \tag{34.79}$$

Newton's equations in their relativistic four-form read

$$\frac{d}{d\tau}p^\mu = F^\mu \qquad \text{or} \qquad m_0\frac{d}{d\tau}\left(\frac{dx^\mu}{d\tau}\right) = F^\mu;$$

hence in our case here

$$m_0\left((\ddot{r}-r\dot{\varphi}^2)\,\mathbf{e}_r+(2\dot{r}\dot{\varphi}+r\ddot{\varphi})\,\mathbf{e}_\varphi,0,ic\dot{\gamma}\right)=\left(\gamma\frac{Qq}{r^2}\mathbf{e}_r,i\gamma\frac{Qq}{r^2}\frac{\mathbf{v}}{c}\mathbf{e}_r\frac{Qq}{r^2}\mathbf{u}\cdot\mathbf{e}_r\right). \tag{34.80}$$

Comparing the components of the four-vectors and taking into account that

$$\gamma\mathbf{v}=\dot{r}\mathbf{e}_r+\dot{\varphi}\mathbf{e}_\varphi\,, \tag{34.81}$$

we end up with the three equations

$$\gamma\frac{Qq}{r^2}=m\left(\ddot{r}-r\dot{\phi}^2\right), \tag{34.82}$$

$$0=m\left(2\dot{r}\dot{\phi}+r\ddot{\phi}\right), \tag{34.83}$$

$$\frac{1}{c}\frac{Qq}{r^2}\dot{r}=m\frac{d}{d\tau}(\gamma c)=m\dot{\gamma}c. \tag{34.84}$$

The dots always denote derivation with respect to the proper time τ.

Equation (34.83) multiplied by r yields as in the nonrelativistic case the angular momentum conservation:

$$m_0\left(2r\,\dot{r}\dot{\phi}+r^2\ddot{\phi}\right)=m_0\frac{d}{d\tau}\left(r^2\dot{\phi}\right)\qquad\Rightarrow\qquad L\equiv m_0r^2\dot{\phi}=\text{constant}. \tag{34.85}$$

The equation (34.84) ensures the conservation of energy:

$$\frac{d}{d\tau}\left(m_0\gamma c+\frac{1}{c}\frac{Qq}{r}\right)=0\quad\Rightarrow\quad E=m_0\gamma c^2+\frac{Qq}{r}=\text{constant}$$

$$\Leftrightarrow\quad\gamma=\frac{E}{m_0c^2}-\frac{Qq}{m_0c^2r}. \tag{34.86}$$

Now we still wish to extract from (34.82) an equation of motion. For this purpose we employ the two conservation laws just obtained.

$$\left(\frac{E}{m_0c^2}-\frac{Qq}{m_0c^2r}\right)\frac{Qq}{r^2}=m_0\left(\ddot{r}-\frac{L^2}{m_0^2r^3}\right). \tag{34.87}$$

We introduce the variable $s=1/r$ and, as in the nonrelativistic Kepler problem, transform to a differential equation for $s(\phi)$.

$$\dot{r}=-r^2\frac{ds}{d\phi}\dot{\phi}=-\frac{L}{m_0}\frac{ds}{d\phi}, \tag{34.88}$$

$$\ddot{r}=-\frac{L}{m_0}\frac{d^2s}{d\phi^2}\dot{\phi}=-\frac{L^2}{m_0^2}s^2\frac{d^2s}{d\phi^2} \tag{34.89}$$

$$\Rightarrow\quad\frac{1}{m_0c^2}(E-Qq\,s)Qq\,s^2=-\frac{1}{m_0}\left(L^2s^2\frac{d^2s}{d\phi^2}+L^2s^3\right)$$

$$s''+s=\left(\frac{Qq}{cL}\right)^2s-\frac{EQq}{L^2c^2}. \tag{34.90}$$

We define the "angular frequency" $\Omega^2 = 1 - (Qq/mLc)^2$ and thereby may at once give the solution of this well-known differential equation:

$$\frac{1}{r} = s = -\frac{EQq}{m_0^2 L^2 c^2 \Omega^2} + A\cos(\Omega\phi). \tag{34.91}$$

There are bound solutions to the problem if

$$qQ < 0 \qquad \text{and} \qquad A < \frac{E|Qq|}{m_0^2 L^2 c^2 \Omega^2}.$$

In equation 34.91 we already have used that we wish to start with $\phi = 0$ at the perihelion. Since Ω differs from 1, there is no closed path, but rather an orbital precession arises (perihelion motion). A closed orbit does only exist for $q = 0$ or $Q = 0$, hence if there is no force. This means that the relativistic Kepler problem always yields rosette orbits, which show perihelion motion. This is plausible if we recall our discussion in Chapters 26 and 28.

The constant A may be determined by inserting our solution in the following relation, which must be obeyed by the four-velocity $u^\mu = (\dot{r}, \dot{\phi}, 0, \gamma c)$ from equation 34.78:

$$\vec{\vec{u}} \cdot \vec{\vec{u}} = u_\mu u^\mu = -c^2 = \dot{r}^2 + r^2\dot{\phi}^2 - (c\gamma)^2 \tag{34.92}$$

Inserting in this equation the definition of angular momentum 34.85, the conservation of energy 34.86, and the relation 34.88 for $\dot{r}$, we get

$$\frac{L^2}{m_0^2}\left(\frac{ds}{d\phi}\right)^2 + s^2\frac{L^2}{m_0^2} - c^2\left(\frac{E}{m_0c^2} - \frac{Qq}{m_0c^2}s\right)^2 = -c^2. \tag{34.93}$$

If we further insert the solution 34.91 for the orbit, we get an equation we can solve for A:

$$\begin{aligned}
-c^2 &= \frac{L^2}{m_0^2}\Big(A\Omega\sin(\Omega\phi)\Big)^2 + \left(A\cos(\Omega\phi) - \frac{EQq}{m_0^2L^2c^2\Omega^2}\right)^2\frac{L^2}{m_0^2} \\
&\quad - c^2\left(\frac{E}{m_0c^2} - \frac{Qq}{m_0c^2}\left(A\cos(\Omega\phi) - \frac{EQq}{m_0^2L^2c^2\Omega^2}\right)\right)^2 \\
&= A^2\frac{L^2\Omega^2}{m_0^2}\sin^2\Omega\phi + A^2\frac{L^2\cos^2\Omega\phi}{m_0^2} - 2A\frac{EQq\cos\Omega\phi}{m_0^4c^2\Omega^2} + \frac{E^2Q^2q^2}{m_0^6L^2c^4\Omega^4} \\
&\quad - \frac{E^2}{m_0^2c^2} + 2\frac{EQq}{m_0^2c^2}\left(A\cos(\Omega\phi) - \frac{EQq}{m_0^2L^2c^2\Omega^2}\right) \\
&\quad - \frac{Q^2q^2}{m_0^2c^2}\left(A^2\frac{L^2\cos^2\Omega\phi}{m_0^2} - 2A\frac{EQq\cos\Omega\phi}{m_0^4c^2\Omega^2} + \frac{E^2Q^2q^2}{m_0^6L^2c^4\Omega^4}\right).
\end{aligned} \tag{34.94}$$

Collecting all the terms yields

$$A = \sqrt{\left(\frac{E}{Lc\Omega^2}\right)^2 - \left(\frac{m_0c}{L\Omega}\right)^2} = \frac{c}{L\Omega}\sqrt{\left(\frac{E}{m_0c^2\Omega^2}\right)^2 - 1}. \tag{34.95}$$

For $A = 0$, that is, for $E/m_0c^2 = \Omega = \sqrt{1-(Qq/m_0cL)^2}$, the orbit 34.91 becomes a circle.

We still note that α^2 in general may also become negative ($Qq > mLc$). Periodic solutions of (34.90) would then no longer exist. This case might occur if one lets, for example, the angular momentum L become very small. When considering this in an atom with a nucleus of charge $Q = Ze$, and taking into account that the angular momentum of an electron ($q = e$) is of the order of magnitude $mL \approx \hbar$ (Planck's quantum of action), we may expect this case for the following charge number Z:

$$Z \geq \frac{\hbar c}{e^2} \approx 137. \tag{34.96}$$

That means that this collapse occurs in atoms with a nuclear charge $Z \geq 137$, due to relativistic effects in the approximation of a point nucleus. The quantity $c^2/\hbar c^1 \cong 1/137$ is known as the Sommerfeld fine structure constant.

Discussion of the solution The solution 34.91 is very similar to the Kepler orbits that we know already from Chapter 26 about planetary motions, see, for example, equation(26.32). We thus can write 34.91 as

$$r(\phi) = \frac{r_0}{\epsilon \cos(\Omega\phi) \pm 1}, \tag{34.97}$$

where

$$r_0 = \frac{L^2c^2\Omega^2}{E|Qq|}, \tag{34.98}$$

$$\epsilon = \frac{L^2c^2\Omega^2}{E|Qq|}\sqrt{\left(\frac{E}{Lc\Omega^2}\right)^2 - \left(\frac{m_0c}{L\Omega}\right)^2} = \frac{Lc}{E|Qq|}\sqrt{E^2 - m_0^2c^4\Omega^2}. \tag{34.99}$$

Here, the upper (lower) sign in 34.97 relates to attractive (repulsive) interaction, that is, unequal ($Qq < 0$) or equal ($Qq > 0$) charges. In the nonrelativistic limit, we have $\Omega \simeq 1$ and $E = m_0c^2 + E_{\text{nr}}$ with $|E_{\text{nr}}| \ll m_0c^2$ and 34.97–34.99 reduce exactly to the conic sections of Kepler motion, where, obviously, the force constant γMm has to be repalced by $|Qq|$. One ends up with

$$r_0 \simeq \frac{L^2}{m_0|Qq|} \tag{34.100}$$

and eccentricity

$$\epsilon \simeq \sqrt{1 + \frac{2E_{\text{nr}}L^2}{m_0|Qq|^2}}. \tag{34.101}$$

This corresponds to the results from Chapter 26. In the nonrelativistic limit, one thus finds again the well-known cicular, elliptic, parabolic, and hyperbolic orbits. The relativistic treatment, however, yields two interesting differences.

1. Because $\Omega = \sqrt{1 - (Qq/Lc)^2} < 1$, the period of the orbit for periodic solutions is larger than 2π. Hence for $0 < \epsilon < 1$ there are no closed elliptic orbits any more, but 34.97 describes a *rosette orbit.* At each period, the turning points preceed by an angle $\Delta\phi$, as shown in the figure (for $\Omega = 0.93$). The constraint $\Omega(2\pi + \Delta\phi) = 2\pi$ yields for this angle of precession for Ω close to 1

$$\Delta\phi = 2\pi\left(\frac{1}{\Omega} - 1\right) = 2\pi\left(\frac{1}{\sqrt{1 - \left(\frac{Qq}{Lc}\right)^2}} - 1\right)$$

$$\simeq 2\pi\left(1 + \frac{1}{2}\left(\frac{Qq}{Lc}\right)^2 - 1\right) = \pi\left(\frac{Qq}{Lc}\right)^2. \tag{34.102}$$

When expressed with the help of the ellipse parameters eccentricity ϵ and major semi-axis $a = r_0/(1-\epsilon^2)$, using the approximation 34.100 yields for the precession angle

$$\Delta\phi = \pi \frac{|Qq|}{m_0 c^2 a\left(1-\epsilon^2\right)}. \tag{34.103}$$

Although our derivation was done for a charged particle in the electric field of a point source, one may be tempted to apply the result 34.103 also on planetary motion, since force laws of Newton for gravity and of Coulomb for electrostatic interaction have the same form. Newtonian gravity with a correction by special realtivity hence makes a prediction for the *perihel motion of planetary orbits* of

$$\Delta\phi = \pi \frac{\gamma M}{c^2} \frac{1}{a\left(1-\epsilon^2\right)}. \tag{34.104}$$

A rosette orbit showing precession of the turning point.

This formula gives extremly small values for the perihel motion of the planets of the solar system. The overall precession of the perihel of the orbit of Mercury, for example, is predicted by this formula to be only 7 arc seconds per century. The observed value not accounted for by perurbations of the other planets, however, is 42 arc seconds per century. This discrepancy can be resolved only within the framework of general relativity. The result obtained there is larger then formula 34.104 by a factor 6. Hence, gravitation and electrostatic interaction differ fundamentally, the similarities in the force law notwithstanding

2. In the case of large charges Qq or small angular momentum L the parameter Ω^2 may become negative. This changes the character of the solutions qualitativly. If we define in this case

$$\tilde{\Omega}^2 = \left(\frac{Qq}{Lc}\right)^2 - 1 > 0, \tag{34.105}$$

the general solution of the differential equation 34.90 is

$$s(\phi) = c_1 e^{-\tilde{\Omega}\phi} + c_2 e^{+\tilde{\Omega}\phi} + \frac{EQq}{L^2 c^2 \tilde{\Omega}^2}. \tag{34.106}$$

From $u_\mu u^\mu = -c^2$, a lengthy calculation along the lines leading to equation 34.95 yields

$$4c_1 c_2 = \left(\frac{E}{Lc\tilde{\Omega}^2}\right)^2 + \left(\frac{m_0 c}{L\tilde{\Omega}}\right)^2. \tag{34.107}$$

Because the right hand side is positive, c_1 and c_2 must have the same sign. Without restricting the general case, one can choose the coefficients to be equal, $c_1 = c_2 = \frac{1}{2}\tilde{A}$, since any difference can be balanced by a rotation $\phi \to \phi + \phi_0$ of the coordinate system. Hence, the general solution is

$$s(\phi) = \tilde{A}\cosh(\tilde{\Omega}\phi) + \frac{EQq}{L^2 c^2 \tilde{\Omega}^2}, \tag{34.108}$$

with the prefactor

$$\tilde{A} = \pm\sqrt{\left(\frac{E}{Lc\tilde{\Omega}^2}\right)^2 + \left(\frac{m_0 c}{L\tilde{\Omega}}\right)^2}. \tag{34.109}$$

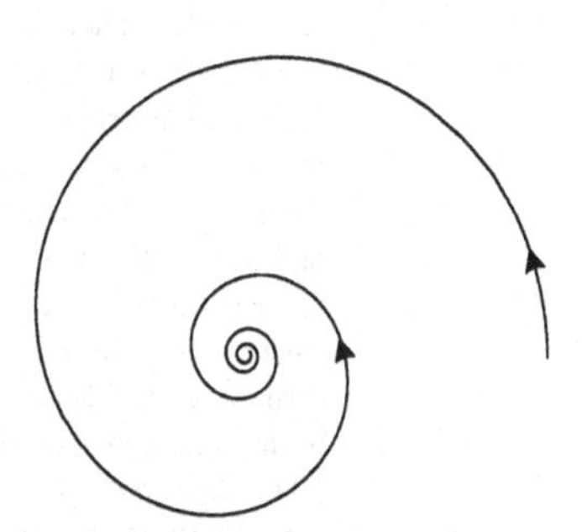

The collapse of the electronic orbit in a logarithmic spiral.

Thus, in 34.91 the trigonometric functions have to be replaced by hyperbolic functions.

With repulsive interaction ($Qq > 0$) there are hyperbolic orbits further on. (Here one has to choose $\tilde{A} < 0$. Positive $\tilde{A}$ together with 34.86 yields unphysical solutions with a Lorentz factor $\gamma < 0$.)

However, when the force is attractive ($Qq < 0$), there are dramatic changes in the type of the orbit. Because the hyperbolic function grows exponentially, $r(\phi)$ goes to zero for large angles ϕ; the orbit has the form of a *logarithmic spiral* as shown in the figure. This "fall onto the center" will happen because the Coulomb force is enhanced by the Lorentz factor γ in such a way that the angular momentum barrier can be surmounted at small distances.

Such a collapse cannot be seen in macroscopic physics, already due to the finite extent of charged bodies. However, the question becomes of interest for (pointlike) electrons with charge $q = -e$ in the field of a nucleus with $Q = Ze$. If one treats the motion of the electron within the framework of classical mechanics, but takes into account that atomic angular momenta are in the range of the Planck constant, $L \simeq \hbar$, the following condition for the collapse of the orbit results:

$$Z \geq \frac{\hbar c}{e^2} \equiv \alpha^{-1} \simeq 137, \tag{34.110}$$

where $\alpha \simeq 1/137.036$ is konwn as the Sommerfeld[3] fine structure constant. Hence, one would expect a collpase of the electronic orbits in atoms with nuclear charge $Z \geq 137$.

Of course, this problem transcends the range of applicability of classical mechanics. But also relativistic quantum mechanics predicts a similar collapse if in an atom the parameter $Z\alpha$ becomes greater than 1. This is closely related to the hypercritical problem of quantum electrodynamics that

[3] *Arnold Johannes Wilhelm Sommerfeld*, b. Dec. 5, 1868, Königsberg, then Prussia (now Kaliningrad, Russia)—d. April 26, 1951, Munich. Sommerfeld attended the Gymnasium in Königsberg (two slightly older pupils at the same school were Minkowski and Wien) and started his studies at the University of Königsberg where he was taught by Hilbert, Hurwitz, and Lindemann. At this time the University of Königsberg was famous for its school of Theoretical Physics, which had been founded by Franz Neumann, but Sommerfeld's interests were in mathematics rather than physics. In 1891 Sommerfeld was awarded his doctorate from Königsberg. In 1893 Sommerfeld went to Göttingen, where he became Klein's assistant. His research there was immediately influenced by Klein, who at this time was involved in applying the theory of functions of a complex variable, and other pure mathematics, to a range of physical topics from astronomy to dynamics. Important work Sommerfeld undertook included the study of the propagation of electromagnetic waves in wires and the study of the field produced by a moving electron. As of 1897 Sommerfeld taught at Clausthal, where he became professor of mathematics at the mining academy. Then, three years later, he became professor of mechanics at the Technische Hochschule of Aachen. In 1897 (first as a professor of mathematics at the mining academy at Clausthal, then, after 1900, as professor of mechanics at the Technische Hochschule of Aachen), Sommerfeld began a 13-year study of gyroscopes working on a 4-volume work jointly with Klein. In 1906 he became professor of theoretical physics at Munich and worked on atomic spectra. He studied the hypothesis that X-rays were waves, which was proved by his collegue Max von Laue by using crystals as three-dimensional diffraction gratings. From 1911 his main area of interest became quantum theory. Sommerfeld's work led him to replace the circular orbits of the Niels Bohr atom with elliptical orbits; he also introduced the magnetic quantum number in 1916 and, four years later, the inner quantum number. It was theoretical work attempting to explain the inner quantum number that led to the discovery of electron spin. In the later part of his career, Sommerfeld used statistical mechanics to explain the electronic properties of metals. This replaced an earlier theory due to Lorentz in 1905 based on classical physics. Sommerfeld's approach was to regard electrons in a metal as a degenerate electron gas. He was able to explain features that were unexplained by the earlier classical theory. Sommerfeld had built up a very famous school of theoretical physics at Munich—among his most famous students are Heisenberg and Pauli—but its 30 years of fame ended with the Nazi rise to power. In 1940 the school closed, but by this time Sommerfeld was 71 years old. He survived World War II and eventually died in a street accident in Munich.

has been investigated extensively by the Frankfurt school and has far-reaching consequences. For example, it leads to a new understanding of the question "What is the vacuum; is the vacuum always empty?" We refer to the volume of the lectures on quantum electrodynamics, and the literature quoted there.[4]

Example 34.4: Gravitational lenses

The demonstration of light deflection at the border of the sun and its correct interpretation and theoretical description in the frame of the general theory of relativity, a few years after its formulation, represented one of the greatest triumphs of the new theory of gravitation by Einstein. According to the general theory of relativity, gravitation manifests itself as a modification of the plane Minkowski space-time geometry. In the vicinity of heavy masses, the space-time is distorted. Light rays that, as everybody knows, propagate along certain geodesic lines, that is, shortest (one may also say "straightest") lines between two world points with $ds^2 = 0$, no longer follow straight lines in the Euclidean sense but in general bent curves (see figure).

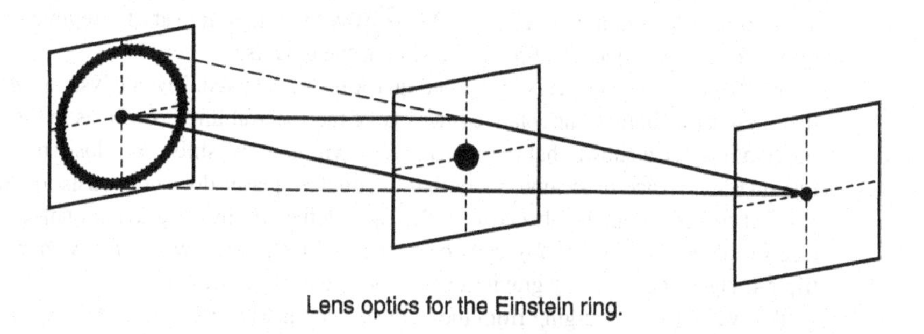

Lens optics for the Einstein ring.

Thus, in a certain sense gravitational fields affect the light propagation just as an optically more dense medium with a definite refractive index does, as is known from geometric optics. We may well imagine that certain arrangements of masses create such a gravitational field, such that light emitted by a far remote object may be deflected when passing this gravitational field, similarly as on passing of an optical lens. Gravitational fields with such properties are denoted as gravitational lenses, analogous to the optical lenses.

Einstein made a calculation on this problem where for the first time the following simple configuration is considered: Let a massive object as source of a gravitational field be positioned between earth and a far remote source of light (e.g., a star), positioned exactly on the optical axis. In such a perfect alignment of successive objects only those light rays from the star may reach the earth which are focused toward earth by the intermediate gravitational field (see figure).

Because of the azimuthal symmetry the star is imaged as a ring visible from earth. Because the star cannot be observed directly because of the intermediate object, one should see only a so-called Einstein ring instead of a pointlike light source. But the configuration just discussed represents a

[4]For popular representations of this domain see, e.g., J. Reinhardt and W. Greiner, *Physik in unserer Zeit*, no. 6 (1976) 171; W. Greiner and J. H. Hamilton, *American Scientist* **19** (1980) 154; J. Greenberg and W. Greiner, *Physics Today* (Aug. 1982) 24. A comprehensive, scientific presentation of the subject can be found in W. Greiner, B. Müller, J. Rafelski, *Quantum Eletrodynamics of Strong Fields*, Springer-Verlag, Berlin, Heidelberg, New York, 1985.

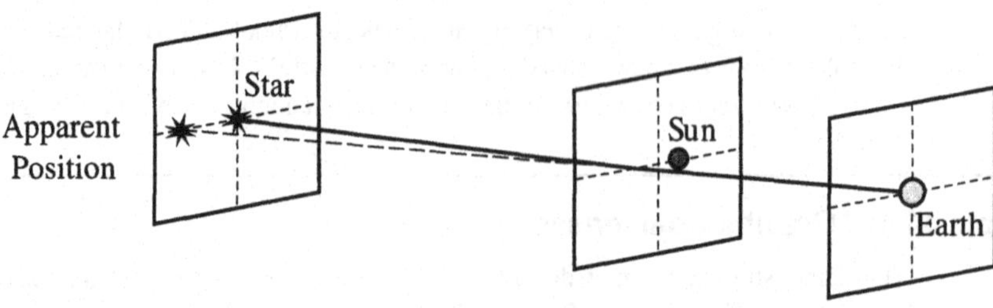

Deflection of light at the rim of the sun.

particularly simple idealized case, and it may be highly unlikely to observe exactly such a situation. In any case one may calculate the imaging properties of general lens systems. It turns out that, besides such ringlike images, there is also a chance for double or triple images of an object.

Particularly interesting objects of astronomy are quasars, known since 1963. Quasars are starlike (i.e., pointlike as seen from earth) light and radio sources displaying a strong red shift in the spectra. Therefore, they cannot be stars of our Milky Way but rather are very far remote objects, the red shift of which is a consequence of the expansion of the universe.

In 1979 a pair of very closely spaced quasars was detected. The analysis of their spectra showed that these agree both in the relative intensity of the spectral lines as well as in their red shift. Further observations have shown that there is a galaxy with low red shift (i.e., closer to us) just between the two quasars. Hence it became likely that the double-quasar does not consist of two distinct objects but that the astronomers—because of the light deflection in the gravitational field of the galaxy—(see Problem 33.3) see two images of a single object.[5] Meanwhile, many more double- and even triple-images of quasars by gravitational lensing have been detected.

The wave fronts emerging from the light source (quasar) are folded in the vicinity of a large mass (galaxy) such that three wave fronts are passing the observer (see figure). On earth one therefore sees three images of the quasar. That one sees only two images in the case of double-quasars may be due to the circumstance that one of the images is very faint or that two images are so closely spaced that they are no longer separated optically.

The gravitational lens effect may be observed only in the radiation from quasars since the deflection angle as seen from earth is proportional to the gravitational potential of the lens at the position of earth. An estimation of the mass distribution in the universe shows that the probability of a gravitational lens effect with a remote galaxy as lens is by about the factor 10^4 larger than the probability of such an effect with a star from our Milky Way as deflecting mass.

The first observations of gravitational lensing very done with arrays of radio telescopes. This technique yielded in 1987 the dicovery of an almost perfect Einstein ring.[6]

Meanwhile, with the advent of the Hubble space telescope and large, modern ground-based telescopes with adaptive optics, the observation of gravitational lenses has become very common in

[5]For an account by one of the discoverers of this first "double-quasar", see F. H. Chaffee: *The Discovery of a Gravitational Lens*, Scientific American, November 1980, 60–68.

[6]Jaqueline. N. Hewitt et al.: *Unusual Radio Source MG 1131+0456—A Possible Einstein Ring* Nature **333**, 537-540 (1988). An overview from this period is given by Edwin L. Turner: *Gravitational Lenses*, Scientific American, July 1988, 26-32.

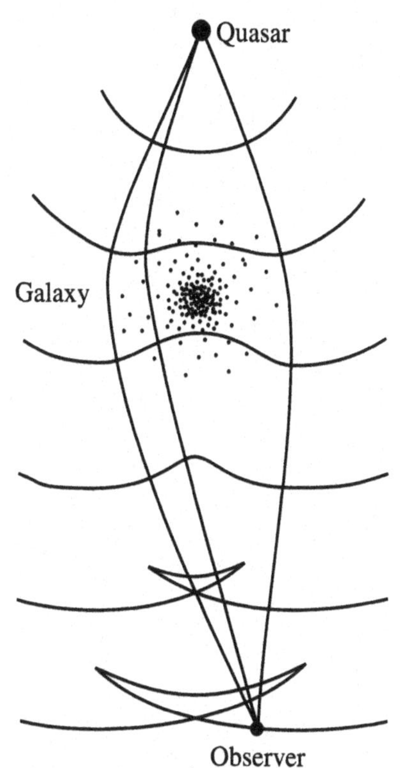

Propagation of a wave front by a galaxy leading to a gravitational lens.

astronomy. It is even used as a tool to extend the range of possible observations and to explore into new issues such as the quest for dark matter or the history of the universe.[7]

Thus, by observing gravitational lens systems, one hopes to clarify a number of highly interesting problems. The light observed has passed cosmic distances. Therefore, gravitational lenses should be affected already by the geometry of space-time as a whole. As the optical properties of a gravitational lens can be calculated exactly, one may take also the influence of the expansion of the universe into account. In principle it will be possible to determine the so-called Hubble constant, which, roughly speaking, connects the extension and the expansion velocity of the universe. A further interesting aspect arises concerning the so-called dark matter as discussed earlier in Section 28. Gravitational lenses must not necessarily be constituted of mass distributions (e.g., quasars, galaxies) that are visible via their electromagnetic radiation. From the relative rate of gravitational lens phenomena, one now might also conclude on the rate of distributions of dark matter in the universe. An estimation of their mass would then have a bearing on the decision between cosmologic models, which all involve a mean mass density as parameter. We shall stop with these remarks and may wait with the expectation for further observations that will make the solution of these problem areas more accessible.

---

[7]Examples for the various modern "applications" of gravitational lenses are give in the article by Joachim Wambsganss: *Gravity's Kaleidoscope*, Scientific American, November 2001, 52-59. See also our remarks in connection with the dark matter problem in Chapter 28.

Index

Printed in the United States
by Baker & Taylor Publisher Services

Printed in the United States
by Baker & Taylor Publisher Services